DUNHAM AND YOUNG'S CONTRACTS, SPECIFICATIONS, AND LAW FOR ENGINEERS

DUNHAM AND YOUNG'S CONTRACTS, SPECIFICATIONS, AND LAW FOR ENGINEERS

FOURTH EDITION

Joseph T. Bockrath, A.B., J.D.

**Member of the Louisiana and
Supreme Court Bar
Professor of Law
Louisiana State University Law Center**

McGraw-Hill, Inc.
New York St. Louis San Francisco Auckland Bogotá
Caracas Lisbon London Madrid Mexico Milan
Montreal New Delhi Paris San Juan Singapore
Sydney Tokyo Toronto

This book was set in Times Roman by Donnelley/ROCAPPI, Inc. (ECU).
The editors were Kiran Verma and Cydney C. Martin;
the cover was designed by Stephanie Blumenthal;
the production supervisor was Marietta Breitwieser.
Project supervision was done by The Total Book.
R. R. Donnelley & Sons Company was printer and binder.

**DUNHAM AND YOUNG'S CONTRACTS,
SPECIFICATIONS, AND LAW FOR ENGINEERS**

7890 DOC/DOC 998765432

ISBN 0-07-018237-X

Library of Congress Cataloging in Publication Data

Dunham, Clarence W. (Clarence Whiting), date
    Dunham and Young's Contracts, specifications, and
law for engineers.

    Includes bibliographical references and index.
    1. Engineering—Contracts and specifications—United
States. 2. Engineering law—United States. 3. Building
—Contracts and specifications—United States. 4. Law
—United States. I. Bockrath, Joseph T. II. Title.
III. Title: Contracts, specifications, and law for
engineers.
KF902.D8 1986   346.73′002462   85-16571
ISBN 0-07-018237-X   347.306002462

CONTENTS

PART 3   SOME LEGAL MATTERS OF CONCERN TO ENGINEERS

16   Agency   263

17   Partnerships   272

18   Corporations   281

PREFACE

This book has been written for students of engineering and architecture, and for practicing engineers, architects, and contractors. The material has been divided into three parts. Part One explains the basic principles of the law of contracts. Part Two is a discussion of the application of contract principles to construction contracts, together with data on the preparation of specifications and other documents which form an essential part of construction contracts. Part Three considers fields of law which are of special interest to engineers, architects, and contractors in their professional lives.

As engineers and architects gain experience and progress in a leadership capacity, they will most likely have to acquire a knowledge of basic legal principles, particularly in the area of contracts and specifications. We hope to introduce the reader to some legal problems which are likely to arise to the end that he or she* will be able to recognize the occasions when a lawyer's aid is needed and will be in a position to cooperate with attorneys with whom he or she is working. We have not tried to teach the law to those who read these pages, but rather to create an awareness and appreciation of more or less typical questions which arise when engineering and law interact.

In an effort to minimize difficulties arising from the use of unfamiliar terminology, we have included a glossary of legal terms which should be consulted when the technical word or phrase employed is not immediately defined. The authors have attempted to state legal principles simply and clearly, rather than to quote extensively from court decisions or complicated documents. Court cases have been used as references and, to some extent, as illustrations. An effort has been

---

*Editors' note: The use of the male pronoun in legal literature is widespread, particularly in the language of standard contracts and, of course, in most of the legal opinions cited. Wherever practical—examples, problems, occasionally in text—we made an effort modify this usage in favor of a neutral or "he or she" approach.

made to clarify points by means of illustrations gathered from engineering practice and from actual construction contracts. Fictitious names have been used in illustrations from fact.

Apart from source material acknowledged as such, it is conceivable that an occasional statement or passage in the text will appear to have been taken directly from another publication. Any such circumstances are entirely unintentional. It can readily be appreciated from the nature of the subject that it is virtually impossible to avoid defining some technical terms, or once in a while describing a rule of law, in language which closely resembles that employed by other writers in the field. There are, after all, a limited number of ways in which to discuss the basic legal principles which have been known, applied, and talked about for many years.

The fourth edition has been completely updated in recognition of the dynamic changes in the law in recent years and every effort has been made to provide current and accurate information. One must be aware, however, that the changes continue, that rules vary from state to state and that every rule has exceptions. Thus, neither this nor any other text is a substitute for a lawyer's help. The expectation of the authors is that a study of this text will enable the reader to recognize when that help is needed.

Joseph T. Bockrath

DUNHAM AND YOUNG'S CONTRACTS, SPECIFICATIONS, AND LAW FOR ENGINEERS

PART ONE

LAW OF CONTRACTS

Chapters 1 and 2 provide a basic introduction to the American legal system. The purpose of Chapters 3 to 7 is to explain the basic principles of law applicable to contracts. The discussion has been condensed so that the reader can secure an overall view of what is involved without having to read too extensively or be burdened with an undue number of details. Illustrations have been given and cases cited in order to convey the ideas as clearly and in as interesting a manner as possible. For a more comprehensive treatment of the subject or any aspect thereof, one may consult any of the standard legal treatises in the field of contracts.

CHAPTER 1

INTRODUCTION

1-1 Law and Engineering Dramatic increases in government regulation in the past few decades have made it imperative that engineers be at least conversant with the basics of law.

Law is a learned profession, and even a modest mastery of it requires years of concentrated study. Of course, mastery is unnecessary for those not going in to the profession. What is necessary for engineers who want to expand their interests is an elementary understanding of how law works. Law is a dynamic discipline, and even lawyers have fewer answers than the public might believe. What they do have are the questions. Learning to answer the questions would require years of study and a restructuring of the thought process, but it is unnecessary; developing a sensitivity for potential legal problems is a relatively short-term and attainable goal. If an engineer, after a study of this work, can recognize situations which give rise to legal questions, the book will have succeeded.

An engineer ordinarily uses a logical method to solve problems, trying generally to identify and control variables and eliminate guesswork from the final equation. Objectivity is important. The result of the process generally has few subjective components or disregards them in the interest of maintaining the integrity of the scientific process.

In law, however, the logic of the process is secondary to the fact of the result. The life of the law, said Holmes, has not been logic; it has been experience. Experience and analogy are most important in the process of legal thought. This ability to analogize and to know when not to do so is critical to the lawyer's training. While many legal analogies may seem absurd to those to whom factual

3

variables are items to be controlled or eliminated, the facts may be legally similar to the questions at issue. It is result rather than method with which the lawyer must be concerned. A number of reasons for this are proffered. Lawyers, rather like engineers, must please clients rather than themselves. It has been said that "the client cares little for a beautiful case." A favorable outcome is what is sought.

Much of the difficulty in this area is inherent in the adversary process. Unlike the engineer, the lawyer is engaged not to find a truth but to present his or her client's position and counter, if possible, the arguments of the opponent. Although a lawyer should not subvert the truth, concern with the correct or "just" resolution of the issue is beyond the province of the individual lawyer. The legal process is one devoted to the resolution of conflicts. Particular laws may be either arbitrary or based on historical irrelevancy or error, but they equally well serve a unitarian purpose if everyone (at least the lawyers) knows what they are and that they will be applied in like circumstances in the future.

To achieve a result in a reasonable time and with at least a semblance of right reason the law regularly utilizes a device which science may find anathema and akin to putting up the solution of a mathematical problem to a vote: compromise. This method of resolving conflicts may be utilized in the law either for expediency or in recognition of the fact that rarely is the side of righteousness easily discernible or all with one party. A jury's finding of guilt based on marginal evidence may be balanced by a light sentence. Likewise, although negligence may in theory be, like pregnancy, a yes or no question and completely separate from damages, a jury unsure of the liability but sympathetic to the victim may compromise its unsureness by finding for the victim but with lower damages. Similarly, the plaintiff's lawyer may settle for less than full compensation if the liability issue is unclear.

The idea of balancing interests in determining a result is deeply rooted in all areas of the law. Injunctions are issued after a balance of equities; custody litigation weighs the interests of the child, the parent, and the community. The "right" answer from the law's view is often fully satisfactory to neither party. The scientist's effort to tolerate if not accept this aspect of legal methodology may determine his or her usefulness as a participant in the making and implementation of policy decisions with a scientific component. It is not rare for the engineer studying law to be appalled at the impotence of the judicial system with respect to seemingly unwise or foolish legislative or administrative action. This view, however, is myopic in two respects.

The legislature, in theory if not in fact, does not have a license to act other than that granted by the voters. The voters' recourse, should they choose to exercise it, is obvious, however slow-working it may be. To think that judges are not subject to political pressures and prejudices is to be naive. Many judges are elected, and most of those appointed are politicians or party loyalists. Appointment to the bench, even for life, cannot displace one's political and social views acquired over a lifetime.

Most significant, however, is the simple fact that judges and lawyers are human, subject to the same prejudices, lapses, and foibles as any citizen. Referring to oneself as "the court" rather than "Judge Smith" does not obscure the fact that a person is deciding questions within the confines of a system designed, or at least accumulated, by human beings.

As Cardozo phrased it, "It is often through these subconscious forces that judges are kept consistent with themselves, and inconsistent with one another. All their lives, forces which they do not recognize and cannot name have been tugging at them—inherited instincts, traditional beliefs, acquired convictions; and the resultant is an outlook on life, a conception of social needs, a sense in James's phrase of 'the total push and pressure of the cosmos,' which, when reasons are nicely balanced, must determine where choice falls. Deep below consciousness are other forces, the like and dislikes, the predelictions and prejudices, the complex of instincts and emotions and habits and convictions, which make the man, whether he be litigant or judge."

Speaking of the process of judicial decision making, it has been suggested that "one wills at the beginning the result; one finds the principle afterwards; such is the genesis of all judicial construction." While this is probably an exaggeration when applied to persons schooled in the value of precedent and continuity, it is true in one degree or another to all persons who face choices. It would be less or more than human for persons not to tailor their logic or procedure in some manner to compel the desired result.

Law should be thought of as a process rather than a set of rules: A process by which disputes are resolved. Oftentimes the dispute has no right or wrong but rather is a contest between competing interests, both of which are legitimate. Rules exist to be sure, but economic and political pressures and considerations are a part of the process, as are the personalities of the litigants, judges, and attorneys. To expect perfect justice from a process of human intervention and administration is to seek disappointment.

1-2   A General Picture of Engineering   The field of engineering involves much more than the utilization of materials and machines. As the engineer advances toward top executive and managerial positions he or she must work effectively with businesspeople, architects, lawyers, economists, and personnel managers, as well as with other engineers.

Much of an engineer's work involves the drawing of construction and other contracts, often with the assistance of a lawyer. He or she must know how to cooperate with others, how to weigh the advice given and the data produced by research, how to make important decisions without inordinate delay, and how to see to it that such decisions are carried out promptly.

In college the principle categories of engineering are necessarily separated from one another. The usual departmentalization is chemical, civil, electrical, mechanical, and industrial, perhaps together with other areas made specialties by particular institutions. In engineering practice these artificial dividing lines tend to disappear. As the engineer gains experience, he or she gradually acquires useful information in fields of endeavor allied to a particular job, becoming a "professional engineer" in the comprehensive sense of the term. Although an adequate college course in a given engineering subject helps tremendously in getting a start in the desired direction, the student must perforce continue study and preparation on the job.

1-3 Contacts with Lawyers Among the ways in which an engineer may have occasion to deal with lawyers are the following:

1 Preparing contract documents and specifications
2 Conducting contractual relationships and the handling of claims and payments
3 Interpreting contract clauses and the settlement of disputes arising therefrom
4 Educating attorneys regarding engineering matters and customs
5 Preparing material for use as trial evidence
6 Serving as an expert witness on technical points at arbitration hearings and court proceedings
7 Settlement of disputes out of court
8 Giving assistance in tax and valuation problems
9 Handling engineering matters connected with the purchase and sale of property and goods
10 Preparing reports on zoning, environmental, and other matters involving governmental agencies or organizations

At one time or another, the engineer may become involved in a variety of legal proceedings, including the following:

1 Regular jury trials
2 Cases heard before one or more judges
3 Hearings before special commissions (such as the Interstate Commerce Commission)
4 Councils of review (for example, special hearings regarding a city's building code or zoning ordinances)
5 Arbitration of disputes

Certainly what you know is important at such times, but so also is your ability to handle yourself expertly under trying circumstances and to cooperate fully with the attorneys with whom you are working.

It is the purpose of this text to give engineers some understanding of the fundamental principles and broad guidelines of the law but not to confuse the issue with the many exceptions and details which should concern only the lawyer. In other words, engineers should know enough about the basics of the law to recognize situations in which the assistance of a lawyer should be sought, preferably before difficulties actually arise.

1-4 Definitions The following definitions will serve for the chapters to follow.

Drawings are sketches or finished line drawings, as well as any notes of explanation or instruction thereon. Prints and other reproductions of drawings are generally deemed to be the equivalent of the originals from which they are made.

Owner denotes the individual or organization for whom something is to be built or furnished under contract. The owner is thus the purchaser, who pays for the goods or services.

Engineer refers to the architect or the engineer (or both) who acts for the owner in the transaction. The term may denote an engineering organization as well as an individual. Sometimes it happens that an individual or organization will occupy the dual role of owner and engineer.

Contractor is the party who undertakes for a stated price to supply goods or to perform a construction job or other project for the owner. In the practice of architecture and construction the contractor not only controls the work of construction but also acts as intermediary between the engineer or architect, who designs the work, and the artisans, who execute it.

The term *engineering* will be used in an extremely broad sense to include the work of all those engaged in design and related activities having a necessary connection with construction contracts.

1-5 Responsibility of the Engineer An engineer may be engaged by the owner (and placed temporarily on the latter's payroll) to handle all the engineering work, including preparation of the contract, for one specific job. Or the engineer may perform all these duties on a fee basis in the capacity of an outside consultant, using his or her own personnel and facilities. Or the engineer may be one of the owner's regular staff members, handling whatever engineering work comes along in the course of business. Whatever the nature of the relationship, the engineer owes the owner a duty of good faith and loyalty.

1-6 Phases of the Development of Construction Projects The engineering endeavors that precede the physical phase of a construction project may be divided into the following stages:

Preliminaries The first steps to be taken are making preliminary studies and cost estimates, selecting a site, preparing a project feasibility report, and justifying the proposed expenditure. After the project is approved and authorization to proceed is obtained, there is arranging for the financing and securing any property, rights-of-way, and permits. Next comes organizing an engineering staff or contracting with an engineering firm to handle the project.

Planning The next essential phase is the making of property and topographic surveys and subsurface explorations, establishing the broad principles and general character of the various parts of the project, and making preliminary layouts of the main features involved.

Designing In a broad sense this phase constitutes the further developing of the general layout and determining just what is wanted. This includes settling of the details of the flow sheet (sequence of operations of an industrial plant); locating the structures exactly; adapting the project to suit the topography; planning the transportation facilities; determining the sizes and types of structures to be built, together with the materials to be used for construction; choosing the architectural features; investigating methods and facilities to be used for construction; deciding upon any provisions for future expansion; and determining many other general features.

Dimensioning   This phase includes developing the detailed plans for all parts of the project, the dimensioning of all elements of the structure, and preparing the drawings.

Preparing the Contract Documents   This phase includes completing the contract drawings, specifications, and all other papers containing data needed to enable contractors to bid on the job. Included also should be preparing the engineer's estimate of the cost of the project.

Securing Proposals   The next phase is advertising for bids and furnishing plans and specifications to prospective contractors.

Awarding the Contract   Included in this phase are receiving proposals from the contractors, comparing these proposals, determining the successful bidder, and signing the contract by both the owner and the builder.

Proceeding with Construction   During this phase plans are converted into reality. Keys to the success of the project are sound management and organization, including the utilization of special skills and know-how and the proper handling of personnel, materials, and machines. A knowledge of the law of contracts and of the general conduct of business is also very important.

1-7   Licensing of Engineers   In most states the law requires a professional engineer or architect to procure a license and a certificate of registration as a prerequisite for practicing in that state. Generally the engineer who is in direct charge of the engineering department of a corporation or of the engineering division of a partnership which carries on engineering work in a particular state will have to have a professional engineer's license issued by that state, whereas those in subordinate positions need not have such a license if the licensed engineer in charge signs all plans or has them issued under his or her seal. A subordinate can normally work for or under the supervision of a licensed engineer without having a license of his or her own because in such an instance the employee does not technically assume responsibility; however, it is advisable for any such professional employee to secure a license as soon as possible. Possession of this document does not make the work any better, but, in the eyes of the public, it carries weight as a sort of recognized standard of ability and accomplishment.

If an engineer has a license in one state and wishes to extend his or her practice to another one where authorization is similarly required, a license from the second state will have to be secured. Holding papers in the first state, through reciprocal agreements or custom, may help in obtaining the license in the second one. The engineer should make advance application to each state in which professional opportunities are anticipated. The initial cost of the license varies according to the regulations of the particular states, and so does the amount of the annual dues to maintain it.

In order to be eligible for licensed status, the applicant must pass examinations of both a theoretical and practical nature prepared by the state's licensing board or other duly constituted authority. Generally, four or five years of acceptable practical experience are also required subsequent to graduation from college before the license will be granted. Many states now permit a graduate of an acceptable college to take the examination in theoretical subjects promptly upon graduation; if this is passed, the graduate becomes an "engineer in training." With the accumulation of sufficient years of experience and the passing of a practical examination, the engineer is ordinarily then eligible for a professional license and certificate. In some states extensive and excellent practical experience may sometimes be excused from the formalized requirements for securing a license.

In designing for the construction of any structure, great care should be taken to guard against features which may endanger the safety of persons or property.

Even if the engineer is an employee, that is, not personally responsible for the project, if something unsafe is about to be done, he or she should call it to the attention of whoever is responsible, doing so not as a critic but as one who is interested in the safety and welfare of all concerned. It goes without saying that whoever approves the drawings and specifications should be personally satisfied with the design and never approve plans which do not meet acceptable standards of engineering.

In a large organization where much of the work has to be delegated, the engineer has to trust the aides who handle many of the details. In no case, however, should he or she sign off on designs made by other persons or organizations without a careful check of their safety.

QUESTIONS

1 Under what circumstances may an engineer or architect have occasion to deal with lawyers?

2 In what kinds of legal proceedings might an engineer be involved?

3 Define "owner," "engineer," and "contractor" as used in contract documents.

4 What are the engineer's responsibilities in handling contract work for the owner?

5 In what ways does the engineer serve the owner?

6 Why may some knowledge of the law be helpful to an engineer?

7 Distinguish between written and printed information.

8 Distinguish between a mistake in design by the engineer and failure to use good judgment.

9 What should be the relationship between the owner and the engineer in the conduct of (*a*) contract work? (*b*) design work?

CHAPTER 2

LAW AND COURTS

2-1 The Law in General *The law* is such a broad term that it is difficult to define. The following statements convey some of the meanings associated with the term:

1 Law means a rule of civil conduct; it commands what is right and prohibits what is wrong.

2 Law constitutes the rules under which civilized individuals and communities live and maintain their relationships with one another. It includes all legislative enactments and established controls of human action.

3 Webster's Dictionary gives the definition "A binding custom or practice of a community: a rule of conduct or action prescribed or formally recognized as binding or enforced by a controlling authority."[1]

4 "Law is that which is laid down, ordained, or established. A rule or method according to which phenomena or actions co-exist or follow each other. That which must be obeyed and followed by citizens, subject to sanctions of legal consequences."[2]

5 Law denotes those rules, standards, and principles which are applied by courts in the decision of controversies.

6 Law relates to such rules as are declared and published by government as a means of developing and maintaining order in society.

Law includes court decisions as well as legislative acts, and it is these applications of rules to facts that often serve as the cutting edge of legal change.

---

[1] *Webster's New Collegiate Dictionary,* G.&C. Merriam Company, Springfield, Mass., 1977.
[2] *Black's Law Dictionary,* West Pub. Co., 5th ed., 1979, p. 795.

2-2   Basic Divisions of Law   *International law* has to do with relations among the countries of the world. It embraces such obligations as one nation owes to another and also the conduct of a nation's citizens toward other nations and their citizens.

• *Substantive law* creates, defines, and regulates duties, rights, and obligations.
• *Procedural law,* on the other hand, prescribes methods of enforcing rights and of obtaining redress in case of their invasion.

• *Constitutional law* has to do with the organization of government, the interpretation of constitutions (which represent the fundamental law of the state and nation), and the validity of statutory enactments with respect to constitutional provisions.

• *Administrative law* deals with rules and regulations established by the executive branch of government through its various agencies, commissions, and the like.

Criminal law defines and prohibits the several types of crimes and provides for their punishment.

Maritime law (or *admiralty law*) includes jurisdiction of all transactions and proceedings with relation to commerce, navigation, and injuries and damages sustained upon the sea. It is a distinct group of traditional practices and rules. The Constitution of the United States has placed this separate jurisprudence in the federal realm, rather than in that of the states. Any nation of the world may adopt the precept of the maritime law as its own, but each nation is free to incorporate whatever modifications and restrictions it deems advisable.

• *Commercial law* is a term having reference to the various rules established, more or less on a worldwide basis, for the control of commercial transactions. The general commercial law is said not to be circumscribed within purely local limits but to partake of an international character. Although the laws of civilized countries are in substantial agreement on many questions affecting the property rights and the relations of persons engaged in commerce, it is also true that commercial law cannot truly be separated from the domain of the particular state or nation whose authority makes it effective.

2-3   Statute Law   A *statute* is a rule of conduct enacted by the duly authorized legislative authority.[3] A statute represents the express written will of the lawmaking power and is rendered authentic by promulgation in accordance with certain prescribed formalities. Court decisions implementing and interpreting a given statute are of great importance. Such decisions, of course, are not themselves statute law; they belong, rather, in the realm of common law (see Art. 2-4).

The Constitution of the United States is the supreme law of the land, and state and federal statutes which contravene it may be declared unconstitutional by the judiciary. Similarly, the constitution of each state serves as the paramount law with respect to state statutes, unless such statutes contravene the United States

---

[3] According to common usage, the term "statute" applies to acts of Congress and to laws passed by the supreme legislative bodies of the several states. On the other hand, enactments at the municipal level (a muncipality being a political subdivision of the state) are referred to as "ordinances."

Constitution or intrude into the federal realm. Apart from constitutions, statutes, and ordinances, the written law in this country can be said to encompass (1) international treaties (trade agreements and the like), (2) orders and decrees of the executive branch of government, and (3) regulations and rulings by state and federal administrative agencies, and latter typified by the Federal Trade Commission and the Environmental Protection Agency.

An enormous quantity of published material is required to handle the ever-increasing volume of the written law. The *United States Code,* whose preface declares that the work "contains a consolidation and codification of all the general and permanent laws of the United States," is subdivided for convenient reference into fifty titles (for example, Title 35, Patents; Title 49, Transportation; Title 41, Public Contracts). Then, too, there are a great number of statute books carrying the laws passed by state legislatures. Thus, the consolidated laws of New York, 1969, consist of sixty-five titles in multiple volumes kept up to date by means of supplements. In addition, there is a wealth of miscellaneous published material embracing such works as administrative regulations, law dictionaries, treatises, and textbooks on various subjects in the law. Although the foregoing publications are the special province of the attorney, the engineer may have occasion to assist the former in researching technical matters.

2-4   Common Law   The *common law* comprises those maxims and doctrines which have their origin in court decisions and are not founded upon statute. The rules thus established through the judicial process are not inflexible but lend themselves readily to such modification as particular facts or situations warrant. Many common-law rules, however, have been codified and are now part of the statutory law of the United States or one of the several states.

A New Jersey case[4] outlines the nature of the common law in this fashion:

> The common law is described by Blackstone as the unwritten law (lex non scripta) as distinguished from the written or statute law (lex scripta), i.e., enacted law. . . . Kent says it includes "those principles, usages and rules of action applicable to the government and security of persons and property, which do not rest for their authority upon any express or positive declaration of the will of the Legislature" (*Kent's Com.,* 471).

Recent cases admit the existence of a field of federal common law which has been given necessary expansion into matters of federal concern where no applicable federal statute exists. However, there is no such thing as a national common law applicable in the state courts of each of the states of the United States.[5]

The United States has primarily a common-law system of jurisprudence,[6] much of it originating in England. The English common law, in turn, drew heavily upon the old Roman law. In addition to its English heritage, our common law has

---

[4] *In re Davis' Estate,* 134 N.J. Eq. 393, 35 A.2d 880, 885 (1944).
[5] *Fiddler v. Rundle,* 497 F.2d 794 (3d Cir. 1974).
[6] The state of Louisiana is a notable exception in that most of its law has as its basis the Louisiana Civil Code, patterned after the Civil Code of France and certain laws of Spanish origin.

developed in line with our own usages and customs, and a number of civil-law principles, such as that of community property, have had widespread adoption.

As circumstances change and the need arises, statutes are enacted which result in modifying or in obviating entirely some particular aspect or rule of the common law. Thus, there is the general principle of common law known as "caveat emptor" ("let the buyer beware"). Continued abuses in the manufacture and sale of food and drugs eventually led to passage of the Pure Food and Drug Act, which was designed to safeguard the public health and welfare. Similarly, questionable practices in the promotion and sale of securities dictated enactment of the so-called "blue-sky laws," aimed at protecting the investor.

To illustrate the operation of judicial precedents and the role of the common law, assume that a dispute arises between an owner and a contractor regarding the meaning of the words "subsurface construction" in a contract. The question is whether this term refers to all construction below ground level or merely to whatever construction is to be finally covered with earth. The dispute is taken to court. The court decides that, under the particular wording of the contract and considering the intent of the parties, justice demands that the dispute be settled in a certain way and that the contractor and the owner fulfill their respective obligations accordingly. This official decision acquires a name and citation by which it may readily be located in the published reports of the cases. Henceforth, if other litigation arises in which all significant circumstances are analogous to those in this previously settled dispute, the prior decision (unless it has been since reversed by a higher court or criticized in other cases) will be treated as a precedent of considerable weight.

It sometimes happens that a particular point in dispute is decided one way by a court in state *A* and otherwise by a court of equal rank in state *B*. It is obvious that these two conflicting decisions may well lead to confusion in the future. The lawyer for one side in some subsequent case (in a third state) involving the same point will cite whichever of the prior decisions tends to support his or her side of the argument, while the opposing attorney, understandably, will cite the contrary holding. The court must then determine whether or not there were any fine points of difference, either legal or factual, between the two cited cases and whether or not either (or both) is truly applicable to the dispute at hand. Assuming that both prior decisions are relevant and that there is no factual basis for distinguishing between them, the court must choose whichever line of reasoning it thinks superior.[7]

A reported judicial decision normally gives the pertinent facts involved, explains the applicable law, and states the court's conclusions. The accumulation of published court decisions from throughout the country is vast. Engineers normally have little need to read the text of an opinion; when the occasion does arise, the librarian of a law library can be of invaluable assistance in finding the relevant decision.

---

[7] Note that three separate states are involved. If, however, the court in the third case happens to sit in state *A* or in state *B,* it would be expected to follow the determination arrived at in that jurisdiction.

● **2-5 Equity**   Literally, *equity* means fairness or equality in dealing, but the term has been employed in a variety of ways. Thus, on the one hand, it is used to refer to a unique system of jurisprudence, and it is also applied to the special doctrines and remedies characterizing that system. The supplementary body of maxims, rules, and forms of action composing equity grew up as a means of filling certain voids in the common law. Suits in equity would be entertained where plaintiffs could show that plain, adequate, and complete remedies at law were not available in respect to their particular grievances. For many years this collateral system of jurisprudence was administered by separate courts of equity. These courts exercised broad discretionary powers in granting relief to aggrieved parties where justice seemed to dictate. The equity court decided what rights the respective litigants had and how such rights could be enforced. Like courts of law, the equity courts were not empowered to *create* rights, nor could they disregard applicable statutes or constitutional provisions.

Today in the federal-court system of this country and in nearly all the states separate equity tribunals do not exist, and the ancient and confusing procedural distinctions between actions at law and suits in equity have been largely removed. In today's typical jurisdiction, legal and equitable principles and remedies have been commingled in one form of action, and litigants are thereby given comprehensive relief in a single suit.

The origin and the rather limited role of equity are well described in the following passage from a 1933 Texas case[8] wherein plaintiff unsuccessfully sought the equitable remedy of injunction:

> The equity system in England grew out of the rigidity and inflexibility of the common law as interpreted by the English judges, and, as modified through the centuries, it has come to America, and its principles, more or less modified, have been adopted in all American courts. Equity, as administered in America at least, is for the purpose of giving aid to the execution of the law according to the principles of justice. It cannot be used to supplant or circumvent the law, but only to give aid and assistance to its higher and better principles. Hence the rule that equity will not interfere when the law provides an adequate remedy for any wrong. Of course, a writ of injunction is an equitable remedy and should not be used except in the execution of law and in the protection of the rights of the individual. It cannot take the place of the law or be made a substitute for it, but must be used in upholding and aiding in the due execution of the law.

Some of the fundamental principles of equity jurisprudence are embodied in the familiar "maxims of equity," a few of which are set forth below:

> "He who comes into equity must come with clean hands."[9]
> "Where one of two innocent parties must suffer, he through whose agency the loss occurred must bear it."

---

[8] *Hinds et al. v. Minus,* 64 S.W.2d 1093, 1095 (Tex. Civ. App. 1933).
[9] In other words, a litigant seeking relief at equity will be denied such if his own conduct has been improper.

"Equity aids the vigilant" and will deny relief to one who has slept on his rights and has been dilatory in prosecuting his cause of action. The undue delay is technically known as *laches*.

The remedies and courses of action which equity makes available are many. Some of the more prominent include (1) injunction, (2) specific performance,[10] (3) subrogation, (4) accounting, (5) rescission of a contract for fraud, and (6) reformation of instruments. Let us consider briefly just one of the foregoing—the injunction. Issuance of a writ of injunction by no means creates any rights but merely protects those already in existence. A preventive injunction requires that the party to whom it is directed refrain from taking certain action. The acts which the injunction is designed to prevent may be only in the stage of anticipation, or they may already have started, in which event their continuation or repetition will be the thing precluded. Whether the injunction sought is of the temporary or of the permanent variety, plaintiff must show that irreparable injury to his or her legitimate interests threatens unless defendant is restrained. Suppose plaintiff seeks an injunction against defendant's continuing trespass on plaintiff's land. The complaint would presumably contain these elements:

1 Statement about the jurisdiction of the court to which the complaint is directed
2 Allegations about the plaintiff's ownership of the realty in question
3 Facts of the trespass
4 Allegations about the inadequacy of plaintiff's remedy at law and statement to the effect that plaintiff's injury would be irreparable if the trespass was permitted to continue
5 Prayer for relief

One frequent utilization of the injunction device arises in cases involving the alleged maintenance of a "nuisance." For example, in a rather exclusive residential section of a Connecticut community a private home was being operated by its owners as a sort of part-time tavern. The neighbors claimed that this situation constituted a nuisance and were able to procure an injunction preventing the occupants from using the structure in any fashion other than as a dwelling.

2-6 Our System of Courts The courts are the last resort for the settlement of disputes. It is therefore desirable for an engineer to have some idea of the setup of the various court systems in the United States and of the kinds of matters which the several categories of courts handle.

---

[10] This decree is appropriate in instances where money damages would be inadequate. Assume, for example, a buy-and-sell agreement concerning a famous racehorse. The prospective purchaser is entitled to that particular horse and not merely to a substitute. The subject matter involved is unique, and, should the seller renege, a court would doubtless compel her to go through with her bargain. For further discussion of the remedy of specific performance, see Art. 6-32.

1 The federal courts deal with those civil and criminal cases which involve (*a*) federal laws or the Constitution, (*b*) disputes between two or more states, and (*c*) disputes between citizens of different states:
  a Supreme Court of the United States. This court, as its name indicates, has the most authority of any in the country. Virtually all its business lies in reviewing decisions of lower courts.
  b Circuit courts of appeals. There are eleven of these courts in addition to the Court of Appeals of the District of Columbia.[11] These "intermediate " courts, sitting in various cities across the country, hear such cases as are appealed from federal district courts. The decisions of the circuit courts of appeals may go to the Supreme Court for review.
  c District courts. There are over a hundred of these courts in the United States, one or more located in each state. These tribunals exercise a broad original jurisdiction of federal matters; appeals may be carried to the proper circuit court or, in unusual circumstances, directly to the Supreme Court.
  d Special courts. Examples of such courts, which are generally created by Congress as the need arises, are the Court of Claims, the Court of Customs and Patent Appeals, and the Tax Court of the United States.
2 The state courts have concurrent jurisdiction with federal courts in most instances and also deal with disputes not within the jurisdiction of federal courts. California will serve as a reasonably typical example. The California system includes the following:
  a Supreme court. This is the highest judicial forum in the state and hears appeals from lower state courts.
  b District Courts of Appeal. As in the federal system, these "intermediate" courts hear cases on appeal from the superior court.
  c Superior courts. These are trial level courts of general jurisdiction, located in each county of the state. Their jurisdiction is both civil and criminal, and their workload may include the trial of felonies, divorce, contract and tort cases, and the like.
  d Municipal and justice courts. Administered on a county basis, muncipal and justice courts have jurisdiction over misdemeanors and civil cases involving small amounts.

QUESTIONS

  1 Define "law."
  2 Briefly define each of these terms: (*a*) muncipal law, (*b*) substantive law, (*c*) constitutional law, (*d*) international law, (*e*) criminal law, (*f*) administrative law, (*g*) maritime law, (*h*) commercial law.
  3 Define and illustrate "statute law."
  4 Define and illustrate "common law."

[11] The number of circuit courts of appeal increased by one in 1980 when the Fifth United States Circuit Court of Appeals was divided creating a new one, the Eleventh Circuit, headquartered in Atlanta.

5 What constitutes the written law of the United States?

6 What is the United States Code?

7 Explain the effect of court decisions upon both statute law and common law.

8 In what form are court decisions recorded? Beyond the opinion itself, what do these reports usually contain?

9 Can statute law change a common-law principle?

10 Describe the setup and general operation of the federal courts; of state courts.

11 Define "equity."

12 What was the origin of equity jurisprudence?

13 State three of the maxims of equity.

14 List several of the remedies available in equity.

15 Define and illustrate "specific performance."

16 What is the supreme law of the land?

17 Is the common law uniform throughout the United States? Explain.

CHAPTER 3

DEFINITIONS

3-1 The Nature of Contract For the practicing engineer, contract law is a most important field of study. Its scope is vast, and to attempt to cover all its ramifications would be incompatible with the purpose of this book. The effort here will be to treat briefly the various kinds of contracts and their interpretation, modification, rescission, performance, and breach.

The law of contracts undergoes a constant process of evolution because of changing customs and practices. Broad contract principles are subject to exceptions and modifications, but the rules and principles discussed here are, for the most part, those that have received general acceptance.

The definitions of "contract" put forth over years of slowly changing usage and convenience are legion. Section 1549 of the Civil Code of California (West 1982) defines a contract as "an agreement to do or not to do a certain thing." The American Law Institute, in its *Restatement of Contracts,* second edition, 1981, section 1, declares that "a contract is a promise or a set of promises for the breach of which the law gives a remedy, or the performance of which the law in some way recognizes as a "duty." As will subsequently appear, the contractual arrangement must involve competent parties and be based upon a legal consideration.

A contract is an understanding enforceable at law, made between two or more persons, by which rights are acquired on the one side to acts or forbearances on the other. To make an agreement which results in a contract, there must be an *offer* and an *acceptance*; and to the promises which stem from the offer and acceptance the law attaches a binding force of *obligation.*

3-2 Express Contracts and Contracts Implied in Fact These differing terms are used to indicate a variation in the character of the evidence by which the contract is proved. The same basic elements are essential in both types.

Technically, an *express contract* is one whose terms are declared by the parties in so many words, either orally or in writing, at the time the agreement is made. An express contract involves an actual promise, while the *implied* type is a matter of inference or deduction from facts and circumstances showing a mutual intention to contract. An implied contract is an *actual* contract, circumstantially proved. To look at the matter from a different angle, the sole difference between an express contract and one implied in fact is that in the former all the terms are set forth by the parties themselves while in the latter the law must imply one or more of the terms from the conduct of the parties. Where one performs for another with the other's knowledge such a service as in the ordinary course of events draws compensation and the recipient of the service expresses no dissent or avails himself of the benefits, a promise to pay the reasonable value of the service is implied.

Both express and implied contracts are founded on the agreement of the several parties and require a *meeting of the minds.*[1]

3-3 Quasi Contracts Obligations created by law for reasons of justice are variously termed *constructive contracts, contracts implied in law,* or, perhaps most popularly, *quasi contracts.*

Quasi contract is a convenient term for a multifarious class of legal relations which possess this common feature: That without agreement—expressed or unexpressed—and sometimes without delict or breach of duty on either side, party *A* has been compelled to pay or provide something for which party *X* ought to have paid or made provision, or *X* has received something which *A* ought to have received. The law in such cases imposes a duty upon *X* to make good to *A* the advantage to which *A* is entitled. Such "contracts" are not real contracts at all but fictions of the law, based on principles of unjust enrichment and without regard to any actual or presumed intent of the party held to be bound—often, indeed, in the face of the party's dissent. There is no "agreement" whatever involved, but these quasi-contractual obligations are clothed with the semblance of contract for the purpose of setting up the remedy. The key difference between quasi contract and true contract is that in the former, the duty defines the contract; in the latter, the contract defines the duty.

3-4 Unilateral and Bilateral Contracts In a *unilateral contract* only one of the contracting parties makes a promise, and that promise is exchanged for an act or

---

[1] Of interest to the engineering profession is the fact that, unless specifically required by statute to be in writing, a contract for construction or repair work may be implied. Where, however, a contractor sublets the contract, the law will not imply any agreement on the part of the owner to compensate the subcontractor for the work performed. The implied obligation of the owner to make compensation to one who improves his or her property is deemed taken away by the special contract between the contractor and the subcontractor.

an executed consideration. In other words, in a unilateral contract there is but one promisor, and he or she is the only party under an enforceable legal duty.[2] A *bilateral contract* comprises mutual promises, with each contracting party playing the dual roles of promisor and promisee. The legal effects of a bilateral contract are reciprocal duties and obligations. To put the distinction another way, the exchange for the promise in a unilateral contract is something *other than* a promise, while in a bilateral contract promises are exchanged and there is something on both sides to be done or forborne.

Suppose Lewis offers to sell Martin a load of topsoil for $12. If Martin accepts by sending the $12, a unilateral contract exists. If Lewis had asked for a *promise* of $12 and Martin had given such promise, the contract would be bilateral. If a baseball fan purchases a reserved-seat ticket for a certain game at Yankee Stadium, the baseball club promises to hold that particular seat for the fan and to let him or her occupy it at the game in question. The foregoing constitutes a unilateral contract, since the baseball club is the only party with a promise outstanding. On the other hand, if the Yankees contract with the General Electric Company for 500 bulbs for the stadium floodlights at a specified price, the arrangement is bilateral since the company promises to deliver the bulbs and the Yankees promise to pay for them.

3-5   Joint and Several Contracts   Whether the rights and duties created by a particular contract are intended to be (1) joint or (2) several or (3) joint and several is sometimes a difficult question of interpretation.[3] Promises of a number of persons (which promises appear in a single instrument) are presumed to be joint unless a contrary intention is evident from the inclusion of obvious words of severance. Some states have laws declaring that contracts in the form joint shall be joint and several.

Where a contract is *joint,* the various obligors must all be joined as parties to any action brought upon the agreement, and release by plaintiff of one obligator discharges the others as well. In contrast, each person bound on a *several contract* has a liability separate from that of any fellow obligor thereunder, and each individual's obligation must be separately enforced.

The distinction between joint and several contracts hinges on the answer to this question: Did all the persons obligated under the agreement promise one and the same performance, or did each one promise only a separate portion of the total? An illustration or two may serve to point up the difference.

Suppose several stockholders employ Y on a commission arrangement to procure a purchaser for their aggregate holdings in a certain corporation. Y's ser-

---

[2] Perhaps the most typical example of a unilateral contract is the *option,* common to real estate transactions. The option is in essence a contract to give another the right to buy, and the party granting the option thereby agrees to sell the property at a certain price at any time within a stipulated period, leaving the prospective purchaser with a clear choice in the matter and (having paid something for the option) under no further contractual obligation.

[3] To simplify the consideration of what is at best a tricky concept, we shall refer only to the duty side of the picture. It should be borne in mind, though, that rights under a contract may likewise be joint or several.

vices are being retained for the joint advantage of all the stockholders involved and not for the particular benefit of any one of them. The contract is joint in the sense that each of the obligor-stockholders is potentially liable for the entire commission. If *Y,* having performed her part of the bargain, needs to bring suit to recover the stated compensation, she must join as defendants all of the obligor-stockholders. Should any one of the latter group be obligated to satisfy the judgment, he would have available a right of contribution as against his fellows.

The case of *O'Connor v. Hooper* [4] will serve as a good example of the opposite type of arrangement. There the contract was entered into between a construction company and certain owners of adjoining property fronting on a street. Each property owner, "contracting severally. . .each for himself and not for the others," promised to pay a stated pro-rata share of the total cost of the construction company's prospective work. The court took the view that the effect of the contract was the same as if the various property owners had written their promises on separate pieces of paper; the obligations created were several and not joint, each signatory owner having expressly disclaimed responsibility for the promises of any of the others.

Finally, there are such agreements as *joint and several contracts* whereby the obligors are bound separately as individuals and are also bound together for the rendition of one and the same performance. Assume an agreement which reads in part, "We, John Smith and Henry Jones, jointly and severally promise to pay Tom Mason $1000." Upon default, separate actions could be instituted by Mason against Smith and against Jones; or one suit could be brought against both defendants. The defendants cannot be sued *both* ways, however—that is, both jointly and separately.

3-6 Entire and Severable Contracts A contract is *entire* when full and complete performance by one party is a condition precedent to the right to require performance by the other party. In such contracts there is no liability for part performance, and failure of one party fully to perform relieves the other of any obligations. Where the drilling of a well is to be performed as a whole and the compensation for such work is to be paid as a whole, the contract is entire, even though the payment is expressed as a stipulated amount per linear foot. Similarly, a contract for the purchase and sale of 1000 head of cattle for delivery within sixty days is entire, although deliveries of 50 or 100 were to be made on various occasions within the time set in the agreement. And where one party agrees to furnish the other a certain quantity of crushed stone for a fixed price per cubic yard at a stated rate per day, the contract is entire, requiring full performance before payment of the consideration is due.

Still another example of this type of contract is afforded by the case of *Kelly Construction Co. v. Hackensack-Brick Co.* [5] Plaintiff in that litigation had a contract for the erection of a high school. It placed a written order with defendant for

---

[4] 102 Cal. 528, 36 P. 939 (1894).
[5] 91 N.J.L. 585, 103 A. 417, 418 (1918).

the furnishing "and stacking on the job of all the common hard brick required by the plans and specifications for the Englewood High School at $7.00 per thousand; brick to be delivered as required by us and sufficient brick to be kept on the job so that we will always have approximately 50,000 brick stacked until completion of the job." Neither the order nor the acceptance fixed any time for payment. Defendant, after delivering some of the brick, refused to fulfill the remainder of its obligation, contending that it was relieved of further responsibility because the brick already delivered in part performance of the contract had not been paid for. The New Jersey court declined to accept this argument, holding that:

> Where, as here, the sale of a specified quantity of brick (i.e., sufficient to complete a building according to stated specifications), the contract is entire, and a failure to pay when a part delivery has been made does not excuse the seller from completing delivery, no time for payment being stated in the contract.

A different problem is deciding whether a building contract is entire which provides for the making of "progress payments" to the contractor. The problem was addressed in *New Era Homes Corporation v. Forster,*[6] where plaintiff had to make extensive alterations to defendants' home. The contract reference to price and payment was:

> All above material and labor to erect and install same to be supplied for $3075.00 to be paid as follows: $150, on signing of contract; $1000, upon delivery of materials and starting of work; $1500, on completion of rough carpentry and rough plumbing; $425, upon job being completed.

Work was commenced and partly finished, with the first two payments made on schedule. Then, when the rough work was done, plaintiff asked for the third installment. Upon defendants' refusal to pay, plaintiff stopped work and sued for the entire balance of $1925. The defendants conceded default but argued that the plaintiff was entitled not to the amount in suit but to the amount of actual loss sustained from the defendants' breach. The court held the contract to be entire, reasoning as follows:

> Did that language make it an entire contract, with one consideration for the doing of the whole work, and payments on account fixed points in the progress of the job, or was the bargain a severable or divisible one in the sense that, of the total consideration, $1150 was to be made the full and fixed payment for "delivery of materials and starting of work," $1500 the full and fixed payment for work done up to and including "completion of rough carpentry and rough plumbing," and $245 for the rest? We hold that the total price of $3075 was the single consideration for the whole of the work, and that the separately listed payments were not allocated absolutely to certain parts of the undertaking, but were scheduled part payments mutually convenient to the builder and the owner. That conclusion, we think, is a necessary one from the very words of the writing, since the arrangement there stated was not that separate items of work be done for

---

[6] 299 N.Y. 303, 86 N.E.2d 757 (1949). See also *S. & W. Investment Co. v. Otis W. Sharp & Son, Inc.,* 170 So.2d 360 (La. 1965), and *M.F.A. Mutual Ins. Co. v. Gulf Ins.,* 445 S.W.2d 829 (Mo. 1969).

separate amounts of money, but that the whole alteration project, including material and labor, was "to be supplied for $3075."

The decision, favoring defendants' position, was that plaintiff-contractor, on defendants' default, could collect only in *quantum meruit*[7] for what had been finished *or* in contract for the value of what plaintiff had lost—that is, the contract price less payments made and less the cost of completion.

In a recent case involving modern construction technology,[8] the contract to deliver and assemble a modular home was also held to be entire, rather than severable, where the contract set a fixed price but provided for periodic partial payments which were not made. A property owner borrowed $18,500 from a bank to finance the purchase of a double-width modular home and procured a fire insurance policy designating the bank as beneficiary. The contract provided that partial payments not to exceed 60 percent of the value of the work in place, less the aggregate or previous payments, would be made at intervals of construction based upon estimates by the contractor of the value of the work in place and approved by the bank. The contractor delivered the two halves of the modular home to the lot for assembly but the home was destroyed by fire the next day. The contractor contended that the contract was severable and that he had provided labor and materials with the value of $17,785. The property owner contended that the contract was entire and, since construction was not complete, the contractor could have no recovery based on the contract.

The court declared that whether a contract is entire or severable is a determination to be made by the court according to the intention of the parties. Such intention should be ascertained from a consideration of the subject matter of the contract, a reasonable construction of the terms thereof, and the conduct of the parties during the negotiations, all of which should be viewed in light of the surrounding circumstances. With these principles in mind, the court found that, the provision for partial payment notwithstanding, the record indicated that the parties intended that the contractor would construct the modular home for a fixed price, $18,500.

> A contract is *severable or divisible* when the part to be performed by one party consists of a number of distinct and independent items and the price to be paid by the other party is apportioned, or is susceptible of apportionment, to each item. An agreement to pay a certain price for every bushel of wheat supplied which corresponds to a given sample is typical illustration of a severable contract. Another is an agreement to build six houses for a stated sum; a proportionate recovery will be permitted for the actual completion of one or more of the six dwellings. In this severable contract, a breach by one party in respect to any one item does not justify the other in repudiating the whole agreement; the latter is still bound to perform, but if he has sustained any injury as a result of the breach, he has a right to recover for such injury in a damages action.
>
> The primary criterion in determining whether a contract is entire or severable is the intention of the parties, regarding the various items as a whole or each contract item as

---

[7] Literally translated, this term means "as much as he deserved."
[8] *LDA, Inc. v. Cross,* 279 S.E.2d 409, (W.Va. 1981).

a separate unit. An illustration of this proposition is afforded by an omnibus insurance policy covering different classes of property, with each class separated from the others and insured for a specific amount; the contract to accord with the apparent intent should be treated as severable and not entire, and breach of the contract conditions for one class of the insured property will not disturb the remainder of the coverage.

3-7   Executory and Executed Contracts   In the executory contract a party is bound for the future to do or expressly refrain from doing a particular thing—as where an agreement is made to build a garage in three months. A fully executed contract, on the other hand, is one in which the object of the agreement has been performed and nothing remains to be accomplished by either party—where an article is sold, handed over, and paid for on the spot.

3-8   Voidable Contracts   The word "void" means null or ineffectual.[9] Frequently the word is construed as having the more liberable meaning of "voidable." Contracts are properly called *voidable* which are fully effectual until affirmatively avoided by some act. Such contracts are prima facie valid but are subject to certain defects of which some party can take advantage. By electing to do so, that party avoids the legal relations which the contract creates or, conversely, by ratification of the contract may extinguish the power of avoidance. Ordinarily, the power in question is confined to one party to the contract, but such is not invariably the case.

Typical instances of voidable agreements are those involving infants or those which are induced by fraud, mistake, or duress.

3-9   Unenforceable Contracts   The law does not enforce such contracts by direct legal proceedings but recognizes them in some collateral way be creating a duty of performance. For example, a perfectly valid contract may become *unenforceable* by virtue of a statute of frauds. If there is nothing in writing sufficient to satisfy the requirements of that statute, the direct judicial remedies at common law are not available to plaintiff if the defendant should choose to take advantage of the statutory provision. The oral agreement, however, is far from being without legal operation. Either party has the legal power to make the contract directly enforceable as against himself by signing a proper written memorandum; he cannot, by such a process, make the contract enforceable in his own favor.

3-10   Subcontracts   Where a person has agreed to perform certain work, for example, to erect a building, and in turn engages a third party to handle all or part of that which is included in the original contract, for example, to install the plumbing fixtures, the agreement with such third person is called a *subcontract.*

3-11   Commercial Paper   Of importance in commercial transaction is the Uniform Commercial Code, a revision and recodification of statutes dealing with

---

[9] Illegal agreements, for instance, are void.

many business subjects, such as negotiable instruments and sales of goods. Adopted on a verbatim basis in virtually every state, the code expressly provides that—except to the extent one or more is displaced by a particular clause therein—general principles of law and equity involved in any commercial transaction shall be deemed to supplement code provisions. The broad scope of the code is apparent from this commentary:[10]

> The Uniform Commercial Code, as drafted and approved by the National Conference of Commissioners on Uniform State Laws and by the American Law Institute. . .would replace the prior uniform acts on Negotiable Instruments, Sales, Warehouse Receipts, Bills of Lading, Stock Transfers, Conditional Sales, and Trust Receipts as well as the nonuniform acts regarding bank collections, bulk sales, chattel mortgages, and factor's liens; would codify the decisional law in relation to letters of credit, and, in addition, would provide statutory authority for the financing of accounts receivable.

When the Uniform Commercial Code was enacted in the typical jurisdiction, it brought relatively few changes in the law of negotiable instruments as applied to "commercial paper"—a term which Article 3 of the code uses to refer to drafts, checks, certificates of deposit, and promissory notes as these items are characterized in Section 3-104. Article 3 was designed to clarify, consolidate, and modernize various provisions of the Uniform Negotiable Instruments Act which it replaced.

A negotiable instrument is readily transferable, being negotiated by simply delivery, if the paper is payable "to bearer," or by endorsement and delivery if payable "to order." A holder in due course who, in good faith and for value, acquires the instrument "without notice that it is overdue or has been dishonored or of any defense against or claim to it on the part of any person,"[11] is not subject to claims and certain defenses of prior parties.

Section 3-104 of the Uniform Commercial Code,[12] setting forth the tests which must be met before an instrument qualifies as negotiable, reads as follows:

1 Any writing to be a negotiable instrument with this Article must:
  a Be signed by the maker or drawer; and
  b Contain an unconditional promise or order to pay a sum certain in money and no other promise, order, obligation or power given by the maker or drawer except as authorized by this Article; and
  c Be payable on demand or act at a definite time; and
  d Be payable to order or to bearer.
2 A writing which complies with the requirements of this section is:
  a A "draft" ("bill of exchange") if it is an order;
  b A "check" if it is a draft drawn on a bank and payable on demand;

---

[10] Taken from Oct. 2, 1961. Report (of the Commission on Uniform State Laws) to the Legislature of the State of New York.

[11] See 302 (1).

[12] See, for example, the same number in *McKinney's Consolidated Laws of New York Book 62½,* pt. 2 (1964).

    c A "certificate of deposit" if it is an acknowledgment by a bank of receipt of money with an engagement to repay it;

    d A "note" if it is promise other than a certificate of deposit.

3 As used in other Articles of this Act, and as the context may require, the terms "draft," "check," "certificates of deposit" and "note" may refer to instruments which are negotiable within this Article as well as to instruments which are so negotiable.

The most common example of negotiable instrument is a *check,* signed by the drawer, reciting an unconditional demand on the named drawee bank to pay a specified sum to the order of a designated person.[13] A *bill of exchange* is a written order (without condition) by which the signing party requires the person to whom the paper is addressed to pay a sum certain to order to bearer upon demand or at a fixed or determinable future time. A *promissory note* is an undertaking in writing by which the maker agrees unconditionally to turn over to the payee a stated amount of money at a given time.

Several types of endorsements play important roles in regard to negotiable instruments. A *special endorsement* names the person to whom or to whose order the instrument is payable; such instrument may be further negotiated only by the endorsement of the person named.[14] An *endorsement in blank* may consist of a signature and specifies no particular endorsee; where an instrument payable to order is endorsed in blank, it becomes payable to bearer and may be negotiable by delivery alone until specially endorsed.[15] A *restrictive endorsement* is defined as follows:[16]

    a It is conditional; or

    b Purports to prohibit further transfer of the instrument;[17] or

    c Includes the words "for collection," "for deposit," "pay any bank," or like terms signifying a purpose of deposit or collection; or

    d Otherwise states that it is for the benefit or use of the endorser or of another person.

The law of negotiable instrument ("bills and notes," as some call it) is complicated and many-faceted. Interested readers should consult the Uniform Commercial Code for a comprehensive picture of the rights and liabilities of parties involved with commercial paper.

QUESTIONS

1 Why may one wish to have work done under a contract rather than hire laborers to perform the work?

---

[13] *Certification* of a check is acceptance, and when a holder procures certification, the drawer and all prior endorsers are discharged (U.C.C. §3-411). A bank certifies a check by deducting in advance the stated amount from the drawer's account so that there is no question about the sum stated on the check being collectible.

[14] U.C.C. §3-204.

[15] *Id.*

[16] U.C.C. §3-205.

[17] Per §3-206, no restrictive endorsement prevents a further transfer or negotiation of the instrument, even though it may expressly purport to do so.

2 What is meant by "contract law"?

3 Distinguish between (*a*) "formal" and "informal" contracts, (*b*) "implied" and "express" contracts, (*c*) "executory" and "executed" contracts.

4 What are some essential elements of a valid contract?

5 What is meant by a "meeting of the minds"?

6 What is a "quasi contract"?

7 Define "unilateral contract" and "bilateral contract." Illustrate each.

8 Distinguish between "joint" and "several" contracts.

9 Distinguish between an "entire" contract and a "severable" contract.

10 What, in contract law, is a condition precedent?

11 What is the meaning of quantum meruit?

12 When is a contract voidable? When is it void?

13 Under what circumstances may a contract be unenforceable?

14 What is a subcontract? Illustrate.

15 What is meant by "commercial paper"? How is it used?

16 What is a certified check?

17 *A* gave *B* an estimate for reshingling the roof of *B*'s house of $450. *B* said, "I want to think it over." Two days later *A* arrived at *B*'s place, put up ladders and, and prepared to shingle the roof. *B* saw what was going on but went off to work, saying nothing. *A* finished the job and sought payment therefor. *B* claimed there was no contract to do the work, and hence no liability. Was *B* correct? Why?

18 *A* contracted with *B* (the owner) to build a warehouse alongside the waterfront as soon as *C* completed the adjoining bulkhead which would retain the ground on which the warehouse was to be built. *C* got into financial difficulties and was unable to finish the bulkhead. What is *A*'s obligation with respect to *B*?

19 Assume that *C* in the preceding example was to build the bulkhead for *D*. Then assume that *C* had finished 50 percent of the bulkhead work before having to give up the job. Would quantum meruit be likely to apply in determining any money which *D* should pay to *C*?

20 At the office one day Russell said, "I wish that I had a couple of sandwiches to take with me this afternoon." Jones overheard Russell's remark, went to the restaurant, and returned with the sandwiches. Is Russell obliged to pay Jones for them? Why?

21 What are the characteristics of a negotiable instrument?

22 What are the customary forms of negotiable instruments?

23 On May 1, 1978, Pratt signed a note for $500 payable to Smith on November 1, 1978. This note, endorsed by Smith to Jones, was given to Brown by Jones in payment for merchandise. On the due date, Brown presented the note to Pratt for payment, but he found that Pratt had gone bankrupt. In the meantime, Jones died, penniless. What recourse does Brown have?

24 Powell wrote a check for $100 payable to Ross. Ross endorsed it "in blank" and told his fifteen-year-old son to cash it at the bank. Can the son do so? How? What could happen to the check if the son lost it on the way to the bank and someone else found it? What would Ross have done if he had wanted to use a restrictive endorsement?

CHAPTER 4

FORMATION PRINCIPLES

4-1   The Essentials of Contract   An agreement, in order to constitute a binding contract, must be entered into by competent parties who express definite assent in the form required by law. Furthermore, such an agreement must be supported by a proper "consideration" (see Art. 4-31), must not at the time it is made be obviously impossible of performance, and must not so contravene principles of law or public policy as to be entirely devoid of legal effect. The present chapter will analyze the prerequisites to the formation of a valid contract.

COMPETENT PARTIES

4-2   The Contracting Parties   There must be a definite promisor and a definite promisee, each of whom is legally capable of playing the intended part in the proposed contractual arrangement.

Although any greater number may be involved, a contract requires a minimum of two parties since it is not possible under existing law for one to contract with oneself. It is possible, nonetheless, for an individual to act in several roles at the same time, such as those of trustee and partner. Regardless of the capacities in which a person may act, he can never contract with himself nor maintain an action against himself.

Moreover, a contract cannot obligate someone who has not the legal capacity to incur at least voidable contractual duties. Certain persons are by law in capable of binding themselves by a promise. Such incapacity may stem from one of several causes, the most significant of which are infancy and lunacy.

4-3 **Contracts of Infants** As a general rule, a minor's contract (unless it relates to so-called "neccessaries") is voidable at the option of the minor, and this seemingly inequitable circumstance explains the reluctance of businesspeople and others to enter into contractual relationships with minors. While an infant may renege on his agreements, the party of full age with whom he contracts has no corresponding option.

The basis for the unusual privilege afforded infants by the law of contracts lies in their supposed immaturity of judgment. Upon exercising this power of disaffirmance, the minor is supposed to return any property or money which may have been turned over by the other party to the repudiated contract; but this requirement in many instances represents but scant satisfaction for such other party, since an infant may disaffirm even though he has previously squandered the consideration received. A reasonably cautious person will, before contracting with an individual under the age of majority, seek protection in the form of a guaranty of performance furnished by a legally competent person.

Theoretically, the privilege given to minors to avoid contracts made in good faith is to be used solely as a shield, and not as a sword; but this is not always realized, and often the minor calls the tune completely. The following example shows how the treatment which the law grants to minors respecting their contractual obligations poses real difficulties for the other party.

Harry Jones, age seventeen, bought a new car from Edwards Motor Company, paying $600 down and signing an agreement to pay the balance on an installment basis. Jones drove away from the showroom and within the hour ran the car over an embankment. The car was demolished, but Jones himself miraculously escaped serious injury and soon was back at the dealer's office, seeking to disaffirm the purchase contract. What is more, he demanded—and succeeded through the ensuing litigation—in getting the return of his $600.

As might be expected, the passing years have brought a number of qualifications upon the minor's formerly absolute right of express disaffirmance. A discussion of these would necessarily be too detailed for this general survey of contract law, but a minor will normally be held liable for the reasonable value of "necessaries"[1] which the supplier actually furnished them on his credit, where the infant's parents or guardians are unable or unwilling to provide them with such items and where they are really in need of same.

An example of a statutory qualification is Section 3-101 of the New York General Obligations Law, which states:

1 A contract made on or after September first, nineteen hundred seventy-four, by a person after he obtained the age of eighteen years may not be disaffirmed by him on the ground of infancy.

2 A contract made on or after April thirteenth, nineteen hundred forty-one, and before September first, nineteen hundred seventy-four, by a person after he has obtained the age of eighteen years may not be disaffirmed by him on the ground of infancy, where

---

[1] Such things as food, shelter, and clothing. *Ragan v. Williams,* 220 Ala. 590, 127 So. 190 (1930).

the contract was made in connection with a business in which the infant was engaged and was reasonable and provident when made.

4-4 Contracts of a Corporation; Ultra Vires Concept The very nature of a corporation—as an entity created by law—imposes some restrictions upon its contractual powers, and the terms of its incorporation may impose others. Persons having dealings with a corporation should satisfy themselves that its activities are in compliance with the requirements of its charter and that the individuals purporting to act on behalf of the corporation are authorized.

If a corporation enters into an agreement which is beyond the scope of its powers, such contract is *ultra vires*[2] and is ordinarily deemed unenforceable. However, many courts have held that a plea of ultra vires will not avail a corporation with respect to a promised act in exchange for which it has received consideration in any case where the status quo cannot be restored. In such instances, the corporation is estopped to set up the defense of ultra vires, and the other party can enforce the bargain.

4-5 Contracts of Mental Incompetents It is essential to the validity of a contract that the parties possess the mental competence to consent. If there is to be a binding contract, the parties must possess sufficient mental capacity to appreciate the import of what they are doing. There is no binding contract where one of the parties was, by reason of physical debility, mental aberration, or other condition, incapable of understanding and appreciating the force and effect of the agreement made. But mere mental weakness, falling short of inability to appreciate the business at hand, will not invalidate the contract. Furthermore, the mental incapacity that affects the validity of a contract is that which exists at the very time the transaction occurs, and prior or subsequent insanity does not enter the picture. Accordingly, a contract normally is held to be binding if it was entered into during a lucid interval. In any event, an insane person, just as a minor, may be liable for the reasonable value of necessaries supplied in good faith.

While the contracts of insane persons are voidable and may be expressly disaffirmed by the lunatic on returning to sanity or during a temporary rational period, the sane party does not enjoy a corresponding option of cancellation. However, where there has been no judicial determination of mental incompetence and where the sane party did not at the time of contracting have knowledge of the other person's insanity, there are decisions to the effect that the contract is binding upon the lunatic if it has been so far executed that the respective parties cannot be restored to their original positions.

---

[2] There is a basic distinction between *illegal* acts, such as those committed in violation of an express statute or those which offend public policy, and acts merely ultra vires. The latter term is used to denote corporate transactions which are outside the object for which the corporation was created and which are beyond the scope of either the express or the implied powers contained in the corporation's charter.

4-6   Contracts of Intoxicated Persons   A contract made by an intoxicated person is invalid if the intoxication has rendered that person incompetent, that is, unable to appreciate the nature and extent of the contract, or if it is accompanied by fraudulent behavior on the part of the other contracting party. The mere fact that one is under the influence of liquor or other drugs at the precise time of contracting does not of itself avoid the agreement into which he has entered.

The enforceability of contracts made by intoxicated persons depends primarily upon the degree of intoxication and whether or not the person was capable of understanding the nature and consequences of the transaction. As is the case with infants and insane people, intoxicated persons will be liable for the value of necessaries received.

PROPER SUBJECT MATTER

A valid contract must have a definite and lawful subject matter. The subject matter should be clearly defined, and the object of the contract must not be illegal.

4-7   Definiteness   An important feature of any contract is that it shall describe what is to be done or what materials are to be furnished with sufficient clarity and enough detail to make evident exactly what is wanted. Courts can neither enforce nor award substantial damages for the breach of contractual agreements which are wanting in certainty. The terms of a contract must be complete, and the various obligations must be described with sufficient definiteness to enable a court to determine whether or not these obligations have been performed.

4-8   Illegal Bargains   All things not forbidden by law may properly become the subject of, or the motive for, contracts; the possible subject matter thus includes such things as future rights and liabilities, pending causes of action, and goods which may be defective or not yet in being.

It is a basic principle that a contractual undertaking must have a lawful purpose and that transactions in violation of law cannot be the foundation of a valid contract. An illegal agreement never attains the dignity of a contract, and such an agreement is void and therefore unenforceable. The taint of illegality may manifest itself in the consideration for the agreement, in a promise expressed in the agreement, or in the purpose to which the agreement is applied. Broadly speaking, a bargain is deemed illegal if either its formation or its performance is expressly prohibited by statute or seeks to accomplish a purpose which is contrary to law, morality, or public policy. It is not necessary that there exist any corrupt intention on the part of the contracting parties in order to render illegal an agreement in violation of law.

There exists no absolute rule by which to determine what contracts are against *public policy*. The term does not lend itself to precise definition, but it may be said that an agreement is against public policy if it is injurious to the welfare of the general public or impinges upon some established interest of society. By way of example, an agreement which contemplates procuring personal or political influence to secure acceptance of a bid on a public contract being let by a governmental department does violence to public policy.[3] Similarly, a contract which will inevitably result in stirring up litigation is contrary to public policy.[4] Likewise, any contract which tends to disturb family relations will be regarded as contravening public policy.[5]

In one case[6] a married man promised to marry a second woman, who was the mother of his illegitimate daughter, as soon as he divorced his wife. Later, asserting that his wife was terminally ill, he again promised marriage as soon as his wife died. The promise to marry was held invalid, even in a state where a breach of promise to marry action is recognized, because the woman knew or should have known that the man was already married. The court said that it would not enforce the contract contravening the public policy favoring stability in marriage, and therefore a promise contingent on divorce or death of a known spouse was not enforceable.

Illegal contracts are infinite in number and variety, but a representative list of such might include an agreement to commit a criminal act, an agreement to commit a tort, an agreement to defraud individuals or the public in general, an agreement in restraint of trade[7] or of marriage, an agreement to encourage litigation, an agreement tending to obstruct the administration of justice, an agreement clearly ignoring a fiduciary duty, and an agreement made in obvious violation of a statute or one which cannot be performed without necessarily violating a statute.

The effect of illegality is quite a different problem. Ordinarily an illegal [8] contract cannot be directly or indirectly enforced by either party, and there can be neither damages for breach nor return of benefits through rescission.

Of real importance from a procedural standpoint is the fact that illegality is one defense a party to the contract is never estopped from pleading as against the other party to the agreement. In other words, the defense of illlegality is always available and cannot be effectively waived, even by means of a written agreement of the several parties.

AGREEMENTS OF THE PARTIES

4-9   Meeting of the Minds   An essential in the formation of a valid contract is mutual assent to the terms of the agreement, a circumstance variously termed *offer*

---

[3] *Glass v. Swimaster Corp.*, 74 N.D. 282, 21 N.W.2d 468 (1946).

[4] *Eco-Phoenix Electric Corp. v. Howard J. White, Inc.*, 1 Cal.3d 2 66, 461 P.2d 33 (1969).

[5] *Daniel v. Daniel*, 222 Ga. 861, 152 S.E.2d 873 (1967); *Goullo v. Goullo*, 361 N.Y.S.2d 769 (1974); *Ladd v. Ladd*, 580 S.W.2d 696 (Ark. 1979).

[6] *Thorpe v. Collins*, 245 Ga. 77, 263 S.E.2d 115 (1980).

[7] There are instances of contracts which clearly restrain trade and yet are held to be perfectly valid; in such cases the restraint involved is either partial or is otherwise limited in its operation (for example, in time or place).

[8] The illegality referred to must inhere in the contract itself, rather than merely being collateral to the arrangement.

and acceptance, or *a meeting of the minds.* The manifestation of mutual assent usually takes the form of an exchange of oral or written words, though occasionally the assent is indicated in some other fashion. It should be remembered that an actual *expression* of mutual assent, rather than the mere existence of the assent, is essential.

4-10 The Nature of Offer An *offer* may be defined as the signification by one person of willingness to contract with another on certain specified terms. It is a statement by the offeror of what will be given or done in return for some desired promise or act on the part of the offeree. By means of such an offer the offeror confers upon the offeree the power to create by acceptance binding contractual relations between them.

An operative offer must be more than simply an expression of desire or hope that a contract can be effectuated. Preliminary negotiations, invitaticns to deal, advertisements, and acts evidently done in jest cannot be regarded as offers because the requisite intent to contract is lacking. Nevertheless, the line of demarcation between preliminary negotiation and operative offer is often difficult to draw. An invitation to enter into negotiations may be just a suggestion advanced with the purpose of inducing offers by others.

Similarly, a mere quotation of price should be distinguished from a real offer. The question of whether a given communication naming a price is meant to be only a quotation or is a full-fledged offer depends upon the prospective vendor's intention as manifested by the facts and circumstances of the particular case or the custom of the trade.

Extravagant terms advanced under circumstances of considerable excitement may reveal the absence of any serious intent to make a firm offer. Not what the apparent offeror actually said but how the recipient of the proposal reasonably interprets the intent provides the real key to the situation.

On occasion the courts have been known to treat a so-called "joke contract" rather seriously. In *Plate v. Durst,*[9] the defendant had promised the plaintiff $1000 and a diamond ring if she would continue certain services rendered in promoting his business. At the trial he admitted these promises, but claimed they were made in jest. The court stated:

> Jokes are sometimes taken seriously by the young and inexperienced in the deceptive ways of the business world, and if such is the case, and thereby the person deceived is led to give valuable services in the full belief and expectation that the joker is in earnest, the law will also take the joker at his word, and give him good reason to smile.

4-11 Communicating the Offer Communication of an offer is necessary so that the intended offeree is cognizant of the proposal and may take appropriate action with respect to it. An acceptance will not convert an offer into a contract

---

[9] 42 W. Va. 63, 24 S.E. 580 (1896). See also the often cited case of *Lucy v. Zehmer,* 196 Va. 493, 84 S.E.2d 516 (1954), in which a contract to sell a farm (which contract was written on a napkin in a bar) was upheld.

unless the existence of the offer is known to the offeree at the time of such acceptance; that is, an offeree cannot "unconsciously" accept an offer (of which he has
not been apprised) merely by happening to take certain action which—unknown
to him—is prescribed in the offer as the mode of acceptance.

In a typical case of the sort just mentioned, Boyle offered by newspaper advertisement a $100 reward for return of her horse. McAdams, without knowledge of
the offer, returned the missing horse and took the magnificent sum of $2 for his
efforts. Later he learned of the advertised reward, but the court denied him any
contractual rights in the matter since there had been no timely communication to
him of the offer in question. In other words, there had been no agreement.

4-12   Imperfectly Transmitted Offers   With respect to errors occurring in the
transmission of offers, it should be noted that, where a person makes use of the
mails to send an offer, the post office will be regarded as the agent. The offer is not
deemed to have been made when the letter is posted but rather when it is received,
and the offeror must suffer any consequences which may arise from delay or
mistake on the part or the postal authorities.

4-13   Duration of Offers   The offeror may fix any time as that within which
acceptance must be made. If no time limit is specified in the proposal the offeree
may assume that a "reasonable" time is contemplated. What constitutes a reasonable time is purely a question of fact depending on the individual circumstances of
the case and on any pertinent business customs involved. For instance, an offer in
connection with speculative stock might logically be assumed to have a shorter
duration than an offer to sell an old rug or an item of used furniture.

Once an offer lapses, it is incapable of being accepted. Similarly, once an offer
is rejected, it ceases to exist and cannot later be accepted without an express
renewal by the offeror.

4-14   Revocation of Offers   In the absence of independent consideration
binding the offeror, the rule in most jurisdictions permits him to withdraw his
offer prior to an unqualified acceptance and prior to the time fixed for its automatic termination. The moving party must, however, bring home to the offeree,
expressly or impliedly, notice of revocation. An exception to this rule is to be
found where an offer has been made through a public medium. In such an instance, revocation is effective if given the same degree of publication as the offer
itself, even though such revocation does not actually reach all those in the position
of offerees.

A sale to *B* of property covered by a previous offer to *A* amounts to a revocation of such offer provided knowledge of the sale reaches *A* before any attempted
acceptance on his part.

In addition revocable offers are generally terminated by the death or insanity of
the offeror or offeree prior to acceptance, through outright rejection or counteroffer by the offeree, and where, after the making of the offer and before acceptance,
the proposed contract becomes illegal.

4-15   Options   Having a specified time for acceptance fixed in the offer represents an option, and the promise to keep the offer open for the period specified is actually a collateral contract to be supported by some independent consideration. Suppose Smith says to Jones, "I will sell you 1000 shares of common stock at $25 per share any time during the next twenty days, you to pay me $100 for the option." If the offer of the option is accepted, the offer of the shares cannot be revoked within the twenty days.

4-16   Acceptance   An offer, until terminated by lapse of time or withdrawn, gives to the offeree a continuing power to create by his acceptance a legally enforceable contract. An accomplished acceptance is necessarily irrevocable, for it is acceptance that ultimately binds the contracting parties. Such acceptance should be absolute, unambiguous, and in harmony with the terms of the offer. It must not fall short of, nor yet go beyond, the terms proposed but must precisely meet them at all points as they stand. The offeror is entitled to impose any conditions he wishes, and if the offeree attempts to effect any material changes in the pattern presented, there is no real acceptance and hence no contract.

4-17   Who May Accept Offers   An offer addressed to a particular person is available for acceptance by that person only. In contrast, a general offer to the public may be accepted by anyone, and an offer to a class or group of persons may be acted upon by any individual in the class described.

4-18   Conditional Acceptance, or Counteroffer   *Qualified (conditional) acceptance* is actually a *counteroffer,* a mere expression of willingness to deal further with respect to the subject matter of the original offer. The submission of a counterproposal by the offeree puts an abrupt end to the negotiations without a contract having been achieved, unless the party making the initial offer renews it or consents to the suggested modifications.

On the other hand, mere inquiry or request by the offeree for more detailed information about certain of the terms of the proposal does not have the effect of rejecting the offer. Similarly, when the offeree embodies in the acceptance terms which were not spelled out in the offer but which the law will imply anyway, the acceptance is not conditional and will effectively bring the contract into being.[10]

4-19   Silence as Acceptance   The silence of an offeree, unaccompanied by other circumstances (such as exercise of a dominion over any of the offeror's property involved in the proposal), is not an acceptance. Indeed, it has been held that silence alone does not operate as acceptance even though the offeror prescribes it as the desired mode of acceptance, and the offeree, knowing this, *intends* his silence as an acceptance. Where, however, the relation of the parties, their

---

[10] For instance, an acceptance might read, "I accept your offer if you can convey good title to the house." Such acceptance is perfectly all right since the offer to sell the property implicitly included a promise to convey valid title.

previous business dealings, or other circumstances are such as to impose an obvious duty to speak, the offeree's failure to do so resulting in harm to the other party can preclude him from later arguing that he never intended by silence as acquiescence.

A famous case illustrating this exception to the general rule is *Hobbs v. Massasoit Whip Co.*[11] This was an action to recover the price of a consignment of eel-skins sent by plaintiff and kept by defendant for some months with no word to plaintiff one way or another regarding acceptance. It seems that plaintiff had made several prior shipments to defendant (which the latter had accepted and paid for) and that there was some sort of vague standing order from defendant for skins of a certain length and quality. It was argued by defendant that one person may not "impose a duty upon another, and make him a purchaser, in spite of himself, by sending goods to him, unless he will take the trouble, and bear the expense, of notifying sender that he will not buy." But the court held that the history of prior dealings created an exception to the general rule advanced by defendant and that, under the circumstances, silence on defendant's part—coupled with a retention of the skins for an unreasonable time—warranted plaintiff in assuming acceptance. The "proposition stands on the general principle that conduct which imports acceptance or assent is acceptance or assent, in the view of the law, whatever may have been the actual state of mind of the party."

4-20 Form and Notice of the Acceptance The offer in a unilateral contract may call for the performance of an act or for forbearance. In a bilateral contract the offer demands a counterpromise, and that is what the acceptance must provide. In case of doubt, the law construes an offer as looking toward a bilateral contract. The explanation for this interpretation presumably lies in the fact there is mutuality of obligation as soon as a bilateral contract is created. With the unilateral type, however, the offeror generally can still revoke even after part performance by the offeree.[12]

Where a bilateral contract is contemplated and acceptance is to be in the form of a promise, notice of the acceptance must be afforded the other party. But where acceptance of the offeror's proposal is by an act, the mere performance of the act—without notice—concludes the contract (always assuming that the offeror has not expressly stipulated that some notice independent of the actual performance be given).

If a particular mode of acceptance is prescribed and the offeree nevertheless elects some other method, the offeror is at liberty to treat the purported acceptance as a nullity. If, however, the text of the offer (or the manner of its communication) does no more than *suggest* a mode of acceptance, the offeree is not strictly limited to such means as the offeror has mentioned. Some other form will be

---

[11] 158 Mass. 194, 33 N.E. 495 (1893).

[12] An exception to this last proposition is posed by the doctrine of *promissory estoppel* in cases involving charitable subscriptions. Once the charity does anything at all in reliance on an offer to make a subscription, the offeror is no longer entitled to withdraw his offer.

effective if it is one that does not cause undue delay and if it successfully brings the acceptance to the knowledge of the offeror. In the last analysis, the propriety of the means adopted to communicate the acceptance, where such method differs from that utilized by the offeror in transmitting the proposal, depends upon what would reasonably be expected by persons in the position of the respective contracting parties in view of the prevailing business usages and other surrounding circumstances.

4-21   Point at Which Contract Is Made   When an offer is sent by mail and the offeree is expressly or implicitly invited to reply in kind, the contract is deemed effective as of the date when the letter of acceptance is put out of the offeree's possession by the act of mailing. This is so even though in the meantime a letter or revocation has been mailed by the offeror but has not reached its destination before the posting of the acceptance.

4-22   Mutual Mistake   The situations in which mistake affects the contract are the infrequent exceptions to an almost universal rule that a person is bound by an agreement to which he or she has expressed a clear assent uninfluenced by falsehood, violence, or oppression brought to bear from some outside source. It will be found that in many instances where mistake is given as the reason for invalidating a contract the trouble is really caused by the act of a third party, or perhaps by the dishonesty of one of the contracting parties. There are relatively few cases of genuine mutual mistake—where parties might contract for a thing which has ceased to exist or are in error about the identity of one another or about the subject matter of the purported agreement. An English case, often cited to illustrate a perfectly innocent misunderstanding of the parties, is *Raffles v. Wichelhaus.*[13] The defendant in that action had agreed to buy from the plaintiff a cargo of cotton "to arrive *ex Peerless* from Bombay." It seems that there were two ships of that name, and both sailed from Bombay—but Wichelhaus meant the *Peerless* which arrived in December. Each party took his respective position in good faith, and neither acted in a negligent or unreasonable manner in arriving at his particular understanding of the matter. The court held that in the light of the varying interpretations no actual agreement had been reached, and there was hence no enforceable contract in existence.

4-23   Unilateral Mistake   The other principal category of mistake which may affect contractual relations is the unilateral type. One party may not avoid a contract simply by showing he erred in what he promised or did. This is so if there has been no misrepresentation, if there is no ambiguity in the terms of the contract, and if the other party has no timely notice of the mistake and acts in perfectly good faith. Subject to certain exceptions, a person must in his contractual dealings shift for himself and exercise ordinary diligence if he is to avoid undue difficulty. The courts will not undertake to relieve a party of the conse-

---

[13] *Hurl. & C.,* 906 (Ct. Ex. 1864).

quences, however unfortunate, of his own improvidence or poor judgment. A buyer, for example, who purchases goods without any statement or warranty of quality from the seller and who subsequently finds the goods to be less valuable than he had thought will ordinarily have no recourse against the seller. The buyer cannot logically expect the other party or the court to rectify any error of judgment he may commit in making his bargain.

Along the same line, is the rule that one who signs an agreement is presumed to know its contents. This presumption very properly aids the other contracting party, who has more than likely shaped his actions and perhaps paid out money in reliance upon the agreement and its enforceability.

The law will not allow a person to enforce a promise if he knew at the time he accepted that the other party understood the terms of agreement in a completely different sense from that in which he understood them himself. In other words, if either party knows that the other clearly does not intend what his words or acts seem to express, this knowledge prevents such words or acts from operating as an offer or an acceptance.

As an illustration, let us take the celebrated case of *Geremia v. Boyarsky et al.*[14] In preparing a bid for work to be done in the construction of the plaintiff's house, the defendants incorrectly added the items contained in their estimate, so that the figure submitted to the plaintiff was $1450.40, whereas it should have been $2210.40. Although the plaintiff had good reason to believe that a mistake had been made, he executed a written contract with the defendants based upon their miscalculation. The defendants discovered the error on the same day and immediately proposed to perform the work at their corrected price or at that offered by any responsible contractor, but the plaintiff refused and, having let the work to others for $2375, which was its reasonable value, brought action for breach of contract. The court held (1) that the mistake was so essential and fundamental that the minds of the parties had never met; (2) that, since the agreement was still executory and the plaintiff had been in no way prejudiced, he would not be permitted in equity to gain an unfair advantage over the defendants, even though the mistake was unilateral, and that the defendants were entitled to a rescission of the contract; (3) that, while the mistake involved some degree of negligence on the part of the defendants, it was not of that culpable character, amounting to a violation of a positive legal duty, which is regarded as a bar to redress in a court of equity.

4-24   Clerical Errors   Where there is an agreement between the parties and the sole difficulty is that a simple error of a clerical nature has arisen in committing the understanding to writing, the mistake is not such as will vitiate the contract. In the occasional instance where the parties cannot correct such an error by common consent and one party seeks unfairly to take advantage of the situation, the other may by court action have the contract enforced in accordance with the

---

[14] 107 Conn. 387, 140 A. 749 (1928). See also *Balaban-Gordon Co., Inc. v. Brighton Sewer District No. 2,* 342 N.Y. Supp.2d 435 (1973).

actual intent and original understanding of the parties. Courts freely exercise the power to correct clerical mistakes when the available proof leaves no doubt that the real contract was meant to be something at variance with what appears in the writing.

4-25 Mistakes about the Law What has been said up to this point relates to mistakes of fact. A mistake of law is something completely different. Though the question is not entirely free of doubt, the general rule is commonly stated to be that an erroneous conclusion about the legal effect of known facts does not destroy the validity of a contract when there is no concurrent mistake of fact.

4-26 Fraud and Material Misrepresentation Fraud, the existence of which renders a contract voidable at the election of the victim, constitutes (1) a reckless or intentional misstatement of a material fact, (2) made with intent to deceive, (3) resulting in misleading the innocent party to contract, (4) to his detriment. Acting as a reasonable person, the defrauded party must have been entitled in the specific situation to rely on the truth of the assertions made. Whether the victim indeed had such a right of reliance will depend on the circumstances of the particular case, such as the precise form of the representation, the relations of the parties, and their respective means of knowledge.

An intentionally false statement of a past or existing fact is the basis for a claim of fraud, and such factual representations must be distinguished from a simple expression of opinion about the future. A misrepresentation may be created just as readily by concealment or suppression of material facts as by affirmative misstatement. Finally, misrepresentation which is perfectly innocent may, if it constitutes a material inducement, render a contract voidable.

4-27 Remedying Fraud Apart from a tort action, the deceived party has available to him a choice of remedies in contract litigation. If he permits the contract to stand (which, of course, he is at liberty to do), he may sue to recover any damages he may have suffered. Or it is possible for the victimized party to rescind the tainted bargain and recover all consideration paid by him, always on the assumption that he is prepared to restore to the wrongdoer anything received pursuant to the contract he is electing to abrogate. Some courts may, at the same time, allow damages resulting from the fraud to flow to the rescinding party in the same action. Generally speaking, the defrauded party may not pursue inconsistent remedies and will be bound by an election once made. Also, such party, upon discovering that he has been defrauded, must act promptly in repudiating the contract.

4-28 Waiver of Fraud A contract induced by fraud is not absolutely void, but only voidable. The fraud may therefore be waived and the contract ratified. Ratification is shown where the innocent party, having acquired actual knowledge of the true situation subsequent to date of the agreement, by acts of commission or omission then evidences a clear intent to affirm the contract despite the fraud.

4-29 Duress and Undue Influence An agreement made under compulsion or threat is not binding upon the person whose assent was involuntarily given. Likewise, undue influence, utilized perhaps by a parent or someone in a confidential relationship to the influenced party, will render a contract voidable. When a person takes advantage of mental weakness, near relationship, or special confidence in order to influence another to make a contract, the second party has not exercised his own volition and will not be liable on the agreement should he choose to avoid it.

In the case of duress, the victim's free choice in the matter is overcome by fear of injury; in the instance of undue influence, it is overcome by importunity. In either event, the resulting contract is voidable, and the victim is empowered to negate the entire transaction.

4-30 Summary on Offer and Acceptance The principles of offer and acceptance are frequently difficult to apply in some of the complicated contract situations which arise in the modern business world. It is often a real problem to determine at just what point a series of acts, promises, and negotiations culminates in a binding agreement. Assume, for instance, that Y says to Z, "I will pay you $1000 to paint my house two coats by the end of this year." If Z promises to take the job and Y thereupon revokes, Z will have no legal recourse, since he should have *painted* rather than *promised* in response to —Y's offer for a unilateral contract.

There are a number of other possibilities which could arise in connection with Y's initial offer. If Z paints one coat complete and then quits, Y will very likely be unsuccessful in any attempt to sue Z, unless the cost of having another painter take care of the second coat proves to be in excess of $1000. Now suppose that Z makes no promise in reply to Y's offer but arrives on the scene with her equipment, looks the job over, and leaves. Quite obviously there is no contract. If, on the other hand, Z gets as far as burning off the old paint before deciding to forget the whole thing, Y would presumably have no recourse—provided there is some other competent painter readily available.

Looking at the same general problem from the other side of the fence, if Z, upon receipt of the offer, arrives at the job site with her equipment, she should be permitted a chance to perform the work despite any belated attempt by Y to revoke the proposal. Certainly Y cannot terminate proceedings without penalty if the painter has, for instance, finished one coat before Y calls a halt; under such circumstances, Y has been unjustly enriched, and Z can recover not merely the cost to her of the work done but also any profit she would have pocketed if the job could have been finished.

CONSIDERATION

4-31 Meaning and Importance Although the minds of competent contracting parties may have met with reference to a common purpose and the other prerequisites thus far discussed may have been fulfilled, the agreement will not be entitled to treatment as a legal and binding contract unless it successfully meets still another test which the law has imposed. A promise, in order to be enforceable, must be supported by a consideration.

Consideration is something of value received by or given at the request of the promisor in reliance upon and in exchange for his or her promise. The consideration may be furnished by the promise or by some other person. Similarly, it may run to the promisor or to a third party. In any event, nothing is deemed to be consideration unless it was so regarded by the contracting parties. It must be something they actually bargained for, something they themselves looked upon as an inducement.

By way of illustration, assume that Smith offers to sell Jones certain property for a stated sum, and, when Jones asks for two weeks' time in which to consider the offer, Smith promises not to sell to anyone else during that period. Under these circumstances, Smith may nevertheless dispose of the property to a third party, since his promise to Jones was unsupported by any consideration. If, however, Jones pays for Smith's promise, there is a contract duly supported by a consideration, and Smith must keep the offer to Jones open for the stipulated period. The payment by Jones is known in law as a consideration, or, as it is frequently expressed, as a quid pro quo.

Consideration, given in exchange for a promise by the other party, may be either (1) a present act, forbearance, or sufferance or (2) a promise to do, forbear, or suffer sometime in the future. In the first case, the consideration is present, or *executed,* and the contract is unilateral; in the second case, it is future, of *executory,* and the contract is bilateral. The fact that the promise given for a promise may be conditional in nature does not affect its efficacy as a consideration. Not every promise, however, is sufficient consideration for a return promise. The test has been stated to be: Where doing of a certain thing will be a good consideration, a *promise* to do that thing will be so too.

When mutual promises serve as consideration one for the other, the law will look to see that there is real substance to each promise. Those without substance are merely illusory promises and of no effect. For instance, an agreement to agree is unenforecable. Similarly, a promise to do what is obviously a physical or legal impossibility is not a true consideration and will not support a binding return promise.

Consideration must be lawful, definite, and certain. A promise which purports to be a consideration may be of too vague and unsubstantial a character to be enforced. For example, a commitment to pay "such remuneration as shall be deemed right" throws upon the courts a responsibility of interpretation which they are not prepared to assume.

The consideration for a contract differs from the motive in that the latter is the underlying cause for entering into the agreement whereas the former is the thing given by one party in exchange for and in reliance upon the other party's promise.

4-32 Forbearance as Consideration   It is well settled that mere forbearance from exercising a legal right, without being asked (the offer) to do so, is not a

consideration which will support a promise. When requested, however, forbearance to sue on an enforceable claim may be sufficient consideration for a promise,
even though there is no waiver or compromise of the right action involved and
even though forbearance is for a short time only. On the other hand, a promise in
consideration of forbearance is not binding if there was originally no cause of
action or if the claim threatened to be enforced is invalid and worthless.

The actual compromise of a suit could without question serve as consideration.
Since days of yore, the law has favored any conduct between disputants that will
preclude litigation and accordingly, looks with great favor on the compromise of
suits.

It frequently happens that the form of the consideration is such that there is no
actual benefit apparent for anybody but merely a detriment to the person furnishing the consideration in question. Such detriment exists where the party affording
consideration agrees to give up something to which he has undoubted right (for
example, the use of intoxicating beverages) or to forbear from some action he has
a legal right to take. Moreover, the surrender of a legal right is rendered no less
valid a consideration by the fact that the right may not come into being for some
years.

For an illustration of the surrender of legal rights as fulfilling the requirement
of consideration, the reader might well refer to *Hamer v. Sidway.*[15] In this case, an
uncle promised his nephew $5000 if the boy refrained from smoking until reaching
age twenty-one. The nephew did as his uncle wished and, when the promised
$5000 was not forthcoming, sued for breach of contract. The defense claimed
there was not consideration for a binding promise, since the uncle received no
actual benefit. The court, however, held that the nephew has given up legal rights
and was entitled to recover.[16]

4-33 Adequacy of Consideration If the parties really intended some thing as
consideration, its apparent "inadequacy," relative to the value of the promise for
which it is exchanged, is of no consequence. If a party gets what has been contracted for, the courts will not look into the question of whether such consideration was of a value equivalent to that of whatever the party was obliged to give
in return. "A cent or a peppercorn, in legal estimation, would constitute a valuable
consideration."[17]

It is the duty of the court simply to interpret and not to redraft the contracts
which come before it in litigation. A shrewd businessperson is entitled to the fruits
of a bargain, and the commercial world, which relies upon the stability of contracts, would be in a bad way if courts could freely set aside contracts for what
seems to them to be inadequacy of consideration. Accordingly, a court of law will
act upon the presumption that parties capable of contracting are likewise capable

---

[15] 124 N.Y. 538 (1891).
[16] Recent cases illustrating this point include *Gill v. Kreutzberg,* 24 Ariz. App. 207, 537 P.2d 44
(1975); *Jemzura v. Jemzura,* 369 N.Y.2d 400, 330 N.E.2d 414 (1975); and *Quayle v. Mackert,* 92 Id.
563, 447 P.2d 679 (1968).
[17] *Whitney v. Stearns,* 16 Me. 394, 397 (1839).

of regulating the terms of their contracts, and judicial relief is granted only when any obvious unfairness in the bargain is shown to have arisen from mistake, misrepresentation, or fraud. There are occasional instances, of course, where the inadequacy or inequality of consideration is so gross as to amount to proof of undue influence of some sort; under such unusual circumstances, the courts will take a hand.

4-34　The Doctrine of *Foakes v. Beer*.[18]　From one of the most famous contracts cases of all time comes the proposition that, because "a lesser sum of money cannot be a satisfaction of a greater," part payment of a liquidated and undisputed debt already overdue is not valid consideration for a promise of the creditor to accept such payment as in full. The debtor, in rendering the part payment, has suffered no detriment he was not already bound to suffer and has surrendered no legal right he was not already obligated to surrender. The end result of the rule in *Foakes v. Beer*—a doctrine which, despite growing criticism, prevails in a great many jurisdictions—is that a creditor can effectively extinguish a $50 debt by accepting in its stead a canary, but he cannot accomplish the same thing by promising to take $49.50 in lieu of the full $50. It is obvious that there are many situations in which application of this rule leads to ludicrous results.

The rule in *Foakes v. Beer* has been repudiated by court decision in several states and by statute in a few more. The modern trend seems clearly away from the *Foakes* proposition. Since the *Foakes* rule is not favored, the courts often exercise ingenuity to avoid application of the doctrine. Seemingly, they have refused to apply the rule whenever they could discover some circumstance which could be regarded as a technical legal consideration. Thus, an agreement to give a smaller sum of money for a greater, if the time of payment is sooner, is binding. One court indicated its reasoning in this connection in rather expressive language: "Peradventure parcel of the sum, before the day, would be more beneficial than the whole sum on the day."[19] Similarly, the giving of a negotiable instrument for a money debt or the giving of virtually anything other than money itself will be deemed sufficient consideration to support a release of the entire debt.

Where the amount due is itself open to dispute between the parties, the rule in *Foakes v. Beer* has no bearing. Under such conditions, if the debtor sends the creditor a check marked "in full payment of my account," which check is for a smaller sum than the amount claimed by the creditor, the entire debt, according to majority opinion, is discharged (by what is technically known as an *accord and satisfaction*) when the creditor negotiates the check without protest.[20] Let us assume that White owes Black $500 and that the debt is past due. White promises that in thirty days she will turn over her speedboat to Black if the latter will accept same in full payment. Black agrees, but the debtor then fails to deliver as sched-

---

[18] L.R. 9 App. Cas. 605 (House of Lords, 1884).

[19] *Chicaro Fertilizer Co. v. Dunan,* 91 Md. 144, 46 A. 347, 351 (19 00).

[20] For a leading case illustrating this point, see *Nassoiy v. Tomlinson,* 148 N.Y. 326, 42 N.E. 715 (1896). It is interesting to note that at least in Georgia a notation on the check is not necessary. See *Souchak v. Close,* 132 Ga. App. 248, 207 S.E.2d 708 (1974).

uled. Under these circumstances, Black can elect to sue for either the $500 or the speedboat. Black's agreement to accept delivery of the speedboat within the stipulated thirty days constituted what is known as an *accord executory*. If White had turned over the boat as promised, there would have been an accord and satisfaction acting as a bar to any attempt on Black's part to get the $500 as well. It should be noted that White's promise to deliver the boat did not of itself operate to discharge the $500 debt, because Black is deemed to have agreed to take the *boat* in satisfaction, not merely the *promise* respecting the boat. This is clearly a distinction of significance.

Moreover, cases in various states have held that the payment of only part of a liquidated debt is a good discharge of the whole where the creditor agreed to accept it as such if it were advanced for the purpose by some person *other than* the debtor. Such payment by a third party of a portion of the debt gives the creditor something different from what that recipient was entitled to demand and therefore operates as a valid consideration for this promise to release the actual debtor from the entire obligation.

4-35   Composition with Creditors   A *composition agreement* between an insolvent or financially embarrassed debtor and two or more creditors permits the debtor a full discharge upon paying each creditor a smaller sum than that admittedly owed to each. Such an agreement operates as a notable exception to the rule that a creditor will not be bound by a promise to accept less than the amount of an ascertained debt. Each creditor acts on the commitment of the others to relinquish, in turn, a pro-rata part of their respective claims, and the benefit which each may derive from the mutual concession is the consideration of inducement which sustains the agreement.

The composition settlement contemplates an understanding not only between the debtor and the various creditors but also among the several creditors. In this respect it differs from the accord and satisfaction since the latter is an agreement between the debtor and a single creditor.

4-36   Performance of Existing Duty   If a promisor gets nothing in return for his promise but that to which he is already entitled, the consideration is unreal and will not support the promise. For instance, where there is an existing contractual obligation to perform certain services for a stated fee, any promise to pay additional compensation for such obligated performance is made without a valid consideration and will not bind the promisor.[21] But where one party refuses to complete his contractual duties by reason of having encountered substantial performance difficulties unanticipated by the several parties at the contract's incep-

---

[21] Let us take another commonplace illustration of this point. Assume that Phelps has borrowed $50 from Dodge and without question owes that amount. Dodge is also asserting an other claim against Phelps, but the validity of this latter claim and extent thereof are in dispute. If Phelps pays $50 in return from Dodge's promise to accept this sum in complete satisfaction of both claims, the promise is unenforceable since it is not supported by a valid consideration. Phelps has paid only what he was under an undisputed duty to pay.

tion and casting an additional and totally unexpected burden upon the performer, a promise by the other party of extra compensation to induce completion of the project is generally regarded as supported by a proper consideration. The presumption in the latter case is that the parties by their subsequent actions waived their respective rights and obligations under the original agreement and substituted the new arrangement.

Instead of having a *contractual* duty, a person might be under a preexisting duty imposed by law to do the very thing he promises to do. In such circumstances, the promise is of no value whatever and is not a proper consideration for a return promise or undertaking.[22] To illustrate, a husband cannot make a contract, enforceable against his estate, to pay his wife for services rendered by her in nursing him during his last illness, she plainly being under legal obligation to render such services in any event.[23] Another example is that of the ordinary witness subpoenaed to appear at a trial. There is no consideration for any promise to pay her something well beyond her expenses, since it is her legal duty to come and give evidence.

The undertaking or doing of anything surpassing what one is already bound to do, though of the same kind and in the same transaction, provides a good consideration. Thus a public officer cannot recover a reward if the services she performs come strictly in the line of her official duty, because a contract to award such additional compensation for the performance in question would be both without consideration and against public policy. But it is also true that a contract to grant a public officer special remuneration or reward for services rendered outside, and not inconsistent with, her official duty is valid.

4-37 Mutuality of Obligation It is often asserted that in executory bilateral contracts there must be *mutuality of obligation*; that is, both parties must be bound, or neither will be. Actually, of course, there are many situations where only one side is really bound, as for instance, where one party is an infant or where one party is guilty of fraud or undue influence.

The proposition which requires mutuality of obligation regarding those contracts to be performed in the future, where the promise of one party constitutes the sole consideration for the promise of the other, stems from the obvious unfairness of enforcing any contract which would require performance without question by one the parties while leaving the other party a free choice in the matter. Thus, where one party alone has an arbitrary right to cancel or terminate the contract at any time, mutuality is absent.

Mutuality of obligation, a most difficult and controversial concept, involves primarily a question of consideration. If the promises which are to serve as consideration for one another are each binding, there is no lack of mutuality, and non-

---

[22] On the same principle, a promise not to do what one cannot lawfully do is an insufficient consideration.

[23] See, on this point, *Foxworthy v. Adams,* 136 Ky. 403, 124 S.W. 381 (1910). With respect to a husband's legal duty to love and support his wife, see *Gutman v. Gutman,* 240 N.Y. Supp.2d 657 (1963).

performance of one of these enforceable promises does not constitute want of consideration nor lack of mutuality, though the breach itself may well afford grounds for a damages action by the innocent party.

4-38   Past Consideration   It is a general principle that past consideration is insufficient to support a promise. That is, an act or forbearance rendered antecedently and not as the agreed equivalent for a subsequent promise is normally to be regarded as in the nature of a gift and is no consideration at all. For example, where a father promises to buy his college student son a car next year because the son was admitted to Harvard last year, the father's promise is supported only by past consideration.

Where, however, an act is done upon definite request, under such circumstances that the law would imply a promise to pay therefor, and an express promise is actually made after the performance has been accomplished, the promise is generally held valid, being supported by what is often called an *executed,* or *past, consideration.* In the absence of the express promise, an action will yet lie to recover reasonable value of the requested services. When an actual promise is made subsequent to the performance, it serves as confirmation of the fact that the promisee's act was completed with the understanding that it was not intended to be gratuitous. A *subsequent promise* (that is, one which follows the other party's performance to which it relates) is binding only when the request, the consideration, and the promise form substantially one transaction. In this view the request can be thought of as virtually the offer of a promise, the precise extent of which becomes apparent only when the promise is actually given, some time after the requested consideration—for example, performance of a personal service—is forthcoming from the promisee.

4-39   Moral Consideration   As a general rule a mere moral obligation will not of itself furnish consideration for an executory promise. Thus, if Smith rescues Jones from drowning, the latter is under some measure of moral compulsion to tender at least a nominal reward, particularly if Smith sustained any loss as a result of his act. But, even if Jones, temporarily overcome with gratitude, were to promise to pay a fixed sum as a reward for Smith's valiant efforts, there could be no legal recovery against Jones for breach of that promise. Such an offer could also fail, as the consideration, or act of saving in this case, is past. By the same token, promise or executory contracts made on the basis simply of love and affection—unsupported by a material benefit for which payment might reasonably be expected—create at most bare moral obligations, and a breach of such promises presents no basis for redress by the courts.

But a moral obligation arising from what was once a legal liability—the latter since barred or suspended by operation of some positive rule of law—will furnish a proper consideration for a subsequent executory promise. Where, for example, a person under a legal obligation to pay a sum of money is released by the running of a statute of limitations and afterward promises the creditor that the sum will nonetheless be paid, the creditor can recover. Some courts reach this result on the theory that the renewed promise is valid without a new consideration, since a moral obligation sufficient to support the promise survives. Other courts, while likewise recognizing the validity of the "released" debtor's subsequent promise, enforce it rather on the ground that the particular circumstances obviate the necessity for a consideration in that the statute of limitations is merely a defense the law has given the debtor to use or not at his pleasure.

We have seen that at common law the contracts of a minor cannot be enforced against him if he cares to claim the defense of infancy. Upon reaching majority the infant may decide to validate previously made contracts, and any new promise to perform in accordance with the original agreement will effectively bind him without the necessity of any new consideration.

As in the case of infancy, contracts induced by fraud or duress are regarded as voidable and not entirely void. If, with knowledge of the fraud, the wronged party elects to ignore the obvious defense and makes a new promise to perform in accordance with the antecedent contract, such promise will be enforceable even though there is no consideration given specifically in return for it.

4-40   Estoppel   A promisor is, under certain unique circumstances, precluded from asserting that a valid consideration is lacking. The classic doctrine is that the promise and the consideration must be given as the contemplated exchange one for the other. From this traditional theory of consideration, the idea of *promissory estoppel* represents a departure in that it renders a promise binding if the promisee has taken action or suffered some detriment in reliance thereon, even though such detriment was not requested as consideration for the promise. A typical example might be the situation where a mortgagee, following a default in payments, promises not to foreclose for a certain period of time. The mortgagor relies upon such freely given promise and, with the mortgagee's knowledge, makes expensive repairs to that part of the property concerned. Under these conditions, the mortgagee's promise, despite the absence of consideration and other normal contract essentials, will be enforced as the logical means of avoiding injustice to the mortgagor.

A long and constantly increasing line of decisions in the United States has established the binding character of most charitable-subscription agreements.[24] It is clearly understood that a subscription—rather than being part of a bargain that provides for the exchange of agreed equivalents—is a donation, an outright gift; and yet promises of this kind are now almost universally enforced. The reason most commonly advanced is that the promisee has given consideration by acting in reliance on the subscription and by proceeding to carry out the purposes for which the pledge was solicited and given. For example, perhaps the beneficiary has expended money in building construction before the subscriber makes any

---

[24] See, for example, *Salsbury v. Northwestern Bell Telephone Co.,* 221 N.W.2d 609 (Iowa 1974); and *in Re Lipsky's Estate,* 256 N.Y. Supp.2d 429 (1965). It is not, however, unanimous. See, for example, *Mt. Sinai Hospital of Greater Miami, Inc. v. Jordan,* 290 So.2d 484 (Fla. 1974), *Maryland National Bank v. United Jewish Appeal Federation of Greater Washington, Inc.,* 407 A.2d 1130 (Md. 1979).

attempt to revoke the promise.[25] In some cases different reasons have been given for the enforcement of the subscription obligation. Occasionally, a court will find an implied promise by the trustee of the charity to execute the work contemplated, such promise operating as a consideration. Or it has been held that the promises of several subscribers mutually support each other. Under this last theory, the subscription agreement amounts to a bilateral contract between or among subscribers, of which contract the charitable organization is a donee-beneficiary. In any event, regardless of the precise theory adopted, the courts of this country will usually treat the ordinary charitable-subscription promise as clearly enforceable at law.

4-41   Illegal Consideration   An *illegal consideration,* which may be defined as a performance or promise which runs counter to law or public policy, will not support an agreement. Despite some difference of opinion on the point, it is ordinarily said that even partial illegality of the consideration will vitiate the accompanying contract in its entirety.

4-42   Failure of Consideration   Some courts subscribe to a doctrine which they characterize as *failure of consideration.* Such failure is deemed to result when, for some reason or other, the anticipated consideration cannot be supplied or performed, and this unexpected state of affairs renders the contract unenforceable. The failure of consideration is predicated upon the happening of events which materially change the respective positions of the parties and which were not within their contemplation at the time of execution of the contract.

4-43   The Statute of Frauds   By statute, some contracts must be written. An English law (passed in 1677), entitled "An Act for the Prevention of Frauds and Perjuries," served as the forerunner and model for similar statutes enacted in the various states in this country. Such statutes in the several states differ from one another and from the original English version, but all regulate the contract formalities necessary to render the agreement enforceable under the judicial process.

There are many parts to a typical statute of frauds. The following classes of contracts, however, should be noted as representative of those agreements which will not be enforceable unless, in the words of one case, "the same, or some note or memorandum thereof, be in writing and subscribed by the party to be charged, or by his agent:"[26]

1 Contract for the sale of goods valued above a stated amount
2 Contracts for the sale of interests in land
3 Contracts not to be performed within one year

[25] See, for example, *Estate of Timko v. Oral Roberts Evangelistic Association,* 215 N.W.2d 750 (Ct. App. Mich. 1974).
[26] Contracts which do not fall within the categories to which a statute of frauds applies are, of course, equally binding whether oral or reduced to writing.

Where the contract is not explicit about the time within which performance is expected to be complete, courts supply a liberal construction and hold that the test is the "shortest possible" and not the "probable" time required for performance. Thus, if the contract *might* be performed in less than a year, it is not within the statute.

QUESTIONS

1 How many parties constitute the minimum needed to form a contract? Why?

2 What feature distinguishes a "minor?"

3 Under what conditions can a minor be released from his or her contract?

4 If a fifty-year-old businessperson contracts to sell an automobile to a minor, can the promise of the former be subsequently avoided? Can the minor avoid his or her part of the bargain? Why?

5 Explain the meaning of "ultra vires" in connection with the contracts of a corporation.

6 In contract law, what significance has (*a*) mental weakness of one of the contracting parties? (*b*) actual mental incompetence of one of the parties?

7 Under what conditions will a person be bound by a promise which was made when intoxicated?

8 Illustrate a case in which the subject matter or promise would be illegal or against public policy and therefore would void a contract involving same.

9 For a $50 commission, *A* agreed to shop around and get a secondhand car for *B*. What is unfortunate about such a contract from *B*'s standpoint? Has *B* any control over what he will receive?

10 *A* agreed to pay *B* $5000 for certain diamonds which *B* was to smuggle into the country. When *B* delivered the jewels, *A* refused to accept them. What can *B* do about it? Why?

11 *A* called a builder, *B,* describing in considerable detail an additional room which she planned to build on her house, and she asked *B* how much he would charge to build it. After lengthy discussion, *B* said, "I think it will cost you $3000." *A* said, "That will be satisfactory." The next day *B* called *A* and said that, after making a careful estimate of the job, he found the addition would cost $3400. *A* claimed that *B* had already contracted to do the work for $3000. Is *A* correct? Why?

12 *X* sent *Y* the plans and specifications for the complete electric lighting system for a new factory. *Y* replied that she would do the work for $4500. *X* announced that he "hereby" awarded the contract to *Y* but that certain fixtures were to be changed and relocated and other revisions made in the plans. Is *Y* bound to go through with the project?

13 Under what circumstances may silence be treated as an acceptance of an offer?

14 Black told White that she would sell all her "do-it-yourself" tools for $300. White replied: "Unless you hear otherwise from me before April 15, I shall accept your offer." On April 12 White wrote Black that she could not afford to purchase the tools, and she gave the letter to Jones to mail. Jones forgot to do so until April 16, whereupon she posted the letter. Is White obliged to purchase the tools? Explain.

15 *X* sold his automobile to *Y* with the statement that all four tires were practically new. *Y* discovered next day (before paying for the car) that all tires were retreads. Can *Y* rescind the contract and avoid any payment whatever? If not, what recourse does *Y* have?

16 *A* had a nephew, *B,* to whom, when *B* was seventeen years and two months old, *A* loaned $2000 for his educational expenses. In return *B* gave *A* a promissory note calling

for repayment (without interest) two years from the date of the note. One year later, *B* quit school and joined the army. Two months after *B*'s nineteenth birthday, on the due date, *A* demanded payment. *B* refused. Can *A* collect? Explain.

17 Murray wrote to Giegler stating that he would install a completely new heating system in Giegler's house, in accordance with the latter's specifications, for $1200. Giegler was away on vacation when the letter arrived at his home, and he had left no forwarding address. The next day, Murray discovered that he had made an error in his estimate. He therefore wrote Giegler another letter, repudiating the first one, and stating that the cost would be $1400. Giegler received both letters simultaneously upon his return home. Can Giegler hold Murray to the latter's first quotation? Explain.

18 Mrs. *S* was hit by a car, and her ankle was broken. Three days later, a paper was mailed to her by the representative of the insurance company who asked her to sign and return it. She did so without reading the paper carefully, believing it to be an initial payment of $500. Later, she discovered that the paper was a release of the company from all liability upon payment of $500. What can she do about the matter?

19 If the offeror makes an honest mistake in his offer, what can he do to remedy the situation?

20 Explain the effect of duress or undue influence upon contractual relationships.

21 Carlson, a builder, promised Pratt that she would engage the latter to install all the hot-air furnaces in houses to be built by Carlson during the year 1984. Pratt ultimately sued Carlson for loss of anticipated profits because Carlson changed all of her house plans and installed hot-water heating systems instead of hot air. Can Pratt collect? Explain.

22 What is the meaning of "consideration" as the term is used in contract law?

23 *A* told *B* that, if the latter would take him along on a ten-day fishing trip in Maine, he would take care of *B*'s yard and flower beds all summer free of charge. *B* accepted the offer. Is there a binding contract? Explain.

24 Can an agreement *not* to do something be a proper consideration for a return promise?

25 Brown was interested in buying Gray's secondhand truck and happened to tell his (Brown's) daughter that he was willing to pay $1000 for it. Gray called at Brown's house when the latter was away, and Gray told Brown's daughter that he would sell the truck for $1000, whereupon the daughter accepted the offer and arranged for delivery next day. Is Brown obligated to pay? Why?

26 *A* had agreed to sell his house to *B* for $20,000 and stated that he would confirm his offer by letter the next day. When *A* wrote the letter, he made a typographical error, stating that the selling price was $2000. What are the rights of the parties?

27 Black agreed to sell certain secondhand machinery to Green for $20,000, and Green accepted the offer. The following week Black learned that the machinery was actually worth $40,000 and thereupon refused to sell at the lower figure. Was Black's position sound? What can Green do?

28 *X* told *Y* that he would buy *Y*'s old tractor at a "proper" figure as soon as *Y* received delivery of a new machine which was on order. Have the parties made a contract? Elaborate.

29 *X* owed *Y* $500 but was unable to pay on the due date. He told *Y* that he could pay only $300 in cash but that he would let *Y* have two pointer pups in addition if *Y* would "call it square." If *Y* accepts, what is the status of the $500 debt?

30 *X* promised to buy *Y*'s house and lot for $25,000 and made a deposit of $1000 to bind the bargain. She was to receive transfer of title on August 10. On August 6 the house burned down. What happens to the contract? What happens to the $1000 deposit?

CHAPTER 5

INTERPRETATION AND CONSTRUCTION

GENERAL RULES OF CONSTRUCTION

5-1 Contract Terminology Given Its Plain Meaning; Exception The words of a contract are to be given their commonly accepted meaning in the absence of some evidence of a contrary intent. But technical terms will be construed in a technical sense, and a special and widely adopted trade meaning afforded certain terminology will be taken into account in any interpretation of a contract containing such terminology. Moreover, in construing technical terms utilized in an agreement, a court will ascribe considerable importance to the meaning given those particular terms in the course of any prior dealings between the parties.

5-2 Restrictions on Court Interpretations A court is not free to disregard contract terminology employed by the parties or ordinarily to insert words which the parties have not themselves utilized. In other words, judicial interpretation of a contract does not include its modification or the creation of a new or different contract.

5-3 Reasonable and Equitable Construction An interpretation which gives a reasonable meaning to all the contract provisions will be preferred to one which leaves portions of the agreement useless or absurd.[1] Likewise, a court will often make some attempt at a construction equitable to each of the parties, but it cannot

---

[1] Somewhat analogous to this is the proposition that courts will, whenever reasonably available, so construe a contract as to uphold its legality and validity.

ignore or change harsh or even seemingly unreasonable terms which are clearly written into the contract, for passing upon the folly or wisdom of a contractual undertaking is not within the province of the court.

5-4   Parties' Intent Important   Perhaps the cardinal rule in the interpretation of contracts is that the court must ascertain and give effect to the intention of the parties as far as this may be accomplished without contravention of legal principles or public policy.[2]

5-5   Clerical Errors   Obvious clerical errors or omissons, which are revealed by a perusal of the instrument as a whole, can be discounted in determining the intent of the parties. In fact, the court will strike out an improper word or even supply an omitted word if from the entire context it can readily and definitely ascertain what terminology was intended and should have been used.

5-6   Preliminary Negotiations   When preliminary negotiations are consummated by a written agreement or when an oral contract is evidenced by a subsequent agreed memorandum, the latter supersedes any previous understandings, and the intent of the parties must be derived from the writing. In case of real doubt about the contract's meaning, however, all the preliminary negotiations between the parties should be taken into account in giving the final contract a proper construction.[3]

5-7   Parties' Own Interpretation   When the parties to an ambiguous agreement have given it a practical construction by their conduct, as by acts in partial performance, such construction will be considered by the court in determining what mutual intention was as of the critical time of contracting.[4]

An excellent example of this point is afforded by the case of *Wiebener v. Peoples.*[5] Involved there was a contract for the construction of a building, which called for the contractor to furnish all materials and labor to erect the structure according to plans and specifications. One rather nebulous clause read: "Excavations and foundations complete to joist line; will be completed to joist line by owners." The litigation dealt primarily with the responsibility for supplying a stairway and

---

[2] The Supreme Court of Georgia recently stated that "a cardinal rule of construction of contracts is to ascertain the intention of the parties. This intention must be enforced unless it contravenes some rule of law." The court then went on to find that a separation agreement entered into by a husband and wife for payment of 50 percent of the net income of an investment was not intended to be alimony but rather a division of property, and therefore was unaffected by the subsequent remarriage of the wife. *Head v. Hook,* 285 S.E.2d 718 (Ga. 1982).
  See also *Texaco, Inc. v. Holsinger,* 336 F.2d 230 (10th Cir. 1964), applying Kansas law, *cert. denied,* 379 U.S. 970 (1965); *Krison v. Tex. Ind., Inc.,* 253 So.2d 614 (La. App. 1971).
  [3] The parol-evidence rule, discussed in Art. 5-15, would preclude the introduction of any oral evidence to *contradict* the terms of the written agreement.
  [4] For example, see *Patterson v. Southwestern Bell Telephone Co.,* 411 F.Supp. 79 (E.D. Okla. 1976); *Ackerman v. Lauver,* 242 N.W.2d 342 (Iowa 1976); and *Occidental Chemical Co. v. Agri-Profit Systems, Inc.,* 37 Ill. App.3d 599, 346 N.E.2d 482 (1976), *Anne Arundel Co. v. Crofton Corp.,* 410 A.2d 228 (Md. App. 1980).
  [5] 44 Okla. 32, 142 P. 1036 (1914).

an iron railing required by the contract to be placed above the joist line on an outer retaining wall of a light and air space outside the basement. The builder argued that such outside railing was part of the basement and, as such, came within the exception quoted. In the course of its opinion, the court took special notice of the parties' mutual understanding of contract language, and said:

> It appears that in the construction of the building the parties to this contract mutually understood and construed it as requiring the defendants [owners] to make the excavation and foundations and to complete the entire basement of the building as required by the plans and specifications to joist line; and, in view of such construction by the parties themselves, . . . we hold that the expression "excavation and foundations complete to joist line," as used in this contract, refers to and means all that portion of the building within the excavations and below the joist line, i.e., the entire basement below that line, *although we might be disposed to hold otherwise but for such construction by the parties* [6] as there is no other evidence of a technical or special meaning of the language used in this exception and it would not otherwise seem to include stairways leading to and from the basement.

5-8 Implied Terms A contract includes not only what is expressly stated but also what is *necessarily* to be implied from the language used. However, courts proceed with considerable caution in supplying by implication provisions omitted from written contracts and will imply an unexpressed term only where the implication arises automatically from the terminology employed in the instrument or is indispensable to effectuate the apparent intention of the parties.

A rather common example of an implied contract term is the case of persons who present themselves as specially qualified to perform work of a particular character. In such instances there is an implied understanding that their product will prove to be of adequate workmanship and possess reasonable fitness for its intended use. Moreover, those who undertake to accomplish a certain result agree by implication to supply all the means necessary to enable them to perform their bargain.[7]

It is said that there can be no implied covenants in a contract in relation to any matter that is specifically covered by the written terms of the agreement itself. In this connection there is a familiar maxim (*Expresso unius est exclusio alterius*) that the expression in a contract of one or more things of a certain class implies the exclusion of all things of that class *not* expressed.[8]

5-9 Contract to Be Considered as a Whole It is a fundamental rule of construction that a contract must be interpreted as a whole, with the meaning taken from the entire context rather than from a consideration of only particular por-

---

[6] Italics added.

[7] Thus, in *McDonough v. Almy*, 218 Mass. 409, 105 N.E. 1012 (1914), it was held that a contract for the removal of traprock by machinery implies an undertaking on the contractor's part to obtain a license as required by a city ordinance for the use of a stationary engine essential to the performance of the work.

[8] Similarly, the fact that a condition is set forth in one of the clauses of a contract is regarded as excluding the idea that the same condition—this time unexpected—was intended to be declared in another clause as well.

tions or clauses of the agreement. A court will not isolate a certain phrase of a contract to garner the intent of the parties but will grasp the instrument by its four corners and in light of the whole ascertain the paramount and guiding intent of the parties.[9]

Similarly, when several instruments are drawn up as part of a single transaction, they will read each with reference to the other or others. Marginal notes on the contract document or matters contained in separate writings referred to in the contract (when clearly intended as a part of the agreement) may properly be considered in the interpretation of the contract. Thus, in construing a performance bond, it should be considered in connection with the underlying contract.[10]

5-10   Construing Installment Contracts   The proper interpretation of installment contracts has proved difficult. Assume that Smith contracted with Jones for the purchase of 500 tons of coal, to be delivered at the rate of 100 tons a month until the entire amount has changed hands, each installment to be paid for on delivery. If the foregoing arrangement is to be deemed a series of separate contracts and not treated as one entire agreement, it follows that a failure by Jones to deliver any particular installment or a failure by Smith to pay for an installment when received would in no way affect the rights of the parties respecting the remaining installments. If, however, the contract is to be treated as an entire one for the delivery of the 500 tons (the stipulations regarding installments being inserted merely for convenience), the question of whether or not a particular installment has been delivered or paid for would have a definite bearing upon the agreement as a whole. What seems to have been the majority view is that a contract of the type described in this paragraph should be regarded as one integrated whole rather than as a series of independent agreements.

Installment contracts regarding the sale of goods represent one topic out of a great many covered by the Uniform Commercial Code. For example, Section 42a-2-612, Connecticut General Statutes, provides as follows:

1 An "installment contract" is one which requires or authorizes the delivery of goods in separate lots to be separately accepted, even though the contract contains a clause "each delivery is a separate contract" or its equivalent.

2 The buyer may reject any installment which is non-conforming if the non-conformity substantially impairs the value of that installment and cannot be cured or if the non-conformity is a defect in the required documents; but if the non-conformity does not fall within subsection (3) and the seller gives adequate assurance of its cure the buyer must accept that installment.

3 Whenever non-conformity or default with respect to one or more installments substantially impairs the value of the whole contract there is a breach of the whole. But the aggrieved party reinstates the contract if he accepts a non-conforming installment

[9] *Rumph v. Dale Edwards, Inc.,* 183 Mt. 359, 600 P.2d 163 (1979).
See also *Bellak v. Franconia College,* 118 N.H. 313, 386 A.2d 1266 (1978); and *City of Jonesboro v. Clayton County Water Authority,* 136 Ga. App. 768, 222 S.E.2d 76 (1975).
[10] See *Turner v. Wexler,* 14 Wash. App. 143, 538 P.2d 877 (1975).

without seasonably notifying of cancellation or if he brings an action with respect only to past installments or demands performance as to future installments.

5-11   General versus Specific Terms in the Contract   Where in an agreement there are both general and special provisions relating to the same thing, the special will prevail. That is to say, the specific language will take precedence over the more general wording. Where, however, both the general and special provisions may be given reasonable effect, both are to be retained.

5-12   Writing versus Printing   When a contract is partly printed or typed and partly written and the printed portion cannot be reconciled with the written portion, the latter will prevail. The reason behind this particular rule of construction (which is resorted to only where absolutely necessary) is that the written words are the immediate language and terms chosen by the parties themselves for the expression of their meaning, and accordingly such written words are deemed worthy of greater weight than printed terminology intended for general use without reference to particular objects and aims.

5-13   Words versus Figures   In the event of an inconsistency between words and figures appearing in a contract, the words will govern.

5-14   Construing Vague Terminology against the Drafter   Ambiguous language in an agreement will be interpreted against the party who employs it. The explanation for this state of affairs is that the drafter is the one responsible for the uncertainties and must suffer any difficulties occasioned thereby. This is most often applied where the contract is on a *printed* form prepared by one of the parties.

5-15   Parol-Evidence Rule   To limit opportunity for fraud and mistake, the law will not allow oral evidence to be introduced to contradict what the parties have agreed to in writing; that is, evidence to *vary* the terms of a written agreement is inadmissible. This rule does not, however, prevent the introduction of testimony intended to negate the very existence of the alleged agreement, nor does it preclude oral evidence regarding "supplementary" terms when the contracting parties have neglected to reduce the entire agreement to writing. In this last instance parol evidence comes in not to vary but simply to complete the written contract.[11] By the same token, certain explanatory evidence may be received orally, as, for instance, when identity of the contracting parties needs to be clarified (for example, when confusion has arisen from similarity of names, or when an agent has signed on behalf of an undisclosed principal).

---

[11] *Smith v. Michael Kurtz Const. Co.,* 232 N.W.2d 35 (N.D. 1975), see also *Burgen v. Pine County Georgia Ltd.,* 382 So.2d 1295 (Fla. 1980), and *Flynn v. Sawyer* 272 N.W. 2d 904 (Minn. 1978), in which the courts permitted the introduction of parol evidence to establish that several individual contracts were intended to be integrated to constitute one contract.

CONDITIONS

5-16 Definition and Classification A *condition* is a proviso upon which the rights and duties of the parties depend. A condition in a contract operates to suspend, rescind, or modify the principal obligation under given circumstances.

Conditions may be *express, implied in fact,* or *implied in law.* A certain fact may operate as a condition simply because the parties intended that it should and indicated as much in so many words; it is then an express condition. Or it may operate as a condition because the parties intended that it should, such intention being a matter of reasonable inference from conduct other than words; it is then a condition implied in fact. Or it may operate as a condition because the court believes that the parties would have intended it to operate as such if they had thought about it at all or because the court believes that by reason of the mores of the time justice requires that it should so operate; it may then be properly described as a condition implied in law.

A few examples of implied conditions include: (1) in employment contracts, the continued life of both parties is implied; (2) in a contract to sell specific goods, the continued existence of the subject matter is implied; (3) the continued legality of any contract is implied; (4) when a contract is made in reliance upon the happening of a particular event, the actual occurrence of that event is an implied condition of the contract. In *Marks Realty Co. v. Hotel Hermitage*[12] defendant had agreed to place an advertisement in a souvenir program to be distributed in connection with international yacht races. War intervened and the races were called off, but the program was published anyway, and defendant refused to pay for the advertisement. The court decided that defendant's refusal was justified since he would never have contracted unless he and the plaintiff had both assumed the races would be held as scheduled. The object in mutual contemplation having failed, plaintiff could not exact the stipulated payment.[13]

5-17 Condition Precedent Conditions are further classified as *precedent, concurrent,* or *subsequent.* A *condition precedent* arises where a promise is predicated on the happening of a certain event; no obligation to fulfill such promise comes into being unless and until the specified event actually does take place. The employment of such contract terminology as "when," "if," or "as soon as" normally points to the inclusion of a condition precedent. Generally in contracts for services, performance is a condition precedent to payment.

A court may also find that a condition precedent is implicit when a contract is read as a whole, even though a specified condition is not referred to. For example, in *Handley v. Ching*[14] the plaintiff and defendant entered into a joint venture to develop a residential condominium project. The plaintiff, a real estate developer,

---

[12] 170 App. Div. 484, 156 N.Y. Supp. 179 (2d Dept. 1915).

[13] The Marks case is still cited as authority in court opinions. For example, see *U.S. v. General Douglas MacArthur Senior Village, Inc.,* 508 F.2d 377 (2d Cir. 1974); and *Nash v. Bd. of Educ.; Town of Islip,* 382 N.Y. Supp.2d 31 (1976).

[14] 2 Ha. 166, 627 P.2d 1132 (1981).

and the defendant entered into an agreement under which a new Hawaii corporation was to be formed which would own and operate the condominium. The agreement contained a provision under which the defendant was obligated to reimburse the plaintiff for funds expended if the defendant defaulted. The defendant did default and the new corporation was never formed. Defining a condition precedent to call for the performance for some act after a contract is entered into, upon the performance of which the obligation to perform immediately is made to depend, the court said that a necessarily drawn implication from the language of the agreement was that the formation of the corporation contemplated by the parties was a condition precedent to any obligation to perform the provisions of the agreement. Specifically, the court observed that the agreement stated that "whereas, Handley and the Allens desire to form a new Hawaii corporation for the purpose of developing the subject property" and that nearly every subsequent section of the agreement referred to rights and duties of the parties in terms of the existence of the new corporation.

5-18 Condition Concurrent Where performance by both parties is designed to be simultaneous, performance by one is often termed a *condition concurrent* to performance by the other. For example, a city may specify in a contract for sewer construction that, when the contractor installs the sewer in Maple Street, the work shall be done one block at a time and may agree that each block will be closed to traffic while work is progressing in it. The closure is then a condition concurrent.

5-19 Condition Subsequent A *condition subsequent* is one which relieves the promisor of a liability which has already attached. If such a condition is violated, the promisor's liability will be annulled. Conditions subsequent are matters of defense to be proved by the party seeking release from the obligation. One example would be a fire insurance policy clause reciting that no suit may be brought on the contract unless within a stated period of time following a loss. The same sort of time limit may be prescribed for claims of a contractor for extra costs occasioned by changes ordered during the course of a job.

Many conditions which appear to be subsequent in nature are in fact conditions precedent. For instance, the recital of obligation in a surety bond will indicate that the obligation attaches only in the event the principal fails in the specified performance. Default by the principal is therefore to be actually a condition precedent to the surety's liability on the bond.

5-20 Personal-Taste Contracts In contracts involving personal taste, an agreement by *A* to perform to the satisfaction of *B* places the latter in a dominant position, and *A* cannot recover his fee until *B* is in fact satisfied. And any dissatisfaction which *B* might express, though it may be capricious by ordinary standards, requires only good faith to be effective. A typical case is the situation where one party undertakes to paint a portrait for another or to fabricate some female attire—in such instances the customer's standard of taste is the test.

Where personal satisfaction is not involved quite so directly, as in a building contract which contains the common provision that no payments are to be made except on certificate of the architect or engineer, the fact that the building was erected in a workmanlike manner and precisely according to specifications will be evidence that the architect was acting in bad faith in withholding the certificate.

5-21   Conditions about Time of Performance   As in the case of contractual provisions generally, stipulations about time must be construed in accordance with the intent of the parties shown by the fair import of the language employed. Completion of a contract within a "reasonable" time is sufficient if no exact time limit has been stipulated. Thus, an agreement to take certain action "as soon as possible" or "forthwith" has been held to require merely that performance be made within a reasonable time under the circumstances of the particular case. The rule is somewhat more strict in connection with those contracts where time is usually deemed to be "of the essence," as, for instance, agreements for the payment of money. In such cases, where no precise time is set, the assumption is normally that the money is payable immediately. Where a contract to render services is silent regarding the time of payment, remuneration is due just as soon as the services have been rendered.

5-22   Conditions about Place of Performance   Where the place of performance has been named by the parties their expression prevails. If, however, the contract is silent in this respect and there are not circumstances showing the parties intended performance to be elsewhere, the place of performance is deemed to be the place where the contract was made.

5-23   Conditions about Amount of the Compensation   A complete absence of price in a contract may render it invalid, although price may occasionally be inferred. The matter of price may be affected by previous dealings between the parties. Where, for instance, the amount of compensation is not specified in an oral contract for work similar to that done under a previous written contract between the same parties wherein the compensation was specified, it may be presumed that payment under the new agreement was intended to be the same rate used for the original work of the same nature.

When there is a contract to perform certain work according to a schedule and that schedule provides for doing the work under the direction of an engineer, variations and additional work ordered by the engineer may be recovered for in an action on the contract. However, when an architect directs extra work to be done and the architect's sole authority is to see that materials and workmanship are in accordance with the specifications, the contractor probably cannot—in the absence of ratification by the owner—recover for the extra work.

5-24   Failure of Condition   The nonfulfillment of a *promise* is called a *breach of contract* and creates in the other party to the agreement a right to damages. On the other hand, nonfulfillment of a *condition* simply excuses the other party from performing his side of the original bargain but creates no right to damages unless fulfillment of the condition has been expressly guaranteed.

5-25   Waiver of Condition   A condition precedent to a contract duty of performance can without question be waived by a voluntary act of the party who is undertaking the duty. This means that by his own free election he has power to transform his conditional duty into an unconditional one and to carry out the terms of the contract despite failure of the condition upon which his obligation was grounded.

For example, when an insured submits a claim for a fire loss on other than the form stipulated by the carrier but the latter ignores this circumstance in replying and simply notifies the policyholder that it disclaims liability on the ground that the fire had been deliberately set by the insured, the contract condition prescribing definite claim forms is waived.

5-26   Estoppel   If the promisor is himself the cause of the failure of a condition on which his liability hinges, he will be precluded from alleging such failure. His conduct in such a case has made it inequitable for him to insist upon fulfillment of the condition.

In *Vandegrift v. Cowles Engineering Co.,*[15] *A* was to build a boat for *B* and to make delivery on a specified date. *B*'s interference prevented the completion and delivery on schedule. It was held that *B* was estopped from asserting failure of the condition as to time, for "He who prevents a thing may not avail himself of the performance which he has occasioned."

Similarly, in a case where defendant hired plaintiff to find a joint venturer to invest in coal properties and where plaintiff was to receive a finder's fee of 5 percent upon conclusion of the joint venturer agreement, it was held that the finder's fee is payable even if the agreement is never concluded when the failure results from bad faith negotiations on the part of the defendant.[16]

5-27   The Requirement of Privity   A binding promise may, of course, be enforced by the recipient thereof, and his rights encompass promises made for the benefit of some third person. Subject to a very significant exception in the case of certain third-party-beneficiary contracts, it is a general rule that only one who is a party to the contract or someone in *privity* with him may enforce the obligation.[17] *Privity of contract* is a term denoting mutual or successive relationship to the same right of property or to the same subject matter. Thus, the assignee is in privity with his assignor, the donee with his donor, and the lessee with his lessor. Similarly, there is privity of contract between an owner and the general contractor who

---

[15] 161 N.Y. 435 (1899). See also *Weniger v. Union Center Plaza Associates,* 387 F.Supp. 849 (S.D.N.Y. 1974); and *Broadstone Realty Corp. v. Evans,* 251 F.Supp. 58 (S.D.N.Y. 1966).

[16] *Nuvest, S.A. v. Gulf and Western Industries, Inc.,* 649 F.2d 943 (2d Cir. 1981), applying N.Y. law.

[17] Justification for the rule lies in the fact that otherwise a person's responsibility for failure to carry out his agreement with another would have not reasonable limits.

is erecting a building for him, but there is no such privity between the owner and a subcontractor on the job. A 1941 Delaware case will illustrate the distinction.[18] Suit was brought by subcontractor against owner to enforce a mechanic's lien for furnishing materials for the construction of a research laboratory. The general contractor was not included as defendant. The court spoke as follows:

> No privity of contract exists between the owner and the subcontractor. The general contractor is the link which connects the owner and the subcontractor. It is manifestly unjust to an owner against whom our statute operates with sufficient hardship, if properly and fairly construed, that, apart from the general contractor, he should be forced to defend against a claim of a subcontractor or of which he may know nothing.

An obvious corollary to the rule rendering contracts enforceable only by parties thereto or their *privies* is that an action on a contract cannot be maintained *against* a person who is not a party to the agreement.[19] There are several variations from this general proposition, one being that a principal is bound by the contract made for him by his agent (if the latter acted within the scope of his authority), even though said principal is not actually named in the contract nor his interest disclosed. The other side of the coin is that one who contracts in a representative capacity is not individually liable where, from the manner in which the instrument is executed, it is apparent that it is intended as the obligation of the person represented and that the principal (and not the agent) is the real contracting party.

5-28   Third-Party Beneficiaries
Right to Sue   One of the controversial subjects in contracts has been whether the common-law rule that only the promisee can enforce a promise should be applied to cases in which the promise clearly was made for the benefit of a third person. The right of a beneficiary who furnishes the consideration to realize the fruits of the contract in which he has invested has long been acknowledged—often on the theory that the promisee acted as agent for the beneficiary in dealing with the promisor—but there has been considerable diversity of opinion about the rights of a beneficiary who is a complete stranger to the contract he seeks to enforce.

The prevailing view in this country now permits a third person for whose benefit an agreement was made to sue on it even though he is a stranger to the contract and to the consideration involved. Indeed, several jurisdictions have provided by statute that one for whose benefit a promise is made may maintain an action upon such promise. The basis for permitting the third-party beneficiary to sue has been stated to be that "the law, operating on the act of the parties, creates the duty, establishes the privity, and implies the promise and obligation, on which the action is founded."[20] The liberal view permitting the beneficiary to sue makes

---

[18] *Westinghouse Electric Supply Co. v. Franklin Institute*, 41 Del. 319, 21 A.2d 204 (1941). More recently, see *Rackers & Baclesse, Inc., v. Kinstler*, 497 S.W.2d 549 (Mo. App. 1973).

[19] Thus a subcontractor has been held not liable to the owner for defective work when the owner's negotiations were solely with the general contractor. *Lumber Products, Inc. v. Hiriat*, 255 So.2d 783 (La. App. 1971). See also *Coburn v. Lennox Homes, Inc.*, 378 A.2d 599. (Conn. 1977).

[20] *Brewer v. Dyer*, 7 Cush. 337 (Mass. 1851).

a great deal of sense because it should make no difference to the promisor—who has received his consideration—whether the enforcement action be brought by the promisee, with whom he dealt directly, or by a third party, who is the intended beneficiary under the contract.

The contract upon which the third-party beneficiary seeks to recover must be a valid and binding one (entered into between competent parties, duly supported by a consideration, and meeting the various other requirements of an enforceable agreement). The beneficiary's rights against the promisor spring from the contract as it was made, and—if it was in its inception void for lack of any essential element—the third party has no rights. Furthermore, one who sues on a contract made for his benefit must accept such contract as is. His rights are affected by any infirmities of the contract as between the contracting parties themselves. The rights of a third-party beneficiary are derivative, and his position is analogous to that of an assignee. Therefore, the promisor may interpose against the beneficiary any defenses which would have been available were the action brought by the promisee. To illustrate: Where the contract is void or voidable, as against the promisee, on grounds of fraud, mistake, infancy, or insanity, it is to likewise against the third party; or, if the duty of the promisor is subject to a condition precedent, the correlative right of the beneficiary is likewise conditional.

Identification of Beneficiary   It is not essential that the person to be benefited by the contract be named therein as long as he is otherwise sufficiently designated as to be readily ascertainable. In fact, he may be one of a *class* of persons if the class is adequately described. Moreover, in order to enable a third person to elect to enforce a contract made for his benefit, it is not necessary that he become aware of the existence of the contract at the time it was made.

Types of Third-Party Beneficiaries   For the sake of convenience, third-person beneficiaries may be divided into three categories: (1) *creditor-beneficiaries,* (2) *donee-beneficiaries,* and (3) *incidental beneficiaries.* Beneficiaries who comprise category (3) may not maintain an action on the contract.

Creditor-Beneficiaries   It is the normal holding that where two parties enter into an agreement whereby one promises to pay the other's debt to a third person, the third person may recover against the promisor. In other words, a *creditor-beneficiary* may enforce a promise to pay a debt owed to him by the person to whom the promise was made.[21] The case of *Lawrence v. Fox*[22] is regarded as the leading authority to the effect that a creditor-beneficiary has an enforceable right. In the case Holly owed Lawrence $300, and Holly was ready to pay. Fox borrowed the money from Holly overnight, promising to pay the debt to Lawrence the next day. Despite the absence of any "privity of contract" between them, it

---

[21] Here is a typical example of a creditor-beneficiary situation: Where a mortgagor who is himself personally indebted sells his interest in the mortgaged property to a grantee who assumed payment of the mortgage debt, the mortgagee is a creditor-beneficiary and is almost universally allowed to sue the grantee to secure a personal judgment against him for the amount of the debt. See, for example, *Albert v. Cuna Mutual Soc.,* 255 So.2d 170 (La. App. 1971). See also *Choate, Hall and Stewart v. SCA Services, Inc.,* 392 N.E.2d 1045 (Mass. 1979), where a creditor-beneficiary was held entitled to bring suit to enforce settlement agreement.

[22] 20 N.Y. 268 (1859).

was held that Lawrence could sue Fox in enforcement of the latter's promise to Holly.

A creditor-beneficiary is not bound to enforce the contract obligation against the promisor but is free to rely instead upon his original remedy against his debtor, the recipient of the contract promise. Thus, the creditor-beneficiary is armed with an enforceable claim against each of two individuals and can select the victim. If he sues both and secures two judgments, entire or partial satisfaction of one judgment will preclude to that extent enforcement of the other judgment.[23]

Donee-Beneficiaries[24]   A third-person beneficiary under a contract is a donee-beneficiary if the purpose of the promisee in obtaining a commitment from a promisor is to make a gift to the beneficiary or to confer on him a right against the promisor to some performance neither due nor asserted to be due from promisee to beneficiary.[25] The donee-beneficiary's ratification of the contract is presumed from the fact of benefit without burden, and his rights arise immediately upon making the contract.

Recovery in donee-beneficiary situations formerly was limited to cases where the third party was a close blood relative of the promisee, but the modern tendency is to permit any donee-beneficiary to recover against the promisor when the contracting parties intended to confer a benefit upon him.[26] Besides being vulnerable to a suit by the donee-beneficiary, the promisor who breaches is answerable to his promisee.

Incidental Beneficiaries   An *incidental beneficiary* is neither a creditor-beneficiary nor a donee-beneficiary but is simply one who happens to have found in some contract a provision which would insure to his advantage. The incidental beneficiary has no enforceable rights under such a contract[27] since the contracting parties had no intention of benefiting him, and it is the intention rather than the effect that governs. To permit recovery on his part, the intent to benefit the third person must clearly appear from the language of the agreement as a whole, construed in the light of the surrounding circumstances.[28]

The case of *Segar v. Irish*[29] affords a good illustration of the incidental-benefi-

---

[23] Not only does the beneficiary have alternate remedies available, but also the promisor finds himself open to two suits at law; both the creditor-beneficiary and the promisee may sue him for any breach of his contractual obligation, although actual recovery by one will bar recovery by the other. Of course, the promisee's damages in an action against the promisor are the amount of the former's indebtedness to the third party.

[24] In life insurance the beneficiary is usually of the donee classification, and in all jurisdictions he can maintain suit on the policy.

[25] *Aetna Life v. Maxwell,* 89 F.2d 988 (4th Cir. 1937). Recent cases upholding the rights of donee-beneficiaries include *Wolfe v. Morgan,* 11 Wash. App. 738, 524 P.2d 927 (1974); and *N.Y. Telephone Co. v. Secord Bros., Inc.,* 309 N.Y. Supp.2d 814 (1970).

[26] In many cases, however, the damages recoverable by the beneficiary of a "gift promise" will be nominal indeed.

[27] Recent cases denying recovery to incidental beneficiaries include *Matternes v. Winston-Salem,* 286 N.C. 1, 209 S.E.2d 481 (1974); and *O'Connell v. Entertainment Enterprises, Inc.,* 317 N.W.2d 385 (N.D. 1982).

[28] If, however, such intention is indicated with reasonable certainty, an action by third party should be maintainable even though the idea of benefiting him was only a secondary purpose of the contract.

[29] 156 Misc. 714, 282 N.Y. Supp. 450 (Otsego Co. 1935).

ciary proposition. It was held in that case that a carpenter employed by the contractor in building a school was at most an incidental beneficiary of the agreement between the contractor and the school district requiring the former to insure materials used in constructing the building. Thus, the carpenter could not recover on any theory of third-party beneficiary for the loss of tools in a fire which destroyed the structure.

In another case[30] a contract between *A* and *B* under which *A* was to sell his house to *B* contained a provision calling for the payment by *A* of a 6 percent commission to the plaintiff realtor. *B* was unable to perform his obligation because he could not sell his own house, and the realtor sued *B* seeking the commission he would have made had *B* performed. The court found the realtor to be the promisee and *A* to be the promisor and that the broker was, at best, an incidental beneficiary under the contract between *A* and *B* and thus not entitled to maintain an action for breach of contract against *B*.

Rescission of Third-Party-Beneficiary Contracts It is now clear that after the third-party beneficiary has become aware of the contract made for his benefit and has taken some action in reliance on it, the promisee no longer has power to release the promisor from the latter's duty to the beneficiary. This is true whether the beneficiary involved comes within the creditor or the donee category. On the other hand, there has always been divided opinion among courts regarding the power of the contracting parties to rescind *before* the intended beneficiary acquires knowledge of the existence of the agreement and changes his position as a consequence thereof. The weight of authority seems to be that the rights of a donee-beneficiary become indefeasible upon formation of the contract, while those of a creditor-beneficiary may be lost or qualified by prompt action of the contracting parties before the beneficiary can bring suit or otherwise materially alter his position in reliance on the promise.[31]

5-29 The Nature of Assignment An *assignment* by the holder of a right introduces a new party (the assignee) in his place. *Duties cannot ordinarily* be assigned. But the expression of intention to pass a right on to someone else is generally given effect. When such is the case, the assignment extinguishes the right of the assignor against the obligor and substitutes a similar right in the assignee.

A contract to assign in the future does not qualify as an assignment, since the term assignment has no pertinence if any further action on the assignor's part will be necessary to complete the right of the assignee. Moreover, *assignment* should be clearly differentiated from *novation*. Any contractual duty can be extinguished with the consent of the party who holds the right. If the party agrees to the substitution of a new party as the one bound to perform, this arrangement may be

---

[30] *Reidy v. Macauley,* 290 S.E.2d 746 (N.C. App. 1982).
[31] A life insurance beneficiary presents something of a special case. Unless said beneficiary has been irrevocably designated as such by the policy owner (who is typically the insured person), the beneficiary may be changed through exercise of a power normally reserved to the owner by the policy contract itself.

effective as a novation.[32] It is definitely not an "assignment," because the latter term is used to denote a substitution brought about by the action of one party only without procuring the assent of the other.

5-30 What May Be Assigned Assuming the parties themselves have not expressly stipulated otherwise, all contract rights are assignable in the absence of a statute to the contrary or unless the contract is of such a nature as to cause an assignment to conflict with public policy or to increase materially the burden imposed upon the obligor. A contract clause purporting to forbid assignment is almost invariably taken as having been intended to prevent the delegation of performance rather than to affect assignment of rights arising under the agreement. Thus, if Smith contracts with Jones to have the latter paint the Smith garage and the instrument states that "this contract shall not be assignable," it should be held to mean that Jones must personally do the work and not that Jones is prevented from assigning his right to the agreed compensation. The right of an assignee under these circumstances would be conditioned upon the satisfactory rendering by Jones of the stipulated personal performance.

An assignment is not ineffective simply because it is for any reason conditional or is revocable by the assignor. Generally speaking, an operative present assignment may be made of rights dependent upon a performance not yet due or rendered. In order for an assignment to be effective as against the obligor's other creditors, there must be at the time of the assignment an existing agreement under which the rights are expected to arise.

5-31 The Delegating of Duties It is often stated that "rights may be assigned but not liabilities." A promisor can never escape a contractual duty by assigning it to another or by delegating another to perform it. But performance of duty is not the same thing as duty of performance. A contract requires of the promisor the production of a specified result. Usually there are several ways in which that result can be achieved, and it is immaterial which method is employed. Therefore, a particular duty can often be performed by a person other than the promisor, and the promisee ordinarily cannot prevent the discharge of his contract right and the promisor's correlative contract duty by such a vicarious performance. Most contracts do not require or envision the personal performance of the promisor. If Y undertakes to do work for Z which requires no special skill and it does not appear that Y was selected with reference to any personal qualification, Z cannot complain if Y has the work done by an equally competent person. But Y is still liable should the work be poorly done, and Y is the only one who can sue for payment

---

[32] A novation is the substitution of a new obligation for an existing one. For example, a court found a valid novation where a lawyer and client agreed to terminate a contingent fee agreement and substitute in its place an agreement whereby the client would pay the attorney $1000. *People v. Metcalf,* 79 Cal. App.3d 144 Cal. Rptr. 557 (1978).

For other illustrations of novation and discussion of the principle, see *United Security Corp. v. Anderson Aviation Sales, Inc.,* 23 Ariz. App. 273, 532 P.2d 545 (1975) and *Hemisphere Nat. Bank v. District of Columbia,* 412 A.2d 31 (D.C. Cir. 1980).

unless he has effectively assigned his right to the compensation stated in the contract.

There are a number of situations, of course, in which the promisor cannot delegate performance of his duty because the promised performance by its specialized nature necessitates his personal attention. If Caruso had promised to sing on a given program, he could not have delegated performance to another tenor. The artistry contracted for was that of Caruso, and the work of a substitute would neither discharge Caruso's duty nor create a right to payment of the agreed compensation. Instead, Caruso would be liable in damages for breach of this assignment.

5-32   The Effect of an Assignment   The assignment of a bilateral contract really consists of two phases: (1) a *delegation* of the power to perform the assignor's contractual obligations, assuming such are not purely personal in nature, and (2) an *assignment* of such present or future rights as the assignor may have under the terms of the agreement. As we have seen, delegation of performance does not of itself relieve the assignor of responsibility for seeing to it that the stipulated functions are properly accomplished.[33]

With respect to the assignment feature noted under (2) above, the assignee's rights can be no better than these same rights were in the hands of his assignor's and if for any reason the assignor's claim was voidable or conditional, so also is the assignee's; that is, the assignee's claim is subject to all defenses which would have been available against the assignor. The assignee may (and by the usual statute must) sue in his own name, since he, and not the assignor, is the "real party in interest."[34]

5-33   Notice to Debtor Regarding Assignment   If, the debtor (obligor under the contract) receives instructions from his creditor to pay the debt to a third person (the creditor's assignee), payment as directed before notice of countermand will discharge the obligation. On the other hand, if the debtor receives no word that the obligation is due to someone other than the party with whom he originally contracted, he is entitled to the benefit of any payment which he may make to his original creditor. Of course, the assignee has certain rights against his assignor when the latter fails to notify the debtor respecting the assignment and wrongfully collects from the debtor. In such a case debtor and assignee are both innocent parties; the debtor consequently is discharged, and the assignor may be thought of as holding his gains in trust for the assignee.

5-34   Successive Assignments   When the holder of a contract right makes a second assignment without procuring a release of the first one, it is generally held

---

[33] If the assignee should fail to perform or should perform in an unsatisfactory manner, the assignor will be liable unless (by novation) the promisee has permitted the assignee to take the place of the assignor as party to the original contract.

[34] Where, however, the assignment was merely as security for the payment of a previous debt due from assignor to assignee, the former retains an interest, and it has been held that in such circumstances suit can be maintained in his name.

that the effective right is in the first assignee. An exception is made, however, when the second assignee in point of time pays value[35] without knowledge of the prior assignment and in reasonable reliance on a document evidencing the existence of the right in the assignor, which document has negligently been left in the assignor's possession by the prior assignee.

QUESTIONS

1 Define and illustrate (*a*) an express condition in a contract, (*b*) a condition precedent, (*c*) a condition concurrent, (*d*) a condition subsequent.

2 Sullivan, an architect, contracted to make the plans for Russell's new warehouse in accordance with the latter's wishes and to meet with the latter's approval. When the plans were finished, Russell did not like them. Sullivan refused to revise the plans unless paid an extra for doing so. Can Russell compel her to make the revisions without an extra? Explain.

3 On April 2, Conners contracted to deliver 50 cubic yards of topsoil to Buck for grading the latter's lawn. In spite of Buck's telephoned requests, Conners had not delivered the soil by May 15, whereupon Buck purchased the topsoil from Perkins (without notifying Conners) and had the job finished by May 21. A week later Conners started delivery and insisted that Buck go through with the bargain. If litigation ensues, what result can be expected?

4 At a fee of $2000, Kinney agreed to make the detail drawings for the reinforcing bars for Kimball's new warehouse. After the drawings were completed, Kimball claimed that the $2000 charge was excessive and demanded a refund of $500. Can he collect it? Why?

5 Ferris bought a secondhand motor boat from Lewis, to be delivered to Ferris's son on June 1 in return for $500. Just prior to June 1, the elder Ferris died. Can the son force Lewis to deliver the boat for the stated sum?

6 An expert sculptor contracted to make a bust of a movie star. She tried to sublet the job to another sculptor, but the actor refused to permit this. Did the latter have this right? Why?

7 Can *X*, who has a contract with *Y*, assign his rights under the contract to *Z*? His duties? Is *Y*'s approval of the assignee (*Z*) necessary?

8 *A* contracted to furnish certain bridge steelwork "complete and delivered to the site" for $30,000. He fabricated all the steel members properly and deposited them at the designated location, but he refused to erect the structure. Was he obligated to erect it? Explain.

9 Black contracted to deliver to storekeeper Brown three hogs per month for twelve months at a stated sum per pound; that was all the contract stipulated. Black delivered the first two months' allotments dressed, and Brown paid for them. Then Black brought the next three hogs butchered but not dressed and claimed payment based on their "live" weight. Was Brown correct in refusing to pay for more than the dressed weight?

10 Will vague terms in a contract be construed against the one who composed them? Elaborate.

11 *A* promised *B* to build a small bridge for $48,000, constructing and maintaining in the meantime a detour that would be satisfactory to the state highway department. Later on

---

[35] That is, gives consideration.

A discovered that the detour would cost her so much that she would lose heavily on the job. She therefore tried to annul the contract on the ground of inadequate consideration. What are her chances of success?

12 Brown, a consulting engineer, agreed to sublet the making of the design drawings for the bridge approaches to Black—provided that Brown secured the contract for designing this certain large bridge project. Assuming that Brown would get the job, Black contracted to employ White at a salary of $800 per month for one year. Brown failed to get the award. Is Black nonetheless obligated to hire White?

13 Does a third-party beneficiary of a contract have the right to demand enforcement of the contract? Explain.

14 Differentiate among creditor-beneficiary, donee-beneficiary, and incidental beneficiary.

15 *A* contracted to deliver "200 pairs of Firestone tires of assorted sizes" to *B* by January 1. *B* intended to send *A* a list of the desired sizes but did not make this fact clear. *A* promptly shipped 200 pairs of tires, determining on his own the several sizes and the number of pairs of each. *B* refused to pay, and *A* brought suit. Will *B* have to accept and pay for the tires in the assortment tendered?

16 In the preceding example, if *A*, expecting word as to sizes, received no pertinent instructions from *B*, and if he therefore made no delivery before January 1, can *B* declare the contract breached by *A*'s failure to deliver? Explain.

17 In quoting her price on some concrete construction, *A* listed the proposal as follows: "500 cubic yards at ninety-eight dollars ($98.00) per cubic yard." How is *A*'s bid price to be determined?

18 *A* sent a check to *B* in payment for a certain list of lumber but gave *B* no instructions as to time for delivery. What time for delivery can *A* demand? Why?

19 *A* ordered 100 pieces of hemlock two-by-fours 16 feet long. *B* had only 76 such pieces so she sent them and included 32 pieces 12 feet long, giving the same total length. What may *A* do about it?

20 What is the meaning of "privity of contract"?

21 On October 3 Smith sent a check to Jones in payment for a TV set to be delivered to his son on December 24 as a Christmas present. He told his son about the arrangement. On November 15, Smith was killed in an auto accident. What action can Smith's son take if Jones fails to deliver the machine? Explain.

22 *X* claimed to be in the business of mowing lawns for suburbanites. *X* contracted with *Y* to mow the latter's lawn for $50 for the season. When *X* appeared at *Y*'s place, he claimed that *Y* had to furnish the power mower to perform the work. Discuss *Y*'s possible course of action.

CHAPTER 6

PERFORMANCE OR BREACH OF CONTRACT OBLIGATIONS

PERFORMANCE

6-1 Performance as Condition Precedent to Recovery   When the contractual duties of parties to an agreement are supposed to be performed concurrently, neither party can recover (for breach by the other) without a showing that he has himself performed, has a valid excuse for nonperformance, or has made an effective tender of performance—that is, has given timely notification to the other party that he is ready, willing, and able to accomplish whatever the agreement may have required of him.

When a contract has been fully performed on one side, the other party's obligation to do his part is apparent. But when a given contract is yet completely executory, much depends—as far as the obligation to proceed with performance is concerned—upon the nature of the respective promises involved. When these are independent of one another, a party must undertake timely performance of his particular obligation, regardless of what has been done on the other side. Should the promise of A be so constituted as to depend upon performance by B, A is not obligated to discharge his contractual duty unless and until B has accomplished his part of the bargain.

6-2 Substantial Performance   Generally speaking, performance of a contract must be strictly in accordance with its terms, and a party cannot be compelled to accept performance at variance with that for which he bargained. This is so even though a tendered substitute might well prove more valuable in a strictly pecuniary sense.

Nevertheless, there is apparent from judicial decisions a trend toward requiring merely substantial performance in lieu of absolutely literal compliance with precise contract requirements. The equitable doctrine of substantial performance stems from a desire of the courts to protect those who have faithfully and honestly endeavored to perform their contractual duties in all material respects and to preserve the rights of such persons to compensation despite unimportant defects in the performance involved.

Substantial performance may be defined as the accomplishment of all things essential to the fulfillment of the purpose of the contract, although there may be insignificant deviations from certain contract terms or specifications. While courts will enforce the rights of one who has thus substantially performed, it is equally logical that the other party, who has received something less than what he should have, is entitled to recoupment of some sort. A building contractor, for instance, having substantially but not strictly performed, can generally recover the sum stipulated in the agreement less any damages occasioned by the minor defects in performance. In the typical case of *Boggs et al. v. Shadburn,*[1] the builder sued the owner to recover the full contract price of a house, and the latter counterclaimed, alleging that the builder had not properly graded the yard and had not waterproofed the basement as stipulated in the contract. In discussing the principle of substantial performance,[2] the court has this to say:

> Where a builder has in good faith intended to comply with a contract and has substantially complied with it, although there may be slight defects caused by a misconstruction of the terms of the contract, and the house as built has been received by the owner and is reasonably suited for the purposes intended, the contractor may recover the contract price less the damage on account of such defects. In such case the true measure of damages is the difference between the value of the house as finished and the house as it ought to have been finished under the contract, plans and specifications.

It is often difficult to determine what constitutes substantial performance in a given situation. There can be no doubt, however, that the concept was never intended to confer on a party the right freely to deviate from the contractual undertaking or to substitute some material or type of operation that may be personally regarded as just as good as what is actually called for in the agreement.

While the tradition has been that where any deviation from the contract is willful or made in bad faith the offending party is not entitled to recover at all, it has also been held that the fact that a contractor is guilty of the willful breach is a factor to be considered in determining whether substantial performance has been rendered and is not conclusively a bar to the contractor's right to maintain the action.[3] It is only where defects are purely unintentional and not so extensive

---

[1] 65 Ga. App. 683, 16 S.E.2d 234 (1941).

[2] For other representative cases dealing with substantial performance, see *Dittmer v. Nokleberg,* 219 N.W.2d 201 (N.D. 1974); and *Cooperative Cold Storage Builders, Inc. v. Arcadia Foods, Inc.,* 291 So.2d 493 (La. App. 1974).

[3] *Vincenzi v. Cerro,* 186 Conn. 612, 442 A.2d 1352 (1982).

as to prevent the other party from receiving approximately what was bargained for that the principal of substantial performance comes into play.

The doctrine of substantial performance also may be available in construction contract situations where the contractor is prevented from completing the work by the property owner. Thus, the failure to completely perform a contract for the erection of a prefabricated luxury home has been held not to bar recovery for the contract price less appropriate allowances where the defects resulted from the purchaser's incessant efforts at varying the prefabricated home as designed and where the contractor was eventually ordered off the premises.[4]

6-3 Partial Performance *Partial performance* is performance which—while sufficient to confer appreciable benefit on the recipient—falls short of the concept of substantial, or virtually complete, performance.

Partial performance of an "entire" contract will not generally support a recovery on the contract.[5] Thus, complete, or at least substantial, performance by one party is a prerequisite to his right to compel performance by the other party, unless the promises of the respective parties are independent of each other.

While the foregoing represents the general rule, there have been a number of cases which have permitted recovery of the fair value of the benefit afforded by the partial performance, particularly when the benefit is of such a nature that it cannot be returned. Naturally, a party who has performed part of his obligation and is prevented or excused by action of the other party from completing the work is entitled to compensation for material supplied and for that portion of the work done. Similarly, if completion of performance is obviated by the intervention of an act of God or is prevented by governmental authority, an equitable recovery for the part performance may be had.

6-4 The Time Element Unless the characteristics of a given contract are such as to render of extreme and obvious importance the performance thereof at the precise time designated (that is, unless time is "of the essence") or unless the contract itself emphasizes the vital nature of the time element,[6] failure of a promisor to fulfill his particular obligation within the stipulated period will not automatically discharge the other party's duty of performance, though any unwarranted delay will normally subject the delinquent party to damages. It is entirely possible, of course, that the delay might be waived by the promisee or excused in some fashion.

Without a contract provision to the contrary, it seems that time will not be regarded as of the essence where a typical building contract is concerned. Many such contracts call for a certain amount per day as liquidated damages, to be

---

[4] *Pilgrim Homes and Garages, Inc. v. Fiore,* 75 A.D. 846, 427 N.Y. Supp.2d 851 (1980).

[5] This is not the case, however, with respect to those contracts classified as severable (or divisible). For a consideration of the difference between entire and severable contracts, see Art. 3-6.

[6] Some contracts will expressly provide that "time is of the essence of this agreement." When the parties have agreed to the inclusion of such language, a failure to perform within the time specified represents a material breach of contract.

deducted from the agreed total price in the event the job is not finished on schedule.

Obviously, it would be unfair to penalize a contractor for delays in performance occasioned through the fault of the other party. Indeed the innocent party is, in such circumstances, usually held entitled to recover damages.

In *Ericksen v. Edmonds School District No. 15*[7] plaintiff general contractor had been authorized to construct certain additions to a school building. A completion date was set, and liquidated damages were provided. One of the contract's general conditions was entitled "Claims for Damages and Extensions of Time" and read in part as follows.

> The Contractor shall not be entitled to any claim for damages on account of hindrances or delays from any cause whatsoever, but if occasioned by any act of God, or by an act or omission on the part of the Owner, such act, hindrance, or delay may entitle the Contractor to an extension of time in which to complete the work. . . .

Contractor failed to finish on schedule and blamed owner for having brought about the delays. In the process of holding that contractor was not justified under the agreement to claim damages but only certain time extensions, the court had this to say:

> It is undoubtedly the rule in this state, as well as in other states generally, that in the absence of any provision in the contract to the contrary, a building or construction contractor who has been delayed in the performance of his contract may recover from the owner of the building damages for such delay if caused by the default of the owner. . .The rule rests on what is generally held to be the owner's implied obligation to keep the work of construction in such a state of forwardness as will enable the contractor to complete his contract within the specified time.
>
> Where, however, the contract expressly precludes the recovery of damages by the contractor for delay caused by the default of the owner, that provision will be given full effect.

6-5 Approval of Performance   The authorities are far from unanimous regarding the effect of a contract provision whereby one party agrees to perform to the satisfaction of the other. It is reasonably clear that, in any event, the latter must exercise good faith in evaluating the performance. The difficult question is whether or not his decision must be reasonable.

A distinction is frequently drawn between cases where personal taste or fancy is a factor and those involving only mechanical utility or operative fitness. In contracts of the former type, where approval depends upon a matter of personal feelings about which there can be no real standard of reasonableness, it has uniformly been held that the party to be satisfied may, simply because of some whim or other, effectively reject what would appear to be a perfectly good performance.

Where no obvious question of personal fancy is concerned, courts have dis-

---

[7] 13 Wash.2d 398, 125 P.2d 275 (1942). See also *Rowland Constr. Co. v. Beall Pipe and Tank Corp.,* 14 Wash. App. 297, 540 P.2d 912 (1975), and *Allen Howe Specialties v. U.S. Construction Inc.,* 611 P.2d 705 (Utah 1980), which contains an excellent discussion of a "no damages for delay" clause.

agreed on whether or not the party who the contract says shall be satisfied has an absolute right to pass on the character of the performance. A number of cases have held that a promise by one party to perform to the satisfaction of the other will be interpreted literally and that approval may be arbitrarily withheld as long as any dissatisfaction is bona fide and not merely feigned for purposes of avoiding payment.[8] The difficulties of proof of what constitutes good-faith dissatisfaction are obvious. On the other hand, many courts have inclined to the opinion that a performance which would be satisfactory to a reasonable person suffices.[9] If a contract aims at a fixed mechanical result or at operative fitness—about which other persons can judge as well as could a party to the contract—and especially if the tests of performance are fully set out in the contract so as to be readily applied, the tendency seems to be to hold (in spite of the typical stipulation for satisfaction) that performance is successful if it attains the intended result. Under such circumstances, dissatisfaction professed is treated as unreal.

6-6   Satisfying Third Persons; Architect's Certificate   Contracts of any sort may, and construction contracts normally do, stipulate that all work is subject to the approval of some third person, typically an architect or engineer with expert knowledge in the field. When a contract requires the certificate of an architect as a prerequisite to payment, the contractor must show (1) that such a certificate has indeed been issued or (2) that the designated architect has withheld his approval fraudulently or in the exercise of bad faith or (3) that the owner has somehow waived his right to insist upon presentation of the architect's written acceptance of the performance.

Where the contracting parties have thus duly constituted an architect, engineer, or some other third person as sole judge of the proper performance of the contract, the ultimate approval or disapproval given is conclusive in the absence of fraud or bad faith.[10] This qualification is an important one. If the architect or engineer fails honestly to exercise his judgment or makes such gross mistakes as to imply bad faith, his decision about issuance of the certificate will not bind the parties.[11] Refusal to execute a certificate of approval is justified only when there is real and substantial failure by the builder to fulfill his contract duties. In other words, the architect or engineer is not authorized to reject work which complies with contract plans and specifications. The power of approval vested in the third person does not entitle him to exact from the builder performance beyond the plain prescriptions of the contract.[12]

---

[8] *Stone Mountain Properties Limited v. Helmer,* 139 Ga. App. 865 (1976); 229 S.E.2d 779. *Stribling v. Ailon,* 223 Ga. 662 157 S.E.2d 427 (1967).

[9] *Meredith Corp. v. Design and Lithography Center,* 614 P.2d 414 (1980); *Aztec Film Production, Inc. v. Prescott Valley, Inc.,* 128 Ariz. 402, 626 P.2d 132 (1981).

[10] *Laurel Race Course, Inc. v. Constr. Co., Inc.,* 274 Md. 142, 333 A.2d 319 (1975). *Lincoln Construction, Inc. v. Thomas J. Parker and Assoc.,* 617 P.2d 606 (Or. 1980).

[11] *Gold v. National Savings Bank,* 641 F.2d 430, applying Tenn. Law, *cert. denied,* 102 S.Ct. 116, 1981. *Savin Bros., Inc. v. State,* 62 A.D.2d 511, 405 N.Y.S.2d 516 (1978).

[12] *Tobin Quarries v. Central Nebraska Pub. P. and I. Distr.,* 64 F.Supp. 200 (Dist. Ct. Neb. 1946), *aff'd,* 157 F.2d 482 (8th Cir. 1946).

Also, the usual certificate of satisfactory performance binds the owner as regards visible or readily ascertainable defects in the work completed, but not as regards latent flaws.

6-7 Arbitration Clauses Often building contracts will outline the procedure to be followed in arbitrating disputes which may arise during progress of the work, or after its completion. The decision of the named arbitrators, where they act within the authority conferred by the contract, is conclusive unless these supposedly neutral persons have, in making their determination, been guilty of fraud, misconduct, or such gross mistake as would imply bad faith or a failure to exercise an honest judgment.[13]

When the arbitrators undertake to pass upon a matter not contemplated by the contracting parties to be within the arbitrator's province, that is, when they exceed their authority as it is prescribed in the contract, any action taken by them under such circumstances is not binding, regardless of the motives involved in the taking of that action.

6-8 Waiver of Imperfections in Performance Not infrequently an owner will consent to certain deviations from contract plans and specifications and will accept incomplete or defective performance as though it were in strict compliance with the terms of the agreement. Whether or not work has been expressly or impliedly accepted and any imperfections waived are questions of fact which often prove difficult to resolve. If the defects are obvious or if circumstances are such that knowledge of the imperfections may be imputed to the owner, his acceptance of the work will obligate him to pay for it. Furthermore, unless he promptly objects to the shortcomings in the work, he will have no right to damages (for instance, an offset against the contract price) for such patent defects in the performance.

Where the deficiencies in a contractor's work are not such as could reasonably have been ascertained when the owner took possession of the structure and made payment under terms of the contract, the latter is clearly entitled to damages upon subsequent discovery of these imperfections within a reasonable time. What constitutes "a reasonable time" will vary widely with the type and size of structure involved and with other facts of the case.

Apart from latent defects, it is not every acceptance which amounts to a waiver of imperfections and entitles the performing party to remuneration. In the absence of substantial performance, acceptance of work for which a contractor seeks recovery must be something beyond merely "keeping and using," where the situation provides no opportunity for the owner to return whatever has been received. In a Wisconsin case involving a defectively constructed roof,[14] use thereof by the owner was deemed not to represent acceptance of the imperfect performance since

---

[13] *Delta South Co., Inc. v. Louisiana and Arkansas Railway Co.,* 394 So.2d 1299 (La. 1981).

[14] *Nees v. Weaver,* 222 Wis. 492 N.W. 266, 268 (1936). See also *O. W. Grun Roofing & Constr. Co. v. Cope,* 529 S.W.2d 258 (Tex. App. 1975).

the contractor's failures should not put the owner in the unfortunate position of having to abandon the building or to remove the roof in order to avoid paying for the defective work. In dismissing the contractor's action to recover for services rendered, the court said:

> A contractor cannot enter upon the premises of another to perform a contract, ignore its terms, and clumsily or carelessly construct a roof falling far short even of a "substantial performance," and impose upon the owner the responsibility of choosing between being charged with the use and acceptance of the roof or going to the expense of removing it. Appellant had the right to use his building, and the breach of contract by respondents could not put upon him the alternative of abandonment of the building or removal of the roof.

Similarly, acceptance of a house as substantially completed has been held not to be a waiver of claims against the contractor where a list of defects was attached to the acceptance by the owner.[15]

EXCUSES FOR DELAY OR NONPERFORMANCE

6-9   Impossibility
Definition and Types   The term *impossibility*, when used in connection with excuses for nonperformance of contractual obligations, refers not only to literal impossibility but also to impracticality owing to some extreme difficulty encountered.

Impossibility may be characterized as either subjective or objective. *Subjective impossibility* is due to incapacity of the party who has undertaken the performance; *objective impossibility* is that type which is attributable to the nature of the thing to be done.

Subjective Impossibility   When impossibility of fulfilling a contractual commitment is due entirely to inability of the promisor himself—perhaps because of physical handicap or lack of skill—the nonperformance will not be excused.

Objective Impossibility   This type may be present at the time the contract is entered into or may occur subsequently. Where the impossibility of performance arises from circumstances in existence when the promise was made, performance is excused in the absence of evidence that the promisor intended to assume the risk of impossibility.

There is a notable lack of consensus about whether performance of an unqualified contract is excused where, by an unavoidable accident or some unforeseen contingency, it becomes impossible to accomplish what a party has promised. One group of cases has taken the position that, in the absence of an express manifestation to the contrary, a contract will be regarded as subject to an implied condition that, if something necessary in connection with the promised performance goes out of existence through some act of God or other unforeseen occurrence not the fault of the promisor, nonperformance is excused. Even these liberal decisions

---

[15] *Lofton v. Don Trahan, Inc.,* 399 So.2d 818 (La. 1981)

recognize that a release from contractual obligation by reason of impossibility is necessarily subject to several qualifications. First, the impossibility involved must be bona fide and not simply a question of hardship or inconvenience. Second the promisor seeking to be excused must not in his contract undertaking have assumed the risk of impossibility. Third, the performance will never be excused if the impossibility stems even to a limited extent from negligence or want of diligence on the part of the promisor.

A second line of decisions has adopted a completely different approach, much less pleasing to promisors caught unaware by supervening objective impossibility. These cases have held that one who unconditionally obligates himself for a certain performance will not be excused by subsequent impossibility brought on by an act of God or other unforeseeable emergency. In other words, a positive commitment to do something possible of accomplishment as of the time of contracting will bind the promisor despite effective interference by some superior agency over which the promisor had no control but against which interference he might have provided, through exception or reservation, in his contract.[16] Where one of two innocent parties must sustain a loss, the law will leave the burden where the contract has left it.

Finally, there has emerged a sort of compromise position somewhere between the two extremes. Cases which take this third viewpoint have excused performance when it has become impossible by an act of God (such as a great storm or an earthquake) but have refused to afford similar relief when the impossibility is occasioned by any lesser accident or unavoidable circumstance. A typical statement in these middle-of-the-road decisions is to the effect that a promisor is presumed, in the absence of an express contrary provision, to have assumed the risk of unforeseen contingencies arising during the course of the work, unless performance is rendered impossible by operation of law, by the other contracting party, or by a true act of God.

Intentions Manifested in Contract A determination of whether an act of God or an unavoidable mishap will discharge a contract duty often depends upon an interpretation of the particular contract itself. Frequently, contracts are so written that it is evident a promisor intended to bind himself to do certain things (or to pay damages for nonperformance thereof) notwithstanding the possibility, or even the likelihood, that the promised performance might subsequently become impossible. There can be no doubt that a person may contract to do the impossible as well as the difficult and be liable for failure to perform.

On the other hand, many contracts provide that such things as fire, war, accident, or "other causes beyond the control" of the promisor will obviate the necessity for performance. Where such a provision is not present, courts are prone to answer a promisor's plea of impossibility or extreme impracticality with the assertion that emergency situations could readily have been guarded against by appropriate stipulation in the contract.

---

[16] Where performance is prevented by the negligent or wrongful action of the other contracting party, the promisor's nonperformance is automatically excused.

Parties to construction contracts should insert clauses designed to impose upon one side or the other (or perhaps to apportion in some desired fashion) responsibility for loss resulting from accidental destruction of the building before its completion.

Financial Recovery for Defaulting Promisor   Where, after partial performance, subsequent impossibility is deemed to excuse completion of the undertaking, the party benefiting from the partial performance will be obliged to return what has been received or to pay reasonable value for the benefit conferred unless the contract clearly provides otherwise. In this respect, cases recognizing impossibility as an excuse for noncompletion of a promised performance are true quasi-contract cases embodying the principle of unjust enrichment.

Take, for example, a contract for all the plumbing work in a building under construction; a fire destroys the structure before the plumber has finished the work or received any payment. The plumber could normally recover in quantum meruit[17] for the value of such work as had been done before the fire.

6-10   Death or Illness of Party   Closely allied to the subject of impossibility as an excuse for nonperformance is death or disabling illness of the would-be performer. Where contracts for strictly personal services are involved, the death, insanity, or incapacitating illness of the party obligated will prevent his failure to perform from operating as a breach.[18] In other words, absent the spelling out of a contrary intention in the agreement itself, contracts of such a nature as to require for performance the continued existence of a particular person will be treated as including an implied condition that the death of such party before any breach on his part will excuse nonperformance of his undertaking.

The above rule is generally limited to contracts of a personal nature and is not applicable where the service involved is of such a character that it may be performed as well by someone other than the promisor or where the contract language indicates that performance by others was contemplated. Thus, the ordinary building contract is regarded as such that the personal representative of the contractor could duly perform all the contractor herself was bound to do. Should death or incapacity of such contractor intervene before the job is finished, the executor or representative will be obliged to see to it that the work is completed.

6-11   Destruction of Subject Matter   The general rule is said to be that, where an obligated performance is such that accomplishment thereof absolutely depends upon the continued existence of a specific thing, the contract must be regarded as implying that, and should the aforementioned thing become unavailable (before time for performance and without fault of the promisor), the duty in question will be discharged.[19] For instance, if a contractor undertakes to repair or remodel an

---

[17] Literally, *quantum meruit* means "as much as he deserved." An action in *quantum meruit* is based on a promise imputed to the defendant (in the absence of an express promise) to pay the plaintiff reasonable value for services rendered or materials furnished by the latter.

[18] The involuntary dissolution by governmental authority of a corporate promisor presents a similar situation.

[19] This rule assumes, naturally, that the promisor has not expressly agreed to be responsible for the continued existence of the subject matter involved.

existing building, the agreement is deemed to carry with it an implied condition that destruction of the building before the work can be done will discharge the contractor's duty, as long as he is not to blame for the mishap.[20]

A Virginia case[21] affords a good illustration of how destruction of the subject matter of an agreement can relieve innocent parties of their obligations under the contract. There was an undertaking by a public utility to supply natural gas to a housing authority. At the outset of the arrangement it was assumed by all parties concerned that the only known gas deposit in the area was extensive enough to afford an ample supply of fuel. Consequently, the written contract contained no statement about the contemplated position of the parties should the source of supply prove inadequate. Eventually the utility was obliged to cease deliveries of the promised quantities because the gas deposit simply was no longer in existence. The court decided that the defendant utility could not reasonably be assumed to have taken the risk of such an eventuality, and accordingly excused it from further performance of its contract. In the course of its discussion, the court said:

> It is, however, fairly well settled that where impossibility is due to. . .the fortuitous destruction or change in the character of something to which the contract related, or which by the terms of the contract was made a necessary means of performance, the promisor will be excused, unless he either expressly agreed in the contract to assume the risk of performance, whether possible or not, or the impossibility was due to his fault. . . .
>
> Where, from nature of the contract itself it is apparent that the parties contracted on the basis of the continued existence of the substance to which the contract related, a condition is implied that if performance becomes impossible because that substance does not exist, this will and should excuse such performance. . . .
>
> Each party to the contract knew that the supply of gas had to come from that specific locality. No other source of supply was available. Each knew that the supply from the field furnished the essential and necessary means of performance, and that the contract could not be fulfilled unless the gas supply continued to exist. They contemplated and assumed its continued existence and contracted with reference thereto as the means of performance.

The terms of a contract may, of course, foreclose any doubt by stipulating that a promisor is bound by his undertaking notwithstanding the untimely destruction of the subject matter through some unavoidable accident. It is not particularly unusual to find a provision, for example, making a contractor responsible for repairing or reerecting the building under construction at his own expense should fire, storm, or other named causes of damage intervene before completion.

6-12   Hardship or Inconvenience   In its normal connotation, *hardship,* or *severe inconvenience,* is not an excuse. It is entirely competent for a party to contract unequivocally to do a difficult piece of work in a certain fashion, and the mere fact that the undertaking proves more burdensome than anticipated will not make the

---

[20] See, in this connection, *Fowler v. Insurance Co. of North America,* 155 Ga. App. 439, 270 S.E.2d 845 (1980).

[21] *Housing Authority, etc., v. East Tennessee L. & P. Co.,* 183 Va. 64, 31 S.E.2d 273 (1944).

contract invalid or relieve the promisor of his obligation. If he wishes to be relieved of the necessity of performing should various contingencies arise, a contractor should have the foresight to assume only a qualified duty, expressly exempting himself from responsibility in certain events.

Almost any contract involves at least a small measure of risk, and in many instances unforeseen difficulties or expense will result in a considerable loss for one party and perhaps for both. A contractor submitting a bid on a construction job has to protect himself against, among other things, possible adverse changes in the cost of labor and materials. If he underestimates on these items and unfavorable circumstances arise, he must nevertheless perform his obligation according to its terms and absorb any resultant loss. It matters not to what degree (short of absolute impossibility) the promised performance is rendered onerous or unprofitable by surprise troubles of various sorts. Even though a contract is found to be extremely burdensome to one party or the other, the function of the courts is not to remake the contract in a more equitable pattern but to enforce it as it stands.

Excavation contracts typify agreements whose breach the promisor attempts to justify on the grounds of hardship.[22] Oftentimes the contractor runs into solid rock where gravel or sand was expected; but this untoward occurrence will not permit stopping short of a discharge of the undertaking.

While not serving as an excuse for nonperformance, severe difficulties which are unforeseen by the parties at the time of contracting and which impose upon one an additional burden not contemplated by either may well support a promise of extra compensation made in order to ensure completion of the contract. Thus, courts have usually held that such a promise is adequately supported by consideration and will therefore be enforceable.

Likewise, where one party extends time for performance by the other in view of hardship conditions unanticipated by either party, the promise of extra time is generally regarded as supported by a consideration.

6-13   Strike and Labor Trouble   Unless a contract clearly provides otherwise, delay on nonperformance is not excused just because progress has been impeded by strikes or labor trouble. This sort of happening is part of a contractor's calculated risk, and his only means of protection is to insert in the contract a stipulation relieving him from the obligation of timely performance if a crippling strike should occur. Agreements which do not contain clauses of this type cannot fairly be construed as if they did contain them.

6-14   Supervening Illegality   Apart from the infrequent instance where there is a contract provision to the contrary, an agreement is normally regarded as carrying with it the implied condition that a promisor's duty will be discharged if, through no fault of his, performance is prevented or prohibited by statutory enact-

---

[22] Three such cases are *United States v. Blauner Constr. Co.,* 37 F.Supp. 968 (D. Mass. 1941); *Miller v. Johns et al.,* 291 Ky. 126, 163 S.W.2d 9 (1942); and *McKee v. City of Atlanta,* 414 F.Supp. 957 (N.D. Ga. 1976).

ment subsequent to the date of the contract. In other words, frustration of contract by governmental action is a good defense to a claim for breach, and nonperformance will create no right to damages when performance was rendered impossible by a change in the law.

As a practical matter, supervening illegality had to be treated by the courts as a complete excuse for nonperformance of affected contractual obligations. To hold parties as bound to anticipate future legislation would be most unrealistic.

Denial of a necessary license or permit will not discharge a contractor's obligation even though performance is illegal without such governmental permission. The reasoning is that the contractor assumed the risk he would be able to procure the needed license.

6-15 Frustration of Object. This is a rather nebulous concept but one which has, in one guise or another, been creeping into the decided cases for many years. where parties contract with obvious reference to some activity governed by outsiders and that activity fails to develop in the fashion anticipated by such contracting parties, this circumstance may well excuse nonperformance of their respective duties.[23] For example, if the prospective owner of a football team in a to-be-formed league signs a player to a contract and the league is never formed, this doctrine may excuse nonperformance. To effect a discharge, the activity must have been so thoroughly the basis of the contract that, as both parties realized, without it their entire agreement was meaningless.

6-16 Waiver by Other Party The doctrine is well established that strict performance of a contract duty may be dispensed with by the other party. Sometimes the waiver, which may be conditional in nature, amounts to an outright dispensation of any performance whatever of the obligation involved, but more often it takes the form of acceptance of performance which differs materially from that called for by the contract. A waiver must be manifested in some definite fashion, though meaningful conduct will suffice just as well as an oral or written declaration.

Waiver requires both knowledge and intention, and is definable as the voluntary surrender of a right. The term waiver is often confused with *estoppel*; while there are similarities, the two are distinguishable concepts. *Intention* to abandon a right is immaterial to estoppel—where the important considerations are (1) prejudice to one party (2) resulting from misleading conduct by the party estopped.

The authorities are not in accord about whether a waiver in order to be finding has to be accompanied by a valid consideration in cases where the elements of estoppel are not present.

6-17 Material Alteration or Fraud by Other Party In the event one of the contracting parties, without consent of the other, substantially changes the terms

---

[23] It is always open to a particular party expressly to assume the risk of all chance occurrences, and where that has been done, duty of performance is not affected by contingencies of any sort.

of a written agreement, the other party at his option may regard the contract as terminated and his duties thereunder as extinguished. Fraud by the party otherwise entitled to performance will excuse the other party's failure to perform.

6-18   Repudiation or Breach by Other Party   Where contractual promises have been exchanged one for the other, a material breach on the part of one party will, unless occasioned by conduct of the adverse party, clearly justify the other's refusal to perform and will entitle the latter party to any provable damages resulting from that breach. If one party is unable or unwilling to perform his part of the contract and thus furnish the agreed-upon consideration for a return performance, such failure of consideration will operate to discharge the obligation of the innocent party. The breach must be substantial if it is to result in excusing subsequent nonperformance by the injured party; an "immaterial" breach will not terminate the arrangement, and the adversely affected person must yet undertake to fulfill his particular promise, though he retains the right to recover for any loss which the partial breach entails. The standards of materiality vary widely and it is often a close question whether a given breach will justify the other party in treating the contract as at an end.

Prospective failure of consideration is fully as effective as actual failure in excusing the injured person from performing. Thus, where one party to an executory agreement repudiates his own contractual obligations and declines to be bound thereby, the other party, upon being apprised of this state of affairs will no longer be obliged to render whatever performance he had promised. While outright repudiation obviates the necessity of performance (or tender thereof) by the other party, the repudiation must be definite and must go to the essence of the contract. Moreover, the repudiator may nullify the effect of his statement by a timely retraction before the other person has accepted the refusal as final and has acted in reliance thereon.

In the event of repudiation before performance by the other party is complete, such other party cannot, by continuing to perform regardless, enhance the measure of damages which the breach has entitled him to recover.

6-19   Prevention by Other Party   Where wrongful conduct of one party has prevented performance by the other, the latter's duty is discharged. Moreover, the wrongdoer has by his inference committed an actionable breach of contract[24] and will find himself liable in damages.[25]

It follows that a building contractor who is delayed or stymied completely by the owner (or by an architect or engineer acting on the job for the owner) cannot be held responsible for failure to complete the work on time, or for outright nonperformance, as the case may be. The owner is in no position to insist upon fulfillment of a contract when he is the party responsible for breach.

---

[24] Every contract contains an implied covenant by each party not to block performance by the other party.

[25] The general rule is that suit to recover these damages may be instituted even before arrival of the time fixed for performance by the guilty party of his own contract duties.

Closely allied to the idea of prevention of performance is refusal of tender. Where one party has by act or word unequivocally indicated his unwillingness to accept the obligated performance of the other party if and when tendered, the latter will be excused from performing. After all, the party who stands ready and willing to satisfy his contractual obligation according to its terms cannot *force* the adverse party to accept performance. Therefore, his inability to complete his undertaking is directly traceable to conduct of the other party and is excusable.

6-20   Failure of Condition   The terms of a contract may make it apparent that a particular duty of performance is conditioned upon the occurrence, perhaps at a stated time and place, of a given event. Failure of such a condition would, of course, relieve the promisor of the necessity of performing that particular duty. To use a rather unlikely illustration, suppose Smith promises a car to Jones in return for the latter's promise to pay $500 if it rains on May 30. Smith has an unconditional duty to deliver the vehicle, but Jones's duty of performance is discharged if it should fail to rain on schedule.

THE NATURE OF BREACH

6-21   Definition and Forms of Breach   An actionable *breach of contract* occurs when a promisor, without sufficient excuse of justification, fails to perform in accordance with the dictates of his agreement.[26]

While failure to perform an obligation in whole or in part is the most common form of breach, there are several other ways in which a contract may be broken: A party (1) may renounce his duties thereunder or (2) may by his own act make it impossible that he should fulfill them or (3) may prevent or seriously hinder the other party's performance.

6-22   Total Breach; Partial Breach   Although every breach of a contractual obligation confers upon the injured party a right of action, it does not follow that every breach will discharge the latter from accomplishing whatever he has committed himself to do under the contract. The agreement may have been broken wholly or only in part. If the breach is merely partial, it may or may not be of sufficient importance to terminate the reciprocal duty of the injured party. Where one party is guilty of total breach (or of partial breach which, in the particular circumstances, is material in its scope), the other person may cease performance on his part and obtain a judgment for damages as well.

6-23   Anticipatory Breach   If a party announces, before his performance is due, his definite unwillingness or inability to fulfill the contract, he thereby admits he is guilty of breach and relieves the other party from further obligation, mean-

---

[26] When time is of the essence of a contract, failure to complete the undertaking within the period specified is a material breach.

while affording the latter an immediate right to sue for damages. The repudiation must be positive and unequivocal; a mere threat to abandon the contract will not suffice. A true anticipatory breach requires that the repudiation antedate the time fixed in the contract for performance by the renouncing party.

Any proceedings culminating in a bankruptcy adjudication are the equivalent of anticipatory breach, apparent on the theory that every contract carries an implied condition to the effect that a promisor will not allow himself (through insolvency, for example) to reach such a condition that he is unable to perform as scheduled.

6-24   Voluntary Disability as Breach   It is well settled that a person cannot avoid liability for nonperformance of an obligation simply by so maneuvering as to place performance beyond his power. Put another way, a party who deliberately incapacitates himself or renders impossible the proper performance of his contract duties, has broken his agreement and is liable in damages.

An old English case[27] well illustrates the point. Franklyn promised to assign to Lovelock all his interest in a certain lease within seven years from the date of the promise. But before the end of the seven years Franklyn assigned his whole interest to a third person. In the course of holding that Lovelock need not wait until the end of seven years to bring suit, the court had these comments:

> The plaintiff has a right to say to the defendant, You have placed yourself in a situation in which you cannot perform what you have promised; you promised to be ready during the period of seven years, and during that period I may at any time tender you the money and call for an assignment, and expect that you should keep yourself ready; but if I now were to tender you the money, you would not be ready; that is a breach of the contract.

6-25   Prevention as Breach   Interference by one party rendering performance by the other party impossible will excuse the resulting nonperformance of the innocent party. Such tactics also constitute a breach of contract on the part of the obstructionist and provide the innocent party with a right of action. Thus, where owner without justification prevents performance by contractor, the latter's duty is discharged, and the owner is liable for breach of contract.

WAIVER OF BREACH

6-26   Definition; Alternatives for Innocent Party   Waiver is voluntary abandonment of a known right, and, when the term is employed in connection with breach of contract, it means the giving up of a right to treat the agreement as terminated by reason of the other party's breach of a duty thereunder. Such right is deemed to have been waived when the injured party continues to treat the

---

[27] *Lovelock v. Franklyn,* 8 Q.B. 371 (1846). Another case illustrating the same principle is *Johnson v. Meyer,* 209 Cal App.2d 736. 26 Cal. Rptr. 157 (1962).

contract as a existing obligation, as by expressing willingness to accept further performance by the wrongdoer (or a fortiori, by actually insisting upon that performance), notwithstanding the breach.

When a particular obligation has been broken in the course of performance of the contract, the injured party is presented with several alternatives. He may elect to treat the entire contract as discharged outright because of the default, or he may waive the breach and continue to carry out the agreement according to its provisions.[28] Acts evidencing an intent to adopt the waiver alternative will operate as a conclusive election, depriving the innocent party of any excuse for subsequent failure of performance on his own part. If the defaulting party has been induced to carry on with performance or otherwise to alter his position in reliance upon the continued recognition of the contract by the innocent party, the latter cannot subsequently switch signals and assert that the contract was indeed discharged in the first place.

6-27   Elements of Waiver   Courts are careful whenever a claim of waiver arises, since abandonment of a right to insist on performance strictly in line with the contract amounts in a sense to modification of the agreement. To render an alleged waiver binding there must be present something in the nature of consideration or an element of estoppel. Moreover, the party electing to waive must have knowledge of all essential facts surrounding the breach.[29] A Massachusetts case[30] is illustrative of this last point. It was there held that acceptance of highway-repair work did not constitute waiver of subcontractor's failure to erect guard-rail posts reinforced by the number of steel rods prescribed in the specifications, since neither the general contractor nor the state authorities knew that the posts as supplied contained fewer than the required four rods apiece. The court said:

> Neither can we agree with the plaintiffs' contention that the acceptance of the work by the Commonwealth and its payment therefor to Hosmer were a waiver of the failure of Russo to furnish and erect posts having the required number of rods. There is nothing in the auditor's report that will support a waiver. There is an express finding that neither Hosmer nor the department of public works knew that these posts contained less than four rods. Hosmer on this record cannot be said to have waived a defect that was unknown to him. A waiver is an intentional relinquishment of a known right.

Waiver may be accomplished by express words or by conduct which is inconsistent with an intention to abrogate for breach, but mere silence will not amount to waiver where one is not bound to speak. Whether there has indeed been a bona fide waiver in a given set of circumstances often presents a close question of fact. For example, in one case[31] a subcontract for installing panes of glass in a building

---

[28] Such a waiver does not necessarily preclude the injured party from recovering for any damages occasioned by the partial breach.

[29] *Intaglio Service Corp. v. J. C. Williams and Co., Inc.,* 95 Ill. App.3d 708, 420 N.E.2d 634 (1981).

[30] *Russo v. Charles I. Hosmer, Inc.,* 312 Mass. 231, 44 N.E.2d 641 (1942).

[31] *Pittsburgh Plate Glass Co. v. American Surety Co. of New York,* 66 Ga. App. 805, 19 S.E.2d 357 (1942).

obligated the subcontractor to clean the glass before final inspection. Upon his refusal to do so, it became necessary for the contractor himself to perform the work, since the cleaning was a prerequisite to final approval of the entire construction project by the architect and engineer. Under the circumstances, it was held that the general contractor's conduct did not present a waiver of the duty on the subcontractor.[32]

6-28 Mutual Desire to Ignore Repudiation As previously noted, when one party renounces his contract and refuses to perform same, the other party acquires an immediate right to abandon the agreement and to secure legal redress for the harm done him.[33] The innocent party is not entitled to ignore the fact of renunciation and enhance his damages by continuing to perform. However, where there is mutual consent to do do, the respective parties may forget all about the repudiation and may resume fulfillment of the contract.

REMEDIES FOR BREACH

6-29 General Considerations It is axiomatic that persons who violate their agreements to the detriment of others with whom they contract must answer in some appropriate way for the injury done. The contract may, and quite commonly does, provide for the precise consequences which shall follow upon breach. Where the parties have stipulated in advance for a particular remedy, that one will be treated by the courts as the exclusive means of redress.

Apart from express contract provisions which might readily alter the situation, breach normally presents the injured party with a choice of several possible alternatives. His principal option lies between compensatory damages and restitution. Under certain circumstances, a decree of specific performance or an injunction will be an available recourse.

The general rule is that material breach of contract frees the injured party from any further duty to render the performance required of him by the contract. At the same time, he is privileged to seek redress for the wrong done him, and usually this recovery will take the form of an action for money damages, measured, roughly, by the value of the performance he had expected to receive. If, on the other hand, he has himself partly performed at the time breach occurs, he may decide to sue for the reasonable value of what he has supplied or accomplished. If he chooses this remedy of restitution and a recovery in quantum meruit, courts are wont to say he has "rescinded" the contract.

While the injured party usually has such an election of remedies at his disposal, it must be remembered that, once he has manifested his choice and the other party

---

[32] Also illustrating situations where a continuation of work was held not be a waiver are *Bd. of Regents, Univ. of Tex. v. S. & G. Constr. Co.,* 529 S.W.2d 90 (Tex. App. 1975); and *John Kubinski & Sons, Inc. v. Dockside Development Corp.,* 33 Ill. App.3d 1015, 339 N.E.2d 529 (1975).

[33] The repudiation may be nullified by timely withdrawal of the "renouncing" declaration before the other party has become aware of the fact of repudiation and perhaps has changed his position accordingly.

has taken any action in reliance upon that manifestation, the choice is binding and will bar recourse to any alternative.

6-30   Damages

Nominal Damages   A breach of contract invariably creates a right of action in the aggrieved party. Even in those instances where no substantial harm has been done or where no loss whatever is traceable to the breach, and even in those very rare circumstances where the innocent party is actually benefited by the breach, the defaulting party is subject of a judgment for *nominal damages.* This will mean a very small sum fixed by the court more in recognition of the default in perform-ance than in compensation for any bona fide loss.

Compensatory damages   The most common remedy for breach of a valid con-tract is the recovery of a sum of money awarded as *compensatory damages.* In ascertaining the proper measure of such damages, the fundamental consideration is to place the innocent party in the position he would have occupied had the contract been performed according to its terms.[34] To accomplish this, there should be recompense for loss suffered and gains prevented in consequence of the default. As it has been expressed in a number of decisions,[35]

> One who fails to perform his contract is justly bound to make good all damages that accrue naturally from the breach; and the other party is entitled to be put in as good a position pecuniarily as he would have been by performance of the contract.

On the other hand, there can be no recovery for purely speculative damages or for elements of damage which were clearly not within the contemplation of the respective parties at the time their agreement was signed. The injured party is not entitled to *better* his position from what it would have been had no default oc-curred. Similarly, damages for breach are intended as compensation, and not as punishment for wrongdoing. The injured party generally recovers nothing in the nature of punitive or exemplary damages, even if the contract itself prescribes such a consequence for breach.

Expenses incurred by a party in preparation for performance of a contract before its abandonment by the other party are proper elements of damage occa-sioned by the breach. Generally, where a building contractor's performance is prevented by the other party's breach, damages are (1) what has been expended toward the performance or (2) profits that would have been realized by full per-formance.[36] Certainly the builder has a right to reimbursement for his out-of-pocket expenditures[37] (as well as to his reasonably anticipated profits), since, had the owner kept his part of the bargain and paid the full contract price, this amount

---

[34] This concept differs from that of restitution, which seeks—to the extent possible—to restore the innocent party to the position he was in before contracting.

[35] See, for example, *Maraldo Asphalt Paving, Inc. v. Harry D. Osgood Co., Inc.,* 53 Mich. App. 324, 220 N.W.2d 50 (1974); and *Martin v. Phillips,* 122 N.H. 34, 440 A.2d 1124 (1982).

[36] *Dialist Co. v. Pulford,* 42 Md. App. 173, 399 A.2d 1374 (1979).

[37] If there remain on hand at the time of breach materials which the builder has purchased for the particular job but can be utilized by him elsewhere, the value of such supplies should be deducted from the builder's total expense outlay when damages are being measured.

would presumably have sufficed to cover the builder's expense and his expected profit as well.

If the contractor is the defaulting party the owner may generally recover the cost of having the job completed. Should the contractor handle the task in its entirety but it develops that the work is defective, the owner is in a position to seek reimbursement of whatever it costs him to make the work conform to the contract.

Duty to Mitigate Damages A party in default on a contract has the right to expect that the opposite party will do everything reasonably possible to minimize damages, and the latter cannot recover with respect to that portion of the loss which he could readily have avoided once the fact of breach became known to him. For example, the measure of damages for default on a contract to deliver goods is ordinarily the difference between the contract price of said goods and the market price at the exact time when delivery should have been effected. If, however, the injured party might have reduced his loss (as, perhaps, by an immediate purchase at a low price of goods to replace those not delivered), his failure to do so should be taken in account in assessing his recoverable damages. What constitutes a reasonable effort at mitigation is a question of fact in each case.

Liquidated Damages Contracts in general, and building and construction contracts in particular, often stipulate that a certain sum per day shall be paid by way of *liquidated damages* in the event completion of the work is not accomplished on schedule. Such clauses will be upheld if the sum provided for is commensurate with the extent of injury which could reasonably be anticipated and if the effect of the provision is compensation for breach rather than exaction of a penalty from the contract breaker. In other words a stipulation for liquidated damages is enforceable where it is not merely a cloak for a penalty.[38] The test is whether the amount fixed bears some reasonable relation to the damages likely to result from breach of the contract.

A typical case[39] illustrating the various aspects of liquidated damages occurred where a contractor agreed to furnish all the material and labor and to erect a store building for the owner. The structure was eventually completed and the owner made payments on the contract, withholding a portion of the total contract price as liquidated damages for contractor's failure to finish the work on time. The agreement of the parties had called for $10 for each day that the time consumed in performance exceeded the allowed time of seventy-five days. The building was to house two retail establishments, for which tenants had been secured in advance and were ready to pay rent starting with the scheduled completion date. One of the stores had actually been leased before such completion date.

---

[38] It is said that courts "abhor a penalty or a forfeiture"; consequently, they are inclined to examine with a critical eye provisions which purport to set liquidated damages. See *Brower Co. v. Garrison*, 2 Wash. App. 424, 468 P.2d 469 (1970).

[39] *Hall v. Gargaro*, 310 Mich. 693, 17 N.W.2d 795 (1945).

See also *Robbins v. Finley*, 645 P.2d 623 (Utah 1982); and *Stonebreaker v. Zinn*, 286 S.E.2d 911 (W.Va. 1982).

Under these circumstances, the court was convinced that the clause setting the figure of $10 a day for delayed completion was a fair attempt to estimate likely damages and was therefore binding. In the course of its opinion, the court abstracted as follows another Michigan decision:[40]

> Courts will disregard the express stipulation of parties as to the damages for breach of a contract only in those cases where it is obvious from the contract before them, and the whole subject-matter, that the principle of compensation has been disregarded.
>
> In cases where it is difficult to accurately determine the damages which one party may suffer by the failure of the other to perform his contract, the parties themselves may agree upon such sum as in their judgment will be ample compensation for the breach.
>
> A provision in a building contract to forfeit $20 per day for failure to complete repairs on a dwelling house within the contract periods is not *per se* excessive so as to amount to a penalty.

6-31 Restitution Where the aggrieved party in a breach situation does not desire compensatory damages but wants merely to recover whatever he may have parted with in the process of satisfying his own contractual obligations, he is said to seek the remedy of *restitution.* In a sense, he unilaterally "rescinds" the agreement at point of breach and demands the return of money he has paid thereunder or the value of goods he has delivered or of such work as he has done. The usual restitution case does not involve the return by the party in breach of a specific item but rather payment of the reasonable value of whatever consideration he has received under the contract up to the time of the default.

In theory, restitution works both ways, and the aggrieved party must himself return or account for any benefits he received before the breach occurred. This requirement is not enforced with unvarying strictness, however, and is presumably never so applied as to make it a shield for wrongdoing. Such restoration by the injured party as is "reasonably possible" and such as "the merits of the case demand" will suffice. For instance, it has been held that the injured party seeking the remedy of restitution need not offer restoration for his own part where what he has received consists solely of money, the amount of which he will be entitled to in any event.

6-32 Specific Performance Under certain rather well-defined circumstances, courts will enforce by a *specific-performance decree* a contract promise to do a given thing and by an *injunction* a promise to forbear.[41] This particular form of redress is by no means available in all breach situations; some of the more significant limitations are hereinafter mentioned.

Where damages provide an adequate remedy, specific performance will not be granted. In personal-property cases, compensating damages will normally suffice,

---

[40] *Ross v. Loescher,* 152 Mich. 386, 116 N.W. 193 (1908).

[41] In either instance, specific performance or injunction, the promisor is being required to live up to the bargain he has made.

creed on the ground that unique chattels of uncertain value were involved.[42] Agreements for the sale of interests in realty, on the other hand, are very commonly enforced by specific performance decrees. A specified piece of land is impossible of precise duplication, and money damages in any amount for breach of a contract to sell that particular plot are (in theory, at least) inadequate.

By its very nature, the remedy of specific performance is largely discretionary with the court before which it is sought. Unless the contract at issue is fair and just and unless its terms are definite and certain, the chances that specific performance will be ordered are slim. The court must be in a position to determine with reasonable assurance what are the contractual obligations of which enforcement is sought and what are the conditions under which performance is supposed to be rendered. It is not necessary that both parties have identical remedies available in the event of breach; consequently, specific performance may be decreed at the instance of one party even though similar redress would not have been open to the other party had he been the injured one.[43] Moreover, it has been held that specific enforcement of a contractual undertaking is not precluded by the existence of a liquidated-damages provision.

Since courts have real discretion in dealing with petitions for specific performance, they are inclined to exercise their prerogative and refuse to grant the requested relief if it appears that the performance entailed is such that extensive and burdensome supervision would be needed for effective enforcement. If, for example, a court endeavored to enforce an agreement for the supply of goods by installment over a long period of time, it is obvious that this would involve the court in a task of superintendence such as it could not conveniently undertake. Partly for this same reason of undue supervisory difficulty and partly because their specific enforcement would amount to involuntary servitude, promises to render personal services will not be handled by affirmative orders of specific performance. Within limits, however, *negative promises*[44] respecting personal services are enforceable by injunction.

With regard to construction contracts it can be said in general that such will not be enforced by specific performance decrees. There are various reasons for this, but the principal one seems to be that damages afford an adequate remedy, regardless of whether the owner or the builder is the party in breach. Also, of

---

[42] For an illustration of this, see *Corbin v. Tracy,* 34 Conn. 325 (1867). By that decision a promise to sell a patent was specifically enforced, the theory being that damages could not be readily measured and were thus an unsatisfactory form of relief to the injured party. See also *R. 20 Kinsey Cotton Co., Inc. v. Ferguson,* 233 Ga. 962, 214 S.E.2d 360 (1975); *Tomb v. Lavalle,* 298 Pa. Super. 75, 444 A.2d 666 (1981). Liquor license was held to be such a unique asset that specific performance was an appropriate remedy.

[43] Suppose that Green agrees to serve as Brown's gardener for the summer in return for Brown's promise to transfer a specified piece of real estate. Green fully performs her part of the contract, but Brown defaults completely. Under the foregoing circumstances, Green would be entitled to specific performance despite the fact that in no event could Brown have secured a similar decree (since obligations to render personal service will not be specifically enforced).

[44] That is, promises not to do the work in question for someone other than the promisee. Thus, a great musician might promise his services to a particular orchestra and at the same time covenant that he would not perform with any other musical organization.

course, superintending performance of a construction contract would impose upon the court an extremely difficult, if not impossible, burden.

QUESTIONS

1 What is a tender of performance?

2 Illustrate the difference between specific performance and substantial performance.

3 A plumbing contract called for certain "Crane" fixtures, but the contractor installed "standard" fixtures, which she claimed were equivalent to the stipulated type. The owner claimed a deduction because of the substitution. Was he justified in so doing?

4 Lewis contracted with the government to dredge 2 million cubic yards of silt from the ship channel in New Haven Harbor. After excavating 1.5 million cubic yards, he quit the job. Should he be paid anything for partial performance?

5 A particular contract stated that "time is of the essence." What legal significance is there in the quoted wording?

6 Illustrate a contract in which personal taste would determine what constitutes satisfactory performance.

7 What interpretation and effect should be given to a clause stating that all work is to secure the approval of the engineer?

8 *B* ordered a 5-ton truck "on approval" from *C*. When the truck was delivered, *B* objected to the color of the paint and refused to accept the vehicle. Did *B* have his right?

9 Who has the right to waive imperfections in the quality of performance? What does "waiver" mean?

10 Impossibility may be an effective excuse for nonperformance; inconvenience is not. Explain.

11 Distinguish between subjective and objective impossibility of performance.

12 What is meant by "act of God?" What effect may such an occurrence have upon a contract? Under what conditions may it fail to relieve the contractor of the obligation to perform?

13 If an act of God makes completion of a contract impossible after part of the work has been done, what arrangements should perhaps be made regarding compensation to the contractor?

14 *X* contracted to do personally certain engineering work for *Y* for the sum of $2,000. When the job was 25 percent completed, *X* fell seriously ill. Can *X* avoid her contract? Can she have someone else finish the work?

15 Farmer Clark contracted with Pillsbury to furnish 10,000 bushels of wheat to the latter on September 1. Lack of rain caused a poor crop and Clark was able to raise only 7000 bushels. Can Pillsbury collect damages from Clark for the shortage of 3000 bushels? Explain.

16 Douglas contracted to drive a well for Tucker for the sum of $500 and to furnish potable water at a flow of 5 gallons per minute. After spending $800 on the work, Douglas was unsuccessful and quit the job. What payment, if any, is Tucker obligated to make?

17 Explain what is meant by "breach of contract."

18 Name several ways in which a contract may be breached.

19 Can one part of a severable contract be breached without breaching the whole contract? Explain.

20 Define each of the following types of damages: (*a*) nominal, (*b*) compensatory, (*c*) punitive, (*d*) liquidated.
21 What is meant by "restitution?" Under what contract circumstances does this concept enter the picture?
22 How does interference on the part of the owner affect the question of liquidated damages where the contractor does not finish the job on time because of such interference?
23 What is meant by a "latent defect" in speaking of a contracting party's performance?
24 *B* was constructing a shopping center for *C*. In the midst of the job, *C* was threatened with bankruptcy. Is *B* obliged to continue with his undertaking?
25 Fulton contracted to furnish Mortimer 50 bushels of apples per week during October and November. She did so the first week of October, then failed the following week. Mortimer then contracted with Evans for the needed apples to December 1, notifying Fulton that the latter had broken her contract. Fulton stated that she was sick in bed during the second week of October, that this circumstance excused her failure, and that her contract with Mortimer remained in full force and effect. What are the rights of the parties?

CHAPTER 7

TERMINATION OF AGREEMENTS

IN GENERAL

7-1   Means of Discharge   Contracts of any sort may be brought to an end in a variety of ways. Full and satisfactory performance by both sides is the usual mode. The second most common way of termination is breach. Upon default, the contractual ties are loosened, but a new factor enters the picture in the form of a right of action accruing to the innocent party.[1]

A third means of contract termination is mutual agreement, the same process which created the relationship in the first place. Fourth, a contract may be closed out because it has become impossible of fulfillment, under circumstances which completely excuse the respective parties from their obligations and duties. Finally, discharge of the contract may be effected through operation of law as, for example, in a bankruptcy situation. The bankrupt party's trustee is in the enviable position of being empowered to pick and choose among the bankrupt's contracts, enforcing those appearing to be valid and repudiating such as appear unprofitable. Contracts "voidable" by reason of fraud, infancy, and the like are discharged through "operation of law" when the party entitled to do so exercises the privilege of terminating the contractual relationship.

---

[1] This right of action (quite apart from the contract which gave rise to it) may itself be terminated; thus, the judgment of a court of proper jurisdiction rendered in a suit brought upon the right is said to merge such right and, in effect, to discharge same. Somewhat similarly, the several judicial remedies available to the party injured by breach of contract will be barred should the applicable statutory period of limitations run its course before appropriate action is taken to invoke the aid of the courts.

TERMINATION THROUGH PERFORMANCE OR EFFECTIVE TENDER THEREOF

7-2 Type of Performance Required In order to bring about discharge of the contract in its entirety, the performance must be complete on both sides and must accord with terms and conditions of the agreement. Obviously, if the contract involved is a *unilateral* one, wherein one party makes a promise in exchange for an executed consideration, fulfillment by the promisor will bring the contract to an end because that was the only obligation outstanding. On the other hand, parties to *bilateral* contracts have exchanged executory consideration—a promise for a promise—and each must do that which is required before the contract can be said to be terminated. Performance by one party only will, of course, discharge that particular obligation, but the contractual duty of the other party is still alive, and the contract remains in force.

Effective tender of performance will do just as well as actual performance where the latter is prevented by the party entitled. An unconditional offer to discharge one's contractual duty in accordance with its precise definition in the agreement plus the actual present ability to carry out such offer will, if rejected, nonetheless relieve the party tendering of any further obligation.

TERMINATION THROUGH BREACH

7-3 Refinements of Terminology Material breach can be a means of contract termination. Agreements which have been broken and, in consequence, brought to an abrupt close are variously spoken of as *abrogated, avoided, terminated, discharged,* etc. These words are used interchangeably and normally are intended to convey the same thought.

7-4 Scope of the Breach an Important Factor Any material default in performance discharges the other party and affords that party the right to treat the contract as at an end. See, for example, *Wagstaff v. Remco, Incorporated,*[2] where a subcontractor sued a general contractor for amounts allegedly due on the construction of apartments. The general contractor denied any indebtedness and counterclaimed that the subcontractor had caused the damage by failure to perform. Ruling in the plaintiff's favor, the court held that where the general contractor's failure to pay an installment as provided for in the contract was such a substantial breach as to materially impair the subcontractor's ability to perform, the subcontractor had the right to consider the contract at an end, to stop work, and to recover the value of the work already performed. But note that the breach, if it is to justify the innocent party in terminating the relationship, must be so material as to defeat or render unattainable the very object of the contract. A

[2] 540 P.2d 931 (Utah 1975). See also *Zulla Steel Inc. v. A&M Gregos, Inc.,* 174 N.J. Super 124, 415 A.2d 1183 (1980).

default which could be characterized as *casual, technical,* or *insignificant* will not, therefore, suffice.[3]

An early Alabama decision[4] contains an excellent discourse on the various aspects of breach as a reason or excuse for contract terminations.[5] The following excerpt is from that decision:

> The principal question by these pleadings is whether the fact that a small quantity of the ore delivered by plaintiff was not free from foreign substance, and satisfactory to the the furnace company receiving it, operated as a discharge of the whole contract, and authorized defendants to terminate it. The effect of a breach of a contract upon the rights and liabilities of the parties depends upon the nature of the agreement. If the contract be entire in the sense that each and all its parts are interdependent, so that one part cannot be violated without violating the whole, a breach by one party of a material part will discharge the whole at the option of the other party; but, if the contract be severable—susceptible of division and apportionment—the amount to be paid by the one party depends upon the extent of performance by the other, the mere failure to perform a part of the contract in strict compliance with its terms will not of itself necessarily authorize the party injured to refuse further performance. Whether a particular contract is entire or severable depends on the intention of the parties, to be determined from the language employed and the subject-matter. In the contract sued on, plaintiff obligated himself to mine and load on the cars all the ore within a given territory, the ore to be satisfactory to the furnace company to which it might be shipped; but the time and amount of the deliveries and the time of the completion of the contract were left unfixed, and necessarily the aggregate price to be paid for full performance was not named. No particular amount of ore was to be furnished each month, and a failure to furnish any ore in any one month would not, of itself, amount to a breach of the contract. The defendant's obligation was simply to permit plaintiff to mine all the ore within the territory named, and to pay on the 20th of each month a specified sum for each ton delivered during the previous month. There is nothing in the contract to indicate an intention of the parties that the right of plaintiff to make successive shipments of ore until the contract was completed should be dependent on the mere fact that each and every ton previously mined and shipped was free from foreign substance, and satisfactory to the furnace company receiving it, and had been accepted by defendants. The contract was, in its nature, severable, and not entire, and the rights and liabilities of the parties are to be determined according to the principles applicable to such contracts. Not every breach of such a contract by the one party will authorize the other to abandon the contract, and refuse further performance on his part. The circumstances attending the breach, the intention with which it was committed, and its effect

[3] For an illustration of this point, see *Illges v. Congdon,* 248 Wis. 85, 20 N.W.2d 722 (1945), *rehearing denied,* 248 Wis. 85, 21 N.W.2d 647 (1946), *modified and affirmed,* 251 Wis. 50, 27 N.W.2d 716 (1947), *certiorari denied,* 333 U.S. 856 (1948). Involved in this case was an alleged breach of triparty contract for the logging, sawing, and sale of timber.

[4] *Worthington v. Given,* 119 Ala. 44, 24 So. 739 (1898). Plaintiff had agreed to mine and deliver ore to defendants at a specified figure per ton. A negligible quantity of the delivered ore was found to contain "foreign substances" in contravention of the contract specifications; defendants thereupon sought to abrogate the entire agreement.

[5] See also *Jacquin-Florida Distilling Co. v. Reynolds, Smith & Hill,* 319 So.2d 604 (Fla. App. 1975) dealing with severability of an engineering contract providing for periodic payments.

on the other party and on the general object sought to be accomplished by the contract, must be considered in determining whether or not the breach will operate as a discharge. If the circumstances are such as manifest an intention on the part of the party in default to abandon the contract, or not to comply with its terms in the future, or if, by reason of the breach, the object sought to be effected is rendered impossible of accomplishment according to the original design of the parties, the breach will operate as a discharge of the whole contract unless waived; but no such result follows from a mere breach of a severable contract unattended with such circumstances or such effect. The right to claim a discharge of the whole contract depends, not on whether the act constituting the breach was inconsistent with the terms of the contract, but whether it was inconsistent with an intention to be further bound by its terms, or whether the breach was such as to defeat the purpose of the contract. The mere fact that a small quantity of the ore delivered to defendants was not free from foreign substance, and satisfactory to the furnace company receiving it, did not give defendants any right to forbid plaintiff to continue mining ore under the contract. . . . Against the natural and ordinary injury flowing from such breach, defendants had an adequate protection and remedy—the refusal to receive and pay for the unsatisfactory ore, in their action for the breach in this respect. If any extraordinary injury arose therefore, the effect of which was to defeat the purpose of the contract, the pleadings fail to show it, and it cannot be inferred from the nature and subject-matter of the contract.

7-5   Cancellation Clause in Contract   Subcontracts in connection with construction jobs typically contain language permitting the general contractor to terminate the agreements upon giving some stipulated notice if, for instance, the subcontractor fails to show reasonable progress in the work. In a Wisconsin case,[6] painting subcontract read in part as follows:

> If any time during the prosecution of the work of this contract, the Subcontractor, not being hindered by causes beyond his control, fails to maintain a sufficient working force, or if it shall become evident to the Contractor that the work is not being prosecuted with proper diligence to complete said work so as not to delay the progress of the building, or if the Subcontractor shows gross carelessness or incompetency, or if the Subcontractor fails, refuses, or neglects to comply with the contract. . .[the Contractor may elect to terminate the arrangement].

During the course of the work, it became apparent to the general contractor that the painting subcontractor was not in a position to supply workers, materials, or equipment sufficient to permit satisfactory completion of the job in a reasonable time. For instance, the subcontracting firm furnished only five rolling scaffolds for an undertaking which required at least twelve, allowed its accounts for paint to become delinquent, and failed to meet two of its own payrolls. Consequently, the prime contractor wrote a letter of cancellation, indicating that, in order "not to delay progress of construction, we are forced to invoke clauses in

---

[6] *Valentine et al. v. Patrick Warren Constr. Co.*, 263 Wis. 143, 56 N.W.2d 860 (1953). This is an interesting and worthwhile case to read. See also *Burgess Constr. Co. v. M. Morrin & Son, Co., Inc.*, 526 F.2d 108 (10th Cir. 1975).

your contract to complete your work with our own forces." The court determined that the subcontractor's breach of contract was a serious one and that the prime contractor's use of the cancellation provision was justified.

TERMINATION THROUGH AGREEMENT

7-6 Mutual Renunciation (Rescission) of the Contract Parties to a contract may abrogate it, either wholly or in part, by their mutual consent,[7] if the rights of third persons have not intervened. So long as it is the joint will of the respective parties to terminate their arrangement, they may accomplish this aim even though the contract itself might contain a provision purporting to preclude its cancellation within a certain period of time or to restrict in some other fashion the right of the parties to cancel by agreement.[8] The mutual consent to terminate need not be express but may be inferred from a course of conduct of the parties which is obviously inconsistent with any intention of proceeding with the contractual relationship.[9] The question of proof is much less troublesome when the understanding of the parties is reduced to writing or expressed orally before disinterested witnesses. Normally, an agreement to end the contract will include provision for restitution, where there has been partial performance on one side or both.

Where it is an entirely executory contract[10] which is being terminated by agreement the consideration received by each party is the abandonment by the other of his rights under the contract. That is, the release of one party from the obligation to perform is adequate consideration for the corresponding release of the other. Where a contract which the parties jointly consent to cancel stands fully executed on one side, some independent consideration must be afforded the party who has performed if his waiver of the other party's obligated performance is to operate as a discharge.

7-7 Substitution of New Contract Closely related to termination by mutual consent is termination of an existing contract simply as the result of the parties

---

[7] There are state laws to this effect. See, by way of illustration, Revised Codes of Montana (1981 ed.) §13-903(5). The same statute lists four situations (additional to "by consent of all the other parties") in which a party to a contract may rescind the same: "(1) If the consent of the party rescinding, or of any party jointly contracting with him was given by mistake, or obtained through duress, menace, fraud, or undue influence, exercised by or with the the connivance of the party as to whom he rescinds, or of any other party to the contract jointly interested with such party; (2) If, through the fault of the party as to whom he rescinds, the consideration for his obligation fails, in whole or in part; (3) If such consideration becomes entirely void from any cause; (4) If such consideration, before it is rendered to him, fails in a material respect, from any cause; or (5) If all other parties consent."

[8] *Resource Engineering, Inc. v. Siler,* 500 P.2d 836 (Id. 1972).

[9] In *San-ann Service, Inc. v. Bedingfield,* 305 So.2d 374 (Ala. 1975), the Supreme Court of Alabama noted that this rule can be given operative effect only when the conduct of the party is found to be positive and unequivocal. See also *Humphreys v. Laile,* 398 So.2d 703, *writ den.,* 398 So.2d 706 (Al. App. 1981).

[10] In other words, where there are mutual existing duties as yet unperformed.

making a new one inconsistent therewith. The new agreement may or may not expressly repudiate its forerunner. The original contract will be discharged if the parties agree to such a substantial alteration in its terms as to amount to the replacement of the original contract with a new one. In a case from Missouri involving an action to recover fees allegedly due under a contract to perform architectural services, it was said that: "Whatever may have been the parties' original agreement, it was competent for them, before any breach of its provisions, to waive, dissolve or abandon that contract and substitute a new oral contract by conduct and imitation, as well as by express words."[11]

Suppose Smith is contractually obligated to pay a stated sum of money in return for Jones's automobile. Wishing instead to perform certain manual tasks which she knows Jones needs to have done, Smith gets Jones to accept the substitute in lieu of the cash to which the original contract entitles the latter. In this situation the new agreement discharges the old and performance by Smith of the stated tasks will relieve her of all obligation.

Short of canceling the old contract and substituting a new one in its place, the parties may agree that the original understanding merits modification, as by incorporation of some additional provision.

Also to be distinguished from the concept of substituted contract is the matter of mere technical reformation of an existing document. *Reformation* is the correction of defects so as to conform the contract to the actual intent of the parties, which intent was, through inadvertence, improperly expressed in the instrument as originally drawn. Among the errors which have been held reformable are obvious mistakes of computation in the proposals submitted by building contractors.

7-8 Release; Convenant Not to Sue A release is a discharge, by the party entitled, of the right of action which has become his as the result of default in a contractual promise to him.[12] The release will fully bind its maker, but in accordance with the elementary rule of contracts, only if there is a proper consideration. *Conditional releases* are perfectly valid and will be effectuated according to their precise terms.

There is also such a thing as a *covenant not to sue.* Here one party promises the other that he will not, perhaps for a stipulated period of time or—less likely— forever, bring suit to enforce a given contract right.[13]

7-9 Accord and Satisfaction There is considerable parallel between the idea of *accord and satisfaction* and that of substituted contract. The principal distinc-

---

[11] *Pallardy v. Link's Landing, Inc.,* 536 S.W.2d 512, 515 (Mo. App. 1976).

[12] *Release* is differentiated from *waiver* in that the latter term normally relates to the abandonment of a stated right before there has been any breach of contract bearing thereon.

[13] For a case attempting to distinguish a release from a convenant not to sue see *Mercantile National Bank v. Founders Life Assurance Co.,* 236 Ga. 71, 222 S.E.2d 368 (1976). See also *Brantley Co. v. Briscoe,* 246 Ga. 310, 271 S.E.2d 356 (1980).

tions appear to be (1) an accord and satisfaction generally starts with an actual default situation, that is, where one party is in breach and a right of action consequently accrues to the other party; (2) the *accord* itself does not ordinarily discharge the original contract right and correlative duty but simply holds these in abeyance pending *satisfaction* of such accord.

Under an accord-and-satisfaction arrangement, the party in default promises to render, and the party who holds the right of action agrees to accept, some performance differing from that which was originally contracted for and which might legally have been enforced. An accord and satisfaction is really a compromise and settlement by the substitution of a new contract, and the several requisites of contract formation must be present. Accordingly, the "party obligated" under the original pact must afford a proper consideration[14] to the "party entitled" in order to support the latter's agreement to accept a different performance in lieu of his previous contract right. As an example, the quid pro quo might be the transfer of a negotiable instrument to take the place of a money payment initially promised.

In an action by a contractor for a breach of contract and for foreclosure of a mechanic's lien the court said:

"An accord and satisfaction has been defined as '* * * a method of discharging a contract, or settling a cause of action arising either from a contract or a tort, by substituting for such contract or cause of action an agreement for the satisfaction thereof and the execution of such substituted agreement. . . . An accord and satisfaction must be * * * accompanied by such acts or declarations as amount to a condition that if the money be accepted it is to be in full satisfaction and to be of such character that the creditor is bound so to understand such offer. . . . When considering the existence of an accord and satisfaction, we should examine the following elements:

1 Did the debtor make an offer in full satisfaction of the debt;
2 Was there an unliquidated or disputed claim which formed the basis of this offer;
3 Was this offer accompanied by acts and declarations which amounted to a condition;
4 Were those acts and declarations such that the offeree was bound to understand them; and
5 Was the offer accepted in full satisfaction of the debt.[15]' "

An illustration of the workings of a rather complicated accord and satisfaction (where both parties were in technical breach of their original contract) is found in *Davis v. Zaban Storage Co.*[16] Plaintiff there was suing for damages sustained on account of the alleged negligence of the defendant warehouseman in not protecting against fire goods stored by him in defendant's warehouse. The evidence disclosed an agreement whereby warehouseman was released from any claim for fire

---

[14] Basic contract principles respecting consideration apply here. Thus, where a liquidated, matured, and undisputed sum of money is due and owning under the prior agreement, there is no consideration for the contract creditor's promise to accept a lesser sum in satisfaction thereof.

[15] *Smith Constr. Co. v. Knights of Columbus Council No. 1226,* 86 N.M. 50, 519 P.2d 286 (1974). See also *Louorn v. Iron Woods Products Corp.,* 362 So.2d 196 (Miss. 1978).

[16] 59 Ga. App. 474, 1 S.E.2d 473 (1939).

damage in consideration of his relinquishment of custody of the goods without payment of storage charges. It was held that this agreement amounted to an accord and satisfaction and that the respective parties were bound by the terms of same.

Apart from those relatively rare situations where the parties stipulate that the accord itself operates as satisfaction, the right to enforce the original contract duty is merely *suspended,* and only so long as there occurs no breach of the new accord. Upon performance of the substituted agreement the previously existing duty is fully discharged. Where, however, the accord is not followed in due course by performance the injured party can sue either upon that accord or upon the no longer suspended original contract duty itself.

7-10   Novation   In an early Washington decision,[17] the court had occasion to discuss at length the subject of *novation,* both as regards the several forms it may take and as regards its essential elements. It said:

> The doctrine of novation is so well understood that it hardly seems necessary to cite authorities to define it. Novation means substitution. It may be either the substitution of a new obligation for an old one between the same parties with intent to displace the old obligation with the new, or the substitution of a new debtor for the old one with intent to discharge the old debtor, or the substitution of a new creditor with intent to transfer the rights of the old creitor to the new. The second class is the ordinary case of novation, and is the case involved in the cause on trial. A novation is a new contractual relation. It is based upon a new contract by all the parties interested. It must have the necessary parties to the contract, a valid prior obligation to be displaced, a proper consideration, and a mutual agreement. If A. owes B. a sum of money, and C. agrees to pay the debt of A. to B., and B. agrees to accept C. instead of A. as payor of the debt, and to discharge A. from his original obligation that is a novation.

It may be seen from the foregoing quotation that substituted contract without change of parties is technically a form of novation but that the ordinary novation arrangement involves a substitution of parties.

7-11   Exercise of Power Reserved   It is not all unusual to find that one of the contracting parties has expressly reserved the option of terminating the contract under certain circumstances[18] or after a stipulated period of time has elapsed. Typical of this class of contracts are agreements regarding domestic service. Not infrequently the servant is specifically empowered to bring the relationship to an end (with or without cause) upon giving several weeks notice. Actual exercise of such reserved power effects a complete discharge of the contract and, in the sense

---

[17] *Sutter v. Moore Investment Co.,* 30 Wash. 333, 70 P. 746 (1902). See also *United Security v. Anderson Aviation Sales Co., Inc.,* 53? P.2d 545 (Ariz. App. 1975) and Chap. 5, Note 32.

[18] Such as the nonfulfillment of a condition precedent or the occurrence of a stated condition subsequent. For example, see *Commercial Contractors Inc. v. U.S. Fidelity & Guaranty Co.,* 524 F.2d 944 (5th Cir. 1975).

that the power was created by joint consent, can be said to be a form of termination by agreement.

TERMINATION THROUGH FRUSTRATION OF OBJECT, DESTRUCTION OF SUBJECT MATTER, ETC.

7-12   Theory of Frustration   There exists in law the doctrine of *frustration* which holds that, where the existence of a specified thing, condition, or state of affairs is essential to the performance of the contract, the entire agreement is dissolved[19] when, by reason of circumstances beyond the control of either party, the subject matter is no longer available. Similarly, where the purpose of a contract is completely defeated and its performance rendered impossible by a supervening event or factor which was not within the contemplation of the parties, the contract is deemed discharged. By this frustration concept the law reads into contracts an implied stipulation designed to regulate a situation which the parties themselves would presumably have taken care of by agreement had the necessity occurred to them.

To illustrate the general principle of frustration consider the case of *Johnson v. Atkins.*[20] The venture involved a purchase of copra in the United States for shipment to named ports in Columbia and resale to a Columbian buyer. The contracting parties realized that goods could not be sent into Columbia without a permit issued under authority of its government and that the entire purpose of their agreement would be thwarted if the permit was not forthcoming. As it happened, the Columbian government did indeed refuse to issue the necessary credentials, and performance of the contract was consequently blocked. In the ensuing litigation, it was held that the whole object of the agreement had been frustrated (by conditions beyond the control of the several parties) and that the contract was therefore discharged.

QUESTIONS

1 In what ways may a contract be terminated?

2 *A* contracted to paint *B*'s house for $600. *A* tried twice to make arrangements to start the job, but *B* kept stalling. Then *A* secured another job, and when *B* finally wanted her to start painting, *A* refused, saying that she was already "tied up." Is *A*'s position defensible?

3 What are the differences among the following terms as used in referring to a contract: (*a*) abrogated, (*b*) avoided, (*c*) terminated, (*d*) discharged, (*e*) rescinded?

---

[19] When frustration occurs, it does not merely provide one party with a defense in an action brought by the other. It kills the contract by removing its very foundation and discharges both parties automatically.

[20] 53 Cal. App.2d 430, 127 P.2d 1027 (1942). See also *U.S. v. General Douglas MacArthur Senior Village, Inc.,* 508 F.2d 377, 381 (2d Cir. 1974), where it was said that frustration of purpose "focuses on events which materially affect a consideration received by one party for its performance. Both parties can perform but, as a result of unforeseeable events, performance by party *X* would no longer give party *Y* what induced him to make the bargain in the first place. Thus frustrated, *Y* may rescind the contract."

4 Under what circumstances may the right to cancel a contract unilaterally be properly reserved to the owner?

5 Cox had contracted to build an addition to Bollard's factory. A sudden flood virtually destroyed the factory before any work was started. What can Bollard do about the commitment?

6 C agreed to purchase from D 100 tons of steel sheet piling. Later on, C found some secondhand piling which he could buy cheaper. What can C do about the situation?

7 In the preceding case, would D have the right to let C substitute a contract to purchase 100 tons of reinforcing bars from D?

8 X made a contract with Y for the benefit of a third party, Z. Under what conditions can either of the contracting parties release the other?

9 As the term is used in contractual relations, what is the meaning of "accord and satisfaction"?

10 What are the essential elements of "novation"?

11 What is the meaning of "recission" of a contract?

12 Under what conditions might a contractor have the right to cancel a contract with a subcontractor?

13 Evans contracted to buy 10 acres of Olsen's land for a factory site. Thereafter, the zoning board rezoned the area which included the property site into a "commercial," or business, section. Is Evans nevertheless obliged to go through with the purchase or answer in damages?

14 Brown contracted to build a house for Gray for $32,000, starting May 10. On May 9, Brown brought his equipment for excavating the basement. Gray told him that she had decided to postpone his project for one year, and that Brown should build the house then at the same price. Is Brown under any legal obligation so to do?

15 Abbott contracted to deliver to Olcott 2000 cubic yards of fill for grading a low area in the latter's property. In return therefor, Olcott was to deliver to Abbott 25 tons of ¾-inch reinforcing bars. Abbot delivered 500 cubic yards of fill, and that was all. What is Olcott's obligation?

16 When and how may it be desirable to have in the master contract a clause empowering the prime contractor to sublet only a limited portion of the work involved in the contract?

17 Where a contract is rescinded by mutual agreement, the concept of "restitution" is often in the picture. What does it involve?

18 Assume that Long contracted to design the superstructure and building foundations for Short's new factory for a stated percentage of the contract price. After the contract was let, Short asked Long to augment their agreement so as to include thereunder (at 0.5 percent remuneration) all foundations for the machines. Can Long decline and enforce the contract as originally constituted?

19 B had a contract with A whereby the latter was to furnish B's restaurant with all the baked goods needed for one year at stipulated unit prices. During the year, B sold her restaurant to C. Does A have to fulfill the remainder of her contract with B, and does B have to complete her contract with A? What if C agrees to assume B's obligation but A does not want any part of the substitution?

20 Moore contracted to sell 100,000 board feet of ¾-inch oak flooring to Ralston at a stipulated price and to make delivery not later than April 15. Moore actually delivered the lumber on April 3, and Ralston paid for it on April 10. What is the status of the contract on April 15?

21 Norton, aged seventeen, contracted to buy a sports car from Jones but failed to get the approval of her father for this expenditure. What can Jones do about the matter?

22 *C* contracted with architect *D* to design the structural work of the latter's school project. However, the job was delayed by the city because of financing. The project was finally cleared, but before *C* started any of *D*'s work, *C* obtained two other large jobs. He telephoned *D* and explained that he now had all the work that he could handle, and he asked *D* to get someone else. Can *D* hold *C* to his bargain?

PART TWO

APPLICATION TO CONSTRUCTION CONTRACTS

The second portion of this text is designed to show how each of the many facets of the law of contracts has its bearing upon engineering and construction undertakings. This part also covers several other subjects with which construction contracts are customarily concerned.

Probably the first step is arriving at an agreement between the owner and the engineer or architect for the preparation of the contract documents and the rendering of other essential services connected with the project.

The documents prepared for the usual construction job consist of (1) drawings, which portray the structure to be built; (2) specifications, which describe the qualities of materials and workmanship to be provided by the contractor; and (3) clauses setting forth specific features which apply to the work and concern the obligations of the parties.

When the planning of a building project is getting under way, it is necessary to make preliminary layouts and studies in order to determine the general features of the structures, equipment, and other facilities that will be needed. Then follow estimates of cost, research to decide whether the undertaking is economically feasible, and, finally, authorization for the engineer to design the works and prepare the contract documents. It is at this last-mentioned stage of the proceedings that the material included in Part Two enters the picture.

After approval of the project and issuance of orders by the owner for the engineer to proceed come the development of the design and the making of the final drawings which will become a part of the contract. All essential features of the project have to be determined, designed, and developed in sufficient detail for the potential contracts to prepare their cost estimates. We do not try to specify precisely how these drawings should be prepared or how the information on them should be presented, because such considerations will vary widely with the nature of the project. However, some important points to remember in connection with any drawing are mentioned.

The next item on the engineer's agenda, after the contract drawings have been worked in sufficient detail, is the drafting of specifications describing the materials and workmanship which the contractor is to provide. Various principles and ideas to be borne in mind in the preparation of this portion of the contract documents are described. A number of sample clauses are included, so that the student can study the wording used as well as the basic propositions involved.

Then it is necessary to draft, for inclusion in the contract documents, a number of provisions spelling out specific requirements pertaining to the conduct of the work and the obligations of the contracting parties. This is done not for the protection of the owner alone. The provisions constitute rules under which all parties are bound to conduct themselves. The considerable importance of such ground rules is apparent when one recalls that the project may involve the expenditure of vast sums of money.

When the drawings, specifications, and miscellaneous contract clauses are well in shape, the owner and the engineer will determine what type of contract is most desirable from the standpoint of the method of payment; for example, lump-sum or unit-price. After this decision is made, the engineer prepares the proposal form which is to be used by contractors wishing to submit bids on the work.

Two other important details that must next be settled relate to the bid bond and the performance bond, and there must be incorporated in the contract documents any special requirements affecting the preparation and submission of bids, general information for the guidance of bidders, and a statement about the procedures which will be followed in making the award.

When the contract drawings and all pertinent documents are sufficiently complete to enable the engineer to determine just when such papers may be released to potential bidders, he or she sets a date for receipt of proposals and advertises for bids.

CHAPTER 8

CONTRACTS FOR CONSTRUCTION AND ENGINEERING SERVICES

8-1 Application to Construction Industry   The principles of contracts can be well illustrated by applying them to the construction field because participation in the preparation and execution of construction contracts is among the major endeavors of engineers and architects. A building or construction contract presents the characteristics common to contracts in general. For example, good consideration is a requisite. The purpose of a construction contract is to govern the rights, duties, and liabilities of the builder who performs the work, the person or organization for whom the construction is to be executed, and the architects and engineers who design and inspect the work.

Organizations which design a project for the owner generally have the duty of compiling the contract documents. The exact wording to be used in any individual case is something that will be dictated by the particular circumstances and subject matter.

If the contract is drawn properly, if the specifications are clear and complete, and if the drawings are both comprehensive and sufficiently detailed, trouble with the execution of the contract can be minimized because the contractor will know exactly what is expected. Should it become necessary, the engineer can resort to the contract's provisions to force any recalcitrant contractor to abide by its requirements. The contract should be so phrased as to ensure that each party has the right and the power to secure the desired performance but that neither can use the contract provisions as a club unless proper performance cannot be secured otherwise.

To summarize, troubles may develop between contractor and engineer or owner when a contract is loosely and poorly drawn, when the exact scope and character of the work are misrepresented or not shown properly, when the requirements are unreasonable or injudiciously interpreted, and when modifications are attempted after the contract is signed.

8-2   Definitions   For the sake of clarity and in order to avoid repetition in the contract documents, it is advisable to define at the outset of such documents the terms which will be used. Such definitions are to apply whenever the terms involved appear in any of the papers that form a part of the contract. The so-called "standard clauses" are useful in this connection.

Here are some illustrative definitions:

1 "Chief Engineer" shall mean the Chief Engineer of the John Doe Corporation, acting personally.

2 "Engineer" shall mean the Chief Engineer, acting either personally or through duly authorized representatives operating within the scope of the authority vested in them.

3 "Contract" shall mean the entire agreement between the parties, including this Form of Contract together with the Specification, the Advertisement, the information for Bidders, the Contractor's Proposal (copies of which are bound with or accompany the other data), and the Contract Drawings, all of which aforementioned items are to be a part hereof as though set forth in full herein or physically attached hereto.

4 "Owner" shall mean the John Doe Corporation.

5 "Contractor" shall mean the individual, corporation, company, firm, or other organization who or which has contracted to furnish the materials and to perform the work under the Contract.

6 "Work" shall mean all labor, plant, materials, facilities, and other things necessary or proper for or incidental to the construction of ———.

7 "Contract Drawings" shall mean those listed in the Specifications under the title "Contract Drawings" and shall include any future changes and revisions of said drawings.

8 "Inspector" shall mean any representative of the Engineer designated by the latter to act as Inspector within the scope of the authority delegated to him. The Engineer, however, may review and change any decision of an Inspector.

9 "Subcontractor" shall mean anyone (other than the Contractor) who performs work at the construction site directly or indirectly for or in behalf of the Contractor, but the term "Subcontractor" shall not include one who furnishes only his personal services.

10 "Permanent structure," and words of like import, shall mean all construction, installation, materials, structures, and equipment required to be left by the Contractor at the construction site after the completion of the Contract.

8-3   Public and Private Construction   Most construction work falls in one of the two following classifications:

1 *Public work* is performed for some division of the federal government (which is thus the "owner"), a state, a city, or a governmental agency. Examples are highways, bridges, schools, harbor improvements, and public buildings. The project is paid for by funds raised by or on behalf of the proper governmental agency. Contracts for such work may be subject to special requirements.

2 *Private work* is performed for an individual citizen, a group of individuals, a corporation, or any organization that is or is to be in private business. Some typical projects are private houses, stores, factories, power plants, and warehouses. The project is paid for by private funds.

There are other classifications which may be encountered, such as the following:

1 *Military work,* which is performed for, and sometimes by, the military forces of the United States.

2 *Civil work,* which often includes all nonmilitary construction.

3 *Quasi-public work,* which is done for organizations that are partially of a governmental nature. For example, the Port of New York Authority is an organization that has been authorized by governmental sanction, but it is a self-supporting one. The George Washington Bridge was built by (or for) the Port of New York Authority as a self-liquidating project.

4 *Architectural services,* which are generally assumed to mean the planning and design of buildings and other improvements and construction in which the services of an architect are of prime importance.

5 *Engineering services,* which are usually assumed to denote such construction planning and design as requires primarily the services of engineers. Examples are power facilities, industrial plants, bridges, highways, hydraulic structures, and port development.

8-4 Contracts for Engineering and Architectural Services If the owner does not have an engineering and architectural staff, he or she must engage someone to take charge of the design and manage the project. Generally the owner makes a contract with whomever is selected, and the terms can be arranged in whatever manner is acceptable to the parties.[1]

In the case of individual houses, stores, apartments, public buildings, landscaping, parks, and other projects where the planning in terms of architectural features is particularly important, an architect may be the one to place in overall charge of the project. The architect then engages and works with engineers in the preparation of the plans and contract.

In the case of industrial projects, bridges, highways, hydraulic structures, public utilities, waterfront construction, and other structures where the engineering features are predominant, an engineer may be placed in general charge, with an

---

[1] The American Institute of Architects and the National Society of Professional Engineers have prepared standard contracts for professional services. These have been carefully written and tested by experience. They are intended for use by architects, engineers, and their clients.

architect in an assisting capacity. Sometimes an individual is competent in both these fields of professional work and can take charge of all phases directly.

When small jobs are involved, and often when the jobs are to be private work, the owner may not wish to contract with individual architects and engineers. He or she may find it more desirable to select a consulting architectural or engineering firm and negotiate a contract with that firm.

When public work is involved, if the agency concerned does not have sufficient architectural and engineering staff of its own, or if the work is too extensive for it to handle, the agency may notify various consulting firms that it is looking for someone to do the architectural and engineering work of the project. It may ask these firms to send reports or brochures showing the works that each has designed in the past, the experience and capabilities of its staff, the workers that it can devote to the project, and any other data which will show the firm's qualifications for performing the technical work desired. The agency will then choose the firm which seems to be the most desirable one. However, once a particular firm has been selected, there still remains the matter of an acceptable fee to be paid for performing the services or work. It may be desirable for the agency to state in advance the total fee or the percentage of the construction cost to be paid for whatever services are to be rendered.

The law and practices of some states, as well as those of the federal government, require that competitive bids be submitted for architectural and engineering services for public works projects. This procedure forces engineers and architects to state publicly a price (or rate) for their services. This means that they may have to skimp on the quality of service rendered if their stated price is so low that hasty performance becomes necessary in order to avoid losing money on the job. The quality of their service and the finished product may therefore suffer.[2]

In some cases, as for certain architectural services, stated fees have been published for the guidance of all those concerned, the fees varying somewhat with the character and magnitude of the proposed project. For example, the fee may be stated as 6 percent of the "cost" of the structure, 75 percent of this fee being payable to the architect upon completion of the plans and the letting of the contract, the remainder being payable upon completion of the construction.

The legality of such fee schedules was questioned in *United States v. National Society of Professional Engineers.*[3] The fee schedules were found to constitute price fixing, in violation of federal antitrust statutes, because members of the society were thereby prohibited from entering competitive bids. The trial court had noted: "Without the ability to utilize and compare prices in selecting engineering services, the consumer is prevented from making an informed, intelligent choice."[4]

---

[2] In Europe the various contractors often make designs of their own and present their proposals on the basis of these designs. Thus the competition involves both the engineering and the construction side. This method certainly brings out very keen competition, and varying ideas on the engineering portion of the work are elicited. In the United States the bidding is generally confined to the construction.

[3] *National Society of Professional Engineers v. U.S.,* 435 U.S. 679 (1978).

[4] 404 F.Supp. 457, 460 (1975).

Therefore, any fee schedule which is intended to be mandatory rather than simply a basis from which to estimate prices and which prohibits competitive bidding practices is suspect. The proper fees for the planning and design of engineering works are difficult to determine and so are many instances of architectural services. It is perhaps unwise to attempt to set up schedules for such services. Each case may best be left for negotiation between the persons involved.

Here are some of the reasons why it is difficult to estimate correctly in advance the cost of engineering-design work and the preparation of contracts:

1 The engineers generally will not have control of the basic requirements that will be presented by the owner. It is therefore practically impossible to foresee all the work that may be necessary.

2 At the start, the basic requirements are usually vague and uncertain.

3 Preliminary engineering studies and the making of alternate plans require an amount of time and cost that is difficult to anticipate. Nevertheless, such studies are usually of extreme importance in determining the best and most economical layout, and they may bring about large savings for the owner. On the other hand, an attempt to minimize engineering costs may result in the adoption of the first design that is at all reasonable, because the engineers cannot afford to use more time and money on this phase of the work.

4 The owner frequently changes his mind about what he wants as the design progresses, or, at least, he finds that he wants various modifications and additions. This is natural, and it is his right. It is also his duty, and that of the engineers, to make sure that all work will be satisfactory because the structure probably cannot be revised after it is built except at considerable cost. The engineers cannot predict how much delay and expense these changes will require.

One can see how inadvisable it may be for an engineer to be paid for an engineering job on the basis of a percentage of the cost of the project. If the work is done with painstaking care and thereby produces a design that can be built with a minimum of cost, the engineer automatically receives less compensation for greater exercise of skill than would be the case for a hastily designed and more costly job.

When a great deal of preliminary planning is to be done, it may be desirable for the owner to engage an engineer or architect on the basis of the cost of labor and materials used in the preparation of the plans, specifications, and other contract documents, plus a percentage fee for overhead and profit. In this manner, adequate studies can be made without an unfair cost to the designers.

An engineer is sometimes required to serve as a consultant in cases where the owner and his staff wish to have the benefit of the knowledge and experience of a specialist in some particular field. This person may contract to do this special work on a per diem basis or may agree to do it for a lump sum which covers all services that are necessary for the completion of the project. Alternatively, he or she may be retained on the basis of a stipulated monthly or annual salary or fee, with the understanding that specified regular hours will be apportioned to the work. In the latter arrangement the specialist may be engaged with the agreement that he or she will spend whatever time is necessary in assisting the client with the work. In short, the parties can make whatever contract they wish for such special services.

Still another arrangement that may have advantages for the owner and the engineer is the selection of a capable contractor to act as a consultant during the planning and design of the project. This applies primarily to work that involves unusual difficulties of construction. The contractor, having superior knowledge of construction procedures and possibilities, is to be an adviser to the engineer and is generally engaged on either a per diem or a fixed-fee basis. Thus the engineer can have the benefit of the contractor's knowledge during the work on the preliminary studies, the development of the design, and the preparation of the contract documents. In this case, the engineer still retains complete control of the development of the project, but, by securing timely advice, may avoid subsequent revision of the plans because of unanticipated construction difficulties.

Even without a formal contract for their services, many manufacturers and contractors are willing to give their advice about specific points provided they are not imposed upon. For example, during the design of the Bayonne Bridge by the Port of New York Authority, a question arose whether or not any steel company could manufacture the proposed heavy forgings that were planned as a part of the bridge bearings. These forgings were believed to be larger than any that had been made previously. Certain manufacturers were consulted about the possibility of producing these forgings. Upon receipt of their affirmative reply, the final design was made accordingly.

8-5 Components of a Construction Contract The contract documents are generally composed of design drawings, specifications, and what are sometimes called the contract clauses. These documents delineate and describe the work that is required, the quality of materials and workmanship to be furnished, and the various conditions surrounding the agreement. The design drawings that are prepared are often called the *contract drawings* because they are the ones upon which the contract is based. They may be amplified or even replaced by *supplementary drawings* if the contract drawings alone are not sufficiently extensive and detailed for use in carrying on the contract work. In some cases *reference drawings* are used to show the general character of work required, even though they are not part of the contract.

The plans, specifications, and other writings constituting parts of a building or construction contract are to be construed together as integral parts of the whole when any question arises. The contract should state that the engineer has the authority to determine the intent of the contract when the parts differ or conflict in their requirements. As a general principle, what is stipulated in the written documents may turn out to have less weight than what is shown on the drawings because the latter are usually made especially for a particular project.

Precedent and the personal opinions of the engineer are likely to have considerable effect upon where and how certain information is shown in the contract documents. Some persons divide the specifications into two parts: general clauses which are covenants or agreements, and technical data. Others place these general clauses, covenants, and agreements under what is typically called the *contract clauses;* then the technical instructions affecting materials and workmanship are grouped in the *specifications.* However, such details are not of great importance provided that the contract documents contain all the necessary instructions and requirements somewhere and that these are clearly and correctly stated.

Lawyers usually prepare the contract clauses, with the technical help of the engineers involved. The engineers are generally responsible for the specifications.

8-6 Lump-Sum Contract   A *lump-sum contract* is one in which the contractor agrees to perform for a stated number of dollars. For this compensation he is to fulfill all his obligations under the contract, even though the cost in doing so may prove to be greater than the stipulated payment. The point to be emphasized is that the contractor agrees to do his part for the stipulated fee no matter what trouble and expense are encountered. The remuneration may be made in a series of partial payments as work progresses or in one final settlement after acceptance of the work.

From the owner's (buyer's) standpoint, such a contract has many advantages. The bids tell exactly what the project will cost unless he later decides upon changes which involve extras. If he is not satisfied with this cost, or if he does not have enough money, he may reject all bids provided that he has reserved this right. Furthermore the owner need not engage a staff to keep account of exactly what work has been done and materials furnished week after week except as this may be necessary in connection with making partial payments. Of course the engineer, as the owner's representative, has to keep watch to see that the contract is fulfilled in its entirety exactly as called for by its terms.

All that is required of the contractor must be made very clear and specific before the contract is let. In the purchase of an automobile, a diesel engine, or some other item that is a more or less standard article in the trade, the contract can call for the article by model number, catalog number, trade name, or some other acceptable and customary designation. In the case of a large construction contract, the owner should have all the necessary plans, contract terms, and specifications complete in every respect. If the extent and details of the job are not shown properly, many features may have to be determined and provided for as the work progresses, probably resulting in additional costs for such extras.

A lump-sum contract can be used successfully for the routine construction of schools, mill buildings, warehouses, and similar structures with which the average contractor has had considerable experience. It is true that these projects involve many different kinds of work, but a capable and experienced contractor can generally estimate the total cost and prepare the bid with a fair degree of accuracy.

By contrast, if the work required by the contract is not readily determinable in all its features at the time of bidding, a lump-sum contract should not be used. It would be unfair to the bidders to expect them to bind themselves to the performance of work, significant aspects of which cannot be outlined in advance. If obliged to bid on such an indefinite project, bidders would have to protect themselves by making proposals high enough to cover the cost of whatever increases and uncertainties may materialize.

Some types of work which are unsuited to lump-sum contracts are the following:

1 Difficult foundations for which no one knows how much excavation will be required, how many piles will have to be driven, how difficult the shoring and unwatering of the excavation will be, and how much time will be required to complete the work.

2 Contracts involving excavation of uncertain quantity and character, the bidder not knowing how much of the work may be the excavation of rock instead of earth.

3 Emergency projects that have to be rushed without time to prepare complete plans prior to obtaining bids.

4 Alteration projects that involve the maintenance of operations while the work is being performed.

5 Projects that are subject to unpredictable hazards, such as floods, cold weather, and the conduct of persons beyond the contractor's control.

From the contractor's point of view, a lump-sum contract may be preferable (unless of course competition or unsound judgments have caused him to make a bid that is too low). The reason is that whatever the contractor can gain by excellent planning and efficient management is his own to keep, and any legitimate saving that he can devise will add to his profit.

8-7   Unit-Price Contract   A *unit-price contract* is one in which payment for the work is to be made upon the basis of the computed quantities of specifically stated items of work actually performed and materials furnished and used by the contractor in the project, each such quantity being multiplied by the contractor's bid price for that unit. For example, the contract may state that the engineer's estimate of the quantity of reinforced concrete is 10,000 cubic yards. If the contractor's bid on that item is $80 per cubic yard, that is the amount to be paid for each cubic yard of concrete placed in the furnished work whether the actual quantity is 11,000 or 9000 cubic yards.[5]

In order to secure unit-price bids it is ordinarily necessary to prepare general designs and drawings together with estimates of quantities so that the bidders may ascertain the magnitude of the project, the character and difficulty of the work, and the probable cost of performing it. Contractors may properly claim adjustments in their compensation if the actual work proves to be considerably different in quantity and quality from that indicated by the data presented for bidding purposes. The fact that contractors will be paid the bid price on whatever quantity

---

[5] Notice that the cost of the project depends upon the bid unit prices and the quantities used. The total cost may or may not be less than it would be if a lump-sum form of contract were used and bids were made on that basis. Furthermore, the contractor's profit does not vary with the form of the contract; profit depends upon how much compensation exceeds expenses.

of that item they actually furnish on whatever service they perform is likely to reduce some of the gamble in bidding.

From the owner's standpoint, one disadvantage of such a contract is that one cannot be absolutely sure of the total cost until the job is finished. On the other hand, the owner can avoid the delay that would otherwise be necessary in making a large number of contract drawings to show in detail everything that will be needed, as should be done in the case of the lump-sum contract. After the contract is let, supplementary drawings can be prepared far enough ahead to enable the contractor to secure all the necessary information in time for obtaining materials, making detail drawing, and doing the work.

Of course, both the owner and the contractor will have to do considerable computation and bookkeeping during the progress of the unit-price job. It is usual practice for the parties to have their own staff make monthly estimates of the amount of materials furnished and work completed during the preceding period, basing their figures upon field measurements or data obtained from the drawings, shipping bills, and similar information. Their estimates are then compared. If there are discrepancies of importance, the computations are checked, and a set of quantities is usually determined and agreed upon as a basis for payment.

It is not customary to have a unit-price item for absolutely everything that goes into a large project. In the preparation of the contract documents, the engineer selects the payment items which will apply to all major portions of the work, and includes as many of the minor items as necessary or desirable. The list of payment items should be sufficiently complete to avoid being unfair to either contractor or owner but without having the number of unit prices so great that it is unwieldy and difficult to use. Furthermore, the units for payment should be definite and readily measurable.

A contract for the foundations and other concrete work of a large industrial project having steel superstructures will serve as an illustration of the selection of the items that are to serve as the basis for payment. The items used in this contract were the following:

1 Earth excavation, per cubic yard
2 Concrete, per cubic yard
3 Cement, per barrel
4 Steel reinforcement, per pound
5 Clearing the site, grading, disposing of excavated material, and cleaning up the site after completion of the foundations (lump sum)

The computation of the quantity of excavation was to be based upon the theoretical excavation from the original ground surface down to the elevation of the bottom of the structure or part (as shown on the plans) and within vertical planes bounding the outline of whatever the part might be. In determining the unit price for this item the contractor was to include whatever the miscellaneous work, shoring, drainage, and backfilling would cost. The payment for concrete was to include the cost of all materials, forms, placing, curing, and finishing of concrete. Since forms are so costly, it is obvious that the contractor should be given enough data from which to estimate the form costs with reasonable accuracy. The contractor was also to be paid for whatever cement was used. Furthermore, from typical details that were shown on the drawings, the bidder was to make an estimate of the expense involved in cutting, bending, and placing the bars. Detail drawings of the bars were to be furnished by the owner.

The use of a lump-sum bid on item 5 illustrates how one portion of a unit-price contract may be treated as a lump-sum item. This is a convenient way to handle work that would be difficult to classify and to keep track of in the field.

Notice that many of the specific hardware items necessary for such a job are not directly measurable in any line of the unit-price contract. For example, all piping and conduits were to be furnished by the owner and placed by the contractor; the same applied to anchor bolts to be set in the concrete. The contractor was supposed to allow for these features. The contract made this clear, and the drawings showed the nature and magnitude of such special arrangements so that the contractor could include their cost in bid prices on the payment items.

By no means do we intend to convey the impression that unit-price contracts are generally preferable to lump-sum contracts. Take, for example, a small bridge job costing approximately $40,000. It involves large variety of materials, but each kind will be used in small quantities. Actual increases in volume of work done when the estimated quantities are small may lead to unexpected increases in the total cost. Reduction in quantities may lead to claims for extra compensation. It may therefore be desirable to use a lump-sum type of contract for bidding such a structure. However, the reader should realize that the form of contract is not what determines the owner's cost or the contractor's profit.

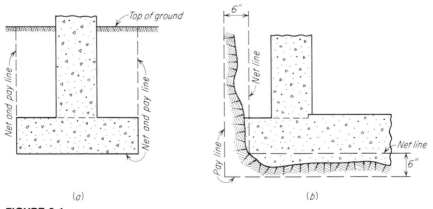

FIGURE 8-1
Payment lines and net lines for excavation in *earth* (*a*) and *rock* (*b*). These are to show the bidder the basis to be used for the measurement of quantities.

8-8   Cost-Plus-Percentage Contract   In an emergency situation requiring construction of something without time to develop plans for it, a contractor should not try to make a proposal to build it on a lump-sum basis, for obvious reasons. A unit-price basis is equally undesirable because it is difficult to forsee all that is required. In such a situation it is possible to engage a contractor to do the work on the basis of the cost plus a percentage of the cost, the owner paying all bills. The percentage of the cost is the remuneration for the contractor.

Materials, labor, rentals of equipment and property, transportation, and everything else except the salaries of the contractor's supervisory staff are generally included in the cost item of such a contract. Sometimes even the wages of the contractor's staff in the field are included as direct costs to which the percentage is to apply, but this depends upon the particular agreement. Ordinarily, the percentage to be paid to the contractor should not be applied to the costs of any portion of the contractor's general office overhead, postage, traveling expenses, insurance (other than that directly applicable to the job), salaries or portions of salaries of company officials who may visit the job, or charges for the use of any equipment that the contractor would not customarily use for the performance of the work. All these points should be made clear in the written agreement.

A cost-plus contract may be very useful as well in situations which are not emergencies. For example, a dam was to be replaced after a washout resulting from the floods in the Naugatuck River Valley in Connecticut during the 1955 hurricane. The owner let the contract to a construction firm on a cost-plus-percentage basis because no one could tell for certain just what troubles would be encountered in the work. In this way decisions could be made as the work progressed.

In the accounting, the contractor generally keeps all records of costs and then presents them to the owner for checking, approval, and payment as the job proceeds.

To make a cost-plus-percentage contract in an emergency, the owner may call in several contractors, find out who can and will tackle the job, make an agreement regarding the percentage to be paid to the contractor, and see that things start moving at once. While the contractor is assembling staff and equipment, the owner will try to determine enough about what is to be done to start the work as soon as the contractor gets the field staff organized; thereafter information is channeled to the contractor on an as-needed basis.

8-9 Cost-Plus-Fixed-Fee Contract   An advantageous arrangement or type of contract for emergency jobs and others involving uncertainties is one based upon the cost of labor and materials plus a fixed fee as a compensation for the contractor. This overcomes the possible weakness of the cost-plus-percentage type of contract. The contractor receives only the stipulated sum for the job of overseeing and running the job no matter what the cost of the project is. If staff salaries are to be paid out of this fee, the contractor will naturally endeavor to expedite the project so as to make as much profit as possible. Even if staff salaries are to be classed as a part of the general costs of the job, the contractor will still want to rush the work so that his workers can be available for another contract elsewhere as soon as possible.

Under this contract this cost-plus-fixed-fee arrangement[6] the owner can select a reliable contractor and engage a capable engineer to design and check the work. All three parties can work in harmony and accomplish amazingly fine results. The gamble is largely removed for both the owner and the contractor: The contractor knows precisely what the profit for the job will be; the owner has confidence that the job will be handled expeditiously and in an economical manner.

8-10 Subdivision into Several Contracts It is sometimes advantageous to split a contract into two or more parts if some portion of the project involves many uncertainties. Here is one example. A large industrial corporation was to make extensive additions to and alterations in the structures and equipment of one of its plants. Foundation conditions were known to be poor and to involve considerable difficulty, and many utilities were to be relocated. All this construction work was to be done while maintaining production in the existing facilities, and its cost would be very difficult to predict. However, as soon as the foundations and subsurface work could be completed, the superstructure and installation of new equipment could go ahead without such uncertainties. Therefore the engineer decided that the best way to handle this job was to make all foundation construction, utility alterations, and subsurface work a cost-plus-fixed-fee contract and to make the remainder a separate unit-price contract. Both parts were to be handled by the same contractor, an arrangement that would assure coordination of the work under one head.

Subdivision may also be advantageous in the case of a very large project. the procedure enables the engineer to make the plans, specifications, and other written material for one portion of the project at a time and to let a separate contract for each portion as soon as the design is completed. This procedure also saves time, keeps the magnitude of each contract within reasonable limits, and spreads the work among various contractors, each of whom may be especially well qualified for a particular part. For example, the work on the Lincoln Tunnel in New York was subdivided into nearly a hundred separate contracts. Typical of these were the following: The caisson for the ventilation shaft on the New York side of the Hudson River, the New Jersey shaft, the under-river portions of the south and north tunnels, each of the three ventilation buildings, various sections of the tunnel between the river and the portals, each of the two plazas, and a series of sections of the approach in New Jersey.

8-11 Contracts for Combined Engineering and Construction Some organizations makes a specialty of both designing and building a project. In other words, they will sign a contract with an owner to make all the preliminary studies and final design and then to go ahead and build the structures involved. This is some-

---

[6] In using the term "cost plus" in a contract one must be sure of its exact meaning, and the contract should be clear in this matter. If it states that the compensation is to be on the basis of the cost of labor and materials or on the basis of time and materials plus a percentage or fixed fee, the contractor probably cannot include in the cost such items as overhead. Such points should be cleared up before the contract is signed.

times called a *turn-key* type of contract. Such a contract may be made on the basis of the cost plus a fixed fee, the cost plus a percentage of the cost, or whatever arrangement is satisfactory to the parties.

Such a combined contract has been used effectively in specialized industrial projects. In some cases, specialists may even assist in getting the plant into operation in order to see that everything functions properly. Or manufacturer may contract to develop the design of an unusual piece of equipment and then to build it. In the field of small buildings, there are some contractors who make a specialty of designing and constructing what are sometimes called *standard buildings.*

The proponents of such contracts claim that work can be greatly expedited in this way because extensive plans and specifications do not have to be prepared and bids do not have to be submitted before work in the field can start. Furthermore, there is not division of responsibility; the contractor prepares the plans and then builds the structures, becoming the final inspector of the initial design work. If handled by a skillful, reputable concern, such methods may be efficient. The owners are, of course, dependent upon the contractor in this situation, just as they are dependent upon the engineer in other cases.

8-12 Preparation of Documents   The need for the exercise of wisdom and skill in the preparation of the contract documents for any construction job should be obvious. This need for special care applies as well to the contractor in interpretating contract clauses and specifications.

It is amazingly easy to make a written statement that can be interpreted in a different way from what was intended. Writers of particular clauses know what they mean, and they think that they say it clearly. However, someone else may derive another meaning from it. The important thing is what the words mean, not what the writer intended to say. It is helpful if one can compose something, then let it rest for a week or two before checking it, because one is likely to be more critical when examining a statement that has not been seen for some time since the writing of it. However, when the contract clauses and the specifications are being prepared, the owner is generally eager to have them completed in a hurry.

The following is a list of some points that will be helpful to one who is writing contract clauses and specifications:

1 Be sure that the wording is clear and precise. Have it checked carefully by some other person. The readers of the documents see only the words that are written. They do not know the thoughts and intentions of the author.

2 Coordinate the written material with the work of those persons who are making the drawings, and have the latter examine all parts relating to their work.

3 Be sure that the grammar and punctuation are correct.

4 Use short simple sentences instead of long involved ones, even if the former do not sound as attractive. You are writing a legal document in which accuracy and clarity are of vast importance. Make sentences complete; do not allow phrases to stand as complete sentences.

5 Do not write on a given point until you are sure of the subject matter.

Obtain competent assistance where necessary. Neglecting the subject may have unfortunate results.

6 Avoid the use of ambiguous words and colloquialisms. Repeat a word as often as necessary rather than using synonyms, because the latter may have slightly different shades of meaning. The vocabulary used should be what is customary in the line of work involved and will be understood clearly by the readers. Complicated, high-sounding words are generally inadvisable.

7 Repeat names as often as necessary. Personal pronouns are likely to cause confusion.

8 Remember that you are giving instructions, not suggestions. You need not explain the reasons behind the instructions.

9 Avoid such all-inclusive expressions as "etc." They have no place in contract documents.

10 Be very careful to make cross-references clear by using title and paragraph numbers.

11 Be sure that the requirements are fair, reasonable, and practicable.

12 Reveal all hazards and give all information which may affect contingencies to be provided for by the contractor.

13 Be careful to examine all the implications of broad, general statements before using such statements.

14 Make an outline of all parts to be covered. Do this before composing the clauses. Divide the documents into whatever sections are appropriate.

15 Avoid omissions, on the one hand, and irrelevant or uselessly repetitious information on the other.

16 Be careful what you say because you must be prepared to enforce your instructions and to abide by them yourself.

17 Establish some kind of numbering system for the various sections and clauses in the documents so that each can readily be referred to later on. It is often convenient for the reader of the papers if titles or labels are added in the margins alongside the various main clauses so that the reader can quickly see what is the nature of the subject matter contained therein.

18 Prepare a table of contents with page references, and also prepare an index.

If revisions of or additions to the contract documents are found to be necessary before the bids are received, issue an addendum to all bidders and demand written acknowledgement of its receipt by all parties who have previously received copies of the contract documents. Be sure that the addendum refers to the specific clauses which are eliminated, revised, or replaced, using definite numbers or labels so that no doubt will remain as to the final instructions. Of course, an addendum takes precedence over the original clause or clauses which it modifies. However, an addendum can, in its turn, be changed or even canceled by a subsequent addendum.

8-13 Examples of Wording The following illustrations, taken from actual documents, are given in order to emphasize the importance of careful writing when preparing contract papers. They show the difference in meaning that may arise from the position of a word, the use or omission of a punctuation mark, the use of a connective, and other seeming trivia. The key words are in italics.

1 "The Chief Engineer *only* has the right to authorize extra work."

This seems to mean that he has no other right. It would be better to say, "*Only* the Chief Engineer. . ."

2 "Within the area of the foundation, the Contractor shall remove all compressible soils *and* blast out all rock above elevation 110."

This seems to mean that the contractor is to remove nothing below elevation 110. If this is not the intent it would be better to write. "The Contractor shall remove all compressible soil within the area of the foundation, and shall also blast out and remove all rock above elevation 110 in this area."

3 "The remaining fifteen (15) percent will be *paid* at the *conclusion* of the job *on* the final approval of the Chief Engineer."

This seems to mean that the chief engineer is to be approved by somebody. It would be better to state that the money will be "paid to the Contractor after the Contractor has completed the Contract, and after the Chief Engineer has approved the completed work."

4 "Brick masonry shall be laid so that all joints are horizontal *and* vertical."

This statement really says that the joints are to be both ways simultaneously— an impossibility. The statement should read, "Brick masonry shall be laid so that the joints between courses are to be horizontal whereas abutting joints are to be vertical."

5 "If an error in the drawings is discovered by the Contractor he is to suspend work until he can *get hold of* the Chief Engineer."

Colloquial or slang expressions like "get hold of" may pass in oral statements but are inadvisable in technical writing. Obviously, the contractor is not to wait until he can physically lay hands on the chief engineer, but rather he is to establish contact with such engineer and obtain the latter's instructions on future action.

6 "The Contractor is to be notified *what to do in writing* by the Chief Engineer."

This sentence implies that the contractor will be told what to write about and how to do it. This is primarily a case of improper sequence of words. It is the notification that is to be in writing. The sentence should read. "The Contractor is to be notified in writing by the Chief Engineer regarding what to do."

7 "The Contractor shall make reparation to the Owner *due to* the mistake."

Reparation is not owed to the mistake. It would be better to say "because of the mistake" or "for damages caused by the mistake." A word like "due" has a variety of meanings, and its use can lead to misinterpretation.

8 "The Contractor is to report any errors discovered *immediately*."

The meaning of this sentence is ambiguous. The contractor is to report immediately regarding any error that he may discover, whether discovered early or late.

9 "The Contractor *should* make a written report to the Chief Engineer *on the fifteenth of each month* showing the estimated quantities of work completed during the preceding month."

The word "should" means that the contractor ought to report but is not compelled to do so. It is better to say "shall" or "is to," which means that he is obliged to do so.[7] It would also be better to start the sentence with, "On the fifteenth of each month the Contractor shall. . ."

10 "The Contractor shall be compensated for any extra time *or* effort involved."

This implied that he will be compensated for one or the other, not for both. This situation leads some persons to use "and/or" as a connective in order to cover one or both items. If the statement is ". . .time and effort involved," it probably will be satisfactory because even though one item is omitted the sum will contain the applicable item.

11 "Upon completion of the installation, the Contractor shall leave the premises in an orderly condition."

This sentence is actually quoted from a contract. A jokester might pretend to wonder whether it is the condition of the contractor or the condition of the premises that is important. The intended meaning is, "Upon completion of this installation, the Contractor shall clean the premises and see that they are in an orderly condition."

QUESTIONS

1 Is it desirable to have construction work done under a contract? Why?

2 What party or parties are obligated by a construction contract? Explain.

3 Who prepares the contract documents for a construction project? What is the objective?

4 Who may suffer if a contract is poorly or unwisely prepared? Explain.

5 For what services does the owner have to depend upon the engineer and architect? The contractor?

6 Define the following as applied to the preparation of construction projects: (*a*) planning, (*b*) design, (*c*) dimensioning.

7 Explain the principal difference between public and private work.

8 What legal problems arise from the use of fee schedules?

9 Why should carefully prepared definitions be included early in the contract documents? Illustrate one likely definition.

10 It is the purpose of the engineer to prepare the contract so that the work will be done properly, expeditiously, and amicably. Discuss this idea and its application in practice.

---

[7] Similarly, the word "may" would be permissive and should not be used when a specific order is intended.

11 How is a court likely to interpret an ambiguous clause in the contract documents? Explain.

12 If the requirements of the drawings and the specifications differ, what should be done about it?

13 Discuss the implications of a "reasonable interpretation" of the contract documents in case some statements in the papers are ambiguous.

14 On what basis is it customary to make contracts for engineering and architectural services?

15 What safeguards should be provided if you are planning to contract for rendering your own personal services as an engineer or architect?

16 What is the primary purpose of requiring an engineer to have a license as a professional engineer? Describe the type of engineering work which is likely to require the services of such a professional?

17 Explain some reasons why it may be difficult for an engineer to predict accurately the cost of engineering planning and the preparation of contracts.

18 Explain the basic features and advantages of the following types of contract: (*a*) lump-sum, (*b*) unit-price, (*c*) cost-plus-percentage, (*d*) cost-plus-fixed-fee.

19 What sort of contracts are unsuited to lump-sum bidding?

20 Why is it sometimes desirable to have certain items of a unit-price contract bid on a lump-sum basis? Explain. Illustrate possible advantages of this system.

21 Under what circumstances might an owner prefer to have a project built under a lump-sum contract instead of under a unit-price contract?

22 Is the contractor likely to make more profit on a unit-price contract than on a lump-sum contract or a cost-plus-fixed-fee contract? Elaborate.

23 Does a unit-price contract generally have a separate payment item listed for every different kind of material and every bit of work required in such contract? Explain.

24 How is the total payment under a unit-price contract determined? Is such a contract likely to be more economical for the owner than a lump-sum contract? For the contractor? Explain.

25 What are the advantages and disadvantages of cost-plus-fixed-fee contracts from the standpoint of the several parties concerned?

26 When a unit-price contract requires certain minor work to be done and material items to be furnished for which there is no stated payment item in the contract, what will the bidder probably do to obtain compensation therefor? Illustrate.

27 Illustrate a situation in which the subdivision of work into several contracts may be preferable to having the entire project performed under one general construction contract.

28 What are the pros and cons in the debate regarding (*a*) engineers bidding for engineering work on the basis of price? (*b*) engineers advertising for receipt of bids for engineering services by other engineers on a price basis?

29 Can an owner have the contract documents prepared by engineer *A* reviewed by engineer *B?* Explain the proper steps to be taken by the parties concerned. If *B* discovers errors in the documents prepared by *A,* what should *B* do? Explain.

30 An engineer who was engaged to design a large building made studies of the soil conditions. As a result, he decided to use piles to support the structure. The owner, without notifying the engineer, called in Ms. Doe, a fellow engineer, and asked the latter to make a separate and secret study of the problem to see if the piles could be eliminated. What should Doe do about it when receiving this request?

31 Upon what logical basis might a contractor determine the percentage to charge in case of a cost-plus-percentage contract?
32 Upon what logical basis might a contractor determine the fee to charge in case of a cost-plus-fixed-fee contract?
33 When and why may it be advantageous for an engineer to have charge of the inspection of work to be performed under contracts which he has prepared?
34 Define subcontracts. When and why may they be used?
35 What is an addendum? When and why may one be used?

CHAPTER 9

ADVERTISING

9-1 Purpose of Advertisement Advertising for bids is usually aimed at interesting a number of contractors in the proposed work to secure the benefit of keen bidding competition. In other cases, the advertisement may be made for the purpose of fulfilling legal requirements or simply as a matter of policy. Again, the purpose may be a combination of the foregoing.

If a large construction job is being designed, it is probable that news concerning it will spread by word of mouth and that interested contractors capable of doing the work will know most of the story in advance of formal advertisement. In the case of smaller projects, however, the published notice may be very effective in reaching prospective bidders who might otherwise fail to learn about the job.

In private work the owner need not publicly advertise for bids but has the right to negotiate directly and to contract with whomever he or she chooses. This procedure has much to recommend it in connection with unusual structures, industrial construction, and other work that requires special knowledge and equipment. There are contractors who have built up an excellent reputation for certain specialized kinds of work, and the owner may prefer to trust one of this elite group to make a good proposal.[1] On the other hand, private work should, in the average case, be advertised, both to secure the advantages stemming from competition and to make sure that no stockholder or other concerned person can with justification accuse those responsible for awarding the contract of favoritism detrimental to the owner's interests.

---

[1] This is especially true if the owner has on previous occasions experienced highly satisfactory results with a particular contractor.

The law generally requires that federal, state, and municipal contracts be advertised publicly. This method opens the way to having the work done at a minimum cost to the public. If the requisite three bids are not received on certain government projects, special authorization will ordinarily be needed if an award is to be made.[2] A scarcity of bids is most likely to occur when the project involves special work that only a few contractors are equipped to do.

9-2   Advertising Media   Since printed advertisements cost money, it is important to utilize expenditures to the best advantage by advertising a contract in whatever media are most likely to bring the message to the attention of the greatest number of potential bidders. A large-city daily newspaper reaches many people, but contractors may not be accustomed to looking into it for data about extensive contract work in the offing. On the other hand, local projects are frequently advertised in small-town newspapers, and those who are interested in such work will look for the advertisements there. Trade journals—the *Engineering News-Record,* for one—make a specialty of advertising for proposals on construction contracts, and many contractors look to that source for information.

Before advertising a contract, one should study the nature of the job and the available means of conveying the information to potential bidders. The object is to get the best results for a minimum of expenditure, and the fact that one advertiser's rates are very low is but one aspect of the picture; despite its attractive price schedules, that advertising medium may not offer the best prospect of success in securing active competition and reasonably low bids.

9-3   Timing of Advertisement   A related question is that of timing. An advertisement appearing several days in succession may be desirable for some purposes. In other circumstances, one that is repeated weekly or biweekly for a month or so may be preferable provided the publication carrying the advertisement appears on a regular time schedule.

In regard to public work, a check should be made to assure that no legal requirements are overlooked in selecting the medium and the timing for the advertisement.

9-4   Attracting Attention   The heading of an advertisement should be such as to attract the attention of the readers and enable them to see at a glance that the text may well contain something of interest. The site of the work and the owner for whom the structure is to be built should be disclosed in the heading or immediately following it. If the mentioned location is within operating distance, the prospective contractor will want to learn more about the job; conversely, if it is far removed, he or she will not take the time to read the details. Likewise, a contractor's reaction to the advertised work may depend to some extent upon the owner's

---

[2] In case of war or other emergency, it may be necessary for the government to dispense with public advertising and to negotiate with someone directly because the public welfare requires that the work be started and executed with all possible speed.

identity. The builder may have had some previous experience with the particular owner or may have heard something which will have a bearing upon his desire to do the work in question.

The following title appeared on an advertisement in the *Engineering News-Record,* July 7, 1983:

Bids August 18, 1983
City of San Jose, California, San Jose/Santa Clara Water Pollution Control Plant, Intermediate Term Improvements, Sludge Processing Facilities.

Thus the reader was quickly informed about the general character and location of the project and the date for bids. Those who are interested in doing that type of work and in that locale are then likely to read the full advertisement, whereas others will merely pass on to something else.

If readers have to look over in detail one involved item after another in a long series without finding anything of particular interest, they are sooner or later going to stop looking altogether and may thus miss the very advertisement you wish them to see. If all the advertisers in a given publication see to it that the titles of their respective advertisements are clear and instructive, readers will be likely to glance at each one, and thus will notice your particular item.

9-5 Information to Be Given in the Text[3] The body of the advertisement should consist of a brief resume of the proposed project and of the special conditions or requirements which distinguish it—for example, regulations affecting labor rates for personnel working under compressed air during the construction of a subaqueous tunnel. Just what a particular advertisement ought to contain will depend upon the circumstances, but the following are points to cover in advertising typical construction work:

1 The kind of job (school construction, sewers, etc.). This matter is of obvious import and must be thoroughly handled, even if the heading of the advertisement—as it should—affords a broad outline of the picture.

2 The locale in which the work is to be performed.

3 The owner for whom the job is to be accomplished. From the identity of the owner and the engineer, many contractors can judge the quality of the work which will be required and the extent of the cooperation they can expect. Other things being equal, contractors are likely to make lower bids on work designed by and under the charge of an engineer whom they know personally (and favorably) or concerning whom they have heard good reports than on work handled by an engineer totally unknown to them or in whom they for some reason lack confidence.

4 The magnitude of the project. The advertisement should give some scale to the job, so that the reader can see whether or not it is too small to be interesting and profitable or, perhaps, so big that it cannot be handled reasonably. It is

---

[3] In most cases, the advertisement is, to a large extent, a condensed version of the instructions for bidders.

usually desirable to avoid stating the engineer's estimate of cost in dollars, although such figures are occasionally included. The engineer's estimate may or may not accurately represent the true cost of the project, but publication of such estimate tends to give the contractors a sort of target at which to aim in making their bids. Each bidder may be inclined to quote as close to the engineer's estimate as possible without risking the loss of the award, provided, naturally, the work can be done at a cost no greater than the engineer's figure. In lieu of a statement of estimated cost, the size and character of the job are often shown by one or more of these methods:

   a Dimensions of width, length, and height
   b General size, number of floors, type of framing, kind of floor and wall construction; for example "School, 75 by 250 feet; 3 stories, steel frame, brick walls, concrete floors"
   c Square feet of floor area (plus general description of the structure)
   d Cubic feet of volume (plus general description of the structure)
   e Estimated quantities of some major items of material or work; for example, "50,000 cubic yards of earth excavation"
   f General capacity and character; for example, "Reinforced-concrete, four-story hospital to accommodate 200 beds"
   g General scope and scale; for example, "5 miles of concrete pavement for four-lane highway," or "200-foot two-lane steel-truss highway bridge"

   5 The type of contract ("lump-sum," "unit-price," etc.). This aspect is of particular significance in the case of large projects.

   6 The date, place, and form of receipt of sealed bids. Note that the date of opening of the bids is the first item appearing in the advertisement example given in Art. 9-6.

   7 The procedure which will be followed in opening and reading of bids. The advertisement should state that the proposals are to be opened and *read aloud publicly*.

   8 The action regarding letting of the contract. If the contract is to be awarded to the lowest *responsible* bidder, that fact should be made clear. Also, a time within which the contract will be awarded after opening of the proposals should be stated—for example, "two weeks after opening of the bids."

   9 The possibility of rejection of any or all proposals. It is generally advisable to state in the advertisement the fact that the owner has the right to reject "any or all" bids (perhaps because the indicated costs may be too high or because all bidders may be "unsatisfactory").

   10 The type and size of deposit required with the proposal. Not only should this requirement be stated, but the procedure for the return of deposits should be clarified.

   11 The performance bond. If such a bond is required, this fact should be set forth, together with the amount expressed in dollars or as a percentage of the bid price.

   12 The time set for starting and completing the work. The reason for detailing these points in the advertisement is obvious. As a corollary, any contemplated bonus or liquidated-damage provisions should be mentioned.

13 The availability of data and forms. State (*a*) where and from whom the contract papers, drawings, and specifications may be obtained; (*b*) at what cost; and (*c*) whether or not a refund will be made upon their return in satisfactory condition.

14 The withdrawal of bids. Indicate whether or not this is permissible. If bids may be withdrawn—which is not customary—state the conditions under which such a maneuver may be accomplished.

15 The date for signing the contract. State any final date by which the selected bidder must sign the contract so that work may progress on schedule.

16 Special conditions. If special hazards exist, if restricted or full traffic must be maintained, if particular labor conditions prevail, or if unusual features are involved, these matters should be mentioned in the advertisement.

17 Further data. It may be desirable to (*a*) state that, at a specific date, place, and hour, the prospective bidders may meet with the engineer to examine the site, ask questions about items which need clarification, and familiarize themselves with the problems involved in the work; (*b*) include a statement as to whether or not the contractor is to reflect all federal, state, and local taxes in his bid price; (*c*) give applicable regulations regarding equal opportunity in employment; and (*d*) state requirements relating to workers' compensation, general liability, automobile and public liability, and property damage insurance.

It may seem that much of the foregoing information is superfluous in a mere advertisement. However, a fairly complete description at this early stage of the proceedings is helpful to all concerned. It aids the engineer by reducing the number of inquiries from persons accustomed to making a full investigation before deciding whether or not a given job is something they can and wish to handle. In many cases—absent a complete advertisement—prospective contractors might too hastily conclude that they are not interested, thus reducing potential competition and presumably working to the disadvantage of the owner. On the other side of the picture, a reasonably detailed advertisement helps, too, by cutting down the number of useless requests for plans. Presumably, only those contractors who are real prospects will bother to apply for supplementary information.

Moreover, a sufficiently comprehensive, accurate advertisement protects the engineer against the possibility of a valid assertion of misrepresentation of the nature and conditions of the proposed contract. In this connection, also, it is advisable to reprint the entire advertisement in the contract documents, particularly when the project involved is a public job.

9-6   Examples of Advertisements   The following advertisement has been copied from *Engineering News-Record* of July 7, 1983. Its content should be compared with that suggested in the preceding discussion.

Bids: August 3, 1983

City and County of San Francisco
Department of Public Works
Westside Pump Station and Force Mains
Specification No. 1114W

Sealed bids will be received in Room 282, City Hall, San Francisco, California, 94102 between 2:00 and not later than 2:30 P.M. local time on August 3, 1983, after which time they will be publicly opened and read. Plans, specifications and bidding forms will be available at the office of the City Engineer, Room 353, City Hall, San Francisco 94102, telephone (415) 558-3676. A nonrefundable fee of $180.00 paid by cash or check will be required for each set. Checks should be payable to "Department of Public Works." Bidders may make their own arrangements to pick up the bidding documents, or, if they make a request in writing, and submit the required fee, such documents will be shipped or mailed to them C.O.D. C.O.D. will only cover the shipping and mailing costs.

All Plan Holders on record as of May 18, 1983 will receive Addendum No. 3 including a new bid proposal package at no additional charge.

The bids for the subject project received on May 18, 1983 are hereby rejected. The average bid amounts for Proposition "A" were $17,458,364.00 and for Proposition "B" $17,471,614.00.

This project includes the construction of a sewage pump station with a nominal wet weather capacity of 140 mgd and installation of force mains, 42-inch, 48-inch, 54-inch diameters and gravity pipelines, 54-inch and 84-inch diameter.

The pump station will be located within the boundary of the Great Highway, south of its intersection with Sloat Boulevard, and immediately adjacent to the San Francisco Zoo.

The work shall consist of building the pump station, force mains, gravity pipelines, 160 feet of connecting consolidation sewer, berm embankment, appurtenant electrical, instrumentation and controls, mechanical, structural, landscape, architectural, roadway, and utility work.

Major components of the pump station will include seven submersible sewage pumps (300 hp each); two mechanically cleaned bar screens, diesel generator, dewatering pumps; piping, valves, sluice gates; ventilation and odor control equipment; auxiliary systems; electrical and instrumentation systems.

To accommodate the equipment, all work required to excavate, demolish existing structures and underground encumbrances, backfill, control groundwater, place embankment and dispose of excess and/or undesirable materials, pave and landscape will be included, as well as relocating any existing interfering utility systems.

The work shall include trenching, demolishing part of two pedestrian underpasses, furnishing materials, installing testing, backfilling, restoring the ground surface to original condition and all other effort required for relocation of existing utility interferences, fences, pavement and the like.

Supporting, working around, and protecting certain utility agency and company facilities in conjunction with the work under this contract, and located in the public street areas, shall be done where shown or where required and in accordance with the pink wages in these Special Provisions, except as otherwise specified.

Upon completion of the construction and installation of equipment, all systems will be tested and their adequacy to perform their functions suitable demonstrated as described herein, and, to the satisfaction of the Engineer, the Contractor shall accomplish all other related and incidental work, as shown on the Plans, and in accordance with the Specifications.

Federal and State funds are involved in this project.

Contract will provide for progressive payments for the value of the work done and materials incorporated based on the unit prices bid. Contract will provide for Liquidated Damages, as fixed in the specifications.

Any contract or contracts awarded under this invitation for bids are expected to be funded in part by a grant from the United States Environmental Protection Agency. Neither the United States nor any of its department agencies or employees is or will be a party to this invitation for bids or any resulting contract. This procurement will be subject to regulations contained in 40 CFR 35.936, 35.938 and 35.939.

In order for the City to receive grant assistance, approval to award this contract must be received from the California State Water Resources Control Board prior to date of award. Therefore, Proposals are subject to the condition that the successful bidder will accept the contract as bid up to a period 120 days after receipt of bids.

Bidders on this work will be required to comply with the President's Executive Order No. 11246. The requirements for Bidders and Contractors under this order are explained in the specifications.

A Pre-Bid Conference will be held on July 11, 1983 at 9:00 A.M. local time in Room 282, City Hall, San Francisco, California. The purpose of the conference is primarily to discuss the required Affirmative Action Programs for equal opportunity in employment and minority entrepreneurship, and any questioned contractural requirements. It is strongly recommended that all bidders attend this conference, but is not mandatory.

Bidders are warned that failure to submit properly completed Affirmative Action Program forms included with the proposal will render the proposal nonresponsive.

A certified check or corporate surety bond of 10 percent of the amount bid must accompany each proposal.

The successful bidder will be required to execute the contract within 10 days after the award of the work to him, and shall furnish bonds to the satisfaction of the City and County of San Francisco for faithful performance and for labor and materials, each in the sum of 100 percent of the amount of the contract as awarded. In case of failure to execute the contract as stated or to furnish bonds the bidder will be considered to have abandoned the contract and the bond or check accompanying the proposal shall be forfeited to the City, not as penalty but as liquidated damages.

Minimum wage rates for this project, as determined by the Secretary of Labor and by the Director of the State of California Department of Industrial Relations, are set forth in the Special Provisions or available for inspection at Room 353, City Hall, San Francisco. If there is a difference among the wage rates predetermined by the Secretary of Labor and those determined by the State of California for similar classifications of labor, the higher wage shall prevail.

Right is reserved to reject any or all bids.

Approved:

Dir. of Public Works & Clean Water Program.

QUESTIONS

1 Why may an engineer advertise for bids on a public project? On a private project?
2 In what ways other than advertising may an engineer request bids?
3 Name an acceptable media for advertising contracts for (*a*) construction, (*b*) machinery.

4 List the kinds of data which should be given in the advertisement for a junior high school. Assume all necessary information.

5 Why should the advertisement specify the procedure to be followed (*a*) for receipt of proposals, (*b*) for opening proposals, (*c*) for awarding the contract?

6 An advertisement for a construction project contained the words "estimated cost is $1,800,000." Is this proper or advisable? Explain.

7 If a project is suddenly canceled after being advertised, what should be done about the request for proposals?

8 Smith's advertisement for bids on the construction of a sewage-disposal plant stated: "$500 per day liquidated damages if not completed in fifteen months after signing of the contract." Reynolds saw Smith, explained the effect upon cost, and secured permission to omit this requirement. Has Smith the right to make this change? What should Smith do about other bidders?

9 Why is it advisable, in a public project, to print a copy of the advertisement in the contract documents?

10 Prepare a list of information which should customarily appear in an advertisement for bids on a typical construction contract, listing items in the preferred order of appearance in the advertisement.

11 Why might a contractor who reads an advertisement of a construction project be interested in and influenced by (*a*) the location of the project, (*b*) the name of the owner, (*c*) the character of the job?

12 What influence may the magnitude of the job have on whether or not prospective bidders will be interested in preparing proposals for performing the work required?

13 Explain the principles to be followed in giving bidders an idea of the character and magnitude of a construction project when advertising.

14 Why state in the advertisement that the contract will be awarded in a specified time after receipt of proposals?

15 Why state in the advertisement that the successful bidder must sign the contract within a specified time after notice of award of the contract?

16 Why might a private owner prefer to negotiate a construction contract directly with one particular contractor instead of advertising for bids?

CHAPTER 10

PROPOSALS

10-1 Significance of a Proposal A *proposal* is an offer made by the bidder to the owner in which the bidder states that he will furnish all materials and perform all work required by the contract documents and that he will do so for the remuneration stated in the proposal. Furthermore, the proposal should be so framed as to constitute an undertaking by the bidder to sign the contract if his proposal is accepted.

A proper proposal and its acceptance together contain the essential elements of a contract:

1 There is an agreement (a "meeting of the minds").
2 The agreement is made between competent parties.
3 The agreement concerns specific subject matter.
4 The parties promise to do certain lawful things.
5 There is a proper consideration.

The preparation of a proposal by contractors is a very important matter, and they should have a great deal of information before bidding upon any major contract. If any of the data given on the design drawings and in the specifications are not clear, definite, and sufficiently comprehensive, the bidders may be unable to estimate the cost properly. Therefore, they may endeavor to protect themselves against uncertainties by entering bids on the high side. Any such maneuver means that the owner will probably end up paying more than he should for the job—even though he may not immediately realize his misfortune in this regard.[1]

---

[1] Uncertain or inadequate information supplied to bidders carries a further drawback in that it increases the likelihood of subsequent disputes between the contractor and the engineer, with the owner frequently being the one to suffer.

From the contractor's point of view, if his estimate is poorly made and if his judgment regarding the character, scope, and conditions of the project is faulty and leads to a proposal which proves too low, he may suffer serious consequences.

It takes time and costs money for a bidder to prepare a firm and proper proposal. Sometimes the owner, once he decides to go ahead with a job, wants everything done with maximum speed, even to the preparation of proposals and the letting of the contract. It is unwise for the engineer to go along with this urgency and to be unrealistic in setting the date for the receipt of proposals. He should allow adequate time for the prospective bidders to study the work at hand and to make suitable, accurate estimates of the cost. Permitting an unduly limited period for this study and calculation work will probably cause bidders to increase their quotations in order to cover possible expenses which they have not had time to estimate properly.

10-2 Purpose of Proposal Forms   In construction contracts it is practically essential to provide all bidders with the same proposal form, prepared or selected by the engineer. This practice should assure that each bid is entered on the same basis as every other bid. Preferably, the form should be incorporated in the pamphlet that contains the written portion of the contract documents. Use of the very same form for all quotations will help to avoid uncertainties regarding just what the respective bidders propose to do and will aid the engineer in comparing proposals.

It should be clearly stated in the contract documents that all proposals not only must be set forth in full on the printed form prepared by the engineer but also must conform strictly with the instructions for bidding. Deviations, omissions, and errors may constitute cause for rejection. Included with the form may be any requirements by the engineer for special information—for example, the bidder's available equipment and organization, other projects completed which are comparable to that in prospect, and proposed time schedule for performance of various parts of the job.

10-3 Standard Proposal Forms   Some engineering offices have developed standard proposal forms for general use. If necessary, various items in the form may be modified in accordance with the peculiarities of a given situation.

Standard proposal forms are most widely used in the case of contracts for services and for the purchase of materials, machinery, and some kinds of equipment.[2] However, large construction projects are likely to involve so much specialized work that they may not lend themselves to use of a standardized proposal form.

10-4 Special Forms   Special proposal forms will vary with the type of contract, the character of the work to be done, the materials to be furnished, and a

---

[2] Such forms should be examined carefully before being put to use. This examination is necessary in order to avoid the inclusion of erroneous or irrelevant material and the omission of data essential to the particular contract in contemplation.

host of conditions that apply to each case. Therefore, the preparation of the form will usually require considerable original composition.

In such writing the author should have some other competent person check the final product. It is also desirable for the author to set aside for a few days what has been written, then reexamine it carefully. It is surprising how many times something that has been misstated or overlooked will surface on this second reading.

In the preparation of special proposal forms it may be practicable (1) to use parts from some previous contract, combining them with whatever new material is necessary, or (2) to use the forms from former contracts as a general guide. However, any "cut and paste" method is likely to be hazardous, since the author may be "reading into" the previously prepared writing something that is not there, thinking that such writing conforms to what is intended at present. Slight differences or shades of meaning may go unnoticed. There may be possible interpretations of the previously written material that are far from what is currently intended. Moreover, it may prove difficult to coordinate new material with old and, in the process, to maintain a similar style in the writing.

Before preparing a proposal form, the engineer must determine the type of contract most suitable for the particular job to be done, that is, a lump-sum, unit-price, cost-plus-fixed-fee, or cost-plus-percentage contract.

10-5 Proposal Forms for Lump-Sum Contracts A lump-sum contract is in a relatively simple category as far as the usual form of proposal is concerned. The latter item is basically a statement to be signed by the bidder, indicating that he proposes to complete the prescribed work for the sum of money stipulated by him directly on the form. Just how much more ground is to be covered will vary with the details of the project, with the personal opinions and wishes of the author of the form, and with customs and legal requirements. Of course, the proposal must be clearly tied in with the drawings, specifications, and other contract documents.

The following is intended to be an illustration of a proposal form for a lump-sum contract for a municipal improvement:

CITY OF X-Y-Z, NEW YORK
WAYNE AVENUE BRIDGE

PROPOSAL

INSTRUCTIONS

Parties submitting proposals shall be very careful to follow all the requirements in connection therewith.

The amount of the proposal shall be written in full and then repeated legibly in figures in the space provided below.

If a check is submitted with the proposal, it must be certified; if a bid bond is submitted, it must be approved by the Comptroller and the Corporation Counsel.

Any proposal not complying with all these requirements may be rejected.

DEPARTMENT OF PUBLIC WORKS
ROOM 10, MUNICIPAL BUILDING

March 12, 1984

PROPOSAL FORM

To the Commissioner of Public Works:

I, the undersigned, hereby declare to furnish all the labor and materials necessary in connection with the performance of the following named work or improvement: FOR THE CONSTRUCTION OF A BRIDGE OVER UNCAS RIVER AT WAYNE AVENUE FOR THE SUM OF _____ ($_____) and to receive in payment therefor such vouchers on the City Treasurer as the laws and regulations of the City provide.

I hereby declare that I have carefully examined the plans, specifications, and contract papers on file in the office of the Commissioner of Public Works of X-Y-Z, New York; that I will provide all necessary tools and apparatus, do all the work, furnish all the materials, and do everything required to perform the above-mentioned work or improvement in strict accordance with the plans, specifications, and contract papers.

Accompanying this proposal is a certified check for twenty thousand dollars ($20,000.00), which shall become the property of the Commissioner of Public Works of X-Y-Z, New York; that I will provide all necessary tools and apparatus, do all of the work, furnish all of the materials, and do everything required to perform the above-mentioned work or improvement in strict accordance with the plans, specifications, and contract papers.

Accompanying this proposal is a certified check for twenty thousand dollars ($20,000.00), which shall become the property of the City of X-Y-Z, New York, if I shall fail to execute a contract with, and give bond to, said city, within seven (7) days after the date of a written notice by the Commissioner of Public Works of the City of X-Y-Z, New York, stating that this proposal has been accepted by the Commissioner of Public Works and that said Commissioner is ready to have the said contract signed.

Promptly after signing of the contract I promise to present to the Commissioner of Public Works of the City of X-Y-Z, New York, a surety company performance bond in the penal sum of twenty-five (25%) percent of the amount of this proposal, in conformity with Article 25 of the contract documents.

I further declare that no officer or employee of the City of X-Y-Z, New York, is or has been directly or indirectly interested in this proposal nor in the labor or materials to which it relates nor in any portion of the profits thereof; that said proposal is made and contract will be executed without collusion with any other person or persons presenting any proposal for the said labor and materials; and that said proposal is in all respects fair and just.

_____

_____

Address _____

It might be entirely satisfactory to replace the paragraphs dealing with (1) the examination of data, (2) the certified check, (3) the performance bond, and (4) the employees of the city by a single paragraph that automatically ties the proposal to the remainder of the contract documents (in which such information and instructions are given). This paragraph might be as follows: "The Instructions for Bidders, and all papers and things required by it, by the Form of Contract, by the Specifications, by the Drawings, and by any papers made a part thereof, are a part of this Proposal."

Notice that the sum for which the bidder undertakes to perform the desired work is to be recorded in writing as well as in figures. In case of any discrepancy between the two, the written sum is considered to represent the correct bid.

10-6 Proposal Forms for Unit-Price Contracts In unit-price contracts, the engineer will prepare a schedule of items together with a list of estimated quantities for each item. Such data may be incorporated as part of the proposal form or may be located in a separate schedule to which reference is made in the form itself. The bidder is to insert in the schedule (1) each bid unit price, (2) each bid unit price times the applicable estimated quantity, and (3) the total of such products. The bidder is to provide any additional data required and then to date and sign the form.

The form is similar to that for a lump-sum contract except for the schedule of prices in lieu of simply a total sum. The aggregate of these estimated quantities times the pertinent unit prices is treated as a lump-sum bid in determining the bid price for the contract award purposes. Actual payment will be made by applying the bid unit prices to the computed quantity of each item actually furnished or performed. The computation of such quantities is usually made by both the contractor and the engineer, and both parties are to agree on the correct figures to be used in computing the payment due. If the parties cannot come to an understanding about the quantities, it may be necessary to have an independent calculation made or to restore to arbitration[3] or other means available for settling disputes.

The following represents part of a hypothetical proposal form for a unit-price type of contract:

CONTRACT NHA-12

TO THE A-B-C CONCRETE PRODUCTS CORPORATION:
The undersigned*

Offer to enter into Contract   hereby offers to enter into a Contract with the *A-B-C Concrete Products Corporation* in accordance with the attached form entitled "Form of Contract." The prices inserted by the undersigned in the clause of said Form of Contract entitled "Schedule of Prices for Classified Work" form a part of this Proposal.

---

[3] See Chap. 28, infra.

Warranties and representations   The undersigned hereby makes each and every representation and warranty required to be made by the bidders in said Form of Contract and agrees that the acceptance of this Proposal shall have the effect provided in the Information for Bidders bound herewith, and that effect only.

Papers forming part of Proposal   The Information for Bidders, all papers required by it and submitted herewith, and the Form of Contract and all papers made a part thereof by its terms, are made a part of this Proposal.

Bidder's Office   The undersigned hereby designates Number _____ Street, City of _____, State of _____, as his office to which notices may be delivered or mailed.

The undersigned further agrees to comply with all Contract requirements about conditions of employment and minimum wage rates.

Dated, _____19__

By* _____

_____

_____

_____

* Insert the bidder's name. If the bidder is a corporation, give the state in which incorporated, using the phrase "a corporation organized under the laws of . . . ." If the bidder is a partnership, give the names of all partners, using also the phrase "copartners doing business under the firm name of . . . ." If the bidder is an individual using a trade name, give the individual's name, using also the phrase "an individual doing business under the trade name of . . . ."

In order to make certain that a proposal signed by a corporate officer will properly bind the corporation, it may be desirable to insert in the proposal form (following such signature) a paragraph somewhat as follows:

CERTIFICATE OF AUTHORITY, IF BIDDER IS A CORPORATION

I, the undersigned, as Secretary of the corporation submitting the foregoing Proposal, hereby certify that, under and pursuant to the bylaws and resolutions of said corporation, each officer who has signed said Proposal on behalf of the corporation is fully and completely authorized so to do.
(SEAL)

In one contract for a portion of the New Jersey approach to the Lincoln Tunnel at New York City, there were fifty-four classified items in the schedule of prices. Many of them contained detailed descriptions which were necessarily lengthy and which had to be prepared with great care. Careful instructions should be given the bidders, and the several items to be priced should be adequately outlined, as in this sample:

PRICES AND PAYMENTS
Schedule of Prices for Classified Work

Prices cover all costs   The following Schedule does not constitute a complete outline of the work to be performed by the Contractor in accordance with the requirements of the Contract Drawings and Specifications in their present form but is merely a list of the items of work to which unit prices are to be applied in computing the Contractor's compensation. It contains all such items, and the compensation computed therefrom is to be the full compensation for all work and materials whatsoever required by the Contract Drawings and Specifications in their present form, whether or not indicated on said Schedule or pertaining to the items of work shown thereof.

Quantities not guaranteed   The quantities shown in the following schedule are estimated quantities and are given solely for the purpose of facilitating the comparison of Proposals. The Owner shall not be held responsible if the stated quantities are not even approximately correct. All computations of the Contractor's compensation shall be based upon the quantities of work actually performed, whether greater or less than the estimated quantities.

Writing controls over figures   In case of discrepancy between the prices stated in writing and those stated in figures, the writing shall control.

Computation of payment   In computing the Contractor's compensation, none of the prices stated in the Schedule of Prices for Classified Work will be applied to any work except that expressly provided therein. Furthermore, in all cases where prices are quoted upon materials, they will be applied only to materials installed by the Contractor and forming a part of the permanent construction.

*If the proposal is signed by an officer or agent, give his title and address.

Application to Construction Contracts

LINCOLN TUNNEL
NEW YORK CITY
NEW JERSEY APPROACH

SCHEDULE

Item	Estimated quantities	Items of work with prices written	Dollar amounts
1	7500 cu yd	ROCK EXCAVATION PER CUBIC YARD	

_____ Dollars

_____ Cents

Payment lines for rock excavation are shown on the Contract Drawings. The above unit price will be applied to the volume of rock between the surface of the rock and such payment lines, whether such volume of rock is greater or less than the volume of rock actually excavated.

In cases where the Engineer hereafter orders rock excavation in addition to that required by the Contract Drawings and Specifications, the above unit price will be applied only to the volume of rock between the surface of rock and payment lines established six (6) inches outside of and parallel to the net lines established for such additional excavation, whether such volume of rock is greater or less than the volume of rock actually excavated.

As used herein, "surface of rock" denotes the surface of rock exposed after excavation of the earth above the rock has been completed, and "rock" means any solid material which normally cannot be removed by hand or mechanical means and the removal of which would normally require both drilling and blasting.

2	20,000 cu yd	EARTH EXCAVATION
		PER CUBIC YARD

_____ Dollars
_____ Cents

Payment lines for earth excavation are shown on the Contract Drawings. The above unit price will be applied to the volume of earth between the surface of the ground and such payment lines, whether such volume of earth is greater or less than the volume of earth actually excavated.

In cases where the Engineer hereafter orders earth excavation in addition to that required by the Contract Drawings and Specifications, the above unit price will be applied only to the volume of earth between the surface of the ground and the net lines established for such additional excavation, even though a greater volume is actually excavated.

As used herein, "surface of the ground" denotes the natural top surface of the earth prior to the commencement of excavation operations, and "earth" means all materials lying below the surface of the ground other than "rock" as defined in Item 1.

3	12,500 cu yd	CONCRETE, PER CUBIC YARD

_____ Dollars
_____ Cents

The above unit price will be applied to the volume computed from the Contract Drawings in accordance with the lines of the structure and/ or the payment lines for concrete shown thereon.

No deductions will be made for reinforcing bars or any rock left projecting within the payment lines, but deductions will be made for the volume of embedded pipes, ducts, and conduits two (2) inches or more in diameter.

10	9600 sq yd	CONCRETE BASE (FOR PAVEMENT), ON SUBGRADE, PER SQUARE YARD

_____ Dollars
_____ Cents

In applying the above unit price, no deductions will be made for the surface area of manholes, catch basins, or other similar structures.

19	1,800,000 lb	REINFORCING BARS, PER POUND

_____ Cents

The above unit price will be applied to the weight of reinforcing bars (computed from the bar lists on the approved working drawings) used in connection with the materials to which the unit prices quoted under Items 3, 7, 8, 9, 10, and 11 pertain. It will be applied in addition to such prices but will not be applied to any other reinforcing bars, nor to the weight of form ties, spacers, chairs, clips, wires, or other devices for holding and supporting the reinforcement or forms.

It is entirely practicable for the engineer to prescribe that some item or items of a unit-price contract are to be bid as an individual lump-sum. Such things as clearing and grubbing of the site, cleaning up and landscaping after other work is finished, interior decorating, and installation of a septic tank and tile field do not lend themselves to the devising of satisfactory units of measurement; any selected payment items are likely to be either too general or too excessively detailed. However, an experienced contractor can usually estimate the cost of such work fairly accurately, and the lump-sum bid on the item stands as a final figure for all the work of that sort which turns out to be involved.

The list of unit prices includes everything that is to be done or furnished under the contract. Many miscellaneous jobs have to be performed by the contractor; payment for these is supposed to be made "indirectly," as will be explained in Art. 10-13. Payment to the contractor will be made only upon the basis of the stated items, the bid unit prices, and the quantities as finally approved.[4]

The engineer should be careful to see that the selection of payment items for inclusion in a unit-price contract represents a wise choice. The items should be definite, readily measurable, of customary type, and of sufficient scope to cover as much as is reasonably possible of the total work required.

10-7　Proposal Form for Cost-Plus Contracts　The proposal form for either a cost-plus-percentage or a cost-plus-fixed-fee contract is similar to that for a lump-sum contract. The form should clarify just what the contractor is to do; for instance, furnish all labor, equipment, material, transportation, and power. Also, perhaps, it should indicate the staff he is to provide, the length of time he is to be in charge, the nature and extent of his authority, and the point at which his services are to terminate. When the bidder signs this form with all the necessary data recorded thereon, he has executed a proposal giving his idea of proper compensation expressed either as a specific percentage or as a lump-sum figure.[5]

Naturally, the contract documents that by reference are made a part of the proposal will cover many details essential to the job. In the case of a cost-plus-percentage contract, the engineer should be careful to clarify what will be included in the cost of the project. Land, easements, taxes, insurance, housing, and the salaries of the owner's supervisory and engineering force are items for which cost responsibility is likely to be regarded as the owner's, and they would consequently be excluded from the list of costs to which the agreed percentage is to be applied.

The contractor, in addition to recouping his costs (as distinguished from those of the owner absorbed from the outset), receives as compensation the "fee" included in his bid, when the contract is on a cost-plus-fixed-fee basis, or the agreed-upon percentage of the final "contractor's cost" total, when the job is under a cost-plus-percentage arrangement.

---

[4] In addition, perhaps, to a lump-sum item or two as above described.

[5] The proposal for the rendering of engineering services in designing a project or in supervising its construction may be merely a letter stating what percentage of the construction cost or what lump sum will be satisfactory to the engineer. No special proposal form is then necessary.

10-8 Lowest Responsible Bidder It is advisable to state in the contract papers that, if awarded at all, the contract will go to the lowest *responsible* bidder. Of course, the owner wants to get the job completed as economically as possible, but this may not necessarily mean awarding the contract to the *lowest* bidder.[6] There are other considerations, such as:

1 The lowest bidder may not have sufficient finances to handle the project. It may therefore be cheaper in the long run to award the contract to the next-higher bidder rather than to chance the trouble and loss for the owner which might ensue if the financially embarrassed low bidder were to get the award and then, by reason of inadequate cash reserves, be unable to finish the work. Even though the posting of a proper bond by the contractor affords the owner a considerable measure of protection[7] (as will be explained in Chapter 28), it is desirable to require the bidder to present with the proposal a statement of financial condition.

2 The lowest bidder may not have sufficient experience in the particular kinds of work involved. The record of past performance is perhaps the best (but not infallible) guide on that score. Limitations therein revealed need not mean the bidder is incapable of accomplishing something previously undertaken. However, it may be reasonable, for example, to question the ability of the bidder to perform a $500,000 contract when the largest one that was previously handled was in the neighborhood of $50,000.

3 The lowest bidder may have an unsatisfactory reputation—at least in the opinion of the engineer. Thus, the contractor in question may have in the past been guilty of careless work, of always maneuvering for extras, of being uncooperative, or of making it necessary for the engineer to inspect the construction work with extreme care. Although rejection on this rather nebulous ground of "reputation" may lead to many disputes, it remains a proper course of action when circumstances call for it.

4 The lowest bidder's staff and equipment may be inadequate. This, too, is difficult to establish. The engineer may ask the bidder to submit an itemized list of his proposed staff and of the equipment he intends using. Justification for such a move is obvious, for it would be most unwise to award to a small building contractor with limited resources the contract for the construction of a large and difficult harbor improvement. Though his bid may be in perfectly good faith, such a contractor is not really equipped to do the job involved, and acceptance of his proposal would likely result in misfortune for him as well as for all others concerned.

10-9 Prequalification of Bidders To avoid the possibility of having to reject a proposal for reasons such as those just described some engineers have adopted the procedure of *prequalification of bidders*. To accomplish this, an individual or group

---

[6] The engineer should be very careful to make sure that rejecting the low proposal is objectively justifiable—that it is not based on personal feelings alone.

[7] There are holdings to the effect that a provision in the advertisement for tenders (stating that the bidders will be required to give security for the performance of the work) may be waived by the owner, and a valid contract will be created by the acceptance of a proposal, although no security is in fact given by the successful bidder.

is empowered to determine whether or not particular general contractors are qualified to handle contracts of a certain size and type.[8] Basing its action upon the results of such research, a state highway department, for example, may prepare a list of contractors who are approved as prospective bidders on its future projects. Compiling such a list is not infrequently a difficult task because (1) omission of a particular contractor is likely to be criticized as unfair; (2) political pressure may be exerted in favor of one or more concerns; (3) a contractor who can readily handle one kind of work might not be equipped for work of other character; (4) a contractor may be able to perform a $50,000 contract but not one of many times that magnitude; (5) a contractor who is not able to qualify at one time may be able to meet all requirements later, or vice versa; and (6) the ability to accomplsih contract work successfully is not always easily determined.

As a general proposition, it seems preferable to disallow unqualified contractors to bid at all—rather than to refuse to award the contract after they have gone to the trouble and expense of putting together proposals. In addition, the prequalification procedure may prevent some builders from becoming saddled with a contract which, because of its scope and complexity, would likely prove disadvantageous for them, notwithstanding their personal beliefs to the contrary.

Private owners are able to use this principle of selection simply by sending inquiries and date to only those whose qualifications have received advance approval. Likewise, governmental bodies (such as certain highway and public works departments) may, unless there are statutes forbidding it, use this method to advantage, though claims of partiality may make its use difficult on occasion.

Prequalification of subcontractors may also be practiced, and if it is, the general contractors should be given the approved list well before bidding, since this information will affect their advance arrangements with prospective subcontractors.[9]

10-10   Deposit for Securing Plans   The bidders will be informed as to how, where, from whom, and for what price copies of the plans, specifications, and other contract documents may be maintained. It is customary to require prospective bidders to make a cash deposit for these papers, the charge representing the estimated cost of printing or an amount that will preclude requests from those who are not seriously interested in bidding. Payment usually must accompany the request for the papers. It is also customary to state that all the deposit (or some stipulated portion of it) will be refunded upon the return of the papers in good condition. Sometimes the refund is limited to those who not only return the papers but actually submit bids as well.

10-11   Miscellaneous Items for Information of Bidders   Some information to be given bidders includes:

---

[8] The investigators will generally ask the prospective bidders for the same sort of financial data and performance records as they would supply were they submitting a proposal on the actual job. The details, of course, will vary with the circumstances.

[9] In every case the owner should retain the right to approve or reject any subcontractor.

1 Familiarity with the work. As stated elsewhere, it should be made clear that, upon signing the proposal, the bidder thereby gives assurance that he has familiarized himself with (*a*) the conditions at the site, (*b*) the requirements of the contract, (*c*) the character of the work; and (*d*) the content of the plans and specifications.

2 The size of the "bid" and "performance" bonds, and any "labor and materials" bond required.

3 The amount of time allowed between the opening of the proposals and the award of the contract. This interval is ordinarily from one to four weeks. The purpose of limiting the interval is to avoid undue delay in signing the contract. If the owner decides to reject all proposals, such action should be taken within this one- to four-week period.

4 The time limit between (*a*) written notice by the owner to the contractor that the latter's proposal has been accepted and (*b*) signing of the contract. A reasonable period may be seven days.

5 Important dates. It is generally desirable to set a date for the start and for the completion of the contract work.

6 Liquidated damages. Data are needed regarding any liquidated damages which will apply in the event of failure to complete the work on schedule. Conversely, any bonus for particularly rapid completion of the work should be mentioned.

7 Taxes. There may be special taxes, license, or permit fees which the contractor must pay. If so, this should be made clear. On the other hand, there may be work which, because of its nature, will be exempt from certain taxes.[10]

8 Corporate charter. In the case of corporate bidders, it may be desirable to require that a copy of the corporation's charter (or other proof of its right to conduct business and perform the work indicated) be furnished the owner.

9 Labor. Any special requirements affecting labor should be given.[11] Thus, it may be required in some places that the rates of pay for labor be published in the instructions for bidders. Likewise set forth should be any consideration that would compel employment of a particular type of labor.

10-12 Preparation of Bid on Cost-Plus Contract Engineers should understand how the bidder arrives at the figures appearing in the final proposal. The determination of a proper fee or percentage for a cost-plus contract is largely a matter of judgment on the part of the contractor and the engineer. They have to

---

[10] Here, for example, is a clause from the instructions for bidders who were estimating the cost of an extension to a hospital:

"Sales Tax:

The Institution is not subject to the State Sales Tax, and the Contractor shall take advantage of this in his bidding."

[11] There may be a stipulation that at least a specified percentage of the workmen on the construction job be residents of the jurisdiction in which the work be performed. On one project in Chile it was necessary for the contractor to obtain 90 percent of his labor from local sources.

consider the particular problems to be met. These are so special in many cases that it is impractical to attempt to state general rules to govern the proper extent of the contractor's compensation. However, in emergency jobs for the federal government, it is probable that precedents have been established which will afford a worthwhile guide. In some instances involving public work there may be laws or "recommended fees" that govern the compensation.

Large private organizations sometimes have a particular contractor to whom they award contracts on a cost-plus-percentage basis in view of excellent work such contractor has done for them in the past. They believe that, in the long run, this is the most advantageous way for them to handle things.

10-13 Preparation of Lump-Sum Bid by a Contractor A logical bid for a lump-sum contract for the purchase of materials is generally based largely upon component costs and competitive factors. In many instances the prices of common materials and products are somewhat standardized or at least known to lie within a narrow range at any particular time. Prices of some articles are given in catalogs or on price lists—with or without discounts. In such cases, the total cost of a product can be estimated with reasonable accuracy, and bids are not expected to vary widely one from another.

Making a lump-sum proposal on a large construction project is a different matter. Here bidders have to estimate various contributing costs carefully since the work entailed requires the use of many different materials, the labor of large numbers of employees, and the performance of a wide variety of operations. Estimating the probable overall costs of such work requires considerable time and expense.

Bidders must estimate the quantities of materials involved, then compute their probable cost. Next they must calculate the approximate cost of all labor that will be required to perform the work. The drawings and other contract documents should give a clear picture of the project itself.

Bidders also have to include many items of cost besides those that are directly associated with materials and labor involved in the construction. Here are some of the other items the expense of which will have to be considered in the estimate:

1 *Supervisory and office staff* in the field. This includes foremen, bookkeepers, and secretaries. In this item there may also be included such things as office supplies, blueprinting, and postage.

2 *Engineering.* Sometimes this may be a considerable item if the contractor's staff is to make detail drawings; to design special features like cofferdams, shoring, and heavy formwork; and to do the surveying to establish lines and grades.

3 *Rentals or construction of* office space, field headquarters, and sometimes housing—if the contractor has to provide temporary living accommodations for some of his employees.

4 *Equipment.* This includes depreciation and repair of the contractor's own equipment, possibly a portion of the cost of new equipment (either additional or to replace that which wears out on the job), the cost of tools that may be lost or damaged, and the rental of machines leased from someone else.

5 *Storage facilities.* Sometimes this includes a storage yard and temporary warehousing at a shipping terminal as well as tool houses and storage facilities at the jobsite.

6 *Power and means of* getting it to the places where it is needed. This may include power lines or small generating plants and the supplying of heat, steam, and compressed air.

7 *Transportation of all materials and personnel.* This may include the building of access roads, transportation of workers to and from their living quarters, demurrage on freight cars not unloaded promptly, and the cost of transfer from one means of transportation to another.

8 *Contingencies.* This topic covers the miscellaneous odds and ends which the contractor may have to take care of and for which he may not have made specific allowance in his estimate. For example, if delays, construction hazards, and liquidated damages are likely, this contingency item is designed to "reimburse" the bidder should these extra expenditures materialize. This item is often estimated as a percentage of the cost of the project or of direct materials and labor; it may range from 5 to 15 percent of this sum, or it may be a lesser percentage if the bidder is convinced that the work is shown completely and clearly on the drawings and in the other contract documents and that no unpleasant surprises are likely to develop.

9 *A portion of the general overhead* of the central or home office. This is because the job will require the services of various persons who are not at the site (or who are there only occasionally) and who may be assigned to the inspection and expediting of materials. "Overhead" might also include a figure to represent the cost of making estimates for *this* job (and perhaps a little extra for other bids that were or will be made by the contractor).

10 *Profit.* This item will vary in accordance with what the contractor thinks is reasonable and can be realized without losing the job.

11 *Interest on money.* The contractor may have to borrow money to finance his operations. If so, he will probably reflect in his bid the estimated cost of interest on loans made to him. He may also wish to take into consideration what his own funds tied up in the job might earn if they were drawing interest at current rates.

12 *Workers' compensation coverage.*

13 *Other insurance on property and personnel.* This item includes, among other things, fire and public liability insurance protection.

14 *Bonds.* This refers to the cost of procuring the bid bond, the performance bond, and any other bonds requried.

15 *Taxes.* In the picture may be sales taxes and income taxes with which the contractor must contend.

16 *Allowances for estimated increases in prices* with the passage of time. This point may be governed by the anticipated duration of the job. The best way to avoid losses caused by rising costs of material and labor is by means of an *escala-*

tor clause,[12] which obligates the owner to pay for these increases occurring during the life of the contract; this is fair because if no increases occur there is no extra charge against the owner. In the absence of an escalator provision in the contract, the bidder must protect himself (against a possible rise in costs) by adding something to his basic figures.

17 *Special expenses.* These may include payments for a temporary right-of-way, the cost of permits of various kinds, fees on the use of patents, fees for consultants, legal expenses, and commissions to agents who may perform certain duties. If the contractor is to run a cafeteria to provide lunches for employees, any expense connected with that operation will be taken into account.

When the estimated costs of all the miscellaneous items are added to those for direct labor and materials, the sum is the total estimated expense. This may or may not turn out to be the actual bid price. If, in the contractor's opinion, the total estimated cost is higher than the sum he thinks others will bid on the job, he may *reduce* his own bid figure somewhat, hoping that he can still make a profit by doing the work more economically than the estimated costs indicate. If he is badly in need of work, he may even seek the contract without allowing for any real profit at all.

Here is an example of such bidding. In a period of depression a steel company had very little work at hand. The drafting room was almost completely out of work when along came an inquiry for a bid on a contract involving approximately 10,000 tons of structural steel. The company estimated the cost, then deliberately made a bid that was considerably below this cost. The contract was secured. The officials believed that this was good business; although a moderate loss could be expected, they did not want to be paying their employees for no work at all, nor did they want to lay off most of their workers. Keeping them busy seemed a good temporary expedient.

It may be, on the other hand, that the contractor believes the estimated costs are too low. He may then *round up* the bid price to what he thinks is adequate. If he is already swamped with other work, he may decide that he will not take any serious chances but will set a price such that, if he gets the job, he will be enabled to augment his staff and organization and still make a profit. If he thinks that there will be little competition for the work, he may likewise boost his figures a little above the estimates, depending upon what he thinks the trade will bear.

It is important to recognize the difference between an *estimate* and a *proposal.* An *estimate* is not assumed to be an accurate statement of the final cost. It is generally based upon information that is not complete and exact, and it involves the making of assumptions. By nature it is an approximation. A bidder estimates cost and will use these or a modification of same as the basis for the *proposal* on the job. The proposal becomes a binding contract when accepted, even though the

---

[12] An escalator clause usually applies in one direction only; that is, it provides for an increase in cost but seldom is concerned with a decrease in cost. Such a clause is most often appropriate in contracts which may require substantial time to complete.

figure quoted does not purport to be an exact mathematical computation of the real cost.

10-14 Preparation of a Unit-Price Bid by a Contractor When a contractor is trying to estimate costs on a contract for which a unit-price proposal is required, he will proceed at first substantially as he would if it were a lump-sum contract. However, he may not have to be so careful in estimating the quantities of material and labor involved. He will investigate carefully such things as the nature of the work, the various construction difficulties that may be anticipated, the expected duration of the job, the availability of materials, and the quality of workmanship needed. He will thus arrive at an estimate of the total cost of the project, including his profit factor.

As shown in Art. 10-6, the list of items in the proposal form usually states the estimated quantities and the units of measurement. The contractor will therefore distribute the estimated total cost of the project among these various items, adjusting the figures until the sum of the bids on all items substantially equals the aggregate number of dollars previously reckoned to be the cost of the job as a whole.[13]

As an illustration of this procedure, assume that the following items are given in the proposal form for a construction contract pertaining to a highway and bridge project:

Rock excavation	2,000 cu yd
Earth excavation and fill	5,000 cu yd
Concrete	3,000 cu yd
Reinforcing steel	300,000 lb
Membrane waterproofing	600 sq yd
Structural steel	800,000 lb

Assume further that the contractor estimates that the total bid price with profit included should be about $850,000. Now he may make a trial summary as follows:

Rock excavation	2,000	cu yd at	$ 20.00	$ 40,000
Earth excavation and fill	5,000	cu yd at	5.00	25,000
Concrete	3,000	cu yd at	100.00	300,000
Reinforcing steel	300,000	lb at	.45	135,000
Membrane waterproofing	600	sq yd at	5.00	3,000
Structural steel	800,000	lb at	.40	320,000
Total				$823,000

[13] Or the contractor may estimate the cost of labor, materials, overhead, profit, etc., for each main item of work if this method seems to be desirable.

This is $27,000 less than the desired total. The deficiency has to be made up by various adjustments. The contractor may add to some of the unit prices as follows:

Rock excavation	add $2.00 per cu yd	2,000 × 2 = 4,000
Earth excavation and fill	add $1.00 per cu yd	5,000 × 1 = 5,000
Concrete	add $6.00 per cu yd	3,000 × 6 = 18,000
Total		$27,000

Therefore, the final figures recorded on the proposal form may read:

Rock excavation	2,000	cu yd at	$22.00	$ 44,000
Earth excavation and fill	5,000	cu yd at	6.00	30,000
Concrete	3,000	cu yd at	106.00	318,000
Reinforcing steel	300,000	lb at	.45	135,000
Membrane waterproofing	600	sq yd at	5.00	3,000
Structural steel	800,000	lb at	.40	320,000
Total				$850,000

Notice that in making the upward adjustments the contractor has added relatively more to the items of work that will be performed early in the course of the contract than those that will be done later. This is called *making an unbalanced bid.* By so doing, partial payments made in the first few months will yield a generous enough return so that less of the contractor's cash will be needed to finance the work.

Another type of unbalanced bid may be arrived at by bidding low on an item that the contractor thinks will underrun the quantities listed in the proposal form and by making a high bid on something that is likely to overrun[14] the stated quantity. Remember that the contractor will be paid the bid price per unit times the number of units actually performed or furnished. Reasonable "unbalancing" is perfectly proper, but it represents something of a gamble by the contractor in most cases.

In one contract that involved a large amount of both earth and rock excavation the borings had been made about 150 feet on centers. The site was in territory where the rock surfaces often varied greatly in elevation within short distances.

---

[14] The word "overrun," as used here, means that the actual quantity work to be done under some particular item exceeds the quantity stated by the engineer in the list of items; of course, "underrun" means just the opposite. For example, the item for earth excavation in the preceding list states that the unit price is to be applied to 5000 cubic yards of material when the amount of the contractor's bid is figured. If the contractor thinks that 6000 cubic yards will actually prove to be necessary, he believes that the work required will overrun the estimated quantity.

One contractor realized this and asked permission to take additional borings at his own expense. This was granted, and the information that he obtained was naturally his own to use. Actually, he found that his own estimates, based on the more complete information, showed much more earth excavation and less rock excavation than stated in the proposal form. He therefore bid a little high on the unit price for the earth excavation and a little low on the rock excavation. He was awarded the contract, and turned a good profit when the final computed quantities of earth and rock turned out to be substantially as he had estimated.

Sometimes the quantity of work and the difficulties to be surmounted in a particular case are uncertain. The owner wants to have a unit-price contract, but also wishes to have a stated limit for the total expenditure. Thus the owner agrees to pay the costs as determined by the unit prices up to this limit. Suppose that the contractor states that the total cost will not exceed $100,000, and the owner agrees to this. Then this limit is binding on the contractor. However, if the latter states that the estimated maximum cost will be $100,000, those words "estimated maximum cost" merely mean an approximation, and the contractor is not bound to the stated figure.[15]

10-15 Improper Proposals If a proposal is not prepared in the manner required, this is proper cause for its rejection. In one case, two steel companies presented nearly equal bids on a contract. The unit prices were supposed to be based upon the "scale weights" of fabricated members—that is, actual weights as the material was loaded on the cars. One contractor added to his proposal a paragraph stating that his bid was based on the "calculated weights"—computed from the detail drawings without allowance for beveled cuts, copes, and other reductions from the overall size of each piece as it was constituted before any cutting.[16] This particular contractor's unit price was slightly under that of the other bidder. However, he had failed to quote on the basis stipulated in the invitation for bids, and the contract was let to the company whose bid was in proper form.[17]

In another case, a contract called for bids on a project which involved pouring large quantities of concrete during the winter season. The papers stated that enclosures and heating would be required to protect the concrete during setting and curing in freezing weather. The lowest bidder added a paragraph stating that the quotation submitted excluded the cost of this protection, and the proposal was rejected because it deviated from the prescribed pattern. If in the case just cited the engineer had accepted the lowest bid, such action would have been binding on the basis of the terms stated in that proposal, because the bidder had made an

---

[15] *J.E. Hathman, Inc. v. Sigma Alpha Epsilon Club,* 491 S.W.2d 261 (1973).

[16] The contract involved considerable steelwork as to which the calculated weight would considerably exceed the scale weight.

[17] It may be that a proposal form will contain a statement to the effect that the owner reserves the right to waive any informality or irregularity in the bids received. This provision is intended to be utilized when the bidder makes some minor mistake or is guilty of an oversight which does not materially affect the actual bid.

offer and the engineer had accepted it. On the other hand, such action by the engineer would have been unfair to the other bidders. They, too, might have submitted lower figures if they had realized that they could violate the specification with respect to the cold-weather protection.

10-16 **Alternates** In some cases proposals may be requested on the basis of two or more alternatives. One purpose for this securing of bid prices on each of several possible arrangements is to make possible a broad choice after the costs have been set forth in the various bids. A second objective may be that of obtaining bids on a basic contract and, with them, securing additional proposals on special items that the owner may or may not decide to add to such basic contract.

The proposals for the construction of the suspension system of the George Washington Bridge will illustrate one function of "alternates." The Port of New York Authority made two different designs—one for wire cables and another for eyebar chains. This bridge was to have a main span that would be twice as long as the then longest existing suspension structure—the Philadelphia-Camden Bridge. There were many arguments among engineers for and against each of the two designs. It was ultimately decided that complete designs and specifications would be prepared for both types and that the matter should be settled on the basis of actual bid prices—or, at least, the decision would be made after the costs became known. Roebling's bid on the wire-cable design was lower than any other of its kind and lower than any on the eyebar-chain proposition. The wire-cable design was adopted, and the contract was awarded accordingly.

The use of alternates for additions or subtractions may be illustrated in this way: Assume that a municipality desires lump-sum proposals on the construction of a large school building for which the appropriation is to be a specific and limited amount. The city wants to have the building constructed in accordance with at least the minimum plan. The estimated cost (as prepared by the engineer) of this minimum structure is well within the appropriation. However, the authorities would like, if financially possible, to include an extension of the wing that is to house shops and manual-training facilities so as to increase their size. Furthermore, although space is provided for a swimming pool in the minimum plan, the pool itself and the locker rooms and equipment are not included in the basic proposition. These items are all desired if their costs will permit their inclusion. The engineer or architect may ask for lump-sum proposals as follows:

1 The basic plan as shown
2 The extension of the shop wing as one alternate (addition)
3 The swimming pool as a second (separate) alternate

Assume that the bids received are as shown in Table 10-1 and that the planned appropriation is $2,100,000. All the bids on item 1 are less than this sum, and A is the lowest bidder. For items 1 and 2 together, B is the lowest bidder, and three of the contractors' bids are within the limit. For items 1 and 3, C is the lowest, and the same three proposals are again under the appropriation figure. However, for

TABLE 10-1
ASSUMED BIDS ON A LARGE SCHOOL

Item No.	Contractor			
	A	**B**	**C**	**D**
1	$1,950,000	$1,965,000	$1,970,000	$2,050,000
2	125,000	100,000	120,000	95,000
3	100,000	95,000	60,000	75,000
1+2	2,075,000	2,065,000	2,090,000	2,145,000
1+3	2,050,000	2,060,000	2,030,000	2,125,000
1+2+3	2,175,000	2,160,000	2,150,000	2,220,000

items 1, 2, and 3 combined, *C* is low, but all bids are beyond the limit of the available funds.[18]

Now suppose that the choice of alternates is open so that the city officials can decide whether to go ahead on the basis of item 1, items 1 and 2, or items 1 and 3. In the first case *A* would presumably get the contract; in the second instance, the award would go to *B*; in the third, *C* would be successful. The combinations of items 1, 2, and 3 is automatically eliminated because of prohibitive cost.

It is obvious that if there are several alternative plans available, it may be possible to award the contract to some preferred builder merely by selecting the arrangement on which that total bid was the low one. Operating in that fashion might well lead to abuse, or at least to accusations of unfairness or partiality. It is therefore desirable to determine in advance, and to state in the proposal form, the order of acceptance of any alternates. In the case of this school, the contract papers might well be so prepared as to call for the basic bid on item 1. Then, if one or more alternates is to be accepted, item 2 is to be first on the list, with item 3 following. With the use of this procedure in the school example given, and with $2,100,000 as the top limit, the authorities must then award the contract to *A* for item 1 alone or to *B* if item 2 is to be included. They have no other choice, assuming both *A* and *B* are responsible contractors.

The same principles apply in case of subtractions. Again, the sequence of the elimination of items should be specified. In the case of this school, the basic design might cover the entire project, and lump-sum bids might be asked for on the complete structure (including pool, etc.). Then the proposal form might call for the statement of a figure by which the contractor would reduce the bid price if the swimming pool and its accessories are eliminated. Still another figure might be requested in order to show the further reduction in the bid price should both the pool and the extension of the shop wing be removed from consideration. Use of

[18] An important item to consider is how much more it would cost to add a portion of the ultimate project later as a second contract than it would to proceed at once with the entire construction. If the total cost of two-stage construction is enough more, the authorities may wish to try immediately to raise funds to complete the structure without delay.

alternates may avoid the difficulty otherwise encountered in revising and readvertising a contract when all bids for the *complete* structure exceed the funds available. The program of reductions, if any, should be clearly described in the proposal form and should be followed in awarding the contract.

10-17   Withdrawal of Proposals   As a general rule, withdrawal of a proposal will not be permitted.[19] In no event should a proposal be withdrawn after the public opening and reading session is under way.

Even where timely withdrawal is allowed, the bidder is expected to reveal an acceptable reason. One might be the discovery of a numerical error in the figures, not to be confused with second thoughts about the wisdom of his proposal. It may be that some major quantity of material or some substantial labor item has been inadvertently omitted from the calculations, or the same item may have been included twice. In these relatively rare instances, the engineer must decide whether or not to allow the affected bidder to rectify the situation either by correcting the proposal or by withdrawing it completely.

The engineer should see to it that every course of action taken is fair to the owner, the bidders, and himself. If the lowest bidder has made a very serious error, one that will obviously cause him substantial financial embarrassment, the engineer should not insist upon punishing him just because a chance to do so presents itself. Fair play in a particular case may demand that the engineer recommend (1) that the contractor be allowed to withdraw the unfortunate proposal, (2) that he enter into a modified agreement satisfactory to him and to the owner, or (3) that his bid be rejected. On the other hand, the mere fact that a contractor makes an error that may cause him to lose some money on the job is not always in and of itself sufficient cause for leniency by the engineer. The decision in any event is to be made by the engineer rather than by the contractor.[20]

In cases where the contractor's calculations have obviously gone astray, the engineer's primary duty is to protect the owner's interests. It may very well not be to the latter's advantage to force the contractor into possible bankruptcy by having him go through with a thoroughly unrealistic proposal and then perhaps end up by having to find someone else to finish the work at considerable additional cost. Often, in such cases, it is preferable to accept the next-higher proposal in the first place and to release the erring low bidder from the potentially disastrous commitment.

Assume that a contractor submits proposals on two different contracts that happen to be scheduled for award at approximately the same date and that he

---

[19] Any prohibition should be publicized in the invitation for bids.

[20] On a large tunnel job, the lowest bid was approximately $1 million (15 percent) *under* both the next-higher proposal and the engineer's estimate. The engineer called in the low bidder and told him that, if he had made an error, he might withdraw his proposal. The contractor stated that his bid was as he intended, and he revealed that it was based upon the use of certain new, patented equipment. The proposal therefore was accepted. In another case, it is reported that a contractor forgot to include a $50,000 stack in his proposal for the construction of an industrial plant. The contractor was not allowed to withdraw his proposal without penalty.

knows he is not equipped to handle both jobs simultaneously. If he proves to be the low bidder on both, his limited capacity will not serve as a sufficient excuse for withdrawing either bid without penalty. He should have been content to take his chances on the acceptance of one proposal or the other rather than bidding on both jobs.

10-18 Opening of Proposals Instructions should be given regarding the procedure to be followed after bids are received. It is desirable to specify that each proposal must be contained in a separate, sealed envelope addressed as directed in the instructions. Each envelope should be clearly labeled "Proposal for Contract." The reason for keeping the contents of each proposal secret is to avoid subsequent claims of collusion; not even the owner should know the contents of any proposal prior to the official opening. At the stated hour the sealed proposals are to be publicly opened and *publicly read aloud.* Reading the proposals in silence, even if done in the presence of others, is not sufficient. Usually a representative of each bidder will be on hand to watch out for his principal's interests. Furthermore, it is desirable to have a disinterested person check the bids as the reading proceeds or promptly after it terminates.

10-19 Rejection of Proposals Information for bidders should state that the owner has the right to reject *any* or *all* proposals. A list of possible reasons for the rejection of a particular proposal was given in connection with the discussion about the lowest responsible bidder. The same objections may apply with respect to the next-higher bidder or to any other one. It may also be desirable to reject any proposal that has not fulfilled all the prescribed requirements and any that seems to contain a flagrant unbalancing of bid unit prices.

The owner should have the right to reject all proposals if the lowest proves so high as to make the project unaffordable. It might happen that sickness, a financial reverse, or some unforeseen happening has occurred and made it inadvisable for the owner to proceed in accordance with the original intention. If so, it is better to abandon the plan than to consummate a contract that is likely to prove unduly burdensome.

Assume that Brown advertised for and received proposals on the modernization of her large warehouse. Before the contract was let she received a very attractive offer from Jones, who wanted to purchase the structure "as is." Being near retirement age, Brown decided to accept this offer and to discontinue her warehousing business. The unexpected opportunity to sell meant that she was no longer interested in the remodeling project; she therefore rejected all the proposals. This example points up the relatively weak position in which bidders find themselves—normally they cannot withdraw their proposals once submitted, whereas the owner is free to drop the entire undertaking if he or she decides to do so.

In government work it is generally required that at least three proposals be received on a project before any can be accepted. This rule is designed to assure what is deemed "adequate competition."

10-20   Readvertisement   If the lowest bid is higher than the amount of money available to pay for the job, the engineer—in the absence of changes in the contract—should not readvertise in the hope of obtaining lower bids the next time around. With the lowest proposal in the original batch known at the time a second set of bids is sought, each bidder is told in effect to beat the figure in order to stand a chance of getting the contract. This "shopping for bids" is a bad practice.

If the first group of proposals is to be rejected and the contract readvertised, the entire design or some major feature thereof should be substantially altered, perhaps by reducing the extent or the quality of the performance required so that the new bids will be submitted on an independent basis.

Assume that all the bids received on a proposed large school building are in the vicinity of $2,500,000 whereas the available funds are limited to $2,000,000; the proposals consequently have to be rejected. Perhaps one entire wing will then be eliminated and the job sent out for new proposals. If the one wing that is to be omitted is about one-fifth of the structure as originally contemplated, the total cost will be proportionately decreased, and the bidders on the smaller structure will be influenced by the knowledge obtained in regard to the former proposals. They are thus given a sort of monetary target at which to aim. It will probably be more desirable not only to omit this wing but also to change the framing to a more economical type, to use cheaper materials, or to make other revisions in materials and workmanship extensive enough so the bidders practically have to reestimate the project in its entirety.

It is possible in such a case to negotiate a revised contract on a lesser structure with the lowest responsible bidder, eliciting thereby a greatly reduced quotation. However, this procedure is somewhat unfair to the other bidders, and the cost to the owner is likely to be greater in the long run than if all bidders are permitted to submit new proposals on the basis of the revised project.

10-21   Awarding the Contract   No particular words are necessary to constitute an acceptance of a bid or tender for building or construction work provided the intention of the owner to accept is made evident. However, it is best to have the form which will be signed by the owner and the contractor, constituting part of the official contract, prepared in advance and included among the contract documents so that all data will be available for inspection before the respective parties sign.

Let us consider the engineer's handling of the proposals. Assume that sealed proposals on a lump-sum contract are duly opened in public and read aloud. The lowest bidder is determined at once, absent the extremely improbable circumstance of a tie. To be investigated, however, is the question of the lowest *responsible* bidder, unless prequalification of bidders has been required, in which latter event all proposals under consideration have been made by acceptable concerns. The engineer should make sure that the successful bid is in order and that the contractor's resources, equipment, experience, and sureties are satisfactory. Thereafter the engineer should give notification that the proposal in question has been accepted and that the contract is to be signed at a specified place and time. This notice should be written and delivered to the successful bidder by messenger, registered mail, or telegram, although proof that it was depositied in an official United States post office or mailbox will suffice.

In the case of a unit-price contract, the public opening and reading aloud of the sealed proposals will similarly reveal who is the low bidder. The procedure for awarding the contract is similar to that used in the case of lump-sum proposals.

QUESTIONS

1 Illustrate a situation in which standard proposal forms might be used advantageously. Explain the advantages of the use of such proposal forms.
2 Why may use of the cut-and-paste method of preparing proposal forms be helpful? When would it be inadvisable?
3 Why is the use of a printed proposal form advisable for construction contracts?
4 Is a printed proposal form essential for the making of a construction contract? Why?
5 Describe how a contractor might go about preparing a bid on a school building to accommodate 400 pupils, assuming that it is to be a lump-sum job. List several of the typical work items involved.
6 For the school mentioned in the preceding question, describe how a contractor might prepare the bid if it were to be on a unit-price basis.
7 Who determines the contingency figure in the cost estimates for a construction project?
8 Who ultimately pays the cost of preparing proposals on a contract in the case of (*a*) the successful bidder? (*b*) the unsuccessful bidders?
9 How does a bidder estimate the profit which she hopes to obtain from the performance of a contract on which she makes a proposal (assuming a unit-price contract)?
10 What are "escalator" clauses as used in connection with proposals on construction contracts? Why may they be used?
11 Does the engineer generally make an estimate of the cost of a project before obtaining proposals? How? Why?
12 What is an unbalanced bid? Why might one be used?
13 On what basis is the successful bidder determined in the case of (*a*) a lump-sum contract? (*b*) a unit-price contract?
14 In what respect may a proposal be improper? What should be done by the engineer in such a situation?
15 Should the engineer specify a time limit after receipt of a proposal before it will cease to be binding on the bidder?
16 When a bidder submits a proposal, is that act something which completes a contract?
17 Who should usually determine the type of contract to be used for a given construction project? On what basis should it be determined?
18 For a unit-price contract: (*a*) How are the quantities which are given in the proposal form determined? (*b*) Are these quantities guaranteed to be accurate? (*c*) What are the stated quantities used for? (*d*) Are these quantities the same as the actual quantities that will probably be used in the job? (*e*) What happens if the stated and the actual quantities differ?
19 How are the quantities determined for a lump-sum contract?
20 Does the fact that a proposal for a construction project is on a unit-price basis mean that the cost of a job to the owner will be less than if the proposal were made on a lump-sum basis? Explain.

21 Can a portion of the work in a unit-price contract be listed as a lump sum? Why? Illustrate.

22 If the plans and specifications are vague and incomplete, what can bidders do to protect themselves?

23 How can a bidder determine the percentage or fixed fee to charge in connection with a cost-plus contract?

24 Explain how and when alternatives might be used when one is seeking proposals on the construction of a large building project.

25 Is it wise to have alternates applied to a main contract as additions? To have them applied as subtractions?

26 Why should proposals for a construction contract be sealed when submitted to the owner?

27 Why forbid withdrawal of any submitted proposal before actual opening of all proposals?

28 What should be done if a bidder changes any portion of the proposal form?

29 Jones's proposal for a building contract was $125,500. After finding that her bid was low, she checked her estimates and found that an item of $25,000 had been incorrectly recorded as $2500 in making the summary of costs. What can she do about the situation? Explain.

30 What is meant by "lowest responsible bidder" in connection with construction contracts?

31 Under what conditions may the lowest bid be rejected (*a*) in private work? (*b*) in public work?

32 Davis's bid of $542,000 was the lowest on a building job. The cost of the largest construction contract which he had performed previously was $78,000. What should the engineer do to determine whether or not to let the contract to Davis?

33 If, in the case described above, the advertisement says, "The contract will be let to the lowest bidder," must this be done?

34 Explain possible procedures for and advantages of prequalification of bidders.

35 Why and under what conditions should the advertisement state that the owner has the right to reject any or all proposals?

CHAPTER 11

CONTRACT CLAUSES— GENERAL CONDITIONS

11-1 Introduction The preparation of the contract clauses entails a great deal of painstaking effort. The arrangement and content of the material will depend on the preferences of the writer and the character of the project; the number of clauses to be prepared will also vary with the scope and complexity of the work to be done. Brevity is desirable—but it should not be attained at the expense of adequate coverage of the essential material.

A first step in the writing of the contract clauses is the preparation of an outline of points to be covered. The author must then organize these points into some logical arrangement. One way to do this is to prepare a table of contents which will become an essential part of the contract documents. It is also desirable to establish a numbering system for sections and paragraphs in order to have the benefit of the numbers in referring to various items.

This chapter and the next have been written for the purpose of showing many of the contract clauses that should be included in a large construction contract. Others will be needed for particular projects, but those given in these two chapters will illustrate most of the general principles involved and the kind of information that has to go into such clauses.

In the preparation of these papers the engineer should have the help and guidance of competent legal advisors. In large organizations it may be that he or she is to assist the attorneys, who will take the lead in compiling the material.

In studying these clauses, remember that a contract is binding in both directions. Although we generally think of the contract clauses and specifications as a means of binding the contractor, they also serve as a vehicle whereby the contractor can hold the owner and the engineer to their side of the bargain.

11-2 Purpose of Written Contract Papers It should be emphasized that a construction contract contains so many important general and technical features that it is essential to have all points clearly shown and described in writing or on drawings. All parties concerned will thereby have the same data, and all future interpretations will be based upon the given information. In this way misunderstandings are minimized, since no one has to depend upon what was thought to be said or implied.

Perhaps the reader will think that some of the matters discussed subsequently are unnecessary in a contract because the contractor can be expected to perform them without specific instructions to do so. Nevertheless, in preparing the contract papers the engineer should include all provisions that may possibly prove important for the protection of the owner, on the one hand, and the contractor on the other.

11-3 Standard Contracts In some lines of business it is practicable to prepare standard contract forms that can be used repeatedly by merely filling in the names, dates, prices, terms of payment, etc., appropriate to the particular fact situation (for example, the purchase of an automobile). It may also be good practice for small construction contracts, such as those for private homes. For example, the American Institute of Architects has prepared "The Standard Form of Agreement between Contractor and Owner for Construction of Buildings." As with standard specifications, these standard contract forms eliminate the need for composing the papers anew each time a similar contract is made. After considerable use, such standard contracts will generally have been proved by experience to be satisfactory, and any errors will have been found and remedied.

In major construction contracts, however, the various projects differ so much that it is generally not feasible to prepare standard contract forms for repeated use.

11-4 Standard Contract Clauses Since there are usually many individual contract clauses that are used repeatedly in construction contracts, it is advantageous to prepare standard clauses which can be incorporated into many different contracts. This plan, to a lesser degree, has the advantages afforded by the use of standard contracts in their entirety. Not only is the saving of time and expense in the preparation of the contract documents an advantage, but so also is the avoidance of omissions and mistakes.

The engineer should be careful to see that each standard clause utilized from a former contract or form a set of standards is exactly what is required. Such care would have proved valuable in the case of a consulting engineer's office which had prepared a clause to describe in considerable detail the contractor's responsibility in the maintenance of water mains, sewers, telephone lines, and electric power lines. This clause was copied verbatim in a subsequent contract involving the maintenance of some gas mains as well as the other utilities just mentioned. No one noticed at the time that reference to the gas mains was not included in the wording of the standard clause. The result was that the contractor in the second case claimed extra compensation for their maintenance.

11-5 Contractor's Warranties It may be that contractors will at some juncture contend that they did not understand a certain portion of the contract, that they did not realize that certain things were to be included in the work, that they did not fully comprehend the local conditions affecting the work, or that they did not realize the extent of the difficulties involved in it.

To avoid subsequent claims based upon such considerations, it is advisable to place in the contract, under the heading "Contractor's Warranties" or "Contractor's Understanding," clauses such as the following:

1 "The Contractor is thoroughly familiar with all requirements of the Contract." Perhaps this appears to be unnecessary or superfluous because a contractor would be foolish to bid on a job with which he did not first become sufficiently familiar. However, addition of the clause should preclude the chance of his subsequently making the lame excuse that he misunderstood what was expected of him.

2 "The Contractor has investigated the site and satisfied himself regarding the character of the work and local conditions that may affect its performance." Again, this is a safeguard and a warning.

3 "The Contractor is satisfied that the work can be performed and completed as required in the Contract." In other words, if something seems to be wrong or some task impossible of performance, it should be brought to the engineer's attention by the contractor *before* the latter signs the contract.

4 "The Contractor accepts all risk directly or indirectly connected with the performance of the Contract." This is stated primarily for emphasis.

5 "The Contractor warrants that there has been no collusion." By this is meant collusion (in preparing the bid) with other contractors, with any of the engineer's staff, or with any other persons.

6 "The Contractor warrants that he has not been influenced by any oral statement or promise of the Owner or the Engineer, but only by the contract documents." This provision is aimed at preventing the contractor from getting special "inside information" from the engineer's staff or anyone else regarding the work and its requirements and from obtaining assurances from the engineer or the owner that certain extracontractual favors will be granted to him.

Other statements that are sometimes required of the contractor include:

1 "The Contractor is financially solvent."
2 "The Contractor is experienced and competent to perform the Contract."
3 "The statements in the Proposal are true."
4 "The Contractor is qualified and authorized to do work in the state."
5 "The Contractor is familiar with all general and special laws, ordinances, and regulations that may affect the work, its performance, or those persons employed therein." These requirements are especially important for out-of-state and even out-of-city contractors to investigate.

6 "The Contractor is familiar with tax, labor, and pay regulations that may affect the work." This warns "foreign" contractors. On large projects it is desirable to publish in the contract any special data regarding wage rates with which the out-of-state contractor may not be familiar. In subaqueous tunneling, for example, state regulations and stipulated wage rates for compressed-air work should be quoted or at least referred to.

11-6   Approval of Contractor's Plans and Equipment   The engineer should not allow the contractor to do "anything and everything" that he wants to do in the name of performance of the contract, even though the contractor is undeniably responsible for performance. If, for instance, the engineer learns that the contractor intends to do something that appears to be unsafe, the contractor should be questioned about the matter and told where the contemplated procedure is unsatisfactory. It may well be that the engineer's knowledge of the project enables him to recognize some important consideration which the contractor has failed to grasp.

It is therefore advisable to state in the document that the contractor is to submit to the engineer for approval[1] prints of any special construction accessories which he intends to use, and he is to secure approval of such prints before undertaking any of the affected work—for example, the falsework for the erection of a large bridge or the cofferdam to be used around the site of a power plant located at the edge of a river. This procedure is designed to give the engineer a chance to see if there is anything in the contractor's plans that the engineer thinks may cause trouble. The move is usually welcomed by the contractor, since he stands to gain if the engineer assists in avoiding subsequent trouble by pointing out some difficulty which had previously escaped the contractor's notice. But it must be remembered that the engineer is not to *dictate* solutions to the contractor because the primary responsibility for the successful completion of the work rests with the latter.

11-7   Defective Drawings   When the engineer prepares the design drawings and specifications, it is taken for granted by the contractor that these are correct and, if carefully followed, will produce the desired results.[2] If it later develops that the design drawings contain errors, the contractor obviously is not at fault provided he did the work properly and neither knew nor was expected to know of the defects.

It is not the contractor's duty to check the design drawings. In the best interests

---

[1] It should be clearly understood that this approval does not relieve the contractor of any of his own responsibility on the premises. In *Appeal of N. MacFarland Builders Inc.,* ASBCA No. 19832, 75-2 BCA, 11,608 (1975), the court held the contractor liable for a cave-in at the excavation because of failure to construct shoring even though the contract stated that the government was to approve the method of protection selected by the contractor.

[2] Therefore, the engineer is considered to be responsible for their accuracy unless contract documents specifically state that the drawings are to show general features only and that they are not in any sense "warranted."

of all parties, however, the contractor should call to the engineer's attention anything that is apparently incorrect. This action may strengthen the contractor's hand if the questioned feature ultimately develops trouble. The contract should state that, if the contractor discovers an error in the drawings or specifications, (1) he is to orally report it promptly to the engineer, (2) he is to confirm it in writing soon thereafter, and (3) he is not to proceed with the affected work until directed to do so by the engineer.

This principle might well be carried beyond the range of apparent errors in the drawings to matters which may involve the best interest of the structure or even the safety of the performance of the work. For example, assume that the contractor notices that the plans do not call for shelf angles to be used at upper floors to support the brick facing of the masonary cavity walls of a multistory building, whereas experience shows that such supports are essential. In such a case, the contractor should recommend in writing to the architect or engineer that such angles be used and state actual cases to illustrate the reasons for the recommendation.

11-8 Approval of Contractor's Drawings The contractor generally has to provide *shop drawings* which are to be used in the actual construction or manufacture. The contract should state that these shop or "working" drawings are to be approved by the engineer before any work to which they apply is done, but these drawings should first be checked for accuracy by the contractor.

For example, assume that a mining company is building a large plant for the crushing and concentration of copper ore. A contract for a large gyratory crusher is to be let to some manufacturer of heavy equipment. The engineer may specify certain basic requirements for this machine, some of which are the following:

1 Size of feed
2 Size of discharge
3 Material for body and liners
4 Capacity
5 Type of spindle at top
6 Arrangement and other details of drive
7 Type and position of discharge
8 Provisions for maintenance and repairs
9 General proportions

The manufacturer who is awarded the contract will prepare shop drawings to show every detail of the crusher. Then prints of these drawings will be examined by the engineer to make sure that the equipment will be suitable. The engineer will then return a copy of these prints to the manufacturer, having indicated thereon approval or desired corrections. The contractor is to make the corrections unless the engineer, after discussion, agrees to let the drawings remain as they are or to accept substituted details devised by the manufacturer. When the latter thinks that the drawings are in satisfactory form, he will submit new prints for the final approval of the engineer. When this approval is received, the drawings will be ready for release for production.

The important point to be emphasized about this procedure, and to be set forth explicitly in the contract, is that although the shop drawings are to be approved by the engineer, the *contractor is responsible* for the accuracy of the work.

There might well be inserted in the contract a clause reading somewhat as follows:

> The approval by the Engineer of any shop or working drawing (or of any drawing of the Contractor's plant or equipment) shall not in any way be deemed to release the Contractor from full responsibility for complete and accurate performance of the Work in accordance with the Contract Drawings and Specifications; neither shall such approval release the Contractor from any liability placed upon him by any provision in the Form of Contract.

The following two hypothetical cases will illustrate this matter of responsibility for the data shown on various types of drawings:

> **1** Assume that one piece of the crusher previously referred to is supposed to be 2 feet 0 inches deep (according to the approved general plan for the equipment) but that the contractor's drafter incorrectly dimensions it 2 feet 1 inch. The checker in the shop misses the error on the shop drawing. The engineer, too, fails to notice it, and he stamps the drawing "Approved." The piece is accordingly made 2 feet 1 inch deep. When the crusher is assembled, the piece obviously will not fit. In this situation the contractor has to stand the cost of revision or replacement of the incorrect part since he was obligated to create the detailed design and to furnish the equipment.
>
> **2** On the other hand, assume that the engineer shows on his design drawings that the overall depth of the crusher is to be a certain dimension. The contractor makes the machine accordingly. Later it is discovered that the depth should have been 1 inch less in order to assure a fit. This is a mistake in the engineer's basic data. The contractor made the machine the size called for in the design; therefore, he is not responsible for the error. The cost of correcting the difficulty falls to the owner.

11-9   Guarantee by the Contractor   Many contracts require the contractor to guarantee that the materials and workmanship of the job will prove satisfactory for one year after completion of the contract. By that time any defects will probably be discovered, and the contractor will have to make them good or arrange some settlement that is acceptable to the owner.

To illustrate the principle involved, assume that a contractor erected a large high-speed forced-draft fan that has a direct-connected drive. The motor was not positioned in exactly the right alignment, and the equipment did not operate properly. Since the trouble was discovered a few weeks after the contractor completed the work, the one-year guarantee required him to remedy the situation.

In another case, a reinforced-concrete underpass or tunnel for a railroad was built under a large embankment. Shortly after the work was completed the engineer discovered that the structure had cracked. The problem then was to ascertain the cause. Considerable study and investigation revealed that the contractor had placed a large amount of fill on one side of the tunnel before placing any whatever on the other side, thus causing an imbalance of loads which distorted and over-stressed the structure. Such inequality of filling was contrary to the specifications. Although the engineer believed that there would be no collapse of the tunnel, the structure was not what the owner paid for; the contractor was held responsible for damages but was not compelled to replace the entire tunnel.

In general, contractors are responsible for their own work and that of their subcontractors. They are to guarantee that the work is free from defects caused by poor performance of construction, but they are not responsible for faulty plans and specifications or for anything that is done under direct orders of the engineer or owner.

11-10 Conduct of Work Although such a course may be ill advised, the owner sometimes may issue instructions directly to the contractor regarding the conduct of work. The *engineer* is the one who should give all such instructions. If the owner nevertheless persists, and especially if the engineer has advised the owner that his orders will cause defective work or other trouble, neither the engineer nor the contractor is to blame for the unsatisfactory results. Proper procedure requires that if the owner believes that something in the plans or specifications should be changed, he should confer with the engineer, who, if he is in agreement, then issues instructions to the contractor. The contractor ought to require that the contractor's operations should be carried on in a workmanlike manner, and it should call for him to expedite the work. He should have a superintendent or other responsible official on duty at all times. The engineer ought to have the power to order that work be speeded up if it is evident that this can and should be done and when the best interests of the project require such action. On the other hand, the engineer must be sure that such action is justified because it may result in overtime work that materially increases the cost of the job.

One occasion upon which the engineer ought to have authority to order that the job be expedited is when there is impending danger. For example, the threat of a flood may imperil completed foundation work of a cofferdam. In some cases, if such orders result in additional costs, the contractor may claim extra compensation provided there has been no negligence on his part. In most cases of emergency the engineer will be acting in the interest of the contractor as well as of the owner. Remember that the progress and safety of the job are essential and that all parties have a real interest in these matters. When the contractor does not move so as to assure both progress and safety, the engineer must have the right to compel such action.

While being careful to avoid dictating to the contractor regarding exactly how the work should be done, the engineer may justifiably recommend and even try to persuade the contractor to do certain things and to conduct operations in what seems to the engineer to be a desirable manner. For his part, the contractor may suggest improvements in the job or the work product as originally outlined, but, when all is said and done, the design is the engineer's responsibility.

Circumstances may develop which make it essential for the owner to take over all or a portion of the work from the contractor and to have the job finished by the contractor's forces or by outsiders. In this connection, note the following clause:

If, in the opinion of the Engineer, the Contractor is or appears to be unable to finish the work, the Owner shall have the right to take over and complete the work as agent of the Contractor. However, the exercise of such right by the Owner shall not release the Contractor or his sureties from any of his or their obligations and liabilities under the Contract or the Performance Bond. The Owner shall also have the right to use or permit the use of any and all plant, materials, and equipment provided by the Contractor for the execution of the Work, and the Contractor shall not remove them from the site without the written permission of the Engineer. The Engineer shall determine the cost of all Work done by the Owner as agent for the Contractor, shall credit the Owner with the cost thereof, and shall credit the Contractor with the compensation which would have been earned thereby; the difference in actual cost above the contract price shall be payable to the Owner.

11-11 Defective Work The contract may contain this clause:

All material furnished and all work done which, in the opinion of the Engineer, is not in accordance with the plans and specifications shall be removed immediately, and other material which is satisfactory shall be furnished and other work which is satisfactory shall be performed.

On the other hand, it is not always practicable or fair to insist upon complete replacement of work which fails to meet the specifications and the requirements of the drawings but is not valueless. Concrete work is a common example of this. The contract may state that test cylinders are to have an ultimate twenty-eight-day compressive strength of 3000 pounds per square inch. What is to be done if, a month after pouring the cylinders, the concrete tests 2800 pounds per square inch? That particular portion of the structure is now covered by many more cubic yards of concrete, and it would be very costly to remove this later concrete plus that which did not meet the test requirements. To deal with such a situation the contract may contain a clause to the effect that although the engineer has the right to demand removal of defective material and work, he also has the right in lieu of such demand to make a deduction in the contractor's compensation if the tests of the concrete cylinders fail to meet the specified twenty-eight-day strength. For example, the contract may specify a reduction of $1 for each 1 percent deficiency times the number of cubic yards of concrete in the portion of the work to which said test was to pertain. The engineer should be careful to specify that he alone has the power to require that repairs, strengthening, or even complete replacement of the work is to be made at the contractor's expense or, in the alternative, that the contractor is to suffer a reduction in payment.

Suppose that the contractor inadvertently erected a light steel beam where the contract drawings called for a heavy one, and vice versa. The error went unnoticed until tons of equipment were installed in the building and caused excessive deflection of the light beam. The contractor was required to bear the expense of making the changes[3] needed, even though the engineer had inadvertently accepted the work.

On the other hand, if the engineer's staff, who designed the beams in the preceding case, had labeled them incorrectly on the design drawing, the contractor could not be expected to notice the error. As long as the work proceeds according to the engineer's plans, the contractor should not stand the expense of any necessary repairs or alterations.

11-12 Relations with Other Contractors and Subcontractors The contract should state whether or not more than one contractor is to work on parts of the project at the same time. It should also specify that the contractor is to conduct operations in such fashion as to cooperate with all others and to avoid interference with their work.[4] The same requirement of cooperation should apply to relations between the general contractor and the subcontractors, even though the former controls the latter and is responsible for the conduct and quality of their work.

Where storage space is required by the general contractor, and by other contractors as well, it is almost essential for the engineer to show on the contract drawings the particular storage areas and the means of access that are to be allotted to each such contractor. A contractor can then arrange with his subcontractors for their use of part of the space reserved for him.

The division of work responsibility should be made clear in advance. It will probably be impossible to delineate everything exactly, but an attempt should be made. For example, in a contract for a large factory building with pile foundations, the project might be divided among three contractors as follows:

1 One contract for all foundations, including piling and concrete work below and including the ground floor

2 A second contract for all of the superstructure above the ground floor

3 A third contract for all the mechanical, electrical, and other equipment, including the erection of same.

As far as it can be done practically, the contract documents should clearly allocate the duties of the respective contractors. For example, here are some questions to be settled:

1 Who provides and installs water supply lines, sanitary and storm sewers, and other utilities? When will material and data be available for this work?

2 Who is to maintain utility services (lights, etc.) during construction?

3 Who is to be responsible for the protection of the public, and of the property itself, and for how long?

---

[3] Since safety is involved here, there is no doubt about the necessity for remedial measures.
[4] Sometimes a construction manager may be in general charge of the project in order to coordinate all phases of the project.

4 Who furnishes and sets the anchor bolts, and when and from whom will information for this work be obtained?

5 Who grouts column bases and bed plates for machinery?

6 Who provides and erects conduits for electric power and light, and when will the material and necessary data be available?

7 Who provides attachments for equipment that is to be connected to the superstructure (including provision of openings for that equipment)?

8 Who is to be responsible for the maintenance of traffic and access, and for how long?

9 If part of the project is to be put to use before completion of the entire project, who is to be responsible for finishing such part?

10 If various articles of equipment are to be furnished under separate contracts, who erects these articles, who provides their supports, and who is responsible for their protection?

11 What is to be the order of construction?

12 When is each contractor to begin and to complete his part of the work?

The following is a paraphrased and abbreviated version of a provision from a contract for a large bridge:

> During the time that the Contractor will be performing the Contract, other persons will be engaged in doing work at the site or in the immediate vicinity thereof, among them being the following:
>
> **1** Under Contract G-1, the John Jones Corporation will perform the work of excavation, filling, and grading in accordance with the information shown on the reference drawings and in conformity with the Specifications for said Contract, all of which papers may be examined at the office of the Owner.
>
> **2** Future contracts will provide for the construction of foundations and superstructures of the east and west approach spans adjoining the main bridge, together with the building of the connecting roadways.
>
> The Contractor shall therefore plan and conduct his operations so as to work in harmony with all others on the project, and he shall not delay, endanger, or avoidably interfere with the activities of such others; he shall at all times maintain a roadway across the area allotted to him for his operations, so that others may have access to the work in which they will be engaged; and he shall cooperate with other contractors in whatever is, in the opinion of the Engineer, for the best interests of the Owner and the public.

11-13   Order of Completion   It is essential that the contract state specifically any requirements there may be regarding the sequence of construction and the order of completion. This is particularly important for large projects that involve various structures or parts, and especially when two or more contractors may be working simultaneously or in relays. Such information not only obligates a given contractor to cooperate with the others but also affords him the opportunity to plan his work more intelligently and to allow in his bid for the extra costs that may be entailed because of delays stemming from the operations of the other contractors or by the effect of the stipulated order of completion.[5]

The engineer must be careful when setting up a schedule and order of completion of the structures. He should be sure that the requirements can be met by the contractors and will not cause unjustified additional expense to the owner through high bids that the contractors are obliged to submit in order to be able to afford to perform the work on time.

Although the engineer is not responsible for the contractor's operations, he may properly require that the contractor prepare in advance an outline of his program showing the contemplated sequence of work. The engineer may also call for the contractor to compose a schedule of expected progress in order to determine whether or not the contractor's plans appear to be adequate in terms of time. The engineer can then use this schedule of operations later on to see whether the work is indeed progressing as anticipated. A clause such as the following may be found in many contracts:

> Within two weeks after the signing of the Contract the Contractor shall submit for approval of the Engineer six (6) black-and-white prints of his proposed construction schedule. This is to show graphically the date on which he expects to start and to complete the various major portions of the project. This schedule is to be so prepared that the *actual* progress of the work can be recorded and compared with the *expected* progress. On the first and fifteenth of each month, and until the job is fully completed, the Contractor shall deliver to the Engineer three (3) prints of the construction schedule with the progress to date shown thereon.

If no time for commencement or completion of the work is specified, a *reasonable* time is to be implied. In other words, the contractor is to begin without unwarranted delay and push forward with diligence. When there is no important reason to set a deadline for completion, the contract might properly state that the "Contractor shall expedite completion of the work, and shall conduct his operations in a workmanlike and efficient manner."

The foregoing concept applies as well to contracts for engineering services. Thus, the engineer is engaged to make plans for a project; if the time for their completion is not stipulated, a reasonable time for this work is implied. Or, the agreement may state that the plans are to be completed "as the Owner may direct." In that case the engineer is entitled to a reasonable time after receipt of the anticipated advices from the owner.[6] Naturally, delays caused by the owner are to be considered in determining what is a reasonable time.

---

[5] Here is a commonplace instance involving a completion date for a portion of the work. A contract for the steelwork of a large industrial plant covered several buildings and was let directly to a steel fabricator. The general excavation, concrete work, roofing, and siding for all structures were to be done under other contracts. All the contracts stated that the foundations, steel framing, roofing, and siding of one of the major buildings were to be completed by Nov. 15, 1965, so that the difficult work of constructing concrete foundations for machines, erecting machinery, and installing electrical equipment in this building could be accomplished under shelter during the winter.

[6] *Giffels & Vallet, Inc. v. Edw. C. Levy Co.,* 37 Mich. 177, 58 N.W.2d 899 (1953).

11-14   Inspection of Materials   The purpose of the engineer's inspection of materials to be used in construction is to see that the contract is properly carried out and that the owner's interests are adequately protected. Such activity by the engineer or his representative generally involves inspection of shopwork as well as of the quality of the materials themselves. For example, in a large bridge project involving important steelwork, the engineer may have an inspector at the steel mill to make sure that the material meets the specifications. He may also have an inspector at the fabricating shop for several months to see that all shopwork is satisfactory. In smaller jobs, the engineer may accept the report of the manufacturer's or the contractor's inspector when this report is submitted with an affidavit vouching for its accuracy. In any event, the contractor should be obligated to notify the engineer in advance so that the latter can, if he desires, have his inspector on hand when work affecting the contract is to be performed.[7]

A large project might involve so much inspection of materials that a separate department in the engineer's office is designated to handle this work. As an alternative procedure, the engineer may engage a firm of specialists to take charge of all inspection of (1) materials, (2) shop fabrication, and (3) to carry out any necessary testing.

The contract should make clear what is to be done regarding inspection and exactly what is expected of the contractor. For instance, if test cylinders of concrete are to be made in the field, the bidder should be told in the contract documents what he will have to do in this regard so that he can reflect the cost in his proposal.

11-15   Inspection of Field Operations   In a large job it is common practice for the owner to have the engineer maintain a resident engineer with a staff of inspectors in the field to see that work called for by the plans and specifications is performed satisfactorily. The inspectors often check such things as the dimensions and adequacy of forms for concrete, the arrangements and sizes of reinforcing bars, the grading of the site, the compaction of fill, the erection of structural steel, the quality of riveting or bolting and welding, the location of anchor bolts, and the erection of machinery.

In such circumstances, who is answerable for an error made in the fieldwork? The contract should be worded so as to make the contractor responsible for his own work. The resident engineer and his staff should try to help the contractor to avoid making errors, but their approval of his work should not relieve him of ultimate responsibility. The inspectors must be careful to see that they help the contractor but do not dictate precisely what he should do or how he should do it.

In one case, some of the footings of a big concrete structure were to be subjected to large overturning moments. The contractor placed the reinforcement in these footings, and an inspector gave his approval. However, in some unexplained manner, only half as many bars as were required by the drawings projected up

---

[7] In *Kaminer Construction Co. v. United States,* 488 F.2d 980 (Ct. Cl. 1973), the court said that "the right to inspect does not imply a duty to inspect."

(from the footings) to withstand the bending moment that would be applied by the future construction. After the concrete of the footings had been poured and set, the men placing the next tier of reinforcement discovered the mistake. The inspector's approval did not relieve the contractor from the very costly task of remedying the error.

In projects of small or medium size it is not usually necessary for the engineer to have an inspector at the site continuously. It may be that the engineer or his representative will visit the job two or three times each week just to supervise the work in a general way. Another arrangement might be the scheduling of regular weekly job meetings for inspection and conferences with the contractor and the owner's "clerk of the works." Thus the engineer will be available to discuss any questions with the contractor and see whether or not the work is progressing satisfactorily.

It is generally advisable for the engineer to make inspection trips on a predetermined schedule so that the contractor knows when the engineer will be available for consultation. The contract, if it does not prescribe a schedule for inspections, should provide for the contractor to notify the engineer in advance of starting on a particular phase of the work so that the latter or his representative will have time to arrange to be on hand at the appropriate time. The contractor should be obligated, further, to have his key operatives available and to otherwise do everything necessary to facilitate inspection by the engineer.

Here is the wording of a clause that was prepared to apply to both the inspection of materials and of fieldwork:

All work, materials, processes of manufacture and methods of construction shall be subject to the inspection of the Engineer (or his duly authorized representative), who shall be the judge of the quality and suitability of them all for the purposes for which they are used. If any of them fail to meet with his approval, same shall forthwith be replaced, corrected, or otherwise made good, as the case may require, by the Contractor at his own expense. Rejected material shall be disposed of as the Engineer may direct. Acceptance of any material or workmanship shall not serve to prevent subsequent rejection by the Engineer if he finds either or both to be unsatisfactory. Unless ordered by the Engineer, no material shall be shipped from its place of manufacture without prior inspection and approval.

11-16 Duties of an Inspector The function of an inspector is that of seeing that the work is done *in accordance with the terms of the contract.* If the contract calls for certain things that are incorrect, it is not the inspector's job to change them, nor is he empowered to do so. If something has been overlooked in the plans or specifications, or if an error has been made in them, the inspector may properly assist the engineer in bringing about remedial action under those terms of the contract that cover revisions, extras, and adjustments. The inspector is representing the engineer and the owner through the engineer.

The inspector may have to keep accounts of time spent, materials furnished, and work done by the contractor. He cannot be everywhere at once, and he may not have to be on the job continuously. He has to learn what to watch for and how things should be done. He should be able to detect developing as well as existing troubles and uncertainties. He should report such matters to the engineer at once, so that the latter can make any necessary decisions promptly. He should remember that delay on his part may cause loss for the owner.

Adopting the middle ground between laxity and fastidiousness is a central concern of the inspector. It involves a balanced approach—protecting the owner's rights without unnecessarily antagonizing those executing the job. Here is an example of undue fussiness. In making the shop drawing of the riveted steel girders for a highway viaduct that was to be on a slight grade, the drafter drew the stiffener angles in a vertical position because it was easier for him to do so than to slope them a tiny bit. The fabricator built the girders with the stiffener angles perpendicular to the flanges. This is the practical and customary procedure because the slight skewing needed to make them vertical would prevent the use of efficient multiple punching of the rivet holes in the web plate. When the girders were ready to be shipped to the site, the inspector demanded, without consulting the engineer, that the girders be dismantled and refabricated so as to have the stiffeners truly vertical. The fabricator sent a representative to explain the situation to the engineer, and the latter agreed at once that the girders were fabricated as he intended.

Much practical experience and knowledge are essential for inspection work. A contractor will generally cooperate well with an inspector whom he trusts, who is trying to be fair, and whose knowledge he respects. The inspector who is unreasonable or who does not understand the problems of performing construction work will arouse a contractor's ire immediately; one who operates in a sloppy manner will arouse the contractor's disdain.

The inspector should never fail to protect the owner's interests within the terms of the contract. In a rather extreme case a contractor persuaded an inspector to do otherwise and gave him a share of the "savings." The job consisted of highway and bridge work, one item being timber piling. The contractor drove short pieces of piles a few feet into the ground so that they looked like pile heads. Furthermore, he made the specified 8-inch concrete pavement 8 inches thick at the edges but 4 or 5 inches thick over the rest of its area. The dishonesty and wrongdoing came to light when the structures and pavement failed. The guilty contractor and inspector received prison sentences.

11-17 Land and Facilities Ordinarily it is the owner's responsibility to acquire such real property as is needed, not only for the job site itself but also for whatever use may be necessary during construction. It may be that an easement has to be secured to provide access to the construction area, that a permanent right-of-way must be purchased, or that some land has to be rented for storage and working space. All these land rights and privileges are to be secured by the owner in advance so that full information can be given to the bidders before they submit their proposals.[8]

---

[8] Any arrangement whereby the contractor undertook to secure the necessary rights, etc., would probably prove more costly to the owner in the long run.

Various facilities will have to be provided by the contractor (perhaps on a cost-plus basis), if they are not furnished by the owner. The contract should be very specific in all such matters. Facilities that typically are needed include:

1 Electric power
2 Access roadway
3 Spur track for railroad access
4 Warehouse or other storage
5 Contractor's and engineer's offices
6 Utilities for these offices
7 Provision for maintenance of traffic and other services

11-18   Permits and Licenses   It is generally the duty of the owner to secure any building permits that are required for the construction of the project. He should obtain any other permits that are called for; examples include (1) a permit to close a street temporarily while some particularly dangerous work is being done, (2) a permit to truck heavy steelwork through the streets, (3) a permit to move an existing house to a new location, and (4) a permit to shut off certain utilities for a while when they have to be relocated.

However, arrangements for procedures that the contractor *chooses* to follow in order to complete the work to his own best advantage are to be made by him. To illustrate this, assume that the contractor prefers to deliver heavy steelwork by boat instead of by railroad, the method of delivery anticipated by the owner. It is then the contractor's duty to obtain any necessary permit. If the contractor wishes to build temporary piers for the erection of a bridge and to place them within prescribed "channel limits," it is the contractor who should secure permission to do so.

It may happen that, through inadvertence, a contract contains a clause that directs or permits the contractor to do something contrary to the applicable building code or some other law or regulation. It is, of course, the engineer's responsibility to avoid such a situation, but the contractor, in bidding, is entitled to assume that he is to quote upon the contract as it is. If later events reveal that some change must be made to meet legal requirements, the contractor may rightly claim extra compensation for any additional costs.

Some states require that a contractor, and perhaps even a subcontractor, must have a license issued by that jurisdiction in order to do any work within its boundaries. If such a requirement pertains it is desirable that this fact be stated in the information provided for bidders so that unlicensed contractors will not waste their time and money in preparing a bid on the project, or so that they can take immediate steps to obtain the necessary license.[9]

11-19   Labor Considerations   It is important to become familiar with federal, state, and local laws regarding eligibility for any conditions affecting employment.

---

[9] Any contractor who attempts to perform work without a required license may find that he cannot enforce collection of payment for any work done by him within the state. The same rule is likely to apply to an individual engineer in respect to services performed without a license.

Just what are the rules regarding whom one can hire and under what conditions? Must the job be closed shop or can it be open shop? What personnel must be obtained locally? What are the maximum working hours and minimum wage rates that have to be met?

Especially when the legal and other requirements are not standard practice, the engineer should give bidders any essential data available so that they can make their proposals accordingly. The following samples of contract clauses deal with labor matters:

> No person under sixteen (16) years of age and no convict labor shall be employed by the Contractor in the prosecution of the work.
>
> All laborers except major supervisory personnel shall be residents of the State of New York.
>
> The Contractor shall not employ for the Work any person whose age or physical condition renders his employment dangerous to his health or safety, or to the health or safety of others. However, this provision shall not be construed to prevent the employment of physically handicapped persons who may be employed for work which they are able to perform satisfactorily.
>
> All employees engaged in the Work under this Contract shall have the right to organize and bargain collectively through representatives chosen freely by them. Furthermore, no employee shall be required to join any union or to refrain from joining any union or labor organization of his own choosing.
>
> The Contractor shall not accept more than eight (8) hours of work each day, to be performed within nine (9) consecutive hours, and he shall not employ any person or persons for more than eight (8) hours in twenty-four (24) consecutive hours, except in case of necessity, in which event the pay of employees for time in excess of said eight (8) hours shall be at the rate of time and one-half.

11-20 Notices All notices, instructions, and information sent to the contractor by the engineer should be in writing. It is often necessary to expedite the completion of the work by use of verbal messages given to the contractor in person or by telephone, but these should be promptly confirmed in writing. This procedure avoids many of the arguments and uncertainties that are likely to develop when messages regarding details of the work are reviewed months later, at which subsequent time the recollections of the parties involved may be dimmed. For the same reason the contractor should, in turn, also put in writing the communications emanating from his office to the engineer.

11-21 Work Done by the Owner Sometimes there are projects in which the owner is to handle part of the job with his own forces or have someone do that part under the direct supervision of the engineer as the owner's representative. Such work should be so conducted as to minimize interference with the general contractors and his part of the overall operations. Naturally, the contractor should likewise cooperate.

This applies to small projects as well as to large ones. For example, Mrs. A plans to build a house for herself and her family. She intends to perform certain work personally in order to minimize the amount of money paid out. The contract with the builder should clarify any such arrangement and describe in detail what work the owner is to perform. Such a contract may seem to be "small," but it is very important for the prospective householder. She should obtain professional assistance in its preparation unless she has had adequate personal experience in this field.

11-22   Lines and Grades   The contract should be specific in its requirements regarding the establishment of lines and grades for a construction project. This function usually falls to the contractor, but some engineers, especially on a large project, prefer to keep direct control by having the establishment of lines and grades handled by their own field organization. If the contract provides for the latter arrangement but further states that the contractor is to *check* all the lines and grades and any other dimensions determined by the engineer, failure by the contractor to do so will mean that he probably will have difficulty collecting from the owner for additional costs resulting from errors in the engineer's determinations.

Whether it is desirable for the contractor or the engineer to make these surveys in a particular case depends upon the size of the staff available to the engineer, the character of the job, and the capabilities of the engineer's personnel. In some ways it is best to have the contractor do this surveying as well as the construction part in order to make him clearly responsible for all phases of the work.[10]

When the lines and grades are to be established by the engineer, the contract documents ought to stipulate that the contractor is obliged to preserve stakes, bench marks, and the like, and is to conform to all lines and grades. The contractor is expected to cooperate with the engineer's men in all this work by avoiding interference with them and by conducting his operations so as to expedite their work as much as possible. In this connection, the contractor should be required by the contract to give the engineer reasonable advance notice of times and places at which he intends to perform certain work so that the engineer may proceed to establish all necessary lines and grades and so that all measurements required for payment and record purposes may be made at the engineer's convenience but without undue delay for the contractor.

Often, although the engineer establishes the controlling lines and grades, it becomes necessary for the contractor to make many minor and detail measurements. The contract papers should reflect this eventually so that the contractor can include the cost of such work in his proposal.

11-23   Underpinning   There are likely to be requirements in municipal ordinances or state laws regarding the procedure and responsibility for underpinning a structure whose safety is likely to be imperiled by a contractor's operations. It is

---

[10] In one instance where the engineer was responsible for lines and grades, his men set the elevation of the foundation of a heavy machine 1 foot too high. The owner had to pay the contractor for all the labor, materials, and incidentals involved in the removal of concrete and the rebuilding of the foundation.

proper, if there are no regulations indicating otherwise, to make the owner of a projected structure responsible for the safety and support of his neighbor's existing building, and sometimes of the adjacent land, when new construction is to be of such a nature that the foundations of existing structures will be endangered.

Underpinning an existing building is both difficult and costly. The owner of the proposed structure should have the engineer so design the new structure that danger to the neighbor's property can be avoided, or at least minimized. If underpinning work is not done well, uneven settlement and cracking of the structure may result, causing the affected property owner to make damage claims against the owner of the building under construction. For protection against unjust claims that his neighbor might advance, the owner of the planned new structure might well have a third party make a careful examination of the condition of the neighbor's property before work is started and then again after work has been completed. Photographs taken by a professional photographer in advance of the performance of any building work may be very helpful in establishing the condition of the preexisting structure at the start of the construction operations.

Occasionally, underpinning work can best be handled as a separate contract, with the owner of the new structure absorbing the expense. Contractors are generally reluctant to make a lump-sum proposal on this kind of work because of the uncertainties accompanying it. In general, it is advisable to arrange for such operations on a cost-plus-fixed-fee basis.

The engineer may show the character of the underpinning work to be done and then require the contractor to follow that pattern. The latter, in presenting his proposal, either simply agrees that the indicated work can be accomplished or proposes modifications of his own. In the alternative, the engineer may ask for proposals in which the contractor is to present both the design of the underpinning and his estimate of the cost.

11-24 Order and Discipline The contractor should be made responsible for the maintenance of order and discipline during the performance of the contract. This is true primarily in regard to the supervision of his own workers, but it should include also a subcontractor's workers who are employed on the job.

In this connection it may be essential that the contract forbid the storage and use of intoxicating beverages at the job site.

For example, the following clause:

> In the performance of this Contract, the Contractor shall exercise every precaution to prevent injury to persons or property; he shall erect such barricades, signs, and lights as the Engineer believes to be suitable; he shall adopt and enforce such rules and regulations as may be necessary, desirable, or proper to safeguard the public, all persons engaged in the Work and its supervision, and all traffic on adjacent streets; and he shall be responsible for the maintenance of discipline throughout the conduct of the Work.

11-25 Performance When the contractor has done all that the contract requires, he is said to have performed his part of the agreement. This is naturally a prerequisite to final acceptance and payment by the owner.

Ordinarily, when one makes a contract to perform stipulated services, he is not entitled to the agreed-upon fee until he has done what he said he would. In the case of construction contracts, "substantial performance" of the specified work or services may suffice to render the contractor's compensation due and payable. Perhaps this liberal policy stems somewhat from the fact that there is, as a practical matter, probably no such things as "perfect performance" of big construction contracts. If the builder has done his job in good faith and with the intent of fully accomplishing what is required of him, even though there may be small defects or omissions, he is generally regarded as entitled to the contract price (minus the cost of remedying any such defects or omissions).

A contract may state that the performance is to be "satisfactory" to the owner, the architect, or the engineer. If the particular subject matter of the contract is something that involves the personal taste or fancy of an individual (as in the case of a contract for interior decorating), the contractor is obliged to satisfy that person before the fee has been earned. In construction contracts as a class, satisfactory performance is normally interpreted to mean performance that *should* be satisfactory to a "reasonable" man.

11-26 Final Inspection and Acceptance Upon written notice from the contractor to the effect that he believes he has performed all the work required by the contract, the engineer is to inspect the job promptly to see whether everything is satisfactory. Much that has previously been inspected and accepted by him or by his staff need not be reexamined in great detail, but if something needing correction is discovered in the work that was approved earlier, the former acceptance does not prevent the engineer from requiring that it be remedied by the contractor.

The "final inspection" may not disclose hidden defects, and the engineer may not become aware at that juncture of an imperfection. Ordinarily, the *final* acceptance of the work by the engineer lets the contractor off the hook unless he has guaranteed his work, in which event, regardless of the approval of the work at the time of the final inspection, the contractor is still bound for the period stated in the contract by his guarantees as to the quality of material and workmanship.

If the engineer finds the job to be satisfactory in all respects, or when any deficiencies have been corrected, he is to notify the contractor in writing that the latter has performed adequately and that the structure is accepted. The engineer will simultaneously advise the owner in writing of this development and will recommend that the owner make the final payment to the contractor.

The contract should outline these steps and should state that final payment to the contractor will not be made until the engineer has inspected and approved the completed work and has accepted it on behalf of the owner. The following clause is illustrative:

When notified by the Contractor that, in his opinion, all work required by the Contract has been completed, the Engineer shall make a final inspection of the Work, including any tests of operation. After completion of this inspection and these tests, the Engineer shall, if all things are satisfactory to him, issue to the Owner and the Contractor a Certificate of Final Completion certifying that, in his opinion, the Work required by the Contract has been completed in accordance with the Contract Drawings and Specifications. However, the Certificate shall not operate to release the Contractor or his sureties from any guarantees under the Contract or the Performance Bond.

Assume that a contractor has completed a new building for a merchant. During the final inspection the engineer finds certain work that does not meet the requirements of the contract, and the defects may entitle the owner to a claim for damages. The owner, however, is eager to occupy the building. If he takes possession at once, will he thereby be prevented from claiming damages for the faculty work? It has been held[11] that:

> An owner is not estopped from claiming damages for the breach of a building contract by taking possession and moving into the building. The failure of the contractor to construct the building in accordance with the plans and specifications was a violation of the contract and the taking of possession of the premises by the plaintiff cannot be considered as a discharge of the defendant's liability.

On the other hand it has been said that the normal rule is that acceptance of work constitutes a waiver of its defects.[12] In some jurisdictions the question of waiver by acceptance depends on whether the defect was latent or was discoverable by reasonably careful inspection.[13]

11-27   Miscellaneous   It is impossible to list here all the miscellaneous items that should be included in a typical construction contract. However, the few considerations mentioned below will serve as a reminder.

Employment   The contract should prohibit hiring during the job period of any of the owner's or engineer's employees by the contractor for his own force. This provision is intended to prevent the contractor from raiding and depleting the owner's and engineer's staffs.

Emergencies   It may be advisable to specify general procedures to be followed in case of emergencies, giving the engineer the right to order overtime work under stated conditions if he considers that this is necessary. Overtime work requirements do not usually apply to supervisory, administrative, clerical, and other personnel not engaged in manual work.

Minimum-Wage Rates   In public work it may be necessary to publish legal requirements. On any job, if applicable rates are clearly stated, it also helps to put all bidders on an equal basis in estimating the cost of the work. When given at all, the tabulation of wage rates should be sufficiently extensive to cover all anticipated classifications of work. The tabulation should take into account fringe benefits as well as hourly wage rates.

---

[11] *Michel v. Efferson,* 223 La. 136, 65 So.2d 115, 119 (1953). See also *Ting-Wan Liang v. Malawista,* 421 N.Y. S.2d 594 (1979).

[12] *Village of Endicott v. Parlor City Contracting Co.,* 381 N.Y. S.2d 548 (1976). See also *Brouillette v. Consolidated Construction Co. of Florida, Inc.,* 422 So.2d 176 (La. App. 1982).

[13] *Forte Towers South v. Hill York Sales Corp.,* 312 So.2d 512 (Fla. App. 1975). *State v. Wilco Construction Co., Inc.,* 393 So.2d 885 (La. App. 1981).

Domestic versus Foreign Materials and Labor Public works may require restrictions regarding the use of foreign materials or foreign labor. If so, this circumstance should be made clear.

Construction Reports The contract is to specify what periodic reports will be required of the contractor. These might include monthly estimates of the quantities of materials furnished and work done, progress reports showing the percentage of completion of various portions of the project, and periodic communications covering the status of materials in the process of manufacture.

Payrolls and Bills for Materials The engineer may require that a copy of the payrolls of the contractor and all subcontractors, together with bills for materials, be made promptly available to him in the case of cost-plus work, extras, and the like.

Cooperation among Contractors In large projects, several contractors may be working simultaneously, or their work times may overlap. The contract should require that each is to conduct his operation so as to cooperate with the others.

Patents Specifying the use of patented devices or materials may be undesirable because this restricts competition and may well inflate costs. If such devices or materials are necessary, however, the contract should indicate who is to secure permission to use these devices and materials and who is to pay any royalties or the like.

Schedule of Securities This is generally a tabulation giving the market value and other data respecting any securities that are posted by the contractor in lieu of a bond.[14]

Guarantee of Equipment If an equipment guarantee is to be required, the details should be clearly stated in the contract. The contractor may be obliged to keep any machinery and other mechanical equipment, electrical work, power generating plants, and hydraulic structures which he erected or supplied in repair and in good operating condition for a year or some other specified time after completion of the contract. Such a clause might be worded somewhat as follows:

> The Contractor shall guarantee that the material, equipment, and apparatus required by the Contract and Specifications shall be free from all defects in material, design, and workmanship and shall give satisfactory and continuous service under all conditions of service required and specified or which may be reasonably inferred from the Specifications, and that all Work performed by him shall be perfect in material and workmanship. The Contractor shall agree to repair or replace, at his own expense, any part of the material, apparatus, or workmanship proving defective within one year from date of acceptance.

Dredging This work may require special instructions regarding sequence periods of work, maintenance of traffic, and the handling of obstructions and of any unforeseen difficulties which arise. Governmental permits to do such work should be secured by the owner unless the dredging is purely for the contractor's convenience, in which case the latter will seek the necessary permission.

---

[14] Usually these securities are deposited with the owner.

Borings Beyond what the engineer has done, explorations of foundation conditions may be required on the part of the contractor, especially in the case of unit-price and cost-plus contracts. If so, the engineer should have the right to determine the extent of the explorations needed and to approve or reject the contractor's plans for such activity. The method of payment for work of this character should preferably be specified on a cost-plus basis or as a stipulated price per foot of depth of boring, per soil sample taken, and per soil test made.

Spare Parts These may be needed in the case of electrical and mechanical equipment being supplied by a contractor as his work product, and detailed requirements for such parts should be given in the contract.

Tests Tests of materials and of mechanisms should, if required at all, be fully described in the contract papers.

Medical Facilities The contractor ought to be obliged to provide at the site of any sizable construction job medical facilities sufficient for the administration of first aid. Moreover, he should have the means available for removing disabled persons to a hospital.

QUESTIONS

1 Explain the principal powers of the engineer in connection with the conduct of work in the field. What is his responsibility in case of an emergency?

2 Who should be responsible for coordination of work done (a) by the contractor and subcontractors? (b) by two different contractors? (c) partly by the owner and partly by the contractor? Explain.

3 Why should the engineer specify that he is to approve all the contractor's (a) drawings, (b) equipment?

4 Should the contractor check the engineer's plans? Why? What should the contractor do if he happens to discover an error in the contract drawings?

5 What sort of things are supposed to be covered (a) by a contractor's warranties? (b) by his guarantees? Why should the engineer require him to warrant and to guarantee these things?

6 Who is responsible for the expense resulting from the correcting of defective work?

7 Ordinarily, who should establish the order in which the contractor is to perform various major portions of a large construction project? Explain.

8 Why is it desirable for the contract documents to contain a clause which requires each bidder to visit (or to have one of his men visit) the site of a construction job before submitting a proposal on the project?

9 A contractor was building a highway underpass in an urban area. The contract stated that she was to maintain traffic and to maintain and relocate all utilities. She encountered an old sewer which was not shown on the plans or mentioned in the specifications. (a) Is maintenance and relocation of this sewer her responsibility even though she had visited the site before bidding, as required by the specifications? (b) Can she properly claim extra compensation for this work? Explain.

10 A contract states that the work is to be completed in thirty days. What does this mean? Is the statement proper?

11 A contract states that the work is to be completed promptly. What does this mean legally?

12 If the contractor guarantees a structure for one year, (*a*) what is the usual starting time of the guarantee? (*b*) What sort of things would the guarantee ususally cover?

13 What is the significance of "substantial performance" in construction contracts?

14 Who is to judge if performance is "substantial" or not?

15 What control (if any) should the engineer have (*a*) over the conduct of a contractor with respect to other contractors engaged in the same major project? (*b*) over subcontractors?

16 The contractor designed the falsework for a large bridge, and the engineer approved the drawing. Subsequently one erection bent collapsed. Who is responsible for the resulting damages? Elaborate.

17 A contract called for the building of a large oil tank on silty soil near a wharf—all in accordance with the design. Under one edge of the area to be occupied by the tank the contractor unexpectedly encountered a group of old wooden piles which had been covered by earth and were not shown on the plans. He cut them off just below the future tank bottom, and he covered them with a little dirt. He did not report the situation to the engineer, who knew nothing about the piles. When the completed tank was tested by filling it with water, the greater resistance of the piles (compared with the bearing value of the neighboring soil) caused uneven settlement; as a result, the side of the tank over the piles buckled. Who is responsible for remedying the situation? Explain.

18 Describe some of the duties and responsibilities of and limitations upon an inspector in the field.

19 Who is to do the inspecting of work done by subcontractors—the engineer or the contractor—since the owner has no direct contact with the subcontractors?

20 If an engineer is not able or willing to send someone from her organization to inspect certain products before shipment from the factory, must she accept these products regardless of deficiencies?

21 If the inspector accepts work done by the contractor, does that mean that the contractor is no longer responsible for proper performance and quality of the work which has thus been accepted?

22 Does the inspector in the field have the right to tell the contractor how to perform his work? Explain.

23 Why make a final inspection when all the work has supposedly been inspected and approved before?

24 What are the principle procedures for and significance of final inspection and acceptance of a job? Does this acceptance relieve the contractor of subsequent responsibility for the quality of the work?

25 If defective work done by the contractor cannot practicably be replaced, what can the engineer do about it?

26 In a construction job, who should provide the necessary land, access, power, sanitary facilities, permits, and licenses? If not specified, what can be done about settling these matters?

27 When may it be desirable for the engineer to establish all lines and grades for a large construction project? When undesirable?

28 Why should the contractor be responsible (*a*) for the protection of the public at a job site? (*b*) for the maintenance of discipline?

29 Explain why it is important to confirm (in writing) telephoned messages between the engineer and the contractor when these messages relate to the construction project.

30 Does the contractor accept all risk when he takes the contract? Why make him state officially that he does so?

31 Why make the bidders warrant that they are familiar with all laws, ordinances, pay rates, etc., affecting the job?

32 Why might the engineer want to specify the order of completion of portions of the contract work? Illustrate a case.

33 Has the engineer the right to require the contractor to prepare a construction schedule? Why may one be desirable?

34 Who should provide the building permit? Why is one necessary in many cases?

35 What sort of contract is best suited for underpinning jobs? Why?

CHAPTER 12

CONTRACT CLAUSES RELATING TO FINANCES

12-1 Suspension of Contract The contract clauses should give the engineer and the owner the right to suspend the contract if, in their opinion, an emergency or other unexpected development makes this necessary to protect the owner's interests. On the other hand, the right to suspend the contract should not be given to the contractor. The right is intended to protect the owner, but it should not be used in such a fashion as needlessly to injure the contractor.

The suspension clause might read like this:

> If at any time the Engineer considers it impracticable to start or to continue performance of the work or any portion thereof (whether or not for reasons beyond the control of the Owner), the Engineer shall have authority to suspend performance until such time as he may believe it feasible or desirable to proceed. However, should such action suspending the work be taken, the Engineer shall take all appropriate steps to minimize the duration of the suspension of the work, and the Contractor shall be entitled to such compensation for the resultant unavoidable expenses to him as the Engineer may believe to be just and reasonable because of the suspension.

To illustrate the need for such a clause, assume that a textile manufacturing corporation has let a contract to build a new and modern addition to one of the old buildings at its plant in Massachusetts. The foundations of the new structure are partially completed when the old structure is destroyed by fire. Obviously, the complete execution of the contract may now be utterly inadvisable. Under such circumstances the owner should have the right to suspend the work until he can determine what to do. He may want to change the design of the structure or to build an entirely new plant. He may wish to rebuild the destroyed structure so as to utilize fully the addition now under contract. He may not have had sufficient insurance to enable him to proceed with the reconstruction until he can secure more money. This financing may cause considerable delay, so that several weeks will elapse before he is sure he can go ahead. He may even decide to go out of business, or to carry out the expansion at another plant.

In case of a contract with a definite time limit for completion, an extension of time should be granted to the contractor to make up for the time lost because of the suspension. The extension should be a liberal one because a protracted interruption of work will almost always cause considerable further delay before the contractor can get adequate production under way again. He cannot keep his personnel on hold indefinitely nor keep the flow of materials to the site all ready to be resumed at short notice.

12-2 Cancellation or Termination of Contract It is possible for circumstances to develop that make it necessary or desirable for the owner to terminate a contract. This right should be definitely stated in the contract clauses. The case described in the preceding article illustrates one way in which the need for the exercise of this right may arise.

The contract might well specify the following additional bases on which the owner shall have the right to terminate it:[1]

1 Bankruptcy of the contractor or assignment by him for the benefit of his creditors

2 Appointment of a receiver or liquidator for the contractor's property (unless such receiver is dismissed within two or three weeks)

3 Failure or refusal of the contractor (after adequate warning by the engineer) to supply enough skilled workers or proper materials to carry on the work in suitable manner

4 Failure or refusal of the contractor to prosecute the work diligently

5 Failure or refusal of the contractor to comply with the instructions of the engineer or with applicable laws and codes

6 Failure or refusal of the contractor to pay for labor and materials

The termination of a contract may lead to a great deal of argument and trouble. Therefore, not only should the right to cancel be specifically reserved for the owner but the details of the procedure to be followed should be stated clearly.

The following is one version of a termination or cancellation clause:

> The Owner shall have the right, if the Engineer shall deem the Contractor guilty of a breach of any portion whatsoever of this Contract, to take over and complete the Work, or any part of the Work, as agent for the Contractor and at the expense of the Contractor, either directly or through the operations of other contractors. Likewise, the Owner shall have the same right if the Contractor shall become bankrupt or if his property and affairs shall be placed under the control of a receiver or trustee.

---

[1] It is customary to require, however, that the contractor be given adequate written notice before the owner acts; for example, ten days.

Whenever the Owner shall be empowered to take over the Work as described in the preceding paragraph, or whenever circumstances or the affairs of the Owner make it necessary for his own financial well-being, the Owner shall have the right to annul, cancel, or rescind the Contract, in whole or in part. If an equitable financial settlement cannot be mutually agreed upon by the Engineer and the Contractor, the settlement shall be arrived at in accordance with the article herein entitled "Settlement of Disputes."

12-3 Transfer of Contract The owner should take care to see that the prospective contractor is capable and well-qualified before the contract is awarded. It follows that the successful bidder should not be entitled to transfer the contract to someone else unless such action is first approved by the owner.

A clause to prevent such transfer might be stated this way:

This Contract is made in reliance upon the qualifications and responsibility of the Contractor, and any advance payments made hereunder are intended to assist him in part in the financing of the performance of the Work. Therefore, the Contractor shall not assign or transfer this Contract or any part thereof without the written consent of the Owner. However, the Contractor may subcontract portions of the Work to be performed hereunder to such persons as the Engineer may expressly approve in writing.

Assume that you as the owner plan to build a house. It will be a major investment for you. Therefore you will wish to make a contract for its design with an architect in whose character and skill you have confidence. Now suppose that, after starting the design of your house, this architect is selected to design two or three large jobs. As a result he wants to transfer your contract to an architect friend of hers whom you do not know. Now you engaged her for her known skills, and you want her to do the design work personally, not have someone else do it.

Again, suppose that your house has been designed and that you have negotiated a contract with a particular builder because you trust him and believe that he will provide for you the best possible structure. You naturally would oppose transfer of the contract to anyone in whom you do not have equal confidence.

There may be conditions, however, in which a transfer of the contract is desirable for the owner as well as for the contractor. Sickness of the contractor, death of some key personnel, financial difficulties that have befallen, labor trouble that prevents proper prosecution of the work are all examples. Transfers, though, are likely to be accompanied by considerable difficulty in the matter of adjusting payments for work done by the new contractor and the old. Transfers are also likely to result in delays, confusion as to responsibility, and difficulty on the part of the new contractor in picking up the work where the other one left off. Transfer of the responsibility for completion of the contract should not be allowed unless circumstances require it.

12-4 Revisions It often happens that revisions of plans prove expedient or imperative after a contract has been awarded.

It is important for an engineer to have the proper attitude toward revisions, and any decisions made should be based upon good judgment. A few principles to bear in mind are the following:

1 Any error should be corrected promptly.

2 Matters involving the safety of personnel, structures, and equipment should receive top priority.

3 When something bearing upon satisfactory operation is discovered to be amiss, especially in the case of a moneymaking or self-liquidating project, the indicated changes should be made unless they entail undue hardship or expense. The structure or machine is built once; it is to be used or operated continuously thereafter. Additional trouble and expense in the building of it may be offset many times over by saving in operation.

4 Modifications that are merely the result of differences of opinion and that will have little real consequence are to be avoided unless they can be made early enough to prevent difficulty in execution.

5 Revisions that involve substantial changes in quantity or quality of work should not be effected until the contractor and the owner have agreed upon what is the allowable additional expense involved. This agreement should be confirmed in writing.

6 If a revision constitutes so major a change that it greatly affects the scope or character of the work, the matter may have to be treated as requiring a negotiated modification of the original contract or the preparation of a new and separate contract covering the proposed additional work.[2]

All revisions should be in written form. A telephone call may be an excellent way to pass along information in a hurry, but it should be followed up promptly by written confirmation of the instructions. This confirmation may be in the form of a letter, or it may be the revision of a drawing, which is also a method of recording information. It is advisable to send a letter of transmittal with any revision of a drawing, calling the contractor's attention to the fact that the drawing has been revised.

It is also desirable for the contract to contain a clause stating that the contractor is not to proceed with the execution of revisions until official data and instructions in writing have been received. Similarly, after having been notified officially that certain revisions are pending, he is not to proceed with work that will be affected thereby because doing so would probably result in even greater cost and trouble in carrying out the revisions later.

Assume that the engineer telephones to a steel fabricator stating that the plates for some coal bunkers in a power plant are to be made $\frac{1}{2}$ inch thick instead of $\frac{3}{8}$ inch as shown on the original contract drawing. The fabricator should not go ahead with further work on these plates or on parts that will be affected by their change in thickness, nor should he go ahead with work on new $\frac{1}{2}$-inch plates until official revisions are received. However, if the engineer telegraphs this informa-

---

[2] *Albert Elia Bldg. Co., Inc. v N.Y. State Dev. Corp.,* 54 A.D.2d 327 (1976).

tion, this constitutes notice in writing, and the contractor may proceed accordingly. The engineer should still confirm the change by a revision of the drawing.

The contract documents should state that the contractor must carry out all revisions required by the engineer, who is the best judge of their desirability or necessity. However, in the usual case the engineer should not order the contractor to proceed until the extra expense involved, if any, is settled, or until the contractor has informed him that the revised work will be done under the unit prices stated in the contract.

Furthermore, the contractor should be obliged to notify the engineer in writing within a stated period (such as seven days) after receipt of the official revisions if he considers them to be such as to require additional compensation. He should submit a statement of how much the extra cost will be and how the amount was calculated, thus allowing the engineer an opportunity to cancel the revision or modify it if the price is not satisfactory.

12-5   Extras in General   The contract should make definite and adequate provisions for the handling of "extra" costs, whether the contract is on a lump-sum or a unit-price basis. This matter is especially troublesome in the case of lump-sum contracts. Extra charges may result from ordinary revisions made by the engineer, from a substantial increase in the quantity of work required, or from a real change in the quality of materials and workmanship from what was specified in the contract. Also, something unforeseen may be encountered, and the performance of unanticipated work may become essential. If so, the additional task simply has to be accomplished regardless of extra costs involved.

One way to handle the problem of extras caused by revisions of a lump-sum contract[3] is to specify the procedure in such a way as the following:

> If revisions made by the Engineer cause an increase of the Work to be done under the Contract or in the cost of said Work, the Contractor shall estimate or keep records of the increase in the cost of labor and materials resulting from the revisions. Upon approval of this estimate or record by the Engineer, this computed extra cost plus fifteen (15) percent thereof for overhead and profit shall be payable to the Contractor as extra compensation because of said revisions. On the other hand, if the revisions of the Contract result in a cost reduction, the Contractor shall estimate such reduction in the cost of labor and materials. Upon approval of this estimate by the Engineer, the stipulated sum shall be deducted from the Contractor's bid price, but this deduction shall not include any portion of the original allowance for anticipated overhead and profit.
>
> The preceding paragraph is to apply to Work performed by subcontractors as well as to that performed directly by the general contractor. However, in the case of extra costs because of revisions made by the Engineer in subcontracted Work, the stipulated allowance of fifteen (15) percent of the approved additional estimated cost or approved actual cost of the extra Work is to be paid as an allowance for overhead and profit to the subcontractors only, no additional percentage being payable to the general contractor.

---

[3] A lump-sum contract may contain a clause setting forth unit prices for extra work done and credits for work omitted at the behest of the engineer; such a clause is often called the *additions-and-deductions provision*.

The above provisions are intended to minimize the likelihood of disputes regarding the contractor's claims for extra compensation by reason of changes made in the plans or specifications by the engineer. Provision for extras is not intended to remedy misjudgment, inefficient management, or other shortcomings attributable to the contractor or the subcontractors and which may have caused the cost of performance to be greater than originally anticipated.

If the contractor goes ahead with extra work, duly ordered by the engineer with the understanding that extra expense for the owner will be involved but before the receipt of written instructions to do so, there is probably a clear understanding that this extra work will be paid for by the owner.[4]

12-6   Extras Caused by Subsurface Conditions   Subsurface conditions on a construction job are a prolific source of trouble. The engineer should be very careful to obtain all the advance information that reasonably can be secured, having adequate explorations made and presenting the results to all bidders. But he should not try to interpret, extrapolate, or guarantee the data given in the logs of the borings, even though he has to prepare his design of the structure on that basis.

In the construction of the caisson foundation for a certain bridge, the record of the borings was given to the bidders. However, boulders caused much more difficulty than the successful bidder expected. Later he brought suit for recovery of the extra cost over and above his estimate. His claim was rejected by the court because he had been given all the information that the engineer possessed. In glacial territory, such as that involved in this case, boulders are not unusual. Miscalculation on the part of the contractor was adjudged not to be the fault of the engineer.

In another case a contractor was to build a four-story reinforced-concrete building on pile foundations at an old waterfront. The owner had some soil borings made and furnished the logs of these borings to the bidders. Later, when the contractor was driving piles, an old sea wall was encountered, the removal of which caused considerable expense. He claimed an extra for this work. The court granted his claim because the owner possessed old plans showing such a sea wall, but he did not show them to the bidders—probably because he had forgotten about the existence of the plans. Nevertheless, the fact remains that he failed to present all the information at his disposal.

In a third situation, the record of the boring taken at the center of the site for a large concrete structure indicated bedrock to be at a given elevation. Later it was discovered that the boring had merely hit a large boulder. The extra excavation to sound rock had to be done under very difficult conditions, and the contractor could not readily estimate in advance how much the extra expense would be. The engineer therefore approved the contractor's proposal that the additional work be paid for on a cost-plus-percentage basis.

---

[4] The contract requirement of a written order for extras would not seem to bar the contractor from seeking payment in such a situation.

The following paragraph shows one suggested way to phrase a clause about borings:

Borings have been made at the site for the Owner. The locations of these borings and the information obtained from them are recorded on the reference drawings. Samples of the soils removed and the records made during the boring operations may be examined at the office of the Owner. However, in transmitting information regarding the materials expected to be encountered, the Owner does not guarantee the accuracy or completeness of the data given, and he assumes no responsibility therefor. The Contractor shall draw his own conclusions from whatever information is available, through the Owner or otherwise.

12-7   Extras Caused by Additional Quantities of Work or Materials   Additional quantities of word ordered during the progress of operations may entitle a contractor to claim an extra on a lump-sum job. This may not be so for a unit-price contract when the contractor receives proper payment through the application of the unit prices.

The borings at the site proposed for a large industrial building showed that pile foundations would be needed along one side. The plans were made accordingly, and the contract was let on a unit-price basis. It was later discovered that one boring taken near the center of the structure had encountered a knob of sand that was surrounded by an area containing very unsatisfactory soils. This necessitated the use of much more piling. Since it was a unit-price contract, the owner and the contractor agreed that the extra piling, concrete, etc., should be paid for at the bid unit prices. It would have been unfair to penalize the contractor for this unforseen additional work, even if the contract had been for a lump sum.

As another example, suppose that one unit-price item in a contract covers the payment for 1000 poured-in-place concrete piles with an assumed length of 60 feet at $10 per foot of pile. The assumed length of piles stipulated in the contract is such that the contractor can use equipment already in his possession. When the driving is started, it is found that in order to attain the specified resistance the piles have to be 10 feet longer than originally expected. Because the piles are too big to be driven by his own equipment, the contractor will have to effect a costly equipment rental. Besides the extra cost represented by the rental, the greater length of the piles will cause the contractor so many additional worker-hours of labor, rendering the bid unit price inadequate. Therefore, the contractor should be entitled to extra payment.

Where a building or construction contract has been fully performed by the contractor, his subsequent promise to do extra work must, to bind him, be supported by consideration. This further promise is supplemental to the original contractual undertakings.

12-8   Extras Caused by Additional Difficulty of Performance   Whether the work is increased in quantity or not, unlooked-for difficulty in the way of performance may be a just cause for the contractor to claim extra compensation. An example of this occurred in the construction of the vehicular tunnel at West 178th Street, New York City. The contract was on a unit-price basis. A deep subterranean ravine was found to cross the site between two of the borings. It contained many heavy boulders, and it was the course of an underground stream. The contractor was paid the regular unit prices on all extra excavation, concrete, etc. The work, however, was more difficult than that called for by the contract papers, and a considerable extra sum was paid to compensate the contractor for the additional expense of surmounting the unexpected difficulties.

Concrete construction is another prime source of claims for extra compensation by reason of more difficult work than anticipated, especially in connection with forms and reinforcement. All too often the owner insists that a few general drawings be rushed to "completion" and that unit-price bids for excavation and concrete be secured just as soon as possible. Naturally, the engineer cannot show on the drawings all the ramifications that will ultimately have to be developed— and he certainly cannot predict them at this early stage of the project's development. The contractor therefore makes his bid upon the basis of certain estimated quantities and the expected degree of difficulty. Later on, the owner may be shocked by the extra costs claimed by the contractor for more forms and steel per cubic yard than the owner had been led to expect. The engineer's position in such circumstances is difficult, to say the least.

By way of example, a large plant required about 2,000 linear feet of subterranean tunnels for electric power distribution. The drawings used for bidding simply showed these tunnels in plan by dotted lines and portrayed one typical cross section. Later on it was found that many manholes were needed in the top or sides for conduits and that branch tunnels were required. So also were a number of stairways for escape hatches and fan houses for ventilation of the tunnels. The contractor presented a claim for additional payment because of this more complicated work. All these extra items were really necessary and had to be provided. the contractor's claim was found to be justified.

12-9 Extras Caused by Reduction of Work A reduction in the quantity of work may also be a proper basis for an extra under a unit-price contract. Thus, a contract had been entered into for some extensive pile driving at a harbor. The engineer later decided to eliminate about 25 percent of the piles. Because of the resulting loss of profit and of a portion of the allowance for overhead built into the quoted unit price for piles, the contractor was able to collect an extra payment above that made in accordance with the bid itself. The decision to honor a claim for an extra was based upon the fact that the reduction in work was substantial. But few piles more or less (say, 5 percent) than the estimated number or the number shown on the drawings generally will not entitle the contractor to an extra if the contract is on a unit-price basis. The contract should state the permissible variation in quantities in order to avoid possible controversy.

Here is a suggested clause dealing with variation in quantities:

> The Contractor's proposal is to be based upon the estimated quantities shown in the "Schedule of Prices." Inasmuch as the quantity of concrete and the number of piles

driven may be affected by conditions encountered in the performance of the Work, or by written instructions to be issued by the Engineer as the Work progresses, the quantity of work done under these two items may vary considerably from the estimated quantities. In this event, payment for any overrun of these quantities shall be made to the Contractor at the bid unit prices. However, if the quantities of either or both of these items actually installed are less than ninety (90) percent of the respective estimated quantities stated in the schedule, the Contractor shall be entitled to such adjustment of the bid unit prices for the affected item or items as the Engineer believes will fairly compensate the Contractor for the loss of profit caused by the decreased amount of Work done under the Contract.

12-10 Responsibility for Extra Costs It is very important to determine whether the owner and the engineer are responsible for unanticipated costs or whether these are occasioned by fault of the contractor.

Undoubtedly the owner is obliged to pay when the information given by the contract documents is inadequate to enable the contractor to determine in advance what work he will apparently have to do. The owner is also obligated (1) when inaccurate data are given by the engineer to the bidders, (2) when the extras develop because of improper design and plans in the first place, and (3) when the engineer makes serious changes after the contract is signed.

If the contractor performs the work as required by the contract but unforeseen conditions arise, through no fault of his, necessitating modifications of the design, and if these modifications are approved or ordered by the engineer, then the extra costs involved are properly to be paid for by the owner. In one case a contractor was to build a 150-foot brick stack at an incinerator. No borings had been taken at the site, and the engineer had designed the footing of the stack as though it were to be supported upon firm ground at the specified elevation. However, under the stack it developed that there was deep clay overlain by a few feet of sand and gravel. The contractor was afraid that the compaction of the clay might cause harmful settlement of the stack. Finally the engineer made an investigation of the soils and decided to support the stack on piles. The new design of the foundation required considerable additional construction. This was obviously to be paid for by the owner as an extra charge.

Conversely, extra costs resulting from the mistakes and ill-advised practices of the contractor are his own responsibility. In one instance a contractor tried to pour some heavy concrete foundations during extremely cold weather without first thawing out the existing concrete. Ice lenses formed at and near the junction of the new and the old work, causing small, open, horizontal cracks after the ice lenses melted, the upper portion of the concrete being supported by the reinforcing bars which acted as tiny columns. The contractor was responsible for the replacement and repairing of this work because the contract stated that no concrete was to be poured against frozen surfaces of any kind.

It may be that extra costs in a given situation are caused by both the owner and the contractor. Such costs should be apportioned by agreement of the parties. Thus, a contract for a large industrial plant could not be completed on time unless at least some of the construction was performed during cold winter weather. This was because of delays in getting started, some of which were attributable to the owner and some to the contractor. The contract had called for discontinuance of work from November 30 to April 1. However, to expedite the project and to enable the contractor to proceed continuously and thus to avoid undue loss (despite expensive winter operations), the owner made a special agreement with him to take certain parts out of the former contract and to place them in the new one, which was to be carried on during the winter. This was a compromise arrangement of sorts between the two parties.

The fact that the price was higher than either of the parties anticipated is not part of the proof in a cost-plus construction contract, nor is reasonableness. Absent come claim indicating fraud or gross negligence, it is not incumbent upon the person providing the service to prove that charges for labor and material were reasonable.[5]

12-11   Errors Discovered by the Contractor   The contract should require that the contractor is to report immediately to the engineer any error discovered in the contract drawings or in any other data furnished by the engineer. The contractor should also report at once any detected error made by his own staff and should cease work immediately on the affected part of the job until the engineer can issue instructions regarding what is to be done in the matter.

On a certain job the contract drawings showed the doors of a factory to be hinged so as to open inward, contrary to the applicable building code. The contractor admitted he had timely knowledge that this was an error but had nevertheless gone ahead on the doors without notifying the engineer. Thus the contractor was later obliged to absorb the cost of reversing the doors.[6]

The following clause shows fairly standard wording used to prescribe the procedure in this matter of correcting errors:

> If the Contractor discovers any errors or omission in the Contract Drawings of Specifications or in the Work undertaken and performed by him, he shall immediately notify the Engineer and the latter shall promptly investigate the matter. If, knowing of such error or omission and prior to receiving instructions from the Engineer regarding correction thereof, the Contractor proceeds with any Work affected thereby, he shall do so at his own risk and the Work so done shall not be considered as Work done under the Contract and in performance thereof unless and until duly approved and accepted by the Engineer.

12-12   Delays   Unless the contract specifically states otherwise, the contractor is generally entitled to recover losses sustained because of delays caused by the owner. The latter is regarded as having an implied obligation to provide all drawings and other information necessary to enable the contractor to complete the job

---

[5] *Romine v. Rex Darnall, Inc.,* 541 S.W.2d 50 (Mo. App. 1976).

[6] Again, the reader should notice that the contractor's knowledge of an error in the plans even prior to signing the contract destroys any reasonable claim for extra compensation for correcting the mistake provided he did not call the error to the engineer's attention at that time. See *Wickham, Inc. v. U.S.,* 546 F.2 355 (Ct. Cl. 1976).

within the specified time. If, on the other hand, the contract states that the "contractor herewith specifically waives claims for damages for any hindrance or delay," it means just that.[7] If such a clause is in the contract, the contractor will probably add to the contingency item in his bid an amount he believes sufficient to cover such eventualities. This may mean that the bidder will submit a higher price than he otherwise would.

A better way to handle this problem is to state in the contract that, in case of any act or omission on the owner's part causing delay, the contractor will be entitled to just recompense and to an extension of time.[8] However, the contractor should be required to give written notice of his claim for extension and of the cause for it within a stated number of days after such cause arises. Furthermore, the engineer should have the specific power to grant what is considered to be a proper extension.

Delays brought about by the contractor are naturally his responsibility, and he can be counted upon to utilize every effort to prevent such delays. The engineer should not set an impractical completion data, for if his requirement is unreasonable, the contractor will have to add a sum to his bid to cover possible penalties or else will have to include extra expenses for overtime work and other costs that arise when a job is to be rushed. Before allowing only a marginal amount of time for completion, the engineer should be satisfied that the advantage to be derived for the owner from speed of execution is worth the extra charge that may well be added in the bid.

Delays caused by "acts of God" are not the fault of the contractor, or the owner, or of any other human being. It is often difficult, however, to establish whether or not a particular happening is truly an act of God. For example, assume that a contractor has constructed a cofferdam around the area in which he is building a bridge pier in a river. A big flood occurs. The water overtops and seriously damages the cofferdam, causing a delay of three weeks in repairing it. Who is to blame? If the cofferdam was not made sufficiently high and strong to withstand floods that are known to occur in that river at least once in three or four years, the contractor will probably be held responsible. However, if he made the cofferdam strong enough so that it would have withstood the worst flood that had occurred during the preceding ten of fifteen years, he would probably be considered to have taken reasonable precautions and would be relieved of any responsibility for the flood damage. The owner (or his insurance company, if involved) would presumably have to pay for the reconstruction in that event.

The contractor is entitled to prompt decisions on the part of the engineer. In a large dredging job in a harbor, the owner decided to make some important

---

[7] *S. L. Roland Constr. Co., v. Beall Pipe & Tank Corp.,* 141 Wash. App. 297, 540 P.2d 912 (1975). See also Chap. 6, note 7.

[8] A contract was let for the construction of a large stack at an industrial plant. It was to be built in the vicinity of a similar stack that emitted sulfurous fumes. The owner realized that work might be delayed by these fumes when the wind carried them across the site of the new structure. Therefore, a clause was inserted stating that any delays resulting from the problem of fumes would mean a corresponding and automatic setting back of the scheduled completion data.

changes after the contract was under way. The job had to be stopped until the owner settled various questions concerning the revisions. Each day of delay cost the contractor an estimated sum of $1000 in rentals, overhead, and unavoidable miscellaneous expenses. It is only right that the contractor should be compensated for the cost entailed by such a delay. If the delay were to continue for a long time, he might well claim in addition the loss of profits that he could have made on some other job.

Here is a sample clause dealing with the subject of delays:

> If the Engineer believes that, on account of any act or omission of the Owner or the Engineer and without fault of the Contractor, any of the plant, equipment or salaried men being used by the Contractor in performance of the Contract are necessarily kept idle for more than seven (7) consecutive calendar days, the Contractor shall be entitled to increased compensation. This increase shall be an amount equal to that which he may pay during such period of idleness for the availability of such idle salaried persons and for other related expenses, plus a sum computed at the rate of fifteen (15) percent per annum on the sum which the Engineer deems to be the actual value of idle plant and equipment, all computations being based upon the total duration of the delay.

In the preceding clause the stipulation of a minimum time of idleness which must elapse before the contractor is allowed to make a claim against the owner for damages caused by delay resulting from an act of the owner may not be entirely wise or warranted. Such a limitation may cause the contractor to increase the contingency item in his bid.

Generally speaking, the contractor or subcontractor should be granted an extension of time not only for delays caused by acts or omissions of the owner or of the engineer but also for delays stemming from occurrences such as the following:

1 Acts of God
2 Acts of the public enemy
3 Acts of the government
4 Acts of another contractor over whom the first contractor has no control
5 Fires
6 Epidemics
7 Quarantine restrictions
8 Freight embargoes
9 Strikes

12-13 Subcontracts The term *subcontract* is used to denote an agreement between one contractor and a second one under the terms of which the second undertakes to handle for the first a particular portion of the job for the completion of which the first has become obligated. By no means is the subcontract of inferior character or importance, and it has all the characteristics of contracts in general.

In large construction contracts it is customary for a general contractor, as the successful bidder, to parcel out parts of the work to others who are specialists in particular lines. The prime contractor will be responsible for coordination and completion of all work by these others, and he should be confident that each one is a concern with which he can work successfully. A general contractor for a structure such as a hospital will quite typically subcontract the heating and ventilating work to one company, the plumbing to a second, the electrical installation to a third, and even the structural-steel fabrication and erection to a fourth.

The contract for a building job on which a great deal of subcontracting was to be expected contained the following clause, intended to limit the amount of work that the general contractor could farm out: "The Contractor will be required to perform with his own organization at least thirty (30) percent of the work."

One contract contained the following clause as a precaution against misunderstandings regarding the precise work to be done by subcontractors: "For convenience of reference, the Specifications are separated into divisions (e.g., 'Plumbing'). These shall not, however, operate to establish arbitrary limits to the contracts between the Contractor and subcontractors." In other words, the subcontracts are to call for whatever work it is desired to have them cover. The arrangement may or may not be based upon subdivisions of the specifications, depending upon the general contractor's wishes.

The general contractor may make tentative or binding agreements with the subcontractors ("subs") before making his proposal, which would be based in part upon their offers to him. A subcontractor normally cannot withdraw from his agreement after the contractor has accepted the sub's bid, even though the principal contract has not yet been awarded.[9]

In other cases, the contractor may shop for bids from subcontractors after he has submitted his proposal and secured the general contract. This practice has been abused sometimes, and it is distasteful to many of the specialists, who generally get their work through subcontracts from others. In some states the law requires that, for state and municipal buildings, the heating, plumbing, air conditioning, and electrical work be let as separate contracts instead of being included under the general construction contract. This enables the specialists to bid for work on a direct basis and avoids the evils of shopping for bids, but it may cause difficulties in the way of coordination[10] of work, extra paperwork, additional costs, rivalries, uncertain division of responsibility, and even the lack of desirable cooperation among the parties involved in the overall job.

Two ways for the engineer to help prevent unfair treatment of subcontractors by the general contractor are the following:

1 The engineer specifies that the bidder is to state the names of his intended subcontractors in his proposal and that no change in this list is to be made without the engineer's approval.

2 The engineer specifies that the contractor is to make binding agreements with his selected subcontractors before the presentation of his proposal and that he is to use their quotations in the preparation of his bid.

[9] *W. M. Heroman v. Saia Electric, Inc.,* 346 So.2d 827, *app. den.,* 349 So.2d 1271, (La. App. 1977). *Montgomery Ind. Inc. v. Thomas Const. Co.,* 620 F.2d 91 (1980), applying Texas law.
[10] In such a case the owner's engineer is compelled to perform some of the duties of a general contractor in coordinating the work.

This approach seems to penalize the owner through higher costs because the bidder cannot adjust his estimate and bid in the expectation that in like manner he will be able to negotiate the quotations received from subcontractors. This disadvantage to the owner may prove out to a limited extent, but such procedures make for better relations with all concerned and tend to avoid lowering the quality of materials and workmanship—a device which the subcontractors might attempt in order to avoid possible financial loss to themselves if their "beaten down" compensation was of questionable adequacy. In addition, either of the suggested contract requirements might prevent the general contractor from trying to increase his profit (after the contract is signed) at the expense of the subcontractors.

The Canadian Construction Association has devised a system called *bid depository* to meet the problem of bid shopping. The idea is as follows:

1 A party (such as a builders' exchange) is appointed to act as depository for all proposals.

2 The architect or engineer specifies in the instructions for bidders that proposals made by bidders for mechanical, electrical, and other work to be subcontracted shall be delivered to the depository at a stipulated time.

3 The subcontractors send their proposals to the depository in separate envelopes, one for each general contractor who is bidding on the job and to whom they are quoting.

4 The envelopes are sealed and dated by the depository, and receipts are given to the would-be subs. No late bids are accepted.

5 Promptly after the closing time for receipt of subcontractors' bids, the depository mails out the envelopes for the architect, engineer, and each contractor; then it puts its own copy in a safe.

6 The general contractors use the bids submitted by the subs in preparing their own proposals, which are to be sealed and to be delivered a specified number of days later.

7 The general contractors' copies of the bids by subcontractors can be checked by reference to the copies retained by the architect, the engineer, and the depository if such action becomes necessary or desirable. The information on the bids of the subs is not made public.

This depository system provides a means for detecting (and thereby practically preventing) bid shopping by general contractors. The procedure can be modified to suit local conditions.

A group interested in this subcontracting problem recommended the following procedure for competitive bidding practices:

Provided this information is requested by the architect or engineer, general contractors shall include in their proposals the names of subcontractors (in each trade) whose bids were used in making up such proposals.

General contractors shall not be allowed to substitute other subcontractors in place of those named without written approval from the architect or engineer.

General contractors shall not be required to give the amounts of the subcontractors' bids they used in making up their basic proposals.

Subcontractors' proposals shall be delivered in writing to the general contractors at least twenty-four hours prior to the closing of bids (on the principal contract) and such proposals shall be held in confidence by the general contractors who receive them.

The receiving of proposals from several subcontractors (in any single line of work) by any general contractor shall be discouraged.

At least five days prior to the closing of the bids from general contractors, architects and engineers shall mail to the general contractors, who are figuring a project, all addenda and bulletins clarifying their plans and specifications.

Unit prices in general shall be eliminated excepting those justified for certain types of work, but (are acceptable) in those cases based upon specific conditions applicable to the particular items involved.

Alternates shall be generally eliminated by architects and engineers.

Plans and specifications shall not be issued to the contractors until the scope of the work is completely defined in sufficient detail to enable accurate bidding.

These recommendations seem worthy of consideration, especially for public work.

A clause such as the following would represent rather poor treatment of a subcontractor by the general contractor: "The subcontractor shall perform all work under this trade and furnish all required materials, whether or not they are mentioned in the Contract *or in any other document.*" The subcontractor who signed such an open-end agreement might suffer severe losses.

Frequently, the owner and the engineer are the direct cause of difficulties in subcontract matters. They rush out a contract and do not prepare the drawings and specifications properly. Both the general contractors and the subcontractors are reluctant to bind themselves on the basis of such incomplete data. This causes the general contractors to make additional estimates of their own or to add some general cost per square foot (or other unit) to cover subcontract work. Then, after the contract is let, the successful contractor tries to make the most advantageous terms possible with subcontractors.

The contract should state that the subcontractors are to be approved by the engineer before the general contractor awards them the contracts for their special work. This is proper because the owner's interests are to be protected in each subdivision of the work. The engineer should be satisfied regarding the ability, resources, and trustworthiness of each subcontractor, but the engineer has no right to interfere in the matter of acceptance or rejection of (approved) subcontractors by the general contractor on the basis of price.

It is desirable to indicate clearly in the contract documents that the general contractor is responsible for the conduct, product, and payment of his subcontractors. Broadly speaking, any contract claims brought up by a subcontractor are the general contractor's responsibility. Even preparation of the form for a subcontract is normally the general contractor's concern. All that the owner and engineer have the right to do is to approve or disapprove, under certain conditions, the letting of a subcontract to a particular organization.[11]

It should be noted that, when a subcontractor agrees to do a part of the contract work in accordance with plans and specifications that are binding upon the general contractor, the subcontractor is bound by these same plans and specifications and is responsible to the prime contractor for compliance. A true subcontractor agreement is made with the general contractor, not with the owner. Therefore, the owner looks to the general contractor with respect to the performance of the job as a whole. If the general contractor abandons the job and the subcontractor nevertheless agrees with the owner to complete the subcontract work, a new pact arises between the owner and the former subcontractor.

One major construction contract in New York State employed substantially this wording regarding subcontracting:[12]

> No consent by the Engineer to any assignment or other transfer, and no approval by the Engineer of any subcontractor, shall under any circumstances operate to relieve the Contractor or his sureties of any of his or their obligations under the Contract or the Performance Bond; neither shall any subcontract or approval of any subcontractor cause or be deemed to create any rights in favor of such subcontractor against the Owner. All assignees, subcontractors, and transferees shall be deemed to be agents of the Contractor. All subcontracts and all approvals of subcontractors shall be understood to be based upon the requisite of performance by the subcontractor in accordance with this Contract; and, should any subcontractor fail to perform his Work to the satisfaction of the Engineer, the Engineer shall have the absolute right to rescind his approval at once and to require the performance of such Work by the Contractor himself or entirely or in part through other approved subcontractors.

12-14 Partial Payments It is difficult for a contractor to finance from his own funds all costs for a large contractual undertaking that extends over a period of several months to a year or two. He will generally have to borrow money. Part of the interest on this borrowed money will be added into the contractor's estimated costs and will be included in his bid. Therefore the owner will ultimately have to pay the contractor's borrowing expense.

To improve the picture, it is good practice to arrange for partial payments to the contractor during the course of the job. One convenient method is provision for payments once a month, each payment being based upon an estimate of the value of the work or portion of the contract that has been completed during the preceding month. The mechanics for the determination of the size of these payments should be specified in the contract. The idea of partial payments is applicable to lump-sum contracts as well as to unit-price contracts. Of course, cost-plus

---

[11] This right is reserved in order to protect the owner in case he and the engineer have what they regard as good reasons for distrusting a proposed subcontractor.

[12] This sort of statement must perforce be spelled out also in the subcontract, since it alone controls the subcontractor.

contracts are financed directly by the owner, but the contractor's fee or percentage may be paid in installments or as a final settlement.

If the contract states that partial payments are to be made to the contractor as the work progresses, this commitment must be honored.[13] Even when the contract fails to specify such a procedure, it may be utilized if full financing by the contractor alone proves unrealistic.

Partial payments should not be based upon elapsed time alone. If a contract is supposed to be completed in ten months but the contractor inexcusably takes twelve months to complete it, he should not be paid in ten equal monthly installments that really prepay him for tardy work. On the other hand, if he completes the work in eight months, he should be paid on that basis. Partial payments should be related to accomplishments, not to intentions.

Assume that a unit-price contract has been signed for a reinforced-concrete bridge and its approaches costing an estimated $1 million, to be completed in twelve months. The unit prices cover such items as excavation, concrete, and reinforcement. At the beginning of a given month the engineer and the contractor may make independent estimates of the quantities of work performed during the preceding month, or the engineer may check the contractor's estimates of the work done.[14] In either case both parties are to agree upon the figures or upon a compromise estimate. The engineer should have the authority to state what estimated figures the payment will be based upon if such decisive action becomes necessary because of disagreement between the parties.

Similarly, if the bridge contract is on a lump-sum basis for $1 million, partial payments might well be determined and made on the basis of work accomplished. To do this, the engineer could establish imaginary unit prices, and he might make his own estimates of quantities, and thus of the payment due, practically in the same way as for a real unit-price contract. Or he might specify that, subject to approval of the several segments of the work, the contractor is to be paid as follows:

1 $50,000 upon completion of the excavation and cofferdams or shoring for the foundations
2 $100,000 upon completion of the river piers
3 $75,000 upon completion of the abutments
4 $100,000 upon completion of the two side spans

---

[13] *Guerini Stone Co. v. P.J. Carlin Constr. Co.,* 248 U.S. 334, 345 (1919): "As is usually the case with building contracts, it evidently was in the contemplation of the parties that the contractor could not be expected to finance the operation to completion without receiving the stipulated payments on account as the work progressed. In such cases a substantial compliance as to advance payments is a condition precedent to the contractor's obligation to proceed." See also, however, *U.S. for the Use of Bldg. Rentals Corp. v. Western Casualty & Surety Co.,* 498 F.2d 335 (9th Cir. 1974), where a delay or deviation in payments withheld were found not to justify rescission of the contract.

[14] These computations of quantities take time. Therefore, the contract might well state that "the Contractor is to submit to the Engineer by the tenth of each month the estimated quantities of the work performed by him during the preceding month."

5 $375,000 upon completion of the three main spans

6 $100,000 upon completion of the approaches

7 $200,000 upon completion and final acceptance of the entire job

In any case, the procedure for making partial payments is to be specified in detail in the contract so as to avoid future misunderstandings.

It often happens that materials are furnished or delivered but that the construction involving them is not completed. Technically, the bid unit prices may not be applicable to such items, but the engineer should estimate the value of these materials and, in fairness to the contractor, include such in the calculation of the partial payment. By way of illustration, suppose the structural steel for a large building is contracted for at a certain unit price as "erected." At the end of a particular month all the material has been rolled, 50 percent has been fabricated, 25 percent has been delivered, but none has been erected. It is proper and logical to make some partial payment on account to the contractor who in turn can renumerate the steel company.

The following clause has been paraphrased from a lump-sum contract and is included here for the purpose of showing a poorly worded clause on partial payments:

> Payments will be made monthly in an amount equal to eighty-five (85) percent of the value of all labor and material incorporated in the work and/or suitably stored on the site during the preceding month until eighty (80) percent of the work has been completed. Monthly payments will then be increased to ninety-five (95) percent of work accomplished.

The following relate to this clause:

1 The engineer should specify the day of the month on which the payment is to be made.

2 The clause should state that the quantity of work completed each month (and its assumed value) is to be determined from the estimates made by the engineer.

3 It is unwise to provide that materials "stored on the site" are to be included in the quantities on which partial payments will be determined. Such a provision might cause disputes regarding the value of material simply brought to the site and left there. Also, labor is not "stored."

4 It would be better to let one established percentage apply until the full job is completed and accepted and until the final payment is made.

5 Each time it is employed, the term "monthly payment" should be amplified by adding "to the contractor."

6 Such a phrase as "work accomplished" is altogether too vague. One cannot afford to be ambiguous in these matters.

12-15 Withholding of a Portion of Partial Payment It is usually desirable for the engineer to have the right to withhold 10 or 15 percent of the estimated monthly payments otherwise due the contractors, as computed from the quantity of work completed, so that the owner will have a little reserve. This gives the owner leverage of a sort, thus discouraging the contractor from quitting near the end of the job. The withheld funds can be used by the owner toward remedying poor work or finishing up if the contractor fails to complete the contract. When performance under the contract is fully accomplished by the contractor, these withheld funds are to be paid as part of the final settlement.

A unit-price contract might well contain a clause such as the following: "On the fifteenth of each month, the Contractor is to be paid ninety (90) percent of the sum obtained by applying the respective bid unit prices to the approved estimated quantities of work completed by the Contractor during the preceding month."

Under a system of partial payments, the contractor may get little compensation during the first month or two whereas during the next six or seven months the payments may be relatively large. Finally, during the last month or two, when odds and ends are being cleaned up, the payments may again be small. However, partial payments, large or small, will be of great assistance in financing his operations.

The money withheld from the monthly payments to the contractor is not a substitute for the performance bond.

12-16   Final Payment   Some suggestions in connection with the final payment are illustrated by the following clause:

> Prior to final payment the Contractor shall obtain and furnish to the Owner satisfactory evidence that the Work is fully released from all claims, liens, and demands and shall secure and furnish written consent of his sureties to final payment hereunder. Furthermore, acceptance by the Contractor of final payment shall release the Owner from all claims and liability to the Contractor for all Work done and materials furnished in the execution of the Contract. Final payment (or any other payment), however, shall not itself serve to release the Contractor or his sureties from their obligations under or in connection with this Contract or the Performance Bond.

After the final inspection and approval of the work by the engineer and after its acceptance by the owner, a general accounting is usually necessary before the size of the balance due the contractor can be determined. Normally, unless a construction contract specifies otherwise, all the work must be substantially completed before a final payment is due.

When there are no extra claims by the contractor and no credits or deductions to be made in favor of the owner, the amount due the contractor on a lump-sum contract is simply the bid price less the sum of all partial payments that have been made. Use of a unit-price contract makes it necessary, first, to compute the final quantities for each payment item, then to multiply each such quantity by the applicable bid unit price, and from these products to calculate the total cost of the contract. This total minus all partial payments is the net sum owed to the contractor. Usually the owner and the contractor can agree upon the calculated quantities and the final computed cost. In case of disagreement, the engineer may determine what the computed costs are to be, a third party may make an independent calculation, or the questions may be settled by arbitration. The contract should state which of these three procedures is to be followed.

The general contractor will usually withhold from interim payments to the subcontractors the same percentage as that held back by the owner,[15] making final payment to them only after he has been completely paid off. This practice can be inequitable. Thus, a subcontractor for foundation work may complete his part of the job and have it accepted early in the life of the contract; nevertheless, he may have to wait months before he can collect any withheld percentage that is still due. It would be more equitable if subcontracted work is sufficiently definite and independent to have the master contract state that, upon approval and final acceptance of subcontractor's work, the owner is to pay the contractor therefor and that the latter is to remit promptly to the subcontractor.

Another way to aid the subcontractors would be a statement in the contract to the effect that, when final payment is made by the owner to the contractor, the latter is promptly thereafter to pay the subcontractors any amounts previously withheld from them and properly due to them.

12-17   Failure to Make Payments   If the owner fails to make payments to the contractor as and when provided in the contract, the contractor ought to have the right to suspend the work, but he should be ready to renew his activity if the payment is delivered before the expiration of some prescribed time limit. However, since such a delay almost always causes additional expense to the contractor, he should be entitled to make a claim for the recovery of all the cost directly chargeable to such interruption. A contractor is well advised to keep appropriate records of costs attributable to those delays which are the fault of the owner.

It seems that continued failure by the owner to make payment ought to entitle the contractor to terminate the contract after a stipulated written notice to the owner of the contractor's intention to do so (assuming payment is not made in the meantime). In the event of such termination, the contractor should be able to recover the cost of all materials actually provided, all work done, and all losses and damages suffered by him as the result of the owner's breach.

12-18   Liquidated Damages   The Invitation for Bids may contain the words "Time is of the essence," meaning that the time of completion of the project is one of the important items for the contractor to consider when bidding on the job. The contractor will be bound to meet a designated completion date or else some specified compensation will have to be made to the owner. Such payments for failure to meet the required date are referred to as *liquidated damages,* not as penalties.

A toll bridge, a factory, an apartment house, a store, or any project that is to be a profit source for the owner will be in the class where relatively quick completion is highly important. Even though the typical owner may have been hesitating for

---

[15] One contract put it this way: "The Contractor shall pay to each of his subcontractors, not later than the fifth day after receipt of each payment made to him by the Owner, the respective amounts allowed the Contractor on account of the work performed by such subcontractors, to the extent of each such subcontractor's interest therein."

a considerable time before deciding to go ahead with the project, once the decision has been made the work should be completed with the utmost speed. The contractor's desire to minimize interest charges (on money which he has to borrow to finance construction) constitutes one reason for speed in converting the project into a profit-making one as soon as possible.

In setting the dates for starting and completion, the engineer might well consult interested contractors about the time necessary to prepare for and to complete the construction. If the time allowed for completion in inadequate, the contractor will have to set a high bid price to cover costs which inevitably accompany overtime work and allow for the possibility of certain liquidated damages if the work does not progress satisfactorily. Perhaps these costs will be included as an indistinguishable part of the bid prices. Nevertheless, they are very real, and the engineer should be sure that the owner realizes that specifying a short period for completion of the work is likely to prove expensive.

The liquidated damages are usually expressed in the form of a stated number of dollars for each day that actual completion of the job extends beyond the data specified in the contract. The amount of this daily figure is supposed to be the owner's estimate of what will be lost in income by reason of the delayed completion.[16] The owner generally will deduct any liquidated damages from the sum due the contractor at the time of final payment.

In the construction field, the liquidated-damages clause is the primary means of avoiding litigation by settling in advance on an agreed figure to be used in computing damages for delay in performance beyond the scheduled completion date. In general, courts have dealt kindly with the true liquidated-damages clause so long as it meets the test of reasonableness. An unusually clear recital of what a court will look for in deciding upon the enforceability of a particular damages provision is this extract:[17]

> Where, as the parties here have agreed is the fact, proof of actual damage for delay in performance of a contract is impracticable or difficult, courts look with favor upon reasonable agreements for liquidated damages made by the parties with full understanding of the situation confronting them; and such contract provisions are uniformly enforced as the parties have written them. . . . Where, however, the amount stipulated in the contract as liquidated damage for the failure of performance bears no relation to the actual damage which may reasonably be anticipated from such a failure, courts decline to enforce the terms of the stipulation. . . . Contracts falling in this latter category are regarded by the courts as contracts for unenforceable penalties. Where the contract for liquidated damages is reasonable in character, it is valid and enforceable although the party complaining of a failure of performance neglects or is unable to prove actual damages by reason of a breach of contract. . . . But it is also established that where one seeking to enforce a provision for liquidated damages is responsible for the failure of performance, or has contributed in part to it, provision will not be enforced.

---

[16] This matter of liquidated damages should not be used as a means of making a profit for the owner at the expense of the contractor. Also, it would seem fair to provide that the contractor will be given a certain bonus if he can so conduct his work as to get the structure into service before the time set as the limit. This bonus arrangement is only occasionally provided, however.

[17] *United States v. Kanter,* 137 F.2d 828, 830 (8th Cir. 1943).

Misunderstandings may arise because of uncertainties respecting the date on which payments for liquidated damages are to begin. Assume that the owner of a store signs a contract with an electrical contractor to have the store rewired and to have new lighting fixtures installed. The work is to be done within three weeks after the signing of the contract (after which period liquidated damages of $100 per day are to be imposed) but with the proviso that the owner examine and approve a similar installation which the contractor has completed in a neighboring city. The contract is not intended to be effective until the owner has made this stipulated examination and until he has found the previous job satisfactory as a sample. Receipt by the contractor of the approval notice should signal the start of the three-week period, and it is advisable to make this point clear in the contract. The notice of official "go-ahead" should be given in writing because oral notification may lead to a misunderstanding.

There may be an argument whether the liquidated damages are to be determined from the number of calendar days of delay in completion or from the number of working days in the period of overrun. If the contract states that a job is to be completed in a given number of working days after the signing of the contract or after the specified date for the commencement of work, this is usually interpreted to mean actual days that the contractor can carry on his work. It generally excludes Sundays, holidays, perhaps Saturdays if the job is to be run on a five-day-per-week schedule, days when bad weather prevents the conduct of work, periods of delay caused by the owner and his staff, and periods of delay caused by acts of God. The contract should state just what is intended in this regard.

When liquidated damages are mentioned in the contract, it is especially important for the contractor to notify the owner promptly and in writing when the former asserts that a delay is being caused by the owner. The contractor should give dates, reasons for the delay, and other pertinent data which will assist in establishing his claim for an extension.

What happens when the contractor quits the unfinished job after the time limit has expired? In one such case[18] the court held that liquidated damages applied from the time fixed for completion until the work was abandoned and for a further period of time reasonably necessary to complete the job. Of course, the owner will not be permitted to increase his recovery either by unreasonable delay in taking over the job or by failure to complete it diligently.

When all parties are at fault, apportionment of fault is appropriate to determine the amount of liquidated damages.[19]

12-19   Damages for Defective or Incomplete Work   The contract should state that the contractor will be held responsible for the quality of the work performed. The owner should make it clear that he has the right to make claims against the

---

[18] *School District No. 3 of Ford County v. U.S. Fidelity & Guaranty Co. of Baltimore, Md.,* 96 Kan. 499, 152 P. 688 (1915). See also, *Austin-Griffity, Inc. v. Goldberg,* 224 S.C. 372, 79 S.E. 2d 447 (1953).
[19] See *Aetna Cas. & Surety Co. v. Butte Meade Sanitary Water Dist.,* 500 F.Supp. 1 (1980).

contractor for defective work. Any such claims should first be approved by the engineer, but in any case caution should be exercised because the validity of the claims may eventually have to be substantiated before a board of arbitration or a court. The damages that an owner can collect for defective work by the contractor are not unlimited, and the claims should be reasonable and consistent with the circumstances.

Assume that Jones let a contract for a central-heating system to be installed in her new house. She specified the type, capacity, and general character of the entire heating system, including the boiler. The contractor warranted the equipment and workmanship to be of top quality. During the first winter of operation the boiler proved to be defective, and it also became apparent also that the capacity of the boiler was inadequate. Jones decided to have the old boiler replaced by a larger one that cost $200 more than the original. Although the contractor can be required to replace the defective boiler with a new one of first quality, Jones cannot expect him to furnish and install a larger and more costly article than that for which she initially contracted.[20] The extra cost of the larger unit is not an obligation of the contractor, and Jones should pay him for the excess value of the larger boiler. To reiterate, the contractor should be held to the performance to which he has agreed under the contract,[21] and there should be no attempt to hold him to anything further without proper compensation therefor. Notice that, since Jones had specified in detail the type and capacity of the boiler and the character of the entire plant to be installed, its inadequacy for the intended use was her responsibility.

Uncertainty as to the amount of damage need not prevent the recovery of some amount. Section 352 of the Restatement of Contracts, 2d, states that "damages are not recoverable for loss beyond an amount that the evidence permits to be established with reasonable certainty."

Comment A to Section 352 adds "courts have traditionally required greater certainty in the proof of damages for breach of contract than in the proof of damages for tort. The requirement does not mean, however, that the injured party is barred from recovery unless he establishes the total amount of his loss. It merely excludes those elements of loss that cannot be proved with reasonable certainty."

It should not be forgotten, however, that a failure to introduce sufficient evidence to prove that damage has occurred will bar recovery. Failure of proof usually occurs in cases in which the certainty of damage is difficult to establish.[22]

Where the contractor is clearly to blame but the measure of damages is not controlled by an express provision of the building contract, the decided cases have not followed any single route in compensating the owner. Many courts have em-

---

[20] *Temple Beth Shalom and Jewish Center, Inc. v. Thyne Const. Corp.,* 399 So.2d 525 (Fla. App. 1981). For a variation on this well-established rule, see *North County School Dist. v. Fidelity & Deposit Co. of Md.,* 539 S.W.2d 469 (Mo. App. 1976).

[21] His responsibility is limited to the execution of the plans unless the contract calls for him to design as well as to construct. *Eccher v. SBA,* 643 F.2d 1388 (1981), applying Colorado law.

[22] *Hubbard Brothers Construction Company, Inc. v. Brackner,* 390 So.2d 648 (Ala. App. 1980). *Design and Corrosion Engineering, Inc. v. Piggly Wiggly,* 408 So.2d 292 (La. App. 1981).

braced the so-called "cost" rule whereby the expense to the owner of completing or correcting the work upon partial breach or default by the contractor is regarded as the proper measure of damages to be applied, and this expense item would normally be deducted from whatever might be due the contractor for such satisfactory work as has been accomplished. Other decisions have adopted what is often called the "reduction in value" theory, under which the damaged owner, by way of balm for his wounds, is credited with an amount which represents the supposed difference in value between the building contracted for and the lesser building which was actually presented.

12-20 Special Claims for Damages against the Contractor Since progress and completion of the work are the objectives of the owner, it may be necessary for him to take special steps to secure the desired results if he believes that they cannot (or will not) be attained satisfactorily by the contractor acting independently. To provide for such a situation, it may be advisable to work the contract in such a way that the owner will have the power to complete any unfinished work himself, but at the expense of the contractor, charging the latter for any losses or extra expenses (damages). Imagine that A has a contract to build a store for B, but A is so busy constructing a shopping center in a nearby town that he has seriously neglected the supervision of B's project. Or, assume that A has contracted pneumonia and will be unable to take charge of the work under the contract. In such circumstances as these, it may be desirable for B to take over direct supervision of the work. If so, he should have the right to do so and to charge A for any loss entailed.

As another example, assume that the contractor has been paid what is due him but has failed to pay his employees their wages for the past week or two.[23] The engineer learns that the men have threatened to strike or to leave the job if they are not paid promptly. In such an event the owner should have the authority (notwithstanding the fact that the project is still supervised by the contractor) to make direct payment to the contractor's employees of all or part of what is due them and to deduct the amount of such payments from the money owed or to be owed to the contractor, thereby keeping the men at work and the job progressing.

One way for the engineer to deal with the questions of potential claims of laborers or suppliers is to place in the contract a clause stating that the contractor must prove[24] that he has paid all bills for labor and materials in respect of work performed during a given month before any partial payments are made to the contractor for that month's work. Thus, if these bills have not been paid by the

---

[23] Under laws which are very common, unpaid material suppliers and laborers are entitled to the protection of a so-called *mechanic's lien*, which when perfected, is attached to the structure being built (and the land on which it stands) and secures for them a priority toward satisfaction of their particular claims. Such laws are grounded on the principle that the building becomes a part of the realty, enhancing the value thereof, and that persons who have contributed labor and material to the improvement of property should be entitled, if the need arises, to look to that property for their compensation.

[24] This proof may take the form of affidavits, copies of receipts, or whatever method is mutually agreed upon.

contractor, the owner can refuse to make further payment to him until the difficulties are remedied, and he can use the withheld funds to pay the creditors himself in a proper case.

12-21 Risk, Liability, and Indemnity   It is usually an advantage to the owner to have the work done by an independent contractor instead of by the owner's employees. In this way the owner need not undertake the burden of devising ways and means of performing the desired work and may turn this chore over to the contractor. The agreement should make clear that the contractor is to assume all risks directly connected with the work[25] and is to carry adequate property and other insurances until such time as the owner accepts the structure or products and the contract is terminated. It may be advisable to have the types and amounts of the insurance coverages specified in the contract so as to avoid disagreements about what is "adequate." The contractor can handle the cost of insurance in his bid.

Further, the contract should state that the builder is to protect not only the owner himself but also the holders of adjacent property from injury arising from the performance of the contract. Construction work is likely to be conducted in places where the general public will be affected. The responsibility for protection of the public should definitely be placed upon the contractor, who of course is the person carrying out the work. Protective measures may involve fencing a site to keep out trespassers, roofing a sidewalk adjacent to a large building operation in a city so as to protect pedestrians, erecting fences or other guards around utilities, such as power ducts or cables, building temporary decking over excavations in streets, and even manually guarding busy construction passageways and areas where blasting or other dangerous operations are being carried on. The contract should require that adequate protective devices be maintained at all times of need. Signs and warning lights alone are generally considered to be insufficient.

As a broad generalization a builder is legally responsible where his negligence during the construction process is the proximate cause of injury to (1) persons rightfully on the scene, (2) persons in lawful use of an adjacent public thoroughfare, and (3) persons who are occupants of premises which abut the project site. If a contractor is known to be careless, incompetent, or inexperienced, he should not be awarded the contract in the first place; the owner could possibly be held liable for resulting injuries to outsiders if he fails to use reasonable care in the selection of a contractor. The owner might charge the engineer with negligence if the latter had advised the owner to let the contract to a bidder known to the enginer to be irresponsible.

Not infrequently found in construction contracts is a clause whereby the builder undertakes to indemnify or "hold harmless" the owner against claims for injury resulting from the builder's own negligence. Such clauses vary in scope.

---

[25] But see *Anderson v. Chancellor Western Oil Dev. Corp.,* 53 Cal.App.3d 235, 125 Cal. Rptr. 640 (1976), in regard to nondelegable duties. See also, *Atlantic Coast Development Corporation v. Napoleon Steel Contractors,* 385 So.2d 676 (Fla. App. 1980).

Here is a provision, taken from an actual contract, which would be characterized as comprehensive:

> The Contractor shall be the insurer of the Owner and of the latter's agents and employees against any and all of the following risk, whether they arise from acts of commission or omission of the Contractor or of third persons, excepting only those risks which result from affirmative, willful acts done by the Owner subsequent to the submission of the Contractor's Proposal:
>
> **1** The risk of loss or damage to the Work prior to the issuance of the Certificate of Final Completion. In the event of any such loss or damage, the Contractor shall promptly replace, repair, and make good the Work without cost to the Owner.
>
> **2** The risk of injuries (including wrongful death) and damages, directly or indirectly, to the Owner, his agents, and his employees, and to his or their property, arising out of or in connection with the performance of the Work. The Contractor shall indemnify the Owner and the latter's agents and employees against and from all such claims and demands, and for all loss and expense incurred by them in the defense, settlement, and satisfaction thereof.
>
> Neither the Certificate of Final Completion nor any payment to the Contractor shall release the Contractor from his obligations. The enumeration elsewhere in this Contract or in the Specifications of specific risks assumed by the Contractor or of particular claims for which he is to be responsible shall not be deemed to limit the effect of the foregoing provisions of this numbered clause not to imply that he assumes and is responsible for only risk or claims of the types enumerated.

Another clause of considerable breadth is the following, as quoted in a Pennsylvania case:[26]

> It is understood and agreed that the party of the second part shall be deemed and considered an independent contractor in respect to the work covered by this agreement, and shall assume all risks and responsibility for casualties of every description in connection with the work, except that he shall not be held liable or responsible for delays or damage to work caused by acts of God, acts of public enemy, acts of government, quarantine restrictions, general strikes throughout the trade, or freight embargoes not caused or participated in by the Contractor. Party of the second part shall have charge and control of the entire work until completion and acceptance of the same by party of the first part. Party of the second part shall be alone liable and responsible for, and shall pay, any and all loss and damages sustained by any person or party either *during the performance or subsequent to the completion of the work* covered by this agreement, by reason of injuries to person and damage to property, buildings and adjacent work, that occur either *during the performance or subsequent to the completion of the work* covered by this agreement, or that may be sustained as a result or consequence thereof, *irrespective* of whether or not such injuries or damage be due to negligence or to the inherent nature of the work. . . . Party of the second part agrees to fully indemnify, protect and save harmless the City, the Director and his subordinates, from any and all liability and from all suits and actions of every kind and description brought or which may be brought against them or any of them. [italics supplied]

---

[26] *Keefer v. Lombardi*, 376 Pa. 367, 102 A.2d 695, 696 (1954).

12-22 Power of Engineer in Settling Disputes Many contracts contain a clause to the effect that the engineer (or the architect) has the power to make final and binding decisions in the settlement of various types of disputes that arise in the execution of the contract. This power should be defined carefully.

The engineer in charge of a construction contract has no *implied* power to bind the parties by his interpretations, but express provisions in the contract that any dispute about the construction of the language used shall be submitted to the engineer (whose decision shall bind the parties) have been upheld. Not infrequently, a contract will stipulate that, in the absence of fraud, etc., the decision of the engineer or some other designated person shall be final on any question submitted. An example of such a question is what constitutes "satisfactory" performance.

Since the engineer is representing the owner in the implementation of the contract, it may seem that the engineer is not in a position to make unbiased decisions in disputes between the owner and the contractor. In practice, however, the engineer's reputation is likely to depend upon his fairness as well as upon his technical ability. Except in the case of the engineer's fraud, gross error of judgment, or actions that indicate bad faith, the engineer's decisions are binding provided that the signed contract contains a clause to that effect. It is advisable, however, to make provision for the disposition of certain types of disputes through resort to arbitration.[27]

QUESTIONS

1 Describe the function of the contract clauses relating to the performance of a construction contract.

2 Differentiate between "standard contracts" and "standard contract clauses." When and how may each be used to advantage?

3 Describe a situation which might justify (*a*) the suspension of a contract; (*b*) its cancellation.

4 Assume that Elkins contracted to build a large addition to Cromwell's munitions plant just before the end of World War II. The war ended before Elkins started work, and Cromwell tried to cancel the contract. Can Elkins do anything about the loss of his anticipated profit (*a*) if the contract contained a proper cancellation clause? (*b*) If the contract contained no clause dealing with cancellation?

5 If work under a contract is revised, what procedure should be followed? Is the contractor obliged to follow the revisions?

6 When may claims by the contractor for extra payment be justified in a lump-sum contract? How may these payments be computed?

7 When may claims by the contractor for extra payment be justified in a unit-price contract? How may these payments be computed?

8 The boring data shown on the drawings for two bridge piers indicated the presence of boulders at the site. The successful contractor included in his bid the cost of cofferdams to surround the piers and to permit removal of these boulders. When work started, the

---

[27] See Chap. 28.

contractor decided to try to get along without the use of cofferdams, and he succeeded. Does the contractor have to make a refund to the owner because of the decrease in the actual cost of doing the work? Explain.

9 What would be the situation if the contractor bid on the previously mentioned piers and assumed that no cofferdams would be necessary but later found that they had to be used?

10 Black contracted with White to build a certain number of large transmission towers for a stated price per tower. White subsequently decided to use longer spans for the wires, thereby reducing the number of towers. Is Black entitled to an adjustment in the price per tower?

11 A contractor designed a cofferdam of cellular type and at the engineer's suggestion strengthened certain details. Nevertheless, the cofferdam failed during a hurricane and flood. Who was responsible for the damages?

12 An engineer telephoned her resident engineer to stop work on some retaining walls which were to be revised. The latter was in a conference at the time. One hour later the resident engineer sent an assistant to notify the contractor. When the assistant arrived, 10 cubic yards of concrete had been poured in the forms and 10 cubic yards more were waiting in trucks at the site. What should be done? Who should pay the bill, if any?

13 What are the advantages from the contractor's standpoint of subcontracting portions of a construction contract? Illustrate a typical case.

14 Describe one way in which the evils of "bid shopping" may be remedied.

15 Cramer subcontracted to Williams the furnishing and installation of all equipment for a large pumping station. The operation of the pumps, said the original contract, was to be guaranteed for one year. When the plant started to operate, certain defects in the pumps were discovered. To whom will the owner look for remedy—the contractor or the subcontractor? Why?

16 What is the purpose in allowing the owner to withhold a portion of partial payments due the contractor? What happens to this money when the job is completed?

17 When and how are final payments computed and made to the contractor for a lump-sum contract? A unit-price contract?

18 Explain the purpose and operation of liquidated damages.

19 What is meant by "the contractor shall indemnify and save harmless the owner in case of any claims for damages"?

20 Brown orally told Black to build him a wooden garage 25 feet long and 15 feet wide, having a concrete floor, overhead doors, and a sloping roof covered with asphalt shingles. No other specifications were stated. The agreed-upon price was $1600. The agreement and details were not recorded in writing. Black built the garage, but she included no windows and no electric lights. Brown claimed that these two items were included in the agreement by implication. Is Brown correct in this matter?

21 If you were running an engineering office, when and why might you have use for standard contract clauses regarding workmanship?

22 If the contract is suspended by the engineer for five days because of changes which the owner wants to make, may this affect the cost to the owner?

23 A contract required the job to be completed July 15. The engineer ordered the contractor to suspend operations from May 1 to May 15, then changed the completion date to July 31. Was the change in completion data appropriate?

24 Under what conditions might the conduct of the contractor be proper cause for the owner to terminate the contract?

25 If no time for starting and completion are given in the contract, is the contractor free to construct his own schedule?

26 Why may revisions in a contract justify extra compensation for the contractor?

27 What compensation should a contractor receive because of delays due to actions of his own personnel?

28 In case of a unit-price contract, (*a*) when will the unit prices be applicable to extra work? (*b*) when inapplicable or insufficient?

29 Should the engineer get and issue data regarding soil conditions at a construction site? Should he interpret them?

30 Explain when and why the contractor may reasonably claim extra compensation when the quantity of work in a unit-price contract is decreased.

31 Who should be responsible for the protection of the public during construction operations? Why?

32 Illustrate a case in which the contractor is evidently entitled to extra compensation because of a change in the quality of the work required in a unit-price contract.

33 If the contractor has completed a unit-price contract, can the engineer compel him to build some added part at the same unit prices?

34 Should the engineer have any control over (*a*) the selection of subcontractors? (*b*) the acts of subcontractors?

35 What specifications apply to subcontracts?

13

SPECIFICATIONS FOR WORKMANSHIP

13-1 Specifying Procedures The purpose of this chapter is to discuss and illustrate some specifications relating to the *quality* of the workmanship which will be required of the contractor who is to build a construction project. *Workmanship* is intended to denote the contractor's *operations* in the shop or field rather than the materials used by him in the performance of the contract.

An *independent* contractor must be free from dictation by the engineer about how the work shall be performed. When the contractor signs the contract, it becomes his duty to perform in accordance therewith, to follow carefully the plans and specifications, and to furnish proper materials and workmanship as required by them.[1] Should the engineer (or the contract documents) specify exactly how the work is to be handled, then the engineer has largely assumed responsibility for securing the desired results. This is something about which the owner and his representatives must be extremely careful.

Assume the specifications state the concrete in a structure is to have an ultimate twenty-eight-day comprehensive strength of 4000 pounds per square inch. They also specify the amount of cement, sand, gravel, and water to be used in the mix, together with the time for mixing and the allowable slump. If the contractor follows these last instructions, as he is obliged to do, it will not be his fault should the concrete in its furnished form fail to meet the designated strength requirement. The specifications should prescribe the quality of the ingredients and the desired final result—4000 pounds per square inch—but they should not attempt to tell the

---

[1] Broadly speaking, and absent negligence on his part, the builder who follows defective or inadequate plans and specifications furnished by the owner, engineer, or architect is not chargeable for any damage or loss which can be traced to the inherent defects of the design or the specifications.

contractor exactly how he should go about attaining that result. On the other hand, it is entirely proper and probably wise to have the specifications require a certain minimum slump for the concrete and state that the contractor is (1) to submit samples of the proposed materials to the engineer for the latter's examination, and (2) to include sieve analyses of the aggregates.

In a case some years ago, the owner had provided specifications which detailed exactly how the walls should be built and waterproofing accomplished. The contractor built the walls in the prescribed fashion, and they proved not to be watertight. The court held that this was the fault of the designer of the plans rather than any fault of the contractor—who had faithfully complied with the mandatory and very precise specifications and plans.[2] The court observed that "where a contractor agrees to produce a certain result, he is responsible if the result is not produced, but, where the owner assumes to specify the manner of construction to produce the result, the owner assumes the responsibility if the work does not turn out as expected." Thus, it would have been preferable merely to (1) specify the quality of materials, (2) show all dimensions on the plans, and (3) require that the structure be watertight. Then the contractor would be obliged to make good if the work failed to meet these standards.

While the rules of law are perhaps not entirely clear in this area, it can be stated as a generalization that a builder need only comply with detailed specifications as prescribed; having done so, he is not answerable for an imperfect result which is not attributable to improper workmanship, negligence, or the use by him of defective materials. On the other hand, if plans and specifications are lacking in detail or were so prepared as to leave open the methods of construction to be employed or the kind or quality of materials to be used, then the contractor—because he has freedom of choice—must so select his methods and materials and so exercise his discretion as to produce a result which is substantially satisfactory.

There may be circumstances that make it necessary for the engineer to specify in detail just what is to be done and how it is to be accomplished, thereby deliberately assuming responsibility. This is generally done in instances where the contractor probably will not be able to determine the desired or proper course himself. Here is an illustration of such a situation. A contract for the construction of an industrial plant included drawings and specifications for a dust-collecting system. All motors, cyclones,[3] piping, and other details were shown on the drawings, and the materials and required workmanship were described fully in the specifications. This was really necessary because the contractor could not tell otherwise what was desired. The contractor agreed to build the system according to the plans and specifications. Whether or not the finished product was adequate for its intended service was necessarily the responsibility of the designers. Of course, if the contractor failed to do the prescribed work correctly, he could have

---

[2] *Kuhs v. Flower City Tissue Mills Co.,* 104 Misc. 243, 171 N.Y. Supp. 688, 690–91 (Sup. Ct. 1918). (Decision subsequently *modified* and *amended* on other grounds, and then *affirmed* at 231 N.Y. 637.) See also *Clark v. Whitener,* 296 So.2d 393 (La. App. 1974); and *Kelly v. Bank Bldg. & Equipment Corp.,* 453 F.2d 774 (10th Cir. 1972).

[3] Equipment utilizing centrifugal force for collecting the dust.

been compelled to make any changes necessary to conform to the requirements of the contract. But that would have been the limit of his responsibility.[4]

13-2   Sample Paragraphs   As respects the measure of the contractor's responsibility, the following contract paragraphs illustrate how two separate items were specified in an actual case:

15   Temporary Structures:

The contractor shall furnish and construct all guardrails, fences, trestles, staging, shaftways, falsework, and other temporary structures in the Work, whether or not of the type enumerated. All such structures shall have adequate strength for the purposes for which they are constructed, and the Contractor shall maintain them in a condition satisfactory to the Engineer.

The designs for such structures are to be prepared by the Contractor; nevertheless, drawings showing their design and details are to be submitted to the Engineer for his approval before being used. Examination by the Engineer, however, will be in a spirit of helpfulness but will not relieve the Contractor of his responsibility for their design, construction, and use, and the Contractor shall make good all injuries to persons or things arising on account of them.

27   Blasting:

Whenever blasting is done, the Contractor shall comply will all federal and state laws and local regulations relating to such work. Explosives used shall be only of such character and strength as may be permitted by such laws and regulations. The Contractor shall provide, at his own expense, proper magazines and storage facilities for the storage of explosives, and these are to be in such locations as may be approved by the local authorities having jurisdiction over such work. The magazines and storage facilities shall be marked with large letters *"Explosives—Dangerous,"* and they shall be kept under lock, the key therefor being kept in the possession of the superintendent or other person designated by him.

Notice that the preceding specifications require certain things to be done by the contractor but do not explicitly state in detail how they should be done.

13-3   Specifying Workmanship   The statement that an independent contractor is not supposed to work under the specific dictation of the engineer means that

---

[4] The danger inherent in specifying how the contractor is to secure the desired results and to do the required work is illustrated further by an instance which was reported to one of the authors. In substance, the specifications for a large housing project stated that, to remove groundwater from the excavations for the foundations, "the Contractor shall use well points," whereas the intent of the engineer had been that "the Contractor *may* use well points." The specifications thus became an order instead of granting a permission. The contractor did use well points and, as a result of the transportation of very fine silty material with the pumped-out water, certain neighboring buildings settled harmfully when they were, in effect, undermined. Since the project owner specified the operation, he was subject to claims for damages amounting to nearly $500,000. He could not blame the contractor for doing what the specifications required. Therefore, the owner had to face the consequences.

the contractor is to use his men and equipment under his own direction to attain the successful completion of the job. On the other hand, this does not mean that the engineer should fail to set up in the contract necessary requirements affecting the quality of workmanship demanded and to include carefully prepared statements regarding certain features of the conduct of the contractor's operations. The nature and scope of such statements and requirements will vary with the character of the work. One who is preparing specifications for workmanship should be thoroughly familiar with the pertinent customs, practices, and possibilities in the particular construction field involved. If the engineer is not well versed in this line he is likely to require unwise or impracticable things of the contractor and to overlook important features for which provision should be made. The result may well be undue difficulty and needless expense for the owner. At the very least, the inevitable complaints made by the contractor because of unwise specifications will make the owner lose confidence in the ability of the engineer.

Assume that the motor, gear reducer, and head pulley of a 60-inch belt conveyor are to be set and grouted on concrete foundations. The equipment is "direct-connected." This means that all parts must be aligned very carefully and accurately. What should you, the engineer, describe in the specifications for this work? The following are some of the significant considerations:

1 Require that the equipment be erected so that the center of the pully is at the elevation and in the position called for on the drawings.

2 Specify that the drive and pulley be aligned accurately and set level.

3 Call for all bearings to be supported on steel wedges temporarily for adjustment (or you might call for double nuts on the anchor bolts to serve as jack screws for adjustment purposes).

4 Describe what grouting is to be done.

5 State the time that must elapse after grouting before the wedges are to be removed.

6 Require that the holes left by the wedges be pointed up.

7 State whether or not anchor bolts are to be leaded in the holes in the bases before the nuts are tightened.

8 State if and how nuts on anchor bolts are to be locked to prevent loosening.

As another illustration of composition, assume that you are to specify how a large elevated steel water tank is to be fabricated. You may state that it is to be riveted and that joints are to be caulked for watertightness. You may specify the size and spacing of rivets and you may show typical details on the drawings. On the other hand, you may specify that the tank is to be welded, show the size and type of weld, and give typical details.[5] Your drawings and specifications must be sufficiently complete amd detailed to enable the fabricator to proceed satisfactorily with his shop drawing and other work.

---

[5] Who will be to blame if the tank leaks? You might better merely call for the type of construction—riveted and caulked or welded—and be careful to state that the tank is to be watertight; the contractor must then make it so.

Or, assume that you, as the engineer, are to specify the finish for the concrete of the second floor of a warehouse. Here are some of the things that you would have to determine and to specify:

1 Is the floor to be poured monolithically, or is it to have a topping applied later?

2 If a topping is to be used, what is the desired material, how thick is it to be, and how soon is it to be laid after the structural floor is poured?

3 What kind of surface finish is to be required—wood float or trowel?

4 Is a surface hardener or an integral hardener to be used? If so, exactly what is it to be and how is it to be applied?

5 Is the floor to be scored?

6 Is any particular color desired?

The difficulty of thinking out and specifying all these details is obvious, but so is their importance. Here the engineer is obliged to make many decisions in composing the specifications, and he has to assume considerable responsibility for the result.

Notice that, while the items shown in the three illustrations call for specific things that the contractor must accomplish, they do not tell him in detail *how* he is to do them.

The following are portions of what appear to be well-written specifications for a large warehouse:

1-16 Depositing Concrete:

Concrete shall be deposited only during the presence of the Inspector and by methods approved by the Owner's Engineer. All concrete shall be placed "in the dry." Should water accumulate in any place where concrete is to be poured, the Contractor shall provide and operate sufficient pumps and do whatever else is necessary to remove the water in an approved manner. Water (other than that used for curing) shall be prevented from coming into contact with concrete while it is setting.

6-3 Installation:

All metal sash and doors shall be erected plumb, level, square, and at the proper elevation and location. All joints shall be tight, and members shall be securely anchored in place. Work shall be adequately braced, and the bracing shall be maintained until its removal is approved by the Owner's Engineer. Moving parts such as sash, operating mechanisms, and doors shall be installed complete with hardware fittings, and accessories. Parts shall be made to operate uniformly and smoothly.

13-4 Specifying the Quality of Results Consider again the conveyor drive that is discussed early in Art. 13-3. Could not this work be described in some other way? For example, the specifications might read as follows: "The Contractor shall erect the head pulley and its drive at the elevation and position shown on the drawings. He shall align and level them accurately, grout the base plates fully, lead the anchor bolts in the bases, and adequately tighten and lock the nuts on the anchor bolts, using double nuts." In a way this requires certain actions, but it allows the contractor to perform the work as he may wish as long as he accomplishes the specified results. This is usually a better way to handle the work than by itemizing and describing every detail of procedure. Of course, the engineer's approval of all completed work is to be required in any case.

Now assume that you are to prepare the specifications for a 500,000-gallon tank for fuel oil to be supported upon firm, level ground. One good procedure is to specify as follows: "The capacity of the tank shall be 500,000 gallons; the material, welded steel; the diameter, 60 feet; the contents, fuel oil; the roof, an approved 'floating' type. All details are to be as approved by the Engineer, and the bid price is to include delivery, erection, all piping connections, and painting." Furthermore, the specifications should indicate that the tank is to be oiltight; that the tank is to be tested by being filled with water; that the foundation material is to be firm sand; that all necessary piping and connections are to be provided and erected; and that all material and workmanship are to be guaranteed to be satisfactory for one year after acceptance of the finished structure by the engineer. In addition, sizes of piping, the positions of connections, the type and material of piping, and the lengths of piping to be furnished and installed by the contractor are all items to be definitely prescribed.

A complete set of specifications for workmanship would allow the contractor to intelligently prepare and submit for approval his own design and the proposed details. He can propose whatever type of construction procedure he thinks is best and the results of which he is willing to guarantee. 0In many cases the engineer can use to good advantage the foregoing method of specifying work more or less in outline form. However, the usefulness of such a method is limited to items or work that the contractor is in a position to detail for himself. More complicated construction makes it necessary for the engineer to tell the contractor in considerable detail just what is required.

Here is a case that illustrates the occasional disparity between what is intended by a specification writer and what is actually said in the specifications. The clause on finishing of concrete stated, "The exposed surface of all concrete walls are to be rubbed." To the specification writer this meant the removal of all projecting fins, the filling of any holes, and the rubbing of the concrete surface with carborundum stones and water until the surface was smooth. However, according to a strict interpretation of the inadequate wording, the contractor could rub the surface a few times with a soft cloth and still fulfill the literal requirement of the specification. The way to avoid such a situation is to define "rubbed finish" early in the specifications. Then the term can be used freely thereafter.

The following acceptable paragraphs specify desired results, but they also state certain procedures that are to be used:

III-5 Concrete:

Every effort shall be made to secure a concrete of maximum density and impermeability and minimum shrinkage during setting and at the same time a concrete which can be worked thoroughly around and made to be in close contact with reinforcement and embedded steelwork.

IV-12   Erection of Steelwork:

The steelwork shall be set in place accurately and the riveting done as the Work progresses. Each tier of columns shall be plumbed, braced if necessary, and maintained in a vertical position. Temporary bracing shall be capable of taking care of all loads to which the structure may be subjected, including wind loads and loads caused by erection equipment and its operation; such bracing shall be left in place as long as it may be required. As erection progresses, all members shall be securely bolted with sufficient bolts to resist all dead, wind, and erection loads.

IX-42   Installation of Fixtures and Wiring:

The Contractor shall properly clean all conduits and remove all burrs and obstructions before snaking in or pulling the wires or cables. Wherever the use of grease for pulling wires or cables is permitted by the Engineer, the grease shall be free from acid. All the wires to be installed in one conduit shall be pulled at the same time, and they shall have no splices between junction boxes. Sufficient slack shall be left at junction boxes and at terminals for splicing the wires and for making connections properly. Joints shall be of approved type, carefully spliced, soldered with noncorrosive flux, and properly insulated.

13-5   Itemizing   There is a maxim of law which reads, "Expressio unius est exclusio alterius." This means that the express mention of one thing implies the exclusion of another. The importance of this principle as applied to contracts is that if there is an itemized list of things or acts that are required, other things or acts not mentioned in the list are, by implication, not required. On the other hand, a very general statement that is supposed to include everything that is necessary may be too broad and cumbersome, and it may be subject to varying interpretations. Sometimes the author of a contract clause tries to cover this difficulty by stating that the contractor is to provide certain things, "such as. . ."

The contract for the structural steelwork of a large industrial plant called for "fabrication and erection," and the unit-price bid was asked for on the basis of "f.o.b. the plant site." The railroad siding entered the plant's property but it was about ½ mile from the site of some of the large buildings. The contractor claimed (and received) extra for hauling the steel from the siding to the points of erection, asserting that the contract should have read "fabrication, hauling, and erection" of the steelwork.

The virtual impossibility inherent in attempting to itemize all the required incidental operations is obvious. In the foregoing case a complete list would have to include unloading the steel from the cars, sorting it, loading it on trucks, unloading it at the various positions, and arranging it for convenience in erection. If one operation is overlooked in such a listing, the omission may open up the possibility of a claim for extras. The words "fabrication, delivery, and erection" might have been a better statement than the one actually used. It can be assumed that all operations incident to the fabrication, delivery, and erection" must necessarily be performed by the contractor—and for the bid price.

The following is another instance in which faulty itemization caused difficulties. Plaintiff had taken a contract for "furnishing and delivery only" of certain machinery for what was then the "War Department," and the list of equipment was itemized in his proposal. Later, the defendant telegraphed the plaintiff that other articles, such as cables, switches, and fuses, needed for *the electrical connection and installation of the apparatus furnished* had not been received. Plaintiff protested that these articles were not in his proposal and therefore not included in the contract. Upon the defendant's insistence plaintiff furnished and delivered the materials asked for, but he stated that he was doing so under protest and that he would ask reimbursement for the cost of $2062. The materials were accepted by the defendant, but he later refused to pay for them. In ensuing litigation the court ruled in favor of plaintiff in regard to this point. The decision involved the following reasoning.

1 Where a contract is for furnishing certain machinery, the contractor is not obligated to furnish cables, wiring, switches, and other materials needed in the installation of the machinery supplied under the contract.

2 Within ten days after demand was made upon it to furnish the materials involved, the plaintiff filed a written protest against the demand, persisted in such protest at every stage of the proceedings thereafter, and, immediately upon delivery, filed claim for reimbursement. This made it clear that *the plaintiff at no time waived his right to claim compensation.*

3 The defendant accepted the materials it had ordered and received the benefit of them. In these circumstances there was an *implied promise* on the part of the United States to pay the plaintiff the cost of extra materials, since delivery was made on that condition. The implied promise to pay, however, is limited to the actual cost of the materials and cannot be construed to include anything beyond that.

Notice that the defendant's acceptance of the proposal signified its concurrence with the list of items therein.

13-6 Standard Specifications for Workmanship Construction contracts vary so much in the character of workmanship required that it is not practicable to prepare truly comprehensive workmanship specifications, such as the ASTM has done in the case of construction materials. There are, however, some workmanship specifications prepared by persons interested in a particular field of activity or in the use of a given product. Where these are applicable, reference to them at the appropriate point in the contract documents may be very desirable.

For example, assume that you are to prepare specifications for the construction of a large store that is to have framing of structural steel. It probably will be advantageous for you to state that the fabrication and erection of the steelwork are to be in accordance with the American Institute of Steel Construction's *Speci-*

fications for the Design, Fabrication and Erection of Structural Steel for Buildings. These specifications are well known, and they are generally accepted as requiring good practice for such work. Naturally, you must provide any special instructions necessary for your job but not included in the set of standard specifications to which reference is made.

In the matter of welding, you may refer to the *Standard Specifications for Welded Highway and Railway Bridges.* These are generally recognized as authoritative and represent some of the best thinking about the requirements for good welded construction. The general engineer might do better to trust them than to attempt to compose an adequate description of such highly specialized work.

These standard specifications as well as the contract drawings, "outside" drawings, building codes, and whatever the engineer may deem advisable for a particular contract should be *incorporated by reference* in the specifications. Thus they can control in various applicable phases of the work. For example, assume that the state highway department has published a very detailed manual containing data and requirements for the construction of highways in the state. A contract for a particular piece of highway construction can refer to this publication and incorporate the same as an official part of the contract. This practice ensures uniformity in various highway projects and saves time and effort in the preparation of the contract papers. Contractors throughout the state soon become thoroughly familiar with the contract requirements of the highway department.

Standard specifications will not contain the answers for everything that is necessary. The engineer must make certain decisions on a particular basis, as illustrated by the following case. In specifying the shopwork for structural-steel fabrication, mere reference to the specifications prepared by such organizations as the American Institute of Steel Construction may not be enough. The engineer should state in the specifications whether the shopwork is to be riveted or welded, whether or not the holes are to be subpunched (or subdrilled) and reamed to metal templates, or whether the parts are to be reamed while assembled. He should also specify whether the field connections are to be bolted, riveted, or welded. The standard specifications may tell how various procedures should be done, but the choice of the type of work is a duty of the engineer.

On the whole, because much of the workmanship required in construction contracts cannot be handled through so-called "standard specifications" but has to be specified or supplemented by instructions prepared for the particular job at hand, the engineer should give this latter task careful attention. It is one of his most important duties; many dollars may be involved in work that is described in a few paragraphs of his specifications. By ensuring excellent preparation of the contract documents the engineer can do much (1) to prevent uncertainties that often lead to difficulties in dealings with the contractor, (2) to avoid mistakes or omissions that may cause the owner unexpected expenditures for the settlement of the contractor's claims, and (3) to ensure avoidance of litigation. Excellent specifications reduce the guesswork for contractors, and thus they may result in lower bids because the figure to be included for contingencies could therefore be relatively small.

Any special features which differ from the so-called "standard specifications" being employed should be described in the specifications for the job or on the drawings. If there are such features, special clauses in the specifications describing them should take precedence over the standard specifications whenever the two conflict. As an example, assume that the engineer wishes to have certain members for a complicated portion of a steel structure subpunched, then assembled, and finally reamed in the shop while thus assembled. If his instructions call for this procedure but AISC specifications do not, then the engineer's written statement governs.

The following paragraphs are illustrations of how portions of some successful specifications were worded:

2-2   Structural Steelwork:

Workmanship for all structural steelwork shall conform to the AISC Specifications for the Design, Fabrication and Erection of Structural Steel for Buildings, latest edition.

6-10   Steel Roof Deck:

Each piece of the steel roof deck shall be securely fastened, in accordance with the manufacturer's standard specifications, directly to all the roof purlins under it, as well as to each adjoining piece at the end and side laps.

10-4   Built-up Roofing:

Roofing shall be a twenty year bonded Barrett "Specification" roof Type "AA," 4-ply. It shall be laid in accordance with the Barrett specification for use over insulated steel-roof decks, and the work shall be performed by a contractor approved by Barrett. The roofing contractor shall furnish Barrett's Surety Bond guaranty for twenty (20) years from date of completion in accordance with Note No. 1 of said specifications.

13-7   Use of Previous Specifications   Contracts prepared for comparable former projects as well as contracts written by others may readily be used as source material for the workmanship requirements of a current job, just as for the specification of materials described in the preceding chapter. Although potentially helpful, they should be used with discretion.

Nothing can replace careful thought and careful composition when one is specifying what is to be done and how it is to be done. Think of all the steps that are involved in securing a result, then be sure that what you copy out of a previous specification covers a step or steps in your particular problem clearly and adequately, add whatever is needed, and eliminate inapplicable and erroneous material.

13-8   "As Directed by the Engineer"   There are all-inclusive statements that are sometimes used in the specifications; for example, "as directed by the Engineer," "to the satisfaction of the Engineer," or "as approved by the Engineer." Such language, if interpreted and carried out properly and fairly, can be useful.

The intent of such statements is to leave many details about the conduct of the work and the quality of the results to be settled between the engineer and the contractor as the job progresses. Their utilization saves time and effort on the part of the specification writer. On the other hand, they should not be used as a bludgeon against the contractor. If the latter knows that the engineer with whom he is to deal is an experienced and just man, he will anticipate real cooperation and understanding in their relationships. If he expects otherwise or is uncertain about it, he is likely to increase his bid to cover in his estimate the cost of any trouble that may arise during the job because of the engineer's opinions and unilateral decisions.

In many situations when the exact conditions are unknown in advance, there may be wisdom in not trying to describe with preciseness what is to be done. For example, assume that a considerable quantity of concrete is to be poured in freezing weather. The character of the structure, the severity of the weather, and the difficulty of carrying on the work greatly influence the details of proper protective measures that should be employed. It may therefore be advisable to say that concrete poured in freezing weather "is to be protected in a manner satisfactory to the engineer." The contractor then proposes a method which the engineer accepts or rejects, or with respect to which he suggests modifications.

The rejection of work by the owner or the engineer as "unsatisfactory," when it is supposed to be done to his satisfaction, must not be based upon whim or bad faith but rather upon genuine dissatisfaction. Furthermore, such action must not be taken prematurely before it can reasonably be determined that the work will not or cannot be made satisfactory.

In any construction contract it is generally advisable to have such a clause as the following, stating specifically the authority of the engineer in the conduct of the work:

> The enumeration herein of particular instances in which decision of the Engineer shall control or in which work shall be performed to his satisfaction and subject to his inspection and approval shall not imply that only such matters and those of similar kind shall be so governed and performed.

13-9 First-Quality Workmanship   It is advisable to include in the specification a statement to the effect that all workmanship is to be first class, or "of the best quality." This is not very specific, but the intention is clear. Persons engaged in any particular activity, whether in the shop or in the field, will generally know what is meant. Third parties, too, can judge fairly well whether or not a given performance meets this standard.[6]

---

[6] "It is the duty of every contractor or builder to perform his work in a proper and workmanlike manner, and he impliedly represents that he possesses the skill necessary to do the job he has undertaken. In order to meet this requirement, the law exacts ordinary care and skill only." *Cantrell v. Woodhill Enterprises, Inc.,* 273 N.C. 490, 160 S.E.2d 476, 481 (1968). On the question of implied warranty of skill and workmanship, see *Pollard v. Saxe & Yolles Dev. Co.,* 115 Ca. Rptr. 648, 525 P.2d 88 (1974); and *Hollen v. Leadership Homes, Inc.,* 502 S.W.2d 837 (Tex. App. 1973).

It is practically impossible to include in the contract documents a description of or specification for all the procedures and work that will have to be done in the providing of materials and in the performance of a large construction contract. On the other hand, a general statement requiring first-class workmanship should not be used as a catchall or as an excuse for inadequate preparation of the drawings and the specifications.

The usefulness of a general clause requiring "first-quality workmanship" is shown in the case of a contract for the steelwork of a large suspended bin. The general shape is pictured in Figure 13-1a. The curved plates were to be spliced as shown in (a) and (b). One contractor was to fabricate the steelwork; another was to erect it. The fabricator detailed the main plates with open holes, as pictured in (c). The engineer who approved the shop details did not notice this deviation from normal practice. All the plates were therefore shipped "loose." The erector had assumed, when bidding on the job, that the splice plates would be riveted along one side in the shop, as shown in (d), which is the customary and more economical practice. The erector sent the owner a bill for $6000 to cover the extra riveting that he had not included in his proposal. Because of the general-workmanship clause, and because the fabricator had to admit that the omission of the shop rivets was not customary and was not first-quality workmanship, the fabricator paid the owner $6000 to meet the extra cost of erection.

FIGURE 13-1
Steelwork for suspended bin.

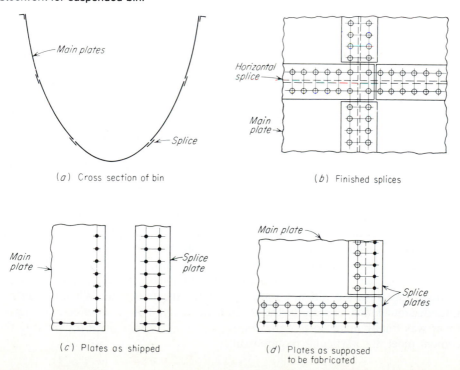

(a) Cross section of bin

(b) Finished splices

(c) Plates as shipped

(d) Plates as supposed to be fabricated

Even though every detail that will be required is not spelled out, a contract for a structure to serve as specified purpose carries the implication that the finished structure shall be reasonably suitable for the intended service. The more fully the requirements are clarified in advance, the less likely are subsequent difficulties between the engineer and the contractor.

A clause typically used in contracts is somewhat as follows:

> All materials and workmanship shall be in every respect in accordance with the best modern practice. Whenever the Contract Drawings, Specifications, or directions of the Engineer admit of a reasonable doubt about what is permissable, and when they fail to state the quality of any work, the interpretation which requires the best quality of work is to be followed.

13-10   Intent of Plans and Specifications   It is often stated in a contract that the contractor is to carry out the "intent" of the plans and specifications. Some say that he is to perform whatever is necessary to carry out this intent, even when the particular item or procedure, though apparently intended, is not shown on the drawings or actually called for in the contract clauses or in the specifications. In some ways this is like asking the contractor to buy something sight unseen. If applied literally it would be unfair to the contractor in many cases. Such all-inclusive clauses generally have only slight value in practice and offer little protection against carelessness and ensuing mistakes on the engineer's part.

A lump-sum contract is generally quite strictly limited to what is clearly and specifically called for in the contract. Minor items that are not shown will usually be taken care of by the contractor without conflict. However, when such things cost an appreciable sum beyond what could reasonably be expected to be covered in his bid, he may properly be entitled to extra compensation.

On the other hand, omissions from the plans and contract documents seldom cause serious argument in unit-price contracts unless the extra materials and workmanship required clearly do not come under any payment item or when the underestimation of a payment item is sizable. A capable contractor is likely to have included in his estimate a margin for contingencies to cover just such omissions. As before, adequate plans and specifications should reduce the size of such cushions.

One contract referred to a certain standard specification for portland cement. The work was to be done in South America. It was the specification author's intention to limit the content of free lime in the cement to $1\frac{1}{2}$ percent, which is accepted as a maximum by many engineers. As it happened, a cement manufacturer in South America received the contract for furnishing all the cement. Tests of the material manufactured by him showed that the free-lime content normally exceeded the desired limit by a considerable amount. When the author tried to hold the contractor to the desired limit, he discovered to his chagrin that the standard specification did not state a limit for the free lime. What the author intended but failed to state in the specifications was not given effect. An agreement was finally made to pay the contractor an extra unit price for making the cement meet this particular requirement.

On the other hand, the matter of intent is considered very strongly when it can be determined with reasonable certainty by a court. Generally a decision that is based upon interpretation of intent will consider the contract as a whole, not just one part of it. Differences between the data shown on the drawings and in the contract clauses or in the specifications may render difficult a decision about what the intent really was. Ambiguity in a language will be construed against its drafter. Even unintentional wording in the contract may cause interpretation difficulties.

13-11   Planning Specifications for Workmanship   This article has been pre-pared to show the reader how to outline and to plan the specifications for the workmanship in two types of construction work. Its purpose is to outline the kind of thinking to be done, and it applies to two construction project phases which the reader can visualize easily: (1) clearing the site, and (2) cleaning up after construc-tion has been completed.

A construction contract is likely to require the clearing of the site, the demoli-tion of existing structures, and rough grading as a part of the work to be done. If so, general instructions should be given regarding such items as the following:

1 Removal and disposition of tress, stumps, shrubs, bushes, and boulders
2 Removal and disposition of existing materials which are to be discarded or salvaged
3 Demolition of existing structures and their foundations
4 Removal and storage of topsoil when necessary. This may include the storage of turf for reuse.
5 Accomplishment of the rough grading—both excavation and fill
6 Disposal of excess excavated material
7 Designation of sources of fill to be brought in
8 Installation of surface and subsurface drainage systems
9 Designation and preparation of storage areas for the contractor's use

Much of this preliminary work may have to be done by the contractor in such fashion as to meet the owner's requirements. Unless the contractor is told what these are, he cannot be expected to fulfill them and, obviously, he cannot properly evaluate them when preparing his bid. For example, consider the item of earth fill. It is not sufficient to say, "The site is to be filled in to the level shown on the contract drawings." Neither is it sufficient to state vaguely that the filling is to be done "in a satisfactory manner." In this particular problem, at least to a certain extent, the engineer should rightfully step in and describe in considerable detail such considerations as the following:

1 Removal and disposal of muck, if necessary
2 Description of the quality and the kinds of materials that will be accepted as fill
3 Deposition of fill in horizontal layers of some stated approximate thickness
4 Compaction of layers of fill by means of sheep's-foot rollers, rubber-tired rollers, or whatever is suitable for the job—or hand tamping if necessary

5 Limitations in the moisture content of the soil—which limitations are to be adhered to during rolling

6 Specification of the density of fill required after compaction

The bid price for clearing the site and for any required demolition and rough grading is often listed as a lump-sum item in a unit-price contract. This is because such work is difficult to measure in detail. On this account, fairness to the contractor necessitates that the engineer be careful to supply the contractor with all essential data regarding both the results to be obtained and any special means to be used in achieving such results.

Now consider cleaning up the site. Assume that you are to prepare specifications to prescribe the work to be done after the rest of a construction contract is finished. What items should be covered by your specifications? You should prepare instructions regarding such matters as the following:

1 Final grading

2 Removal of equipment

3 Removal and disposal of debris and excess materials

4 Demolition and removal of temporary structures

5 Removal of temporary facilities used for access to the site

6 Removal of temporary power lines and other utilities that are for construction purposes only

7 Provisions for surface drainage

8 Completion of landscaping

These matters, if not clarified, may cause disagreements when two or more contractors or the contractor's and the owner's forces are involved in the work. Disagreement may arise even in the case of subcontracting of portions of the job. This "cleanup" is the responsibility of the general contractor, but the requirements should be made clear in the contract wording.

13-12   Sample Paragraphs   The following additional paragraphs from specifications are given in order to enable the reader to study the wording and to see how various pieces of information may be presented in specifications. A number of comments are included to call attention to special features. The examples are purposely taken from a variety of work.

32   Sequence of Work; Increase of Forces:

The sequence of work done under this Contract will be left to the Contractor. Should the Engineer order the Contractor to increase his forces in order to ensure completion of the Contract within the time specified herein, the Contractor shall promptly comply with such order without additional cost to the Owner.

Notice that names are repeated in the preceding specification, the personal pronoun being used only once. This repetition of names (that is, the designations "Contractor," "Engineer," and "Owner") is for the purpose of avoiding ambiguity.

The last portion of the foregoing paragraph indicates to the contractor that, if he believes overtime work may materialize, he had better make allowance in his bid for such contingencies. It would seem more desirable, for all concerned, to specify that the owner shall compensate the contractor for the extra cost involved in such ordered overtime work unless the contractor is at fault, that is, unless the tight situation is the result of his negligence or delay. In any event, the engineer should have the right to order overtime work if he believes that an emergency exists or is pending, as shown in the following example.

A contractor was building a bridge pier in a shallow stream which was subject to flash floods. The forms for the pier were completed but the concrete had not been poured when a weather report came over the radio to the effect that a hurricane was likely to strike that area within two days. The engineer ordered the contractor to work overtime until the concrete was poured and danger to the structure was thus minimized.

61 Sanitary Regulations:

The Contractor shall provide and properly maintain, for the use of all employees on the work, adequate sanitary conveniences, properly secluded from public observation. Such conveniences shall be provided at such points and in such manner as the Engineer shall approve, and their use shall be rigidly enforced.

The Contractor shall effectively prevent the committing of nuisances upon the site and upon adjacent property.

Notice that this is a performance type of specification (specifying results rather than means).

16 Cofferdams:

The Contractor shall construct, maintain, and subsequently entirely remove suitable and efficient cofferdams in order that all work, except when expressly permitted to be done in water, may be performed by the Contractor and inspected by the Engineer in the dry.

The Contractor shall design all cofferdams, and he shall submit drawings of them for the approval of the Engineer when approval is necessary before the Contractor proceeds with the construction of the cofferdams. The Contractor shall be completely responsible for the strength, safety, watertightness, and sufficiency of all cofferdams, and he shall defend and save harmless the Owner from any and all damage suits due to the construction or failure of any part or parts of the cofferdams or sheeted work. The engineer will be concerned primarily with the excellence of the workmanship for the substructure of the bridge.

173 Field Connections:

The movable span trusses, the tower trusses, the lifting frames, and the bridge portals shall be assembled in the shop, the parts adjusted to line and fit, and the holes for field connections drilled or reamed while so assembled. Holes for other field connections, except those in lateral, longitudinal, and sway bracing, shall be drilled or reamed in the shop with the connecting parts assembled, or else drilled or reamed to a metal template.

An expense allowance for this connection work will be charged to the owner through the contractor's bid. The engineer endeavors to see to it, through the specifications, that the structure will fit together properly when erected in the field; otherwise the owner may have to accept delay and an unsatisfactory structure if it is necessary to apply makeshift remedies at the site.

VI-I Machinery Workmanship:

All workmanship shall equal the best practice of modern machine shops. The finish shall be confined to rotating, bearing, and sliding surfaces and to wherever required to produce adequate clearance, accurate fit, and precise dimensions.

Again the contractor is made responsible for the results; the details of workmanship are to be determined by him.

204 Operation Tests:

The complete electrical installation shall be carefully tested by the Contractor in the presence of the Engineer and as directed by the Engineer. All necessary corrections and adjustments shall be made promptly by the Contractor. The Contractor shall furnish an adequate supply of current at the proper voltages for making such tests, and he shall furnish the necessary instruments and appliances therefor.

Notice that such a test is one of performance, not one of materials. In the usual situation, even though the engineer may have tests of the materials used or may have had such tests made by others for him, the contractor still has to prove that the installation functions properly.

38 Erection:

Before starting erection, the Contractor shall submit complete and detailed erection drawings and his program for erection of the steelwork, all of these being subject to the approval of the Bridge Engineer. Whenever, in the opinion of the Bridge Engineer, the proposed plans interfere with, or involve avoidable difficulties or delays in the subsequent erection of other parts of the structure by other contractors, the Bridge Engineer may order such modification either in the drawings or the program as, in his opinion, will best expedite the construction of the bridge as a whole.

The engineer thus keeps in his own hands the authority to see that the contractor coordinates his work with that of other contractors on the same master project.

140 Field Test of Mains:

The work of laying the pipes and special castings, and the work of setting valves and hydrants, shall be of such character as to leave all the pipes and connections watertight. To ensure these conditions, the Contractor shall subject all mains twenty (20) inches or more in diameter, and their appurtenances, to a proof by water pressure of not less than one hundred and twenty-five (125) pounds per square inch,[7] unless the conditions under which the main is installed make this pressure test impracticable.

Such proof of the adequacy of an installation is generally necessary before the construction is approved and accepted.

V-26   Sewers, Drains, and Their Appurtenances:

The Contractor shall provide for the flow of (1) any water courses, (2) closed flumes, (3) open-joint or blind drains, and the like, which flow is interfered with during the performance of the work. He shall immediately remove and dispose of all offensive matter. A flow throughout the entire length of such sewers and appurtenances as are to be relocated or reconstructed shall be maintained until it can be transferred to the new facilities. Flows in house connections shall be maintained, properly restored, and reconnected to the old or connected to the new sewers, as required.

The phrase "as required" in the preceding paragraph is supposed to mean "as necessary to keep the facilities in service"; in fact, it would be wiser to say so. If the clause stated "as required by the Engineer," this would shift the task of determining precisely what was needed from the contractor to the engineer.

8-43   Setting Brick:

Face brick shall be laid in courses of headers and stretchers, every third course being a header course with every sixth course bonded through. Face bricks shall be laid true to line with end joints not exceding one-quarter (¼) inch in width. Horizontal joints shall be two and five-eights (2-⅝) inches center to center. Each brick shall be well parged on the back with mortar and shall be completely embedded in mortar under the bottom and on the sides in one operation. Bed mortar shall not be furrowed. All mortar joints of the face bricks shall be tooled slightly concave.

Here the engineer specifies the results in considerable detail, but he does not specify *how* the contractor is to secure them. However, it is recognized that these matters sometimes involve a fine line of distinction.

10-3   Marking Floors:

At the time the finish on the concrete floors is being floated to proper slopes, the surfaces, except in those places where the Contract Drawings show the floors to be covered with terrazzo or tiles, shall be marked off in squares of approximately four (4) feet in a pattern to be determined by the Engineer. The markings shall be cuts one-half (½) inch in depth, and they shall be properly rounded or chamfered with an approved edging tool.

The engineer has to determine only where the markings shall be. The contractor is to obtain this information before he can proceed with his work.

---

[7] Notice that these dimensions and figures are carefully spelled out, not given simply as "20" or "125 psi."

29   Location of Machinery:

The Contractor shall erect each machine in the location indicated on the Contract Drawings.

Such a statement as this is superfluous because the drawings show where each machine is to be erected, and the drawings are in themselves instructions to that effect.

58   Erection of Anchor Bolts:

All anchor bolts for machines are to be located accurately. The bolts are to be held in place by means of wooden templates to prevent displacement during the pouring of the concrete of the machinery foundations. The location of each bolt is to be approved by the Inspector before any concrete is poured around it.

This clause is necessary because only the locations and elevations of the bolts are usually shown on the drawings. The engineer requires the use of templates in order to make certain that the bolts are held in place properly during the pouring of the concrete. The engineer insists upon inspecting for approval the locations of the bolts as a check on the contractor's work.

13-13   Typical List of Items   Before attempting to prepare the specifications for workmanship for a project, one should compile a list of items to be covered and the types of work to be done. This will help in assuring that all necessary features and workmanship requirements are included. The following is a checklist of work and workmanship items that might have to be described in the specifications for the construction of a building. Each item may, of course, involve a number of features.

1   Demolition and clearing of site
2   Protection of site—fences, lights, etc.
3   Subsoil explorations, if not made in sufficient detail in advance
4   Drainage of the site
5   Rock excavation
6   Earth excavation
7   Filling, backfilling, rough grading
8   Cofferdams
9   Piling or caissons
10   Concrete and reinforcement
11   Masonry
12   Waterproofing and dampproofing
13   Structural steel
14   Fireproofing
15   Walls, siding, and partitions
16   Flooring
17   Roofing

18 Insulation
19 Stairs and railings
20 Elevators, dumbwaiters, escalators
21 Windows, trim and glazing
22 Doors and trim
23 Miscellaneous and ornamental metal
24 Furring, lathing, plastering
25 Carpentry and millwork
26 Acoustical treatment
27 Flashing and caulking
28 Hardware
29 Window screens and weather stripping
30 Window shades and venetian blinds
31 Painting and finishing
32 Roof drainage
33 Plumbing
34 Heating and ventilation
35 Power and lighting
36 Sprinkler system
37 Drainage of foundations and basement
38 Landscaping
39 Order of erection (if important)
40 Maintenance of any services

This itemization is given to show the reader how comprehensive the preparation of specifications must be. The list will serve its purpose if it helps to emphasize the importance of adequate specifications as a means (1) of showing the contractor what is required in the line of workmanship and (2) of enabling the engineer to have data at hand for use in seeing that each part of the work is done properly by the contractor.

QUESTIONS

1 What is meant by "independent contractor"?
2 Illustrate a situation in which an independent contractor is not actually independent.
3 What is the danger of itemizing in the specifications the procedures which the contractor is to follow in order to attain the required results?
4 When may the expression "as directed by the engineer" be helpful if used in the specifications? Illustrate.
5 A specification stated: "All work is to be done to the satisfaction of the engineer." (*a*) What does this statement mean? (*b*) How does the bidder know what will be required? (*c*) When is such a statement in the specifications useful?
6 What direct control should the engineer have over the method of construction adopted by the contractor? Under what conditions might he advise the contractor?
7 If the contractor follows specified procedures in the field, who is responsible for the results?

8 Illustrate a particular case in which it may be necessary for the engineer to specify procedures to be followed in the field.

9 The specifications stated that all joints of a steel oil tank were to be scarfed and butt welded, and that the tank was "not to leak." Is this a satisfactory specification? Explain.

10 A contract for certain machines specified that all bearings should be finished to a tolerance of 0.01 inch. Is this a proper specification?

11 A contract specified that the parts of a certain large rectangular steel box should be connected by means of ¼-inch fillet welds. When the contractor finished the job, the box was found to be badly distorted (out of shape). The contractor claimed that this result was entirely due to the required welding. What defense can be made by each party?

12 The drawings showed the positions and details of the anchor bolts for a machine but nothing was said in the specifications about how these bolts were to be erected. During pouring of the concrete, the bolts were displaced so that they were 1 inch too far apart. Can the engineer do anything about the situation? Would it have been proper if the specifications had required that the bolts be held by metal templates during erection? Explain.

13 Do the statements "as directed by the engineer" and "as approved by the engineer" mean the same thing? Illustrate.

14 The specifications stated that certain long girders for a bridge were to be braced laterally during erection. The contractor proposed using cables attached to the girders and to large stakes driven into the ground. If the engineer approved the idea, who should be held responsible in case the girders buckle sideways during the erection of the bridge?

15 The drawings called for sheet piling to be driven 35 feet into the ground to form a cofferdam around a deep excavation. It proved to be impossible to drive the sheet piling to the required depth. Who should pay the expense of the attempts to do so? Whose responsibility is it to devise some other method to make the excavation possible?

16 A large fan was to be supported upon steel beams in the top of a power plant. The contractor performed the work as intended. When operated, the fan caused annoying vibrations. Who is responsible for the unsatisfactory results?

17 Concrete pavement for a pier was to be screeded level and then broomed. When the work was partly done, a violent thunderstorm occurred, spoiling the finish of the completed portion. The engineer discovered the situation the next day. What possible courses of action can the engineer take?

18 A specification states that the contractor is to "carry out the intent of the plans and specifications." (*a*) When and why may such a statement be useful? (*b*) When is it not useful? (*c*) Does it make any difference whether the contract is lump-sum or unit-price?

19 What does the statement that "all work is to be of approved quality" mean?

20 What is the meaning of "best quality workmanship"? What value has this statement in a set of specifications?

21 Name some standard specifications for workmanship. Illustrate their use and explain the advantages of employing them.

22 Is it proper to specify that "the roofing is to be applied in accordance with the manufacturer's recommendations"? Explain.

23 If standard specifications, like those of the AISC, are used as reference, can any deviation from them be specified? Explain.

24 How is a court likely to determine the intent of the plans and specifications relating to workmanship?

25 A specification states that all work "shall conform to the best modern practice." Is this of value? Why? Illustrate.

CHAPTER 14

SPECIFICATIONS FOR MATERIALS

14-1 Introduction The specifications are to give detailed information regarding *each* and *every* thing and operation within the scope of the contract, to serve as a single and definite basis for competitive bidding, and to constitute a book of instructions for all concerned with the work. Some basic requirements for proper specifications are (1) technical accuracy and adequacy, (2) definite and clear stipulations, (3) fair and equitable requirements, (4) format such as to permit easy use during operations, and (5) careful preparation to the end and they will be legally enforceable.

The differing functions of the contract drawings and the specifications may be illustrated by referring to Fig. 14-1. Assume that this is a portion of a contract drawing for a large industrial plant. The parts we are interested in are the pipe tunnels. These are shown by the dotted lines in the plan view; A typical cross section is pictured in the lower view. Notice that the drawing shows the following:

1 Length of tunnels
2 Elevations of inverts (center line of floor surface)
3 Sizes for the two types of tunnels
4 Thicknesses of roof, floor, and walls
5 Reinforcement required
6 Waterproofing
7 Open-joint pipe drains
8 Spacing of contraction joints

It is left for the specification to describe the following:

1 Quality of the concrete, aggregates, and cement

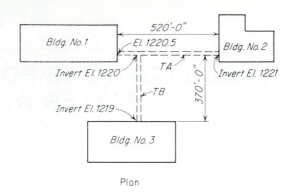

Plan

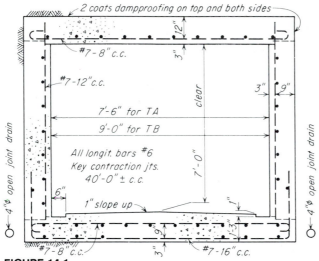

FIGURE 14-1
Pipe tunnels.

2 Workmanship required for mixing, placing, and curing the concrete, including forms

3 Finish required for floor, roof, and wall surfaces

4 Quality of reinforcement

5 Workmanship required for placing and embedding reinforcement

6 Quality of material for dampproofing

7 Workmanship required for application of dampproofing

8 Material and construction requirements for open-joint pipe drains

9 Materials and workmanship required for contraction joints (for example, keyways, waterstops), and painting of surface with asphalt

10 Excavation and backfilling requirements

11 Preparation of soil foundation

It is obvious that the specifications as well as the drawings should be prepared with great care. The engineer and the owner should never forget that if the specifications fail to show all requirements adequately, then the contractor's obligation is limited to performance of what is actually called for in such (incomplete) specifications (and/or on the drawings).

Somewhere in the contract, preferably in one of the contract clauses, there should be a statement describing the procedure to be followed if it should develop that there are discrepancies between the specifications and the drawings. In the haste with which contract documents sometimes have to be prepared, such inconsistencies unfortunately are likely to occur. They may also arise because of changes which are made in the drawings but which are not made correspondingly in the specifications, or vice versa. Again, discrepancies are likely to be caused when one statement in the specifications is changed but the called-for revision of a parallel clause relating thereto is overlooked. This is a danger which arises when the requirements for something are specified in more than one place. It is best to state definitely in the contract clauses that, in case of discrepancies between the specifications and the drawings, the engineer is to determine which of the two is to govern.

When something is called for in the specifications but is not included in the drawings, and vice versa, the contractor should furnish or perform that particular item as though it were covered in both. The contract documents must make this point clear.

At the beginning of the specifications, or perhaps at the beginning of the contract clauses, it is desirable to have a clear statement of exactly what work is to be covered by the contract. The following illustrative material is taken from the specifications for a bridge project.

This Contract shall include:

1 One (1) complete superstructure, including asphalt-paved movable span, towers, counterweights, power equipment, hydraulic jacks, signals, safety gates, barriers, machinery house, operator's cab, floormen's houses, and all miscellaneous equipment

2 Two (2) complete abutments with pavements

3 Two (2) complete approaches with pavements

4 Complete removal of the substructure and superstructure of the existing bridge

5 Removal of the piling which supports the substructure of the existing bridge and removal of all existing fender piles

6 Dredging of the channel of the river in the vicinity of the bridge

7 Protection of the sixteen- (16) inch water main which crosses the river at this location

8 Maintenance of navigation and compliance with all requirements of the local United States Engineer's Office

9 Provision for access during construction to all properties (on the site of the contract) for both pedestrian and vehicular traffic

10 Operation of the bridge after completion for seven (7) days under traffic

11 All miscellaneous items of work shown by the Contract Drawings and/or described in the Specifications

14-2 Discretionary Power of the Engineer   There are many materials that can be used for the building of structures; for example, steel, aluminum, concrete, wood, bricks, stone, tiles, and cinder block. Similarly, there are many manufactures of electrical, mechanical, and other equipment. These materials and products may be substituted for one another to some extent. On the other hand, some may be more suitable than others for use in specific instances. It is the engineer's duty and right to specify the one—or the alternative ones—that he believes to be best for each particular use, article, or structure. This means that he is to be the one to make the selection when the design is made. The responsibility is his alone, just as it is in the determination of the other features of the design of the structure, the machines, or whatever else the project may involve.

Without this discretionary power of the engineer there would be no end of confusion. What would happen if manufacturers of suitable materials that could be used in a project insisted that their products be included as alternatives in the contract and that competitive bids be taken before a decision were made? The choice might then be made upon the basis of the lowest bid price. Such a procedure would be very unsatisfactory. It is usually advisable and often necessary to plan a project in detail so as to accommodate it to the use of a particular desired product or material. It would generally be impracticable for the engineer to try to make the drawings for a project so comprehensive as to adapt it to the use of all materials that manufacturers would like to have employed.

The engineer should make thorough studies to ascertain that the final design is the best that can be devised. Naturally, he should consider basic concepts, arrangements, and workmanship. He should also consider the relative costs of materials when making the design. Low first cost alone, however, is not always the feature that should determine selection of materials, or of anything else for that matter. Longevity, maintenance charges, operating costs, efficiency, power requirements, safety, and general adaptability for its purpose—all these are to be considered as well.

A threat to this discretionary power of the engineer occurred in connection with the design of the Ohio Turnpike.[1] The commissioners, upon the recommendation of the engineers concerned, specified portland-cement concrete pavement because they believed it to be best under the circumstances. Persons who were interested in having asphalt paving included as an alternate bid brought legal action to force the commission to include this option. The Ohio Supreme Court overruled the lower court and decided that the commission (and its engineers) had the right to specify what it considered to be best, provided only that its action be taken "in good faith and not in abuse of its discretion."

14-3 Standard Products It is often both efficient and desirable to specify a standard product by trade name, catalog number, or any other suitable reference that is definite enough and is customary usage. For example, one may specify an electric motor as follows: "X Electric Manufacturing Company Type CSP, open, 10 hp, 1200 rpm, frame 326, 440 volt, 3 phase, 60 cycle." Thus, all concerned know exactly what is required, or they can readily find out about it.

On the other hand, this designation of one specific product is not usually permissible for government work. In such contracts it is normally a requirement

---

[1] *State v. Ohio Turnpike Commission,* 159 Ohio St. 581, 113 N.E.2d 14 (1953).

that at least three alternate suitable articles be designated, any one of which will be acceptable. The purpose of this regulation is to avoid what might appear to be favoritism. It is also intended to secure adequate competition in the bidding. That is why in such contracts it is generally necessary to avoid specifying a patented product that only one producer can furnish.

Such restrictions do not apply in private work, but it is often advisable to avoid criticism and to promote competition by specifying "so-and-so or approved equal." This tells other manufacturers the kind of product or service that is desired, and it enables them to propose a similar product of their own that they consider to be just as good or better. It sometimes happens that, unknown to the engineer, some new and better article has been developed. In such an event a substitution will be welcomed. However, the engineer's approval is essential; otherwise great confusion would result.

The following clauses illustrate the use of references to manufacturers' products in specifying plumbing and heating equipment:

A Minneapolis Honeywell T-418 B outdoor reset control with outdoor weather bulb and water bulb in a separate well shall be provided to operate the oil burner. Control shall provide 210°F water when outdoor temperature is 0°F. Provide and install the following fixtures and equipment:

Lavatories: Standard F 329 A
Tubs: Standard P 210 5T and P 2100T

This specification is intended to cover the complete installation of one (1) Electric Oil-hydraulic (electrically operated hydraulic elevator, using oil as the fluid medium) Passenger Elevator, as manufactured by Rotary Lift Company, Memphis, Tennessee, or equal as approved by the Architect.

It is important to give brand names, pattern numbers, and enough other data to make the requirements very clear and definite. It is also advisable to be sure that the article is a standard one that can be readily secured and that the manufacturer is a reputable concern.

Another way to specify an article without resort to the use of the words "or approved equal" is to list the brand names and other information of a series of articles, any one of which will be acceptable. This procedure is more difficult, and the list, even then, may not be truly complete.

The use of these references to a standard commodity or product is, of course, much easier and safer than the preparation of detailed specifications describing all the desired qualities of the article. On the other hand, when patented or proprietary products are desired but cannot be named in the specifications, it is possible to word the specifications so that the desired product is the only one which completely meets the requirements.

The use of the phrase "or approved equal" may not be strictly correct because articles manufactured by various sources may seldom be exactly alike even though each may serve the desired purpose satisfactorily. The words "or approved equivalent" may be more technically correct. The word "substitute" does not necessarily mean the same quality, and its use is not recommended.

"Or approved equal" is a good illustration of the meaning of words as "used in the trade." A court will construe the former phrase as having its customary meaning unless some contrary statement appears in the contract. The important thing is the requirement that an article differing from the specified one may be substituted only when approved by the engineer.

On the other hand, there are cases where, for one reason or another, a specific article is alone acceptable. An example of this might be the equipment for an addition to a factory. The owner already has certain machines, motors, etc., in the existing plant. For the pending annex he wants more of the same ones because his maintenance people and operators are familiar with that particular type of equipment. Furthermore, he does not want to carry replacements and spare parts for a variety of equipment types and sizes. In such circumstances he is well advised to specify "more of the same."

It is frequently satisfactory to say in the specifications that a product is to be made "according to the manufacturer's standard." For a motor of certain type and horsepower, for a truck of a specific type, model and capacity, and for a bridge crane of some particular span and capacity, such a specification may be perfectly all right.

Such may not be the case, however, if the article to be purchased has to meet various conditions of special nature. For example, a gantry crane of a particular span, height, speed, and capacity was required, and the specifications used the words "according to the manufacturer's standard" in describing it. When the drawings for the contractor's design were received, the engineer objected strenuously to certain basic features and to several details of the design because some parts were not sufficiently strong. However, he was at a great disadvantage in securing remedial measures because the manufacturer said: "This is our standard. We have built similar designs for. . . ." In this case, as in many others, the opinion of the engineer was based upon personal knowledge and experience of what would be best for long-time service. He could not prove that the manufacturer's design was unsafe, but he wished to have certain parts of the construction detailed in accordance with what he had previously found to be eminently satisfactory. The specifications should have reflected this.

It is essential that the engineer be the one to approve or disapprove any substitution. However, if the specifications state that alternative materials will be considered by the engineer, a contractor cannot be certain about what the engineer will accept. Therefore, if the bidder is to play safe, he will have to base his costs upon what is specified because if he states that the bid is based upon certain stipulated substitutions, he will have changed the terms of the invitation for bids. The bidder's proposal is then likely to be rejected because it is not in proper form and because it cannot be readily compared with the proposals of other bidders. Substitutions (and any corresponding price adjustments) will usually have to be arranged after the contract is signed, if they are to be resorted to at all. After the contract is signed, it is unfair in a way to other bidders for the engineer to accept cheaper products as substitutes for those specified, even if the contract cost is adjusted accordingly. If a contractor proposes a substitution prior to bidding, and if the engineer finds that such substitution will be in the best interest of the project, the engineer should issue an addendum to all bidders advising them that said substitution will be acceptable.

14-4 Standard Specifications for Materials It is a great help to refer to standard specifications—where applicable ones are available—as a means of specifying materials. The American Society for Testing Materials (ASTM) has prepared volumes of specifications for a wide variety of materials—in fact, for most of the materials used in practice. These specifications are the result of long and intensive study by experts in each field involved and are generally accepted as authoritative. The ordinary engineer cannot hope to sit down and quickly write a perfectly satisfactory specification for even a few basic materials, nor would it be advisable when such excellent ones are readily available.

For example, assume that specifications are needed for the steel to be used in a number of large castings. The engineer may state that the material is to conform to ASTM designation A 27-82, standard specification for Carbon Steel Castings for General Applications. Thus, manufacturers who wish to bid on these castings will know exactly what is required as far as quality of material is concerned. In using such a reference the engineer may add "or the latest revision thereof," if he is not sure that the reference volume designated contains the very latest revisions of the pertinent specifications.

The title used in the foregoing ASTM refereence is generally descriptive; the designation is the specific reference. In this particular example the number 82 designates the year of issue of the specifications by the ASTM organization.

As the name implies, the ASTM specifications, in addition to specifying qualities, cover or set up standard procedures for acceptance tests of materials. This is an important matter. One can easily imagine the confusion that would arise if excellent specifications of the chemical and physical properties of a material were issued but organizations or users were left to devise their own methods of testing this material to ascertain whether or not a particular lot would be acceptable. Just a few of these tests as issued by ASTM are the following:

> ASTM designation: E8-82 standard method of Tension Testing Metallic Materials.
> ASTM designation: C151-77 standard test method for Autoclave Expansion of Portland Cement.
> ANSI/ASTM designation: C88-76 standard test method for Soundness of Aggregates by use of Sodium Sulfate or Magnesium Sulfate.

Notice that the ASTM specifications do not, and should not, prescribe what material to use for a particular purpose or when to use it. These are matters for the individual engineer to determine. If he wants to make an article of cast iron instead of cast steel, the proper reference to the appropriate ASTM specifications will ensure that he is calling for good cast iron. On the other hand, if the brittleness of cast iron should prove to be a disadvantage in service, this is not the fault of ASTM. Obviously, the ASTM cannot guarantee proper design use of its data and materials.

It is clearly the responsibility of each individual engineer to be sure that he knows the contents of the section of the ASTM specifications that he uses as a reference. Anything he needs that is not so covered should be specified directly. Thus the dimensions and finish of castings are generally to be given on the drawings, while requirements for annealing and any special methods to be used in making the castings are to be stated by the engineer in specifications he personally devises or are to be given through references to proper standard specifications.

The ASTM will specify only such items as the proper ultimate tensile strength, the minimum yield point, and the required minimum percent elongation. Furthermore, the ASTM does not undertake to specify unit stresses to be used for design purposes. For example, if the engineer is using structural steel, he (or some governing code) must determine what unit stress and safety factor are to be used. Again, if the engineer is using concrete, he is the one to determine whether that concrete is to have a twenty-eight-day ultimate compressive strength of 2500 or 4000 pounds per square inch, etc.

Besides the ASTM specifications, there are many other materials and workmanship specifications of real help to the engineer as references. These latter are generally prepared by manufacturers of materials or by groups that are interested in particular materials. For example, the *Standard Grading and Dressing Rule for Lumber*[2] prepared by the West Coast Lumbermen's Association, is useful when one is to prepare specifications for a timber structure; *Specifications for the Design, Fabrication and Erection of Structural Steel for Buildings,* prepared by the American Institute of Steel Construction, is very helpful in connection with building jobs; and the Building Code Requirements for reinforced Concrete (ACI 318-77) of the American Concrete Institute serve a useful purpose in certain phases of concrete work.

The fact that these specifications are prepared by those who are interested in the sales of a particular material does not mean that the specifications are poor or untrustworthy. The manufacturers who advocate the use of their materials obviously want them used properly so that the results will be satisfactory in the present case and that credibility for future sales will be established.[3]

Naturally, standard specifications must be correctly used if they are to prove valuable. It is generally better to use them in their entirety than to select pieces of them for quotation. For example, the *Steel Products Manual for Hot Rolled Carbon Steel Bars,* Section 8, published by the American Iron and Steel Institute in August 1952, contains, in substance, the following specification for the straightness of reinforcing bars to be used in concrete:

> **1** The offset of any bar from a straight line shall not exceed $\frac{1}{4}$ in in any length of 5 ft
> **2** In no case shall the offset of any bar from a straight line exceed $\frac{1}{4}$ in times one-fifth of the length in feet.

Both these requirements are pictured in Fig. 14-2*a*. Specifications were prepared which copied only the first of these items. Fig. 14-2*b* shows how, if only this

---

[2] This rule applies to Douglas fir, Sitka spruce, West Coast hemlock, and Western red cedar.
[3] Design specifications and building codes do not concern the contract as such, although the structure to be built under the contract may have to be designed in accordance with their provisions.

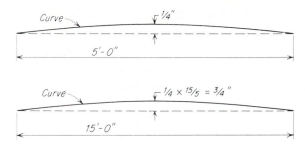

(a) According to American Iron and Steel Institute

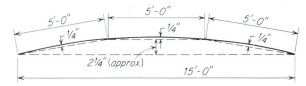

(b) According to specifications as written

FIGURE 14-2
Partial use of a specification for straightness of reinforcing bars.

one requirement is prescribed, the supplier may comply with it and yet ignore the equally important second item. The bars may then be curved excessively but still meet the specifications.

Notice the following advantageous uses of references to standards for electrical work:

The following industry standards shall be considered minimum requirements under this specification:

1 The standard rules of the American Institute of Electrical Engineers
2 The rules and regulations of the National Board of Fire Underwriters (National Electrical Code)
3 The rules and regulations of the local utility company and local Electrical Inspection Department
4 All electrical materials and appliances shall have the listing of the Underwriters Laboratories, Inc., or shall conform to their requirements and shall be so labeled
5 All work is to be done in accordance with all state and local regulations, and the Contractor is to obtain and pay for any charges for all permits, licenses, etc., in connection with his work.

A paragraph in another specification stated:

All work shall be performed in accordance with the National Electrical Code, American Standard Safety Code, and such state and local codes as may be applicable. The Contractor shall obtain all necessary approvals, and he shall pay for all permits.

The following is an illustration of the use of a reference to a standard specification for plumbing work:

All work shall meet the requirements of the National Plumbing Code, unless otherwise noted.

Standard specifications are generally prepared for use in average or ordinary conditions. When special requirements are to be met, the engineer should not trust the adequacy of standard specifications. It is his responsibility to prepare complete and accurate specifications that meet the requirements of the particular job.

14-5 Use of Previous Specifications Many products that are considered as raw materials for a construction contract are really the finished product of some manufacturer, and they cannot be described adequately, if at all, by reference to standard specifications. However, use may be made of appropriate paragraphs that have been previously prepared by the engineer himself or by others. If no such paragraphs are available, original descriptive data will have to be compiled and utilized.

The use of paragraphs or clauses that have been prepared for the specifications of a preceding job may have the advantage of prior experience and testing in practice; however, previous compositions may not be completely applicable to the present case. They may contain errors or inadvisable statements, and they may fail to cover all necessary features. Often there is a statement in a former writing that refers to some feature of the previous contract that is not relevant to the present one. Again,, there may be some different feature in the present contract that the copied clause does not cover at all or, at least, for which it does not prescribe the requirements correctly.

The principal lure of previous compositions is that their use relieves the engineer of the necessity of doing unnecessary original thinking and also helps to avoid overlooking some important feature. But excessive reliance upon the cut-and-paste system may lull the engineer into a false sense of security so that he does not really check to see that all necessary points are covered properly and accurately.

One way for a large organization to handle this matter of the preparation of specifications would be to compose a paragraph or series of paragraphs covering all the subjects that are likely to be applicable to its future contract, to be studied carefully by all concerned. Copies of the master document can be prepared for incorporation in future specifications, with blank spaces left for the insertion of any figures or statements that are not constant for all cases. (This procedure is likely to be safer than merely cutting paragraphs wholesale from previous specifications.) If the master set of specifications is kept up to date and if anything discovered in the course of contract work which affects these specifications is passed on to the one in charge of the master copy for correction the occurrence of mistakes should be minimized.

14-6 Specification in Terms of Service In some cases, such as purchase orders, it may be satisfactory to specify materials and equipment in terms of the service required. The reason may be to secure competition in circumstances such that each of several different products could serve the purpose. The engineer may use this method when he is not sure which particular material or article will be the best and the cheapest and when he wants the recommendations of, as well as prices quoted by, various manufacturers.

It seems that this *performance type* of specifications, which states what is required in the way of performance, qualities, etc., is attaining greater popularity than one which attempts to set down all the detailed requirements for a particular article. This is an especially suitable form of specification when more than one method or commodity may be used satisfactorily for attaining a particular purpose. In this case the manufacturers and contractors are permitted to make their proposals upon the basis of what each one thinks is suitable and best to meet the general requirements. The engineer can then accept or reject any or all proposals if he believes that such action is desirable. If the several producers guarantee that their products will meet the requirements, then the engineer may select one upon the basis of price, desirability, or other characteristics, relying upon the guarantor to make good if necessary.

Let us assume that an engineer is to specify the insulation for the flat roof of a new and very large industrial plant. This will be a separate contract that will include the material and its installation. He may specify the desired resistance to the transfer of heat or may say that the insulation is to be equivalent to a 2-inch layer of foam glass. He may also say whether or not the material is to be rigid, fire-resistant, moisture-proof, and/or nondecaying. The manufacturers can then propose what they recommend, what they will guarantee, and what the cost in place will be. The engineer then has to determine which one of the several possibilities proposed is the best, considering both price and quality.

Again, assume that a mechanical engineer is preparing a requisition for a bridge crane for an industrial plan. He might prepare a sketch showing the clearance diagram, the span, side and vertical clearance, distance of rails above the floor, capacity in tons, auxiliary hook (if any) and its capacity, speed of bridge travel, speed of trolley, voltage to be used for power supply, and position of cab. He might also state the date of delivery required. Various manufacturers have their own particular designs and details for such equipment, or they will make up special designs. Each may submit his proposal for the bridge crane. The engineer can then accept whichever proposal he thinks is best. The details of the proposed equipment may vary, but the guarantee of service and quality given by reputable manufacturers can be trusted.

The following specification is quoted from a contract for some pumps which constituted one part of the equipment to be supplied:

> Close-coupled centrifugal raw water booster pump, complete with motor, of capacity and head as given in "Design Conditions" herein, fully assembled and ready for installation.
>
> The pump shall be designed for the following conditions and in accordance with the standards of the Hydraulic Institute.
>
> Capacity, 20 gpm
> Static head at suction, about 10 ft
> Discharge head, 195 ft
> Water temperature, approx. 35 to 70°F

The final pump suction head, discharge pressure, and rotation will be given at time of approving the overall dimension drawings.

Notice that the preceding specification permits each manufacturer to propose one pump (or to design one) that is suitable for the required service without tying him down to minor mechanical details, although these must be satisfactory to the engineer.

Naturally, this method of specifying service requirements cannot be used on a wholesale basis for a large construction contract that includes a great many items. The contractor must have a fairly definite knowledge of the requirements of the contract. If the specifications have too many materials and products open to choice, the contractor cannot tell in advance just what to bid upon unless he can determine precisely and easily what is to be used to fill each need. This sort of freedom tends to cause confusion.

On the other hand, innovation should not always be stifled. Someone may develop a new way of making something or performing some operation. It may be possible to make use of it by wording the specifications so this method may be considered provided the engineer believes it to be in the best interest of the project. Again, one should encourage participation but without making the specifications too vague.

14-7   Planning Detailed Specifications   It is apparent that one must know a considerable amount about a material in order to write a good specification for it. One should know what qualities to expect in the material, what characteristics are needed for the special purpose, and whether or not (and, if so, how) it can obtain the desired result without undue trouble and cost.

For purposes of illustration, assume that you are to prepare specifications for the piles and timbers to be used in a wooden pier for docking colliers at a power plant located at tidewater. Assume further that you cannot use references to standard specifications for the purpose. How would you go about accomplishing the work? Here are some of the things that you should think about and cover in your specifications:

A Piles
  1 Kind of material; that is, Southern yellow pine, oak, spruce, Douglas fir, etc.
  2 Variation allowed from the stipulated lengths
  3 Straightness
  4 Size range, or at least minimum sizes, of tip and butt
  5 Removal of bark
  6 Limit on knots allowed and trimming of knots
  7 Seasoning
  8 Creosoting—method to be employed and quantity of creosote to be used
  9 Limitations on cracks, shakes, and other imperfections allowed
B Timber for deck
  1 Kind of material
  2 Variation allowed from specified lengths, especially underrun

3 Straightness or warp

4 Limit on knots—number, size, and position

5 Seasoning

6 Creosoting

7 Amount of splits, checks, and shakes allowed; length, depth, direction

8 Grade

9 Chemical treatment, if any

10 Kind of finish—rough or dressed

11 Full trimming—no skips

12 Character of stock—fresh, clean, no decay

13 Grain limitations

14 Limit on pitch streaks or pockets and on stains

15 Allowed variation from nominal sizes

16 Limit on worm holes

17 Requirements if tongued and grooved

Actually, especially with timber, the limitations generally vary somewhat with the size and general character of the material. For example, the specifications for 8 by 12 timbers will differ in some details (other than just size) from those for 1 by 6 sheathing.

As a further example of specifying a material, examine the following which applied to gravel fill at a school building:

GRAVEL FILL: Gravel Fill shall consist of sound, tough, durable particles of crushed or uncrushed gravel, free from soft, thin elongated or laminated pieces and vegetable or other deleterious substances. It shall meet the sieve analysis given below, and the requirements for plasticity and the test for soft particles indicated in Connecticut State Highway Department Form 809, "Standard Specifications for Roads,, Bridges and Incidental Construction," dated, January, 1963, with "Supplemental Specifications" thereto dated October, 1964, Article M-02-07, paragraphs 2 and 3.

The reader can easily realize what a task it is to compose all these detailed specifications. He will soon look for some standard references that he can use in preparing the required data. It must be remembered that a misstatement or an omission may cost money and may compel one to accept material that is not as good as he intended.

SIEVE ANALYSIS FOR GRAVEL FILL

Square mesh sieves	Percent passing by weight
Pass 3½ in	100
Pass ¼ in	30–65
Pass #40	5–30
Pass #100	0–10

14-8   Special Items It is seldom that any engineer or architect has to compose specifications covering some material that is entirely new or some new combination of materials. However, he may desire some special qualities in, or uses for, materials. If so, he must specify them by original composition.

To illustrate, suppose that an architect wishes to use a certain unusual type, texture, color, and finish of stone for the exterior portion of the walls of a monumental building. Under the circumstances, it is necessary for him to make a study of exactly what kinds of stone are available and suitable. He must also determine whether or not the stone of his choice can be finished in the fashion he desires. This necessitates the preparation of special specifications for this material. The architect, of course, must assume responsibility for the results of his choice.

Sometimes it is desirable to require the contractor to prepare samples of his work or product and to submit them for approval. The specifications should state that the materials furnished under the contract are to match the approved samples.

When it is necessary to specify a material in detail or to specify some commodity, it is essential for the engineer to have a thorough knowledge of the physical, chemical, and other properties of that material. He must then be careful to prepare his specifications so as to include the desired properties but to exclude or limit anything which may be undesirable.[4]

Provision should also be made for acceptance tests of materials. Who is to test them? Who is to pay for this work? What are the tests to be? Who is to approve or disapprove the results of these tests? All these questions should be answered in the contract clauses or in the specifications.

The engineer should also be very careful to see that his specifications are practical and that they can be met without excessive cost and trouble.

When confronted with special cases, the engineer or architect might well go to the trouble and expense of having a competent specialist check his specifications before they are published. A misguided sense of professional pride should not be allowed to prevent this sensible extra step of precaution.

QUESTIONS

1 What are the basic requirements of proper specifications?
2 Discuss the difference in the kind of information shown in the drawings and that given in the specifications.
3 What are standard specifications? Illustrate what use can be made of them.
4 What is the special function of material specifications? Do they in any way protect the contractor?
5 When and why may one use copies of previous specifications when composing a set of contract documents? Discuss the advantages and disadvantages of this procedure.

---

[4] For example, sulfur may be an impurity in steel. If an excessive amount will be harmful for the intended use, the engineer should specify the maximum percentage which will be acceptable.

6 What is the ASTM? How can its work product be utilized? What advantages are there for the engineer in using such available material?

7 What is a "performance" or "service" type of specification? Illustrate its use. Who is responsible for results where this type is employed?

8 When and how can the specifications use to advantage a reference to a standard product of a particular manufacturer?

9 Why specify "or approved equivalent"? Who does the approving?

10 The drawings showed that a steel structure was to be field-bolted, but the specifications stated that it was to be field-welded. The contractor bid on the bolted work. Was he justified in so doing? What can the engineer do about the situation?

11 The specifications called for the coating with linseed oil of all "finished" surfaces of machinery. The drawings did not indicate any coating. What should the inspector require of the contractor? Explain.

12 A specification stated: "The insulation shall be 2 in thick and of an approved type." Is this a satisfactory way to put it? Why?

13 Why is it desirable to include in the contract documents a description of the project?

14 A specification stated: "Substitutions may be submitted for the approval of the engineer." Is this a satisfactory statement? Elaborate.

15 Why is it generally inadvisable to allow bidders to make substitutions (in their proposals) for specified structural materials? Is such substitution satisfactory in the case of certain equipment?

16 Has the engineer the right to specify what he wants regardless of the opinions of the contractors?

17 Does an engineer act properly when he seeks the advice of contractors and manufacturers before preparing his specifications for some article or process? Why?

18 A specification states: "Suitable pumps shall be installed to prevent flooding of the excavation." Has the engineer control over what pumps are to be provided? If flooding occurs, who is responsible?

19 What are the advantages and disadvantages of specifying products by name and catalog number?

20 What is the purpose of public works regulations which require that at least three proposals be sought?

21 What advantages and disadvantages may there be in specifying a list of acceptable alternatives for a given purpose?

22 Under what conditions might it be desirable for the engineer to specify one and only one named product?

23 An engineer specified that a bridge crane should "have a capacity of 50 tons, with a 10-ton auxiliary, a span of 75 feet, an acceptable speed of travel, and be driven by 440-volt, 60-cycle power." Is this specification satisfactory under ordinary circumstances? Why?

24 Can a contractor propose substitutions for specified products *a* before the bids are received? *b* after the contract is awarded?

25 Can an engineer specify a material having different properties than those given in the ASTM specifications? Why?

26 Does the ASTM specify how and when a particular material (like Portland Cement, Type II) is to be used? Why?

27 As references, are standard specifications always to be used? If standard specifications are thus employed, can the engineer specify anything which differs from them?

CHAPTER 15

DRAWINGS

15-1 Introduction This chapter is intended to show some of the fundamental requirements for the preparation of the drawings for construction contracts. The drawings, like the printed material, will become a very important part of the contract. Between them, these items are supposed to give the contractor the essential information regarding the work. The preparation of his bid will be based upon his interpretation of these papers. The details will vary with the project, but the basic principles are common to most construction contracts.

In general, the drawings made by the engineer and the architect—who are often referred to as *design professionals*—may be called *design drawings* when they are prepared for bidding purposes. When the contract is let, these drawings become known as the *contract drawings*. These drawings should show enough of the extent and character of the work to enable the contractor to know what is to be expected. For a lump-sum contract, the design drawings should be much more comprehensive and explicit than they need to be for other types. The work to be done under a unit-price contract may be such that the engineer, in order to show various details of the project, has to make many additional drawings after the contract is let. In almost all construction contracts it will be necessary for the contractor to provide *shop drawings* in order to make all the details clear for construction purposes.

15-2 Purpose of Design Drawings Design drawings are used to give the contractor such needed information as can be transmitted more effectively by picture language than by written description. Such data as shapes, sizes, detailed dimensions, relationship of parts, and desired locations for special features would be almost impossible to convey satisfactorily by any means other than drawings.

The preparation of design drawings is an art in itself, but in this text we are primarily interested in the functional aspect of these drawings as an important part of the contract documents. The contractor will have to provide and perform what such drawings require. When arguments arise (most likely during the course of construction) regarding contractual requirements, the key to resolving issues is what the contract drawings really show, not what they *should* have shown or what the engineer wanted. Considerable trouble and expense may well follow drawings that fail to do the job for which they are intended. The adequacy, clarity, and detailed accuracy of the drawings—rather than their beauty—are the really important features. A drawing may consist simply of an amateur's freehand sketch made on a piece of wrapping paper. If so, and if made a part of the contract, the drawing would be just as binding as though painstakingly put together in ink on tracing cloth by an expert drafter.

It is ordinarily assumed that the drawings and the specifications, although prepared by the engineer, are to become the property of the owner for whom they were made and who pays for the engineering work involved. If any other "ownership" is intended, it should be agreed upon in advance between the owner and the engineer.

15-3 Tabulation of Drawings When they are not physically attached to the rest of the documents, the design drawings should be made an integral part of the contract through what is known as *incorporation by reference*. It is customary to have these drawings done on tracing paper or tracing cloth, then to have prints made from these originals. Drawings may be of any selected dimensions; 24 by 36 inches is a common size. In view of their considerable size and bulk, the prints are usually bound together and kept separate from the other contract and specification papers.[1]

It is desirable to have the incorporation-by-reference clause—which states that the design drawings are an integral part of the contract—accompanied by a table setting forth the date of the drawings, their individual numbers, and their individual titles. For example, the following hypothetical list is illustrative of one possible arrangement for the pertinent portion of a unit-price contract dealing with major highway construction:

Contract Drawings

The Contract Drawings which accompany and form a part of this Contract and of the Specifications are dated Feb. 15, 1986. They have the general title "The Connecticut State Highway Department—Route 46, Expressway—John Doe Street to City Line— Bridgeport, Connecticut—Contract No. EX-3."

The drawings are numbered and separately entitled as follows:

---

[1] On the other hand, reproductions of *small* sketches may be bound directly with the contract itself, but suitable small-size photostats or photographic reproduction of ordinary large drawings are unduly costly, and, when made the proper size for convenient binding, may prove too small to be easily read.

Drawing no.	Title
1	Location Plan and Index
2	Plan and Profile—John Doe Street to Robert Roe Avenue
6	Cross Sections
7	Robert Roe Avenue Bridge
8	West Abutment at Robert Roe Avenue
9	East Abutment at Robert Roe Avenue
14	Connection to Maple Street

The contract drawings do not purport to show all the details of the work; they are intended merely to illustrate the character and extent of the performance desired under the contract. Therefore, they may be supplemented or revised from time to time, as the work progresses, by the engineer. Revisions and additional drawings are to be made by the engineer as the proper illustration of the work requires. All such supplementary and revised drawings automatically become part of the contract.

One should be very careful to see that the tabulation of pertinent drawings is complete and correct. If a drawing involving a considerable amount of work is omitted from the tabulation, this omission is as serious as the failure to put some specific written material in the contract and specifications. If the drawing is issued belatedly, the contractor can contend that such work was covered nowhere in the contract as signed, and, therefore, the performance of the work involved is not required. If the contractor then proceeds to perform this additional work, he will be justified in making a claim for extra compensation unless the applicable unit prices cover the cost adequately.

Notice that, in the preceding quotation, the engineer's right to revise the contract drawings and to make binding supplementary drawings is specifically stated in the contract. The need to thus alter and amplify the original set of drawings usually arises when a unit-price contract is let on the basis of general design drawings which do not show the required work completely. It often happens that such a contract is let on the rather imprecise basis of several drawings which picture the general scope and character of the project[2] but which do not show

[2] For example, the unit-price contract for the construction of a $20 million industrial plant was let on the basis of data shown by the following design drawings only (plus, of course, specifications):

1 A general layout of the site and of the plant showing structures, transportation connections, grading, etc.

2 A topographic map of the property

3 A foundation plan showing footings and locations of piles, plus a few typical details

4 A floor plan showing the size, shape, and general features of the main plant

5 Sample framing plans and sufficient details to picture the character of the work

6 Enough architectural details to show the typical roof, wall, and floor construction

7 Layout drawings to picture the character of the plumbing, drainage, heating, and electrical installations

After the contract was awarded, the engineer prepared over 300 supplementary drawings for the contractor's use.

sufficient detailed information to enable adequate performance of the work. After a contract is so awarded, the engineer may have to make from several to a few hundred supplementary drawings in order to portray all the information essential in the construction process.

15-4 Preservation of Contract Drawings Because revisions of the original tracings are not unlikely, it is desirable for the owner to retain at least one *record set* of the contract drawings (in the form in which they were initially issued for bidding purposes) for possible use in the event of future disputes. Reproduced tracings made on cloth directly from the original ones (before any revisions are permitted) would serve as an ideal record set of the contract drawings. Possession of a record set of these drawings may prove to be of great value if, for example, the contractor should present claims for extra compensation. Repeatedly, this question arises: "What did he bid on?" The answer can be gleaned from reference to the record set; by comparing these drawings with what was actually built, the validity of the contractor's claims may be ascertained.

In one case the contract drawings showed a detail for a special expansion joint in some heavy-duty crane rails. The drafter inadvertently omitted the word "thermit" from the note referring to the welding pictured. The contractor consequently assumed that ordinary electric arc welding would be satisfactory. After the contract was signed, the drawing was revised to correct the note. Since the contractor had no facilities for thermit welding, he had to engage someone else to do this extra work. He claimed payment for the extra cost. Reference was made to prints of the original contract drawings, and the contractor's claim was thereupon approved.[3]

Besides retaining a set of the original contract drawings, it is often desirable for the owner to require the contractor to provide a set of tracings of "as built" drawings after the work is completed, showing each part as actually constructed. They should cover all dimensions after the several parts are measured in detail in the field. These tracings may prove very useful in case of future alterations or additions to the structure. On the other hand, it should be made clear that the owner is not allowed to use any reproduced tracings of the contract drawings or tracings of the "as built" drawings for any other purpose or project without the written permission of the design professional to do so.

15-5 Sources of Trouble Carelessly prepared or inadequate drawings obviously invite trouble in the performance of a contract and are a fruitful source of claims by the contractor for extra compensation. The engineer, by way of making sure the owner understands the situation, would be well advised to demand sufficient time and personnel with which to prepare really suitable drawings. In practice, the engineer is generally faced with the problem of getting the drawings and other contract documents completed for bidding purposes as fast as possible,

---

[3] However, this can work both ways, and numerous instances might be cited in which the record set of drawings has enabled the engineer to prove that extras claimed by the contractor were unwarranted.

usually under pressure from the owner and with limited resources available for performing the job.

Sometimes unit-price proposals must be based upon preliminary drawings purporting to show in a very general way the nature and extent of the project. The advisability of this widely used approach is open to question. Subsequent development of the design almost always results in the addition of many essential features which add appreciably to the cost of the work. Some claims by the contractor to recover this excess are practically inevitable and usually justified. The owner should realize this when he lets the contract to a bidder who has had to base his proposal upon drawings, which, taken as a whole, are not thorough and comprehensive.[4]

One principle to remember is that the function of drawings is to make the subject clear and definite by whatever means the drafter can devise. It is obvious that the engineer cannot afford to make the contract drawings so extensive that they show every detail of the project. Many parts may be similar in general character but different in various details. It is essential to show on the drawings enough data to describe the work sufficiently for a knowledgeable contractor to understand what is required. The engineer might well label such illustrative information "typical details" or the equivalent. He should state in the specifications, or by notes on the drawings, that the data shown on the contract drawings are typical and are to describe the general nature of the work, and that the contract clauses and specifications, as well as the contract drawings, constitute the contract as a whole. For economic reasons, efforts should also be addressed to keeping the preparation of the drawings from becoming unduly laborious on the one hand or excessively costly on the other.

15-6   Development of Details   Basically, the design drawings are intended to show the general outlines and enough detail regarding the structure, materials, and equipment to enable the contractor to bid on the job. However, seldom are these drawings entirely adequate for every need in connection with the construction work; it is necessary to enlarge upon the information which they show or imply, supplementing the engineer's drawings with what are variously called *detail drawings, shop drawings, construction drawings,* or *working drawings*—made by the contractor.[5] These drawings of details are intended to picture for the contractor

---

[4] One large unit-price job for heavy concrete construction was undertaken on the basis of a few general drawings together with a set of prints from a previous contract—similar in a broad sense, but very different in detail. The old prints were used for the limited purposes of (1) showing the bidders some sample drawings of the type turned out by the engineer's office and (2) picturing the general nature of the construction which had been required under the earlier contract. The emerging demands of the project at hand naturally revealed many and extensive differences between the old drawings used for bidding and those "current" ones ultimately issued after the contract was awarded and actual construction was under way. The contractor's list of claims for extras reached a monetary total that was staggering. The owner blamed the engineer for the unfortunate situation, saying, "Why didn't you convince me that this would be the result?"

[5] Thus, in some cases the contractor will have to make construction drawings showing designs and details for such elements as cofferdams, heavy shoring, arch centering, and other essentials for actual

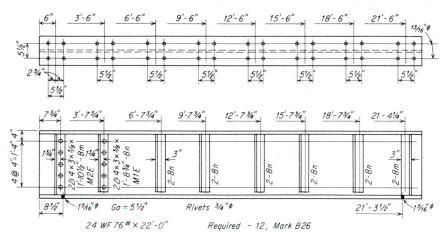

FIGURE 15-1
Shop details of a beam.

everything needed in order to execute the work. The shop drawing for a steel beam, shown in Fig. 15-1, is as it might be made by the drafter of a steel-fabricating shop. Notice that all materials and dimensions are given so that the workers can manufacture the finished product.

The design drawings (as supplemented) might show plans for the steel framing of a building. They may consist in part of drawings portraying the various members by single lines, as illustrated in Fig. 15-2, which is assumed to be a small portion of a drawing which depicts a steel roof truss. Each member is labeled to show the size and material required. These drawings should also reveal necessary information regarding the prescribed lengths and positions of members. Still, it is obvious that a steel-fabricating shop could not build this truss unless it was given more complete information than is available from this design (or "contract") drawing. Therefore, the contractor for the steelwork will have to put together additional drawings, giving in minute detail whatever data are needed to enable the shop to fabricate each one of the members and to build the truss. In addition, the fabricator will have to prepare erection diagrams, bills of material for computing weights, shipping bills, and any other miscellaneous papers which may prove necessary. On a large project the shop drawings and related items may be numbered in the hundreds.

conduct of the work. Preparation of all these papers involves a considerable amount of labor and of engineering ability. When the engineer examines the drawings showing such designs prepared by the contractor for the latter's use, the engineer should make it clear that he does so only in a spirit of helpfulness and to see that, in his opinion, these plans show what seems to be suitable to accomplish the final construction required by the contract. The engineer may make suggestions (and even criticisms) if he believes that such will aid the contractor in safely performing the work involved. However, by no means should the engineer assume any responsibility for such designs and drawings by the contractor.

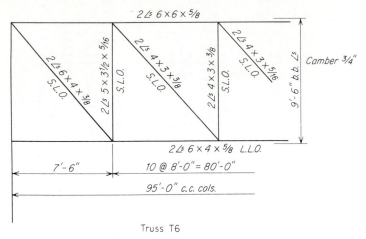

FIGURE 15-2
Truss diagram.

Going back now to the design drawings, in addition to the simple line diagram shown in Fig. 15-2, the engineer may indicate a few typical details as to the character of the work required—as illustrated in Fig. 15-3; this shows how some members are to be connected to each other. Where concrete work is involved, the engineer might prepare a design drawing somewhat like the hypothetical one in Fig. 15-4; this illustrates the general character of concrete piers and steel beams for supporting heavy equipment in an industrial plant. A careful study of this illustration will demonstrate that the contractor's men cannot build this pier without having more information, even though the size and shape of the concrete are given, the character and sizes of reinforcement are indicated, and the sizes and spacing of the steel beams are revealed. A possible shop drawing and bar schedule for this pier are represented in Fig. 15-5.

Of course, it is possible for the engineer to prepare the shop drawings for construction purposes, but the contract should be specific on this point. If the contract contemplates that such drawings are the responsibility of the contractor, it may be desirable to clarify the following points as guidelines:

1 The standard size of drawings

2 Instructions about whether paper or cloth is to be used for the tracings

3 Instructions about whether pencil or ink is to be used in making the final drawings

4 Any necessity for the use of printed forms for the drawings

5 The general title to be used for all drawings pertaining to the project

6 The printed forms to be used on the drawings for recording revisions and approval by the engineer

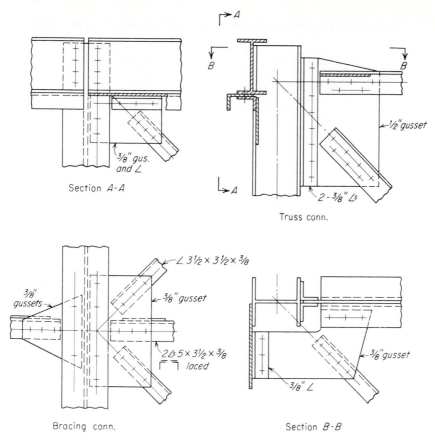

FIGURE 15-3
Details of connections to exterior columns (Scale 1 in = 1 ft).

7 Statement as to responsibility for the accuracy of drawings[6]

8 Any necessity and procedure for the submission of preliminary layouts for approval by the engineer

9 The number of prints required and the mechanics of submitting shop drawings for the approval of the engineer

10 The number of prints required for filing purposes by the owner and the engineer, and the number to be sent to the inspectors and to the field staff of the engineer

11 The number of prints of erection diagrams, shipping lists, shop bills, and the like to be furnished to the engineer and his field men

---

[6] It is desirable to provide that the contractor's drawings are to be checked by him before submission for the engineer's approval.

FIGURE 15-4
Portion of a contract drawing of piers and beams supporting equipment in an industrial plant.

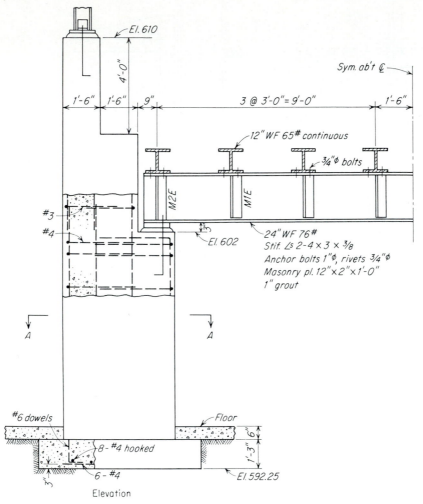

Elevation

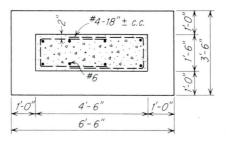

Section *A-A*

Shop drawings are ordinarily to be prepared by the contractor or by someone who is responsible to him. They are not part of the contract papers as such. In examining and approving the shop drawings, the engineer is to do so in order to see that what they show seems to be in accordance with the contract's requirements. By no means does he (or should he) assume any responsibility for their accuracy. His goal is to examine the shop drawings with reasonable care in order to ascertain that they show suitable information.

Of course, it is possible that an owner (such as an industrial concern or a governmental agency) may have a small project to perform from the planning stage clear through to completion of construction, using its own design professionals and other employees. Then any shop drawings which are made by its staff (or by others for them) are still the responsibility of the owner.

15-7   Revisions   Whenever an original contract drawing is changed, new prints of it should be sent to the contractor and to all other persons concerned. It is important to enable the recipient to see easily just what has been revised, and for that purpose, one of the following systems might be used:

1 Have a form printed on the original tracing for use in the recording of any revisions, perhaps as shown in Fig. 15-6. As a part of this form have a space available for inserting the number of the revision, the date on which it was made, the name or initials of the persons who made and checked it, and a note or two regarding the nature of the change. In addition, on the drawing itself, show in a small circle at the pertinent location the number of the revision. This enables the users of the drawing to determine in what respect it has been revised and to compare the altered version (as described in the printed form previously mentioned) with the original. The encircled number should remain as a permanent record on the tracing.

2 It is possible to identify the revisions directly on the tracing (preferably using pencil), and to mark on the *back* of the drawing so that what is added will not interfere with the original penciled or inked information. However, this general procedure is unsatisfactory when the drawing has to be changed more than once, since earlier changes encircled on the back of the tracing become difficult to distinguish from later ones.

3 Revisions may be encircled in colored pencil on the prints of the changed drawings. However, this procedure is likely to lead to errors unless all current prints are correctly marked. Also, this system does not provide an automatic aid in locating the alterations when additional prints are taken from the tracings at some later date.

15-8   Notes   *Notes* on the drawings are integral parts of the drawings themselves. The notes are really "specifications" of a sort and are as effective. These notations give certain information that cannot be shown advantageously, if at all, by the pictures alone. The line work and the notes together are to describe fully

FIGURE 15-5
Details of concrete pier and reinforcement.

BAR SCHEDULE FOR ONE PIER – 24 required

Type	Location	Mark	No. req'd	Bar No.	Type	A	B	C	Length	Total lin. ft.	Remarks
	Footing	S1	6	4	Str.				6'-0"	36'	
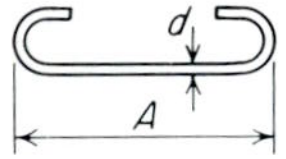	Footing	S2	8	4	I	3'-0"			4'-0"	32'	
	Footing	S3	8	6	II	3'-0"	6"		3'-4"	27'	A vertical
		S4	2	6	II	8'-3"	2'-9"		10'-10"	22'	A vertical
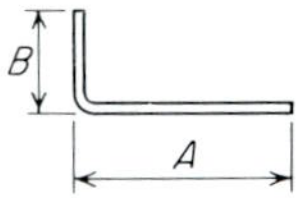		S5	2	6	Str.				12'-3"	25'	
		S6	4	6	Str.				10'-6"	42'	
		S7	4	6	Str.				7'-9"	31'	
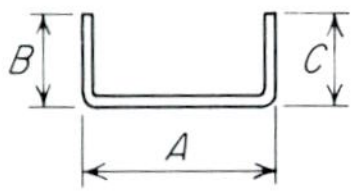		S8	12	4	III	1'-2"	2'-8"	2'-8"	6'-6"	78'	
		S9	6	3	III	1'-2"	1'-9"	1'-9"	4'-6"	27'	
		S10	8	3	III	1'-2"	1'-0"	1'-0"	3'-0"	24'	

Type I, II, III (bar bend diagrams)

Note: Standard hooks and bends: pin diam. = 6d

FIGURE 15-5 (*Continued*)

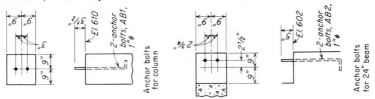

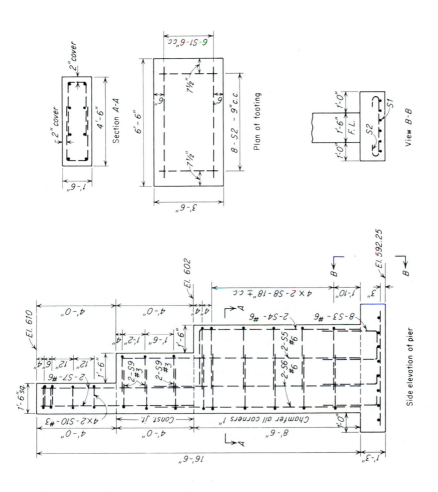

REVISIONS	NO.	MADE BY	DATE	DESCRIPTION		REVISIONS	NO.	MADE BY	DATE	DESCRIPTION
	6						1			
	7						2			
	8						3			
	9						4			
	10						5			

FIGURE 15-6
Form for record of revisions of a drawing.

the work required. Notes should be clear, specific, complete, and as carefully worded as though they appeared in the printed specifications.

In the preparation of data for any contract, questions will arise about whether particular instructions should be placed in the printed specifications, shown in the form of notes on the drawings, or both. The author of the contract documents should bear in mind that, when the field personnel or those in the shop are using prints of the drawings in order to detail or construct anything, it is advisable to have pertinent notes right on the drawings so that the users will be sure to see the information; such persons might not think to search through the printed specifications to see whether there is some instruction therein which will affect the particular job upon which they are working.

In general, if certain instructions relate to what a drawing portrays and if these instructions are to apply to some details of that particular drawing only, it is well to place the information in the form of notes directly upon the drawing in question.[7] On the other hand, if the instructions are common to a number of drawings, it is probably desirable to place the information in the specifications, where it can be made applicable to any and all affected drawings. Thus, if certain bolts in one particular connection (alone) are to be "shouldered bolts with double nuts," an appropriate note should be placed only on the drawing where details of that connection are shown. But if "all field connections are to be high-strength bolts" tightened to produce a stipulated unit stress, this information need not be repeated on each drawing that shows connections but should appear in the specifications. However, it is often advantageous to repeat this information in a set of general notes which is recorded on one drawing but is equally applicable to all if the other drawings are each inscribed: "For General Notes, See Drawing No. ———."

QUESTIONS

1 What kinds of information are best given by way of contract drawings?
2 Is any particular size essential for contract drawings?
3 Which are more important, the line drawings or the accompanying notes?

---

[7] For example, the note about cambering the roof truss shown in Fig. 15-2 is placed on the drawing because it refers to this truss only.

4 Why is it important to have on the drawings a permanent record of the various revisions made in them?

5 What is a "record set" of contract drawings? What is its purpose? Where should it be kept?

6 Can contract drawings be supplemented by (*a*) the contractor? (*b*) the engineer?

7 After preparation of the drawings and specifications and issuance of all contract documents to the bidders, but before receipt of any proposals, the engineer discovered that one of the drawings contained a serious error. What should he do about the matter?

8 Why should the engineer state in the contract documents that he has the right to revise and to supplement the contract drawings? If he fails to make such a statement in these documents, will he be unable to revise or supplement the drawings?

9 What relation have "shop drawings" to "contract drawings"?

10 A dimension on a drawing showed the desired length of a truss to be 75 feet, but, according to the scale of the drawing, the truss measured only 70 feet. Which (if either) can the contractor assume to be correct?

11 The drawing of a basement showed no drain around the outside, but a note in one corner of the drawing said: "Provide 4-inch drain and discharge line around foundation." Is the contractor obliged to install such a drain?

12 Why might an engineer want to have in his files a copy of each of the contract drawings for a factory which he has designed?

13 What is meant by "incorporation by reference"?

14 Why is it desirable to list the contract drawings in the specifications or in some other part of the contract documents? How should the listing be done?

15 After preparation of the specifications is completed but before the contract documents are issued to the bidders, the engineer decides to add two more drawings to the list of design drawings. Can he do so? Explain how.

16 What are "preliminary drawings," and how may they be used in connection with actual construction projects?

17 Differentiate between "design drawings" and "contract drawings."

18 Does the contractor generally prepare the "shop (or detail) drawings"? Can the engineer do so if he wishes to? Explain when and why the engineer might prefer to make them.

19 When the engineer revises the contract drawings, does he have to secure the contractor's approval?

20 Jones sent a freehand pencil sketch to Smith, a builder, showing the dimensions and other pertinent data for a garage which the former wanted to have built. Smith quoted Jones a figure of $1500 for the complete job. Jones accepted the offer. Does the freehand sketch constitute a contract drawing in this case? Explain.

THREE

SOME LEGAL MATTERS OF CONCERN TO ENGINEERS

This final group of chapters is designed to acquaint the reader with some of the highlights of selected legal subjects. Of course, limitations preclude anything approaching comprehensive treatment of these topics. Since the purpose of this section is to give engineering professionals a look at the principles of law in those fields which would seem to hold the most interest and significance for them, a number of areas of legal concern, for example, domestic relations, criminal law, and wills and estates, are not touched upon at all or are considered only incidentally.

CHAPTER 16

AGENCY

16-1 Definitions and Distinctions   Agency can perhaps be well described as acting through another. One party, called the *principal,* authorizes another, called the *agent,* to represent the former in certain business dealings with outsiders. Agency has to be a consensual relationship in that the agent cannot be a pure volunteer. His authority must come in some fashion from his principal.

An agent differs from a servant or employee in that the agent has been appointed to act for his principal in transactions with third parties; the servant or employee normally exercises far less discretion than an agent and is under more detailed orders from the employer about exactly how the assigned work is to be performed. Thus, in the usual situation, the servant or employee is hired to do a specified job, typically of a "ministerial" nature (action performed under a superior with little reliance on discretion of doer), under well-defined restrictions, and is not entitled, for example, to enter into contracts on behalf of the employer. It is possible, of course, for an employee to be authorized also to undertake the separate function of an agent. By way of illustration, a gardener regularly employed to keep a homeowner's grounds in shape could well be granted the right to purchase any needed supplies directly. When exercising the latter prerogative, the employee-gardener is filling the role of agent, and his employer-principal is responsible on the contractual obligations undertaken. The true agent acts for another and under such other's general control; when such an agent proceeds within the scope of his authority, the principal is bound.

A true *independent contractor* is neither an agent nor an employee, though his status bears resemblance to each. The independent contractor is hired to do a given job in return for a set charge. His modus operandi is his own, and the person who did the hiring has no right to interfere in any way during the progress of the work. As long as the end product of his efforts accords with the specifications agreed upon, the independent contractor has done all that can be asked of him. Since the party employing the independent contractor has no right to control performance details, he is not liable to outsiders for harm to them stemming from the negligence of the contractor or his employees.[1]

Various factors must be considered in determining whether a person acting for another is an independent contractor or an employee. One consideration is the degree of control retained by the individual doing the hiring. Another is the intent of the parties who created the relationship in question. Still another is with the method of payment, if "by the job," that is some evidence of independent-contractor status. Ownership of the tools utilized in the work may also be significant.

16-2   Trustee Distinguished from Agent   The essential distinctions between trust and agency are that the trustee acts in his own name and has legal title to the trust property,[2] while the agent merely represents and acts for his principal. A trustee normally is given wider discretion than an agent, the latter being under more or less constant supervision by his principal.

16-3   Importance of Agency   The subject of agency is of tremendous importance in the modern business world and affects virtually every form of commercial undertaking. Corporate affairs are necessarily handled by agents, since a corporation, by its very nature, can act only through its officers and employees. A contract of partnership is a contract of agency, differing from what could be termed a *pure agency* only in that in the latter the agent binds his principal only, while in a partnership arrangement each of the partners doubles as principal and agent, and the acts of each are binding on all.

16-4   Creating the Agency Relationship   Agency is created by express or implied[3] contract or, less frequently, by operation of law, estoppel, or ratification.

Where the relationship rests upon contract, the principal must intend appointment of the agent, and the latter must intend to accept the appointment and act upon it. Similarly, the other requisites of an ordinary contractual arrangement are needed. Thus, the agency must not have been created to further an unlawful purpose and contractual capacity of the parties may have to be demonstrated.[4]

---

[1] Exceptions to the rule do exist. Where, for example, the task that the independent contractor is asked to handle is extra hazardous in nature, the owner cannot divest himself of potential tort responsibility. Furthermore, even where the activity involved is not inherently dangerous, if the owner contracts with someone whom he knows or should know to be incompetent, he may very well be held liable for injury sustained by third parties in consequence of the contractor's faulty execution of the work.

[2] The trustee is expected, of course, to manage and utilize the trust estate in accordance with the terms of the trust instrument and to the advantage of the named beneficiaries.

[3] The implication arises from words and conduct of the parties and from circumstances of the particular case.

[4] A person may handle through an agent whatever business he has the legal capacity to conduct himself.

Some jurisdictions have statutes requiring under certain conditions that the agent's authority to act be given in writing.[5] Illustrative is this New York provision, which reads in part:[6]

As estate or interest in real property, other than a lease for a term not exceeding one year. . .cannot be created. . .unless by act or operation of law, or by a deed or conveyance in writing, subscribed by the person creating. . .the same, or *by his lawful agent, thereunto authorized by writing.* . .[italics supplied].

One familiar form of written authority to an agent is, of course, the *power of attorney.* An offshoot of this is the *proxy* so frequently used in the handling of corporate affairs.

16-5 Agency by Estoppel, etc. An agency relationship may be created *by estoppel.* When Smith so conducts her affairs as to lead outsiders to believe that Jones is her accredited agent, Smith will be estopped from denying the agency when the rights of third persons become involved through acts and behavior of Jones. Agency by estoppel rests upon representations of the principal and a change of position by some third person in reasonable reliance upon such representations. In the true estoppel situation the agent will be found to have no "real" authority (express or implied), but is deemed, rather, to possess "apparent" authority which the principal has knowingly permitted him to assume. Agency by estoppel is an artificial creature which justice imposes upon the unwilling principal. It should be distinguished from *implied agency,* which is as much actual agency as if it were created by express words. The principal is fully as liable to outsiders in the case of agency by estoppel as he is in the case of implied agency, but only in the former must third persons show actual reliance on the principal's conduct.

When the word "ratification" appears in connection with a discussion of agency, it is usually employed to identify the principal's decision to endorse certain acts by the agent in excess of the latter's authority.[7] Occasionally, however, mention is made of a principal-agent arrangement being effectuated by *ratification.* Thus, it might happen that Black, without any prior commitment from White, undertakes as agent to represent White in a series of transactions. Obviously, White is free to repudiate Black's actions. If, on the other hand, she retroactively assents or, saying nothing accepts the benefits of any contracts Black has made on her behalf, it can be said that an agency by ratification has arisen. Like agency by appointment, but unlike agency by estoppel, agency by ratification requires consent of the principal.

[5] The usual instance involves a transfer of land interests, where, for example, the agent is to sign the deed.

[6] *General Obligations Law,* §5-703 (McKinney 1978).

[7] There is a marked distinction between (1) an estoppel to deny the validity of a contract and (2) ratification in fact of such contract. In the latter instance, the principal is bound because he intended to be; in the former, he is bound against his will in order that justice may be done in regard to the innocent third party.

16-6   General and Special Agency   A *general agent* is one empowered to transact all the business of his principal or all the business of a designated type or, all the business at a particular location. Thus, a partner is a general agent for the partnership as regards all matters which come within the scope of its activities.

The authority of a *special agent* is confined to one or more specific acts. It is often said that an agency is special when both the end and the means are specific and when no continuity of service is involved. Typically, the special agent is retained to handle a single transaction and does so in accordance with definite instructions or within restrictions necessarily implied from the nature of the act to be performed.

The terms general agent and special agent are relative, and the line of demarcation is often difficult to draw. In addition to the powers actually conveyed to him, a general agent has, by implication, incidental authority to do whatever is customary and is reasonably necessary to accomplish the purpose at hand, and outsiders dealing with a general agent are entitled to rely upon the full scope of his apparent authority. The rule does not apply when actual limitations upon the powers of the general agent are known to the third person or are contained in a writing which he realizes is intended for his inspection. Misleading conduct on the part of the principal can alter the situation at any point.

Third parties who contemplate doing business with a special agent must at their peril determine the precise extent of his authority, since the boundaries of that authority are by no means as vague as in the case of a general agent. A bank teller illustrates a common type of special agent with familiar and limited powers. He operates under a specific grant or authority, his duties are outlined with exactness, and his position is not such that any implication of undefined powers can legitimately arise.

In summary, neither general nor special agent can bind his principal unless such agent is acting within the confines of his authority. The principal is responsible for all proper undertakings by his agent, regardless of whether the authorization in a particular instance is actual or apparent. Many times an agent will bind his principal through the exercise of apparent authority and, at the same time, will have violated secret instructions given him by the principal. Such conduct by the agent may render him answerable in damages to the principal.

16-7   Position of the Agent   Every agent has certain obligations toward his principal. He must display the utmost loyalty and good faith. He must obey instructions and attempt not to exceed his authority. He is expected to utilize reasonable care and skill in fulfilling his duties, and he is obliged to account to his principal whenever the latter insists and when the agency is terminated. All profits arising from the undertaking are deemed to belong to the principal.

The agent is expected to communicate any and all information which he acquires in the course of his agency and which affects the subject matter of that agency. By no means may the agent "compete" with his principal. Thus, he may not profit at the principal's expense by using knowledge of the principal's business acquired while acting in a confidential capacity. Similarly, the agent is not allowed to indulge in "self-dealing." For example, an agent who is authorized to make a certain type of purchase cannot secretly buy the commodity from himself. Should this rule ever be violated the principal can rescind the transaction for fraud and recover the purchase price.

In order to bind his principal, the agent must have acted within the scope of his authority. When a particular act of the agent's is not an authorized one, the principal may yet elect to ratify, thus rendering the situation the same as though the agent had possessed sufficient authority. Should the principal decline to ratify the agent's overreaching, the latter is potentially liable to any outsiders misled by his claim of authority. The principal cannot repudiate his agent's action as unauthorized and yet retain for himself the fruits of such action.

The agent is generally answerable to injured outsiders for affirmative acts of negligence[8] (that is, misfeasance), but not for nonfeasance (that is, passive negligence). An agent guilty of intentional torts of any sort is liable in damages to the victim.[9]

16-8   Position of the Principal   It is a fundamental proposition in law that a principal is liable on all contracts made by his agent while that individual is acting within the scope of his actual or apparent authority.[10] Unauthorized acts of an agent may nonetheless be ratified by the principal and thus become binding upon the latter.

Generally speaking, the principal is charged with constructive notice of facts acquired by the agent in the course of his duties. Knowledge will not be imputed if its revelation would bare the agent's guilt as when the agent is perpetrating an independent fraud or is otherwise acting adversely to his principal's interest. Information received by the agent long previous to his undertaking the agency relationship will not be imputed to the principal.

The principal has certain basic responsibilities toward his agent, perhaps the most significant of which is the duty to give reasonable compensation for services rendered. Also, the principal is expected to indemnify the agent for injury suffered in the course of conducting the principal's business in line with his instructions. An agent must assume for himself the more commonplace risks, such as that of loss of personal funds through theft.

In the field of torts, the principal is liable for acts of negligence committed by the agent while the latter was operating within the scope of his authority and in furtherance of the principal's business. For nonwillful torts, it is immaterial that the agent was performing his function in an irregular manner or even contrary to explicit instructions.

---

[8] In most cases both agent and principal are liable.
[9] If the situation is reversed, naturally, and a third party commits a personal tort against the agent, a remedy in damages is available to the agent.
[10] Conversely, when the agent makes such contract in the name of his principal, the principal can hold the third party to the bargain.

Where the agent's misconduct is willful, a third party seeking to hold the principal must show that the tort involved is reasonably incidental to the performance of the duties assigned to the agent. This normally proves a difficult burden for plaintiff, since there is a presumption that the agent was not employed with the expectation that he would commit intentional torts.

16-9   The Undisclosed Principal   Whenever a duly authorized agent enters into a contractual arrangement without disclosing his true role or the identity of his principal, the third party, upon discovering the facts of the situation, can hold the agent[11] or, in most instances, may elect to hold the *undisclosed principal.* The doctrine works in reverse, also, and with a few exceptions the undisclosed principal may at some point in the proceedings decide to reveal his interest in the contract and enforce performance by the third party. One of the exceptions comes about when the third party has previously refused to deal with the principal and the latter thereupon attempts to achieve his ends surreptitiously through an agent, who is told not to indicate for whom he is acting. In such a situation, the contract cannot be enforced by the undisclosed principal against the deceived third party.

16-10   Subagents   A principal selects a particular person to act as his agent, and the selection is presumably made on the basis of the agent's known competence and trustworthiness. The principal, accordingly, is entitled to have his affairs receive the personal attention of the person chosen for the job. Thus, it is said that authority delegated to an agent cannot be further delegated by him to a subagent.[12] The prohibition, however, relates only to matters involving the exercise of discretion and does not extend to so-called "ministerial acts." For example, the agent is normally free to utilize office workers to help him with such things as bookkeeping.

16-11   Termination of the Agency   An agency relationship may end by (1) death of either principal or agent, (2) destruction of the subject matter of the agency, (3) fulfillment of the particular purpose for which the agency was formed, (4) expiration of a time period set in the contract, (5) supervening illegality of the agency's purpose, (6) mutual consent, and (7) unilateral termination by one party or the other.[13]

Agency is a voluntary relationship and, as such, may as a usual thing be terminated at will by either party, without such party incurring damages for breach. If the contract between principal and agent has a stipulated duration, it may yet be freely renounced for cause prior to the expiration of the period set. However, lacking justifiable cause, the party breaching would be liable in damages. "Specific performance" of personal-service contracts will not be granted and the innocent

---

[11] An agent is likewise liable in contract to outsiders when he is pretended to represent a principal actually nonexistent.

[12] This assumes that the principal has not expressly empowered the agent to delegate.

[13] A principal who is terminating the authority of his agent will do well to see to it that interested third parties receive some notification.

party in a breach situation can secure damages but cannot force the other party to fulfill the precise contract obligation. But, if an agent whose services are unique fails to carry out his contract with the principal, the agent may be enjoined from working during the remaining period of the breached contract for one of his principal's competitors.

A true irrevocable agency is specifically enforceable on the ground that otherwise the injured party would be irreparably damaged. If the agency is "coupled with an interest," then it is deemed irrevocable. The "interest" referred to is an interest on the very subject matter of the agency, and not merely in the right to earn compensation or commission stemming from the exercise of the agency.

The phrase *"coupled with an interest"* does not mean an interest in the exercise of the power, but rather an interest in the property on which the power is to operate. The test is whether or not the agent has an interest in the subject matter of the agency independent of the power conferred.[14]

The Restatement of Agency, 2d Section 138, sets forth the requirements for the creation of an agency or power coupled with interest. They are (1) that the agency be held for the benefit of the agent not the principal, (2) that the agency is created to secure the performance of a duty to the agent or to protect title in the agent, and (3) that the agency is created at the time that the duty or title is created or is created for consideration.[15]

Let us assume that *A* has an idea of running a trucking business and gets *B* to put up the money. *A* secures a five-year contract as manager, with base salary plus a percentage of the profits. Is the arrangement irrevocable? Probably so. *A*'s argument would be that she will not make sufficient profits from the undertaking unless she is personally allowed to handle the selling end as manager. Thus she has power coupled with an interest.

The case of *French v. Kensico Cemetery*[16] will serve as a further illustration. For some years, plaintiff had paid defendant cemetery various amounts for the current care and maintenance of a plot and mausoleum. To avoid the necessity of further annual payments, plaintiff turned over to defendant $5000, obtaining a receipt which spelled out defendant's obligation to invest the principal in securities and to use the income for perpetual care of plaintiff's plot and mausoleum. Eventually plaintiff attempted to cancel the arrangement and get back her "deposit." The Appellate Division, in holding for defendant, spoke as follows:

> Our conclusion is that the delivery of the money and the giving of the receipt constituted an agreement which created an agency (or a power) coupled with an interest in defendant and that if the agreement was founded on a sufficient consideration, it is irrevocable by plaintiff. The consideration moving from defendant was its undertaking to relieve plaintiff of the trouble and annoyance of attending to the care and maintenance of the property. . . .

---

[14] *Peacock v. American Agrinomics Corporation,* 422 So.2d 55 (Fla. App. 1982).
[15] See also *Becket v. Welton Becket & Assoc.,* 39 Cal. App.3d 815, 114 Cal. Rptr. 531 (1974).
[16] 264 App. Div. 617, 35 N.Y.S.2d 826, 828–29 (2d Dept. 1942), *affirmed,* 291 N.Y. 77 (1943).

Defendant's interest in the performance of the agreement is that if it is not performed as long as there is need for care and maintenance, defendant will be obliged to carry that burden, since no well-managed cemetery could depreciate its property and that of other lot owners by having on its premises a mausoleum in process of falling into decay and ruin upon a considerable plot, with grass uncared for and running to weeds. . . .

In our opinion, the delivery of the money and the giving of the receipt created in the defendant an agency coupled with an interest, irrevocable so long as the need for care and maintenance of the plot and mausoleum continues to exist. That need is manifestly a permanent one. If plaintiff should be permitted to revoke the agency and to obtain the return of the money, defendant would be compelled to assume the duty of care and maintenance in its own interest. An agency may not be revoked where revocation would subject the agent to liability without fault of his part.

QUESTIONS

1 Describe the agency relationship and how it is created.
2 What are the duties, rights, and obligations of (*a*) an agent? (*b*) a principal?
3 Distinguish between agency, on the one hand, and employer and employee (or master and servant) on the other.
4 Explain the differences between an agent and an independent contractor.
5 Distinguish between an agent and a trustee.
6 Distinguish between the rights and duties of (*a*) a general agent and (*b*) a special agent. Illustrate both types of agency.
7 What are "implied powers" in connection with an agency? Illustrate.
8 List several purposes for which an agency may be created.
9 When and how may an agency be terminated? Give illustrations.
10 May an employer discharge his employee when he wishes to do so? May a principal do likewise with his agent?
11 Does an agent have authority to use his own discretion?
12 Can a servant or employee make contracts on behalf of his employer?
13 Can a person be an employee and at the same time be appointed to act as agent?
14 How may one determine whether a given person acting for another is an employee, agent, or independent contractor?
15 In the usual case, which has more in the way of discretionary powers, an agent or a trustee?
16 Can an agency be created by implication as well as by express contract? Explain.
17 Must an agency always be created by means of a writing? Elaborate.
18 Explain and illustrate how an agency may be established through estoppel.
19 An agent does something beyond his authority. Has the principal the power to ratify the unauthorized action? What would be the effect of ratification?
20 Distinguish between the principal's being bound (*a*) by estoppel and (*b*) by ratification.
21 Explain and illustrate "agency of necessity."
22 Is the principal as much bound by acts of a general agent as he is by those of a special agent?
23 Under what conditions may an agency be irrevocable?
24 Whipple is an engineer in private practice. She appoints Monroe to act as her agent in securing business. Has Whipple this right? Can Monroe properly hire Appleton to assist? May Monroe appoint Appleton to represent Whipple as Monroe's agent? As Whipple's agent? Explain.

25 *A* is an agent for *B,* a realtor. *A* agrees to sell *C*'s house but, unknown to *B,* for a "bargain-rate" commission of $400. After the house is sold, *B* demands that *C* pay the standard commission—$500. Must *C* pay the extra $100? Give reasons.

26 Perkins, an agent for Quade, agreed that his principal would design a large store for Dunford for the sum of $2000. It soon became obvious that the job could not be done properly for anything like this figure. Is Quade bound by the arrangement made for him by Perkins? Explain.

27 If Carter has some very hazardous blasting to do, can she, by letting a contract for this work to Dawson, shift to the contractor all responsibility for possible damages to persons and property? Explain fully.

28 Suppose the agent, exceeding his authority, has made certain contracts on behalf of the principal which the latter declines to accept. Has the third party any recourse against (*a*) the principal, (*b*) the agent?

29 Blodgett is on a trip for her principal, Franklin. Just before Blodgett, as agent, signs an agreement with Gordon regarding the purchase of 20,000 tons of coal, Franklin dies. What is the legal situation; that is, is there a binding contract?

30 If an agent quits without cause and before expiration of the contract of agency, can the principal (*a*) compel specific performance; (*b*) collect damages?

CHAPTER 17

PARTNERSHIPS

17-1 Nature of a General Partnership The Uniform Partnership Act[1] (Section 6) defines the ordinary partnership as "an association of two or more persons to carry on as co-owners a business for profit." The persons concerned must, of course, have the capacity to contract.[2] A corporate unit, restricted as it is by its charter and by the laws of its state of domicile, cannot in the usual case find the authorization to join a partnership. Partnership arrangements, which normally are geared for a considerable period of time, feature community of interest in profits, losses, and the capital employed, as well as joint control of the operation. When a person has contracted for proprietary rights in a business, he is an *owner* thereof, and, if there are other owners of the same business, he is by definition a *partner*. Apparent intent of the parties, profit sharing, and mutual control are three factors bearing upon the existence of a partnership.[3]

Each member of an ordinary partnership has certain rights and privileges,[4] chief among them are:

---

[1] See, for example, New York Partnership Law (McKinney, 1948).

[2] A minor may join a partnership. but his contractual undertaking is voidable by him. His investment in the business, however, is subject to the claims of creditors. In many jurisdictions it is also subject to the claims of his fellow partners, unless the minor was induced by fraud to enter the partnership in the first place.

[3] Not infrequently an individual who is not really a partnership member represents himself as a partner or permits firm members to create the impression that he is one of them. Third persons, relying upon the individual's apparent association with the enterprise, may thereupon extend credit to the partnership or enter into other dealings with same. As against outsiders thus deceived, the ostensible member will be deemed a *partner by estoppel* and subjected to the same liability as though he were indeed a full-fledged partner in the firm.

[4] The partnership agreement will govern such things and must be consulted in a particular case.

1 Privilege of sharing in the management and control of the firm's operations

2 Right to share in profits or losses

3 Right of co-ownership of specific partnership property. Each partner is entitled to use the property for any *partnership* purpose. He is, however, essentially a fiduciary as respects his associates and must account to them for any personal profit realized through his private use of the firm's assets.

4 Power to act as agent for the partnership.

5 Right of contribution from his associates in case he makes payments from his own pocket on partnership accounts

Although it has its own assets, carries on its own business activities, etc., a partnership, unlike a corporation is not treated as a legal entity separate from the individuals composing it.[5]

A most significant aspect of a partnership is the unlimited liability of each of the partners for debts of the enterprise,[6] as compared with a stockholder of a corporation, whose liability is limited to his capital investment.

Each partner acts as principal for himself and as agent for his fellows. He may enter into any agreement which is within his actual or apparent authority, and the firm will be bound to honor such agreement.[7] In other words, each partner has full power to act for the group in entering into obligations in the course of business. It can readily be seen that partnership carries with it some elements of danger. Consequently, one should exercise caution in selecting partners, and the partnership agreement should be drawn with great care. Delectus personae ("choice of the person") is of the essence of partnership as a special form of business association; no outsider can join a partnership without first securing the consent of each of the members. It follows that, while a partner may assign his right to profits and his right to share in any surplus available upon dissolution of the business, he may not on his own effectively convey his partnership interest in the sense of entitling the assignee to a role in management of partnership affairs.

17-2 Articles of Partnership Unless a statute states otherwise, it is possible to form a partnership by oral agreement, but written articles of partnership, signed by the parties and setting forth the terms and conditions of the agreement, provide the customary method of formalizing the agreement. Moreover, an oral contract of partnership is under the usual statute of frauds, unenforceable if it is to last more than a year.[8] Oral contracts for periods of less than a year are outside the purview of the statute of frauds; these understandings are binding and, lacking unanimous consent, can be abrogated during the contract term only at the risk of damages for breach.

Quite apart from any statutory considerations, it is advisable as a matter of

---

[5] But the Federal Bankruptcy Act treats partnerships as legal persons, and the firm itself may be adjudged a bankrupt, either separately or jointly with one or more of its general partners.

[6] That is to say, the personal fortune of a partner is vulnerable in its entirety.

[7] See *Fliegel v. Associates Capital Co. of Delaware, Inc.*, 537 P.2d 1144 (Or. 1975).

[8] However, if the parties commence operations under a long-term oral contract, a *partnership at will* is deemed in effect. Such an arrangement is perfectly legal but may *without penalty* be terminated by any party at any time.

common sense for the partners to have a written memorandum of agreement, fixing among themselves their respective rights, duties, and liabilities and thus reducing the likelihood of controversy. The partners may include in their formal articles any agreement they wish respecting such things as sharing of profits and losses, priorities of distribution on winding up the business, etc. Articles of partnership vary widely in the subjects they cover and in the degree of detail involved; much depends upon the nature and scope of the enterprise. A typical list of contents for a partnership agreement might include:

1 Name of the firm and names of the copartners

2 Nature and location of the business to be carried on

3 Date the enterprise will commence operations and intended duration of the partnership

4 Statement of capital, the "what" and "how" of each partner's contribution

5 Designation of services to be rendered by each member and statement about compensation, if any,[9] and drawing accounts.

6 Management of the business—duties and restrictions

7 Statement of the respective shares of the several partners in profits and losses

8 Statement about banking matters, keeping books of account, making periodic inventories, etc.

9 Provision for arbitration of disputes, frequently "in accordance with the rules of the American Arbitration Association"

10 Designation of the rights of the parties upon dissolution[10] and statement about the method to be employed in the final winding up

At any point any or all of the provisions in the partnership agreement can be amended, waived, or abrogated by consent of all the partners.

17-3 Each Partner as Agent of the Firm By virtue of the partnership relation, each member has the capacity to bind his firm by all acts and representations ostensibly within the scope of its business. The point is thus expressed in Section 9(1) of the Uniform Partnership Act:

> Every partner is an agent of the partnership for the purpose of its business, and the act of every partner, including the execution in the partnership name of any instrument, for apparently carrying on in the usual way the business of the partnership of which he is a member binds the partnership, unless the partner so acting has in fact no authority to act for the partnership in the particular matter, and the person with whom he is dealing has knowledge of the fact that he has no such authority.

---

[9] Partners are not entitled to charge each other or the firm for personal services unless there is a special agreement permitting them to do so. However, the Uniform Partnership Act, §18(f) allows a surviving partner "reasonable compensation" for his efforts in winding up the firm's affairs.

[10] It is particularly important to indicate what result is to follow upon the death of a partner, how the decedent's interest is to be evaluated, etc. Many agreements provide the surviving associate an option between purchasing the decedent's interest and liquidating the enterprise forthwith.

The articles of partnership may purport to place definite limitations on the powers of a member of the firm, but such restrictions are ineffective as respects outsiders who deal with a partner unaware of any curtailment of what appears to be his full and complete authority to bind the partnership. On the other hand, limitations thus imposed by the partnership agreement *are* effective with such third persons who are aware of the true situation and among the partners themselves.

Since each member of the partnership acts as agent of the whole, it follows that notice of anything relating to affairs of the enterprise duly given to any one partner constitutes communication to the partnership itself. Similarly, information which a member has acquired in a partnership transaction is considered to be also in the possession of such member's principal, the partnership.

17-4 Overreaching by a Partner While a partner, in his role of agent of the firm, has power to bind his fellows by any acts within his apparent authority, it is possible for him to go too far. Thus, for example, he may not properly give the firm's note in payment of a personal debt nor charge the partnership for property purchased by him for his individual account. Nor is he entitled, without the express consent of each of his copartners, to make a general assignment for the benefit of the concern's creditors or to take action of any sort which would make it impossible to carry on the ordinary business of the partnership.

Where a partner has breached his obligations to his associates, the ordinary and appropriate remedy is an accounting in the course of dissolving the firm.

17-5 Tort Liability Responsibility for a partnership tort is *joint and several* in the sense that every member of the firm is individually liable and that the victim may proceed against all the partners jointly or against such of the partners as he chooses. Under these circumstances, recovery is obviously not dependent upon the personal wrong of the particular member against whom the claim happens to be asserted.

Whether the tort involved is truly a partnership tort, and thus one for which the various members will be answerable, depends upon whether the wrong was committed within the ordinary course of the partnership's business.[11] Typically, the tortious conduct with respect to which the partnership is sought to be charged stems from the negligent operation of a motor vehicle by one of the partners. It is often a close question whether, at the crucial time, the partner was engaged in an undertaking related to the firm's business as distinguished from his own purposes.

17-6 Contract Liability Unlike the situation in regard to partnership torts, the liability of members for partnership contracts is said to be *joint* only. That is, where the alleged obligation is contractual, all partners can be said to be necessary parties. Although for most purposes a partnership is not an entity separate from

---

[11] That a given tort is willful in nature does not necessarily preclude its having been committed in the course of the firm's business, though the willfulness is a factor of significance in this regard.

its members, statutes in many jurisdictions provide that persons carrying on business as partners may sue or be sued in the partnership name.[12] Such laws amount to a recognition of partnerships as legal entities for procedural purposes.

A contractual obligation running *to* the partnership is regarded at common law as the joint right of the various partners. An action to enforce such must consequently be brought in the names of all the partners unless statute permits suit to be filed in the firm's name. Suits on partnership claims must be brought on behalf of the firm. A partner may not in his individual capacity recover upon obligations owed his firm.

17-7   Incoming and Retiring General Partners   One who is brought into a going business as a new partner is not personally liable on preexisting obligations of the firm unless he expressly assumes such personal liability.[13] On the other hand, under the Uniform Partnership Act his share in the partnership property is clearly available for the satisfaction of these old debts.

Unless there is an understanding by which the creditors agree to substitute the credit of an incoming member for that of a retiring general partner, the latter remains liable for the firm's obligations outstanding at the time of his withdrawal. Furthermore, as for subsequently incurred obligations, the former partner may find himself answerable to third parties who are unaware of his retirement and have continued to deal with the concern. To be fully effective, a retiring general partner's notice of withdrawal requires (1) personal notification to those who dealt with the firm during this tenure and (2) notice by publication to others in the area.[14]

17-8   Termination of the Partnership   Termination is a two-step process: first comes *dissolution* of the firm, and then *winding up* of its activity.

There are a number of situations which would occasion dissolution of the ordinary partnership. Perhaps the most common of these are (1) death,[15] bankruptcy, or withdrawal of a general partner; (2) mutual consent of the firm members; and (3) expiration of the time for which the partnership was formed.[16] Notification regarding the dissolution should go to those with whom the firm has had

---

[12] See, for example, New York Civil Practice Law and Rules, §1025 (1976).

[13] In return for admission to the firm, the new member may have promised his copartners or a retiring member that preexisting claims would be paid in full. Such a promise could presumably be enforced against the incoming partner by the creditors concerned, using a third-party-beneficiary approach.

[14] If the associate who is retiring was a "dormant" partner, he need not give notice to creditors as a prerequisite to escaping liability for partnership debts arising in the future. A dormant partner, by the way, is so far inactive in partnership affairs that his connection with the firm is generally unknown. Nevertheless, he is in most respects a full-fledged partner and, as such, is liable for all partnership obligations incurred during his association with the enterprise.

[15] In the absence of an express agreement to the contrary, a partnership is dissolved by death of one of the partners, *Timmermann v. Timmermann,* 538 P.2d 1254 (Or. 1975); *Wagner v. Atoll,* 362 N.Y.Supp.2d 278 (1974).

[16] The partners, of course, are free to ignore the fact that the winding-up date originally agreed upon has been reached and may simply carry on indefinitely. Continuance under these circumstances really amounts to an implied agreement for a *partnership at will* under the same terms as initially set.

dealings. Dissolution of itself has, of course, no effect upon existing liabilities of the partnership.

A dissolved partnership retains its existence in the eyes of the law until the winding-up process is completed. This entails such actions as payment of debts, collection of accounts receivable, and fulfillment of commitments undertaken prior to dissolution. Under the Uniform Partnership Act, the death of one partner means that his right in and to specific assets of the firm vests in the surviving partner or partners.[17] When a partnership is dissolved by the death of a partner, the legal representative of the deceased has a right to an accounting from the surviving partner or partners from the date of death.[18]

The surviving partner, engaged in liquidating the business,[19] will find his powers and duties governed largely by the probate and partnership laws in the particular jurisdiction. If it had been intended that he should exercise any extraordinary powers in the premises, the partnership agreement should have so indicated. Thus, unless the articles of partnership expressly authorize a surviving partner to purchase the interest of his deceased associate, he is not automatically entitled to do so.

In reducing the partnership property to cash and distributing the proceeds, certain priorities must be recognized:

1 In the vanguard come creditors of the firm other than partners.[20]

2 Next are the claims of partners for loans and advances beyond the amount of their agreed capital contributions.

3 After the foregoing categories of claims have been satisfied, under the usual partnership agreement members are entitled to a return of their respective capital contributions. If there is insufficient property available for this purpose, the loss is normally to be shared by the solvent partners in the proportions in which they were supposed to share profits. By way of illustration, assume that partners X, Y, and Z have subscribed, respectively, $20,000, $10,000 and $4,000 of the concern's $34,000 capital and that the articles of partnership called for sharing profits equally. Assume further that, upon dissolution and the settlement of debts, there remained $10,000, which result represented a capital loss of $24,000. Equal sharing means a debit to each partner of $8,000. Therefore, X would get $12,000 and Y $2,000, while Z was paying in an extra $4,000 to meet the deficiency.

4 If the dissolved partnership has been a very successful enterprise and some undistributed assets remain after the various demands outlined above are taken care of, the balance will be disposed of among the partners in the form of profits.

In the marshaling of assets, partnership creditors have priority over partnership

---

[17] Unless the deceased was himself the last survivor, in which event title passes to his legal representatives.

[18] *Hansel v. Hansel,* 446 A.2d 1294 (Pa. Super. 1982).

[19] Normally, either by statute or through provision in the partnership agreement, the surviving partner will be entitled to reasonable compensation for his services in closing out the enterprise.

[20] Should the firm's assets prove inadequate to meet liabilities, the individual partners, must, under certain circumstances, themselves contribute to make up the difference. See the subsequent discussion in this article.

assets, but creditors of an individual partner are generally recognized as having priority with respect to that partner's individual assets. If the latter creditors' claims are not fully satisfied from an individual partner's personal assets, they can reach their debtor's share of whatever surplus *firm* assets might be left undisturbed after the firm's business has been finally settled. They have no legitimate claim against such firm assets until an accounting of the debtor-partner's interest has been made. This cross-application principle works both ways. If the partnership lacks the wherewithal to satisfy its creditors and one of the partners has personal assets on hand after paying his individual creditors, these assets will be available to meet the firm's need.

17-9 Limited Partnership A limited partnership, which must be formed strictly in accordance with applicable statutes,[21] has at least one *general* partner and one or more *limited* (or *special*) partners; the latter contribute cash or property as investors but do not contribute services and have no powers of management.

The principal object of limited partnership is to protect the special partner by exempting him from general liability and placing only his invested capital in jeopardy. The limited partner has certain rights by virtue of his position. Among these are the right to share in the firm's profits and to seek return of his capital contribution upon dissolution of the business. In the absence of a definitive provision in the limited-partnership agreement, the interest of a limited partner is assignable, with the assignee acquiring the right to receive the share of the profits to which the assignor would otherwise have been entitled. As long as he shuns active conduct of firm affairs, the special partner's liability for obligations of the firm is limited to the amount of his investment. Any show of management interference, however, will lose him his cloak of partial immunity and render him fully liable as a general partner.[22]

Like corporations, limited partnerships were unknown to the common law and are purely creatures of statute; the statutory requirements must be closely followed if the special partner would achieve and retain his privileged status. The law in many jurisdictions requires each limited partnership to file a certificate setting forth all substantive provisions of the limited-partnership agreement.

---

[21] The Uniform Limited Partnership Act has been enacted into law in a great many states.

[22] Section 303 of the Limited Partnership Act (1976) says that a limited partner who is not also a general partner does not participate in the control of the business solely by doing one or more of the following: (1) Being a contractor for or an agent or employee of the limited partnership or of a general partner; (2) consulting with and advising a general partner with respect to the business of the limited partnership; (3) acting as a surety for the limited partnership; (4) approving or disapproving an amendment to the partnership agreement; or (5) voting on one or more of the following matters: (*a*) the dissolution and winding up of the limited partnership; (*b*) the sale, exchange, lease, mortgage, pledge, or other transfer of all or substantially all of the assets of the limited partnership other than in the ordinary course of business; (*c*) the incurrence of indebtedness by the limited partnership other than in the ordinary course of its business; (*d*) a change in the nature of the business; or (*e*) the removal of a general partner. Further, this enumeration does not mean that the possession or exercise of any other powers by a limited partner constitutes participation by him in the business of a limited partnership.

In various respects a *lender* is in a better position than a limited partner. The lender often finds it possible to exercise some measure of control over the firm which is indebted to him. If he treasures his limited liability the limited partner must leave the reins of control to others. Moreover, upon dissolution of the partnership and distribution of its assets, the lender (as a general creditor of the firm) enjoys the highest priority.

17-10   Subpartnership   A contract of subpartnership is an agreement between a partner and an outsider, called a subpartner, whereby the latter, typically, is to share on a fifty-fifty basis (with the partner) profits and losses arising from such partner's association with the firm; that is, the subpartner agrees to participate in the contracting partner's share only.

The subpartner performs no active function, nor has he any power of control over the partnership affairs. Since his contract is on a purely personal basis with an individual partner, the "sub" is not directly liable to partnership creditors as a co-obligor on the firm's commitments. He is in no sense a member of the firm and has no contractual relationship whatever with partners other than the one with whom he has made the profit-and-loss arrangement.

QUESTIONS

1 What are the fundamental rights and privileges and the potential liability of ordinary partners?
2 Explain the difference between a general partnership and a partnership. Illustrate.
3 What is the maximum number of members in a partnership?
4 What are the rights and obligations of an underage partner?
5 Are partnerships subject to the same governmental supervision as corporations?
6 How can the existence of a true partnership arrangement be proved or disproved?
7 List the several points which should be covered in the average partnership agreement.
8 What is the usual procedure for bringing new partners into a firm?
9 How and for what reasons may a partnership be terminated?
10 If a partnership is liquidated, how are its assets disposed of? What are the respective priorities of creditors of the firm and of creditors of an individual partner?
11 What are the powers, rights, and liabilities of (*a*) a general partner? (*b*) a limited partner? (*c*) a subpartner?
12 When several persons are contemplating formation of a partnership, how important is it for each to consider the character and personality of his prospective associates?
13 If one has entered into a partnership, is he bound by the acts of his copartners?
14 What is the purpose of the Uniform Partnership Act?
15 Is a partner an agent for his copartners?
16 Why is it desirable to have written articles of partnership?
17 What is the procedure in connection with a partner's retirement?
18 What is a "dominant" partner?
19 What is a "subpartner," and what in general are the rights and obligations which go along with that role?
20 Black, Brown, and White were partners and carried on a small business manufacturing precast-concrete products, cinder blocks, etc. Black, without the knowledge of her part-

ners, agreed to sell the business to Gray. Can Brown and White block the sale, and why? If Black's deal were to sell Gray $10,000 worth of blocks for $8000, what could Brown and White do about the matter?

21 Wallace, who was seventeen years old, went into partnership with Yeager and Smith, who were of age. Wallace invested $5000 (from a legacy) in the business. Within three months, the partners found that they were losing money in their small contracting concern. Wallace wanted to drop out and withdraw his investment. Could he do so? Explain.

22 If, in the preceding case, Yeager and Smith were deeply in debt but had told Wallace they were making handsome profits (in order to induce him to join them and contribute the $5000), could Wallace withdraw his funds when he learned the truth?

23 The partnership of Duncan, Howard, and Kearny was in debt for $100,000. What courses of action could the creditors pursue? With firm assets insufficient, can individual assets of the several partners be moved against?

24 A, B, and C formed a partnership to conduct certain business for five years. What steps, if any, are necessary to continue the relationship beyond the period originally stipulated?

CHAPTER 18

CORPORATIONS

18-1 Types of Corporations Corporate organizations are of different types and are formed to serve a variety of purposes. The basic distinction lies between *public* and *private* corporations. The former are founded by a governmental unit and belong to it. Both a municipal corporation (that is, a city or a town) and an incorporated school district typify this category. Upon such corporate entities the creating authority confers whatever powers and imposes whatever restrictions the public benefit may require.

On the other hand, corporations are *private* which are not truly public, but there are also corporations popularly called "quasi-public," examples of which are the telephone, water, and power companies. Quasi-public corporations are subject to particularly strict regulation because their operations directly affect the public interest. Normally, however, public utilities are conducted for private gain and are properly classed as private rather than as public corporations despite the fact that they are commonly vested with certain powers of a public nature, such as that of eminent domain.

Within the broad category of private corporations are those with and without capital stock. The latter are known as *membership corporations* and include the bulk of the charitable, social, and educational organizations.

Professional Corporations Until relatively recently, anyone, subject to age and residence requirements, could form a corporation for any lawful purpose *except for conducting a profession,* as it was felt that the highly personal and confidential relationship between a lawyer and client or doctor and patient imposed a standard not consistent with the ordinary commercial relationship. Today most jurisdictions allow persons engaged in the practice of professions to form professional corporations. Some states require only a license for professional activity to qualify to form a professional corporation while others confer "professional" status on vocations not ordinarily considered professions and not usually requiring a license.

The majority of states require that the shareholders be licensed to render the service offered by the corporation. A few states also require the stockholders to be employed by the corporation.

Aside from tax and retirement advantages available in some circumstances, professional incorporation may offer attractive limits on professional liability, although not always to the extent of the ordinary corporation. The majority of states provide that the shareholder is liable for his own torts and those of employees directly under his supervision. A minority of states provide that all shareholders of a professional corporation are personally liable for the torts of all members, thus nullifying all aspects of limited liability enjoyed by ordinary corporations.

The majority rule with respect to contract liability is that liability is unchanged from that in a partnership. Some jurisdictions impose contract liability on the corporations for corporation debts and obligations, thus limiting the shareholders' liability to the capital invested. A minority view holds the shareholder personally liable in the corporation's contract obligations if he participates in the transaction.

Subchapter S Closely held corporations, those in which the stock is held by a small number of persons, may find it advantageous to take advantage of Subchapter S (Sections 1371–1379) of the Internal Revenue Code, which allows a corporation to avoid the double taxation normally entailed by doing business in the corporate form. Under Subchapter S, profits are taxed to the shareholders at their respective individual tax rates, whether or not such profits have actually been distributed. Also, most net operating losses may be passed through and deducted at the shareholder level, so Subchapter S election may be useful where a corporation anticipates substantial losses in its early years but desires to retain the advantages normally associated with doing business in the corporate form, such as limited liability.

With the consent of all of its shareholders, a corporation may elect under Subchapter S if it has no more than thirty-five shareholders, has no trust or corporate shareholders, has no nonresident alien shareholders, has no more than one class of stock, and is not a member of an "affiliated group," which means one or more chains of certain types of corporations connected by common parentage.

There are no statutory limits on the size of a Subchapter S corporation, although some limitation may be in existence as a practical matter by virtue of the limitation to one class of stock.

Small Business Stock: Section 1244 The small business section is intended to provide a type of investor's tax insurance against losses frequently incurred by small businesses, particularly losses on the sale of securities or for worthless securities. Losses on small business stock may be deducted as ordinary losses to the extent of $50,000 annual individual returns and $100,000 on joint returns. In the ordinary case, losses on the sale of securities are capital losses.

Unlike Subchapter S, Section 1244 places no limit on the number of shareholders or classes of stock. However, the company must qualify as a "small business corporation," and the stock eligible for special treatment must be common stock issued pursuant to a written plan. Also, persons seeking to deduct the loss on such shares must have acquired them from the corporation directly and have held them continuously since that date. Thus, one who acquired shares by purchase, gift, or legacy from another person is not entitled to these advantages.

For a corporation to qualify as a small business corporation the amount which may be offered under the plan plus the amount of money and other property received by the corporation for stock, as a contribution to capital, and as paid-in surplus, may not exceed $500,000, and the amount which may be offered under the plan plus the equity capital of the corporation may not exceed $1,000,000. The first limitation is intended to limit the total amount of common stock which may be qualified under Section 1244, while the second test limits the tax advantage to shares issued by relatively small corporations at the time the plan is adopted.

Professional corporations, Subchapter S corporations, and small business corporations, are all types of corporate enterprise which small or new engineering and contracting firms may wish to consider, with the advice of their attorney and tax adviser, for a possible limitation of liability, taxation, or revenue-raising advantage.

18-2 "Foreign" Corporations Frequently a corporation organized in and authorized by state A desires to extend its activity to states B and C as well. In that event, it must secure a "certificate of authority" or other document of permission from each of these other jurisdictions and must meet whatever requirements they impose upon such a "foreign" corporation in return for its enjoying the privilege of doing business within their confines. Normally, this will entail such things as filing a copy of the articles of incorporation, designating a local resident (typically the secretary of state) as agent for service of process, and paying fees.

18-3 Defectively Formed Corporations When a corporation is created in accordance with any and all applicable requirements of law, the corporation is referred to as one of de jure. It sometimes happens, though, that compliance with the pertinent laws, while substantially complete, is defective in some regard or other. Where the deviations are relatively minor, the corporation will be treated as one "in fact," or de facto, and, its right to operate can be attacked only in a direct proceeding brought by the state. Four prerequisites are commonly cited to test a defectively formed corporation claiming de facto status:

1 Existence of an enabling act or valid law under which the particular type of corporation involved might be organized
2 "Colorable compliance" with such law, meaning a bona fide attempt to incorporate thereunder (as by filing the necessary papers with the Secretary of State; etc.)
3 Good-faith belief on the part of the prime movers in the enterprise that incorporation has indeed been achieved
4 Actual operation of the corporate franchise

In the eyes of the law, a de facto corporation differs from a de jure one only in that the former alone is vulnerable to a direct ouster proceeding at the instance of the state.

18-4   Characteristics of the Private Stock Corporation as a Form of Business Association   While the impetus comes from the organizers themselves, a corporation is technically a creature of the state and operates within the tight framework of its own articles of incorporation or charter and of applicable laws. The corporation's powers are limited to those enumerated in the incorporating document (or necessarily implied therefrom) approved by the appropriate government authority.

The corporate device enables individuals to pool their invested funds on a limited-risk basis in a profit-seeking business operation to be conducted by management of their choice. The corporation as a distinctive type of business association has several significant advantages, not the least of which is the fact that shares in the enterprise are readily transferable. Another plus is the perpetual duration which the average corporation enjoys in contradistinction to a partnership, which may be subject to dissolution on the death or withdrawal of a partner. Since a corporation is an entity unto itself, the death of any of its directors, officers, or stockholders need not affect its existence or, in most cases, even cause a temporary disruption of its operations.

Perhaps the limited liability afforded its stockholders is the primary advantage of the corporate form of undertaking. Subscribers are responsible to their corporation[1] for payment in full of stock subscriptions, and satisfaction of this debt can be compelled. But the stockholder, generally speaking, has no individual liability for any indebtedness of the firm and, in the event of corporate failure, stands to lose only his investment. One exception should be mentioned. When it can be shown that a corporation has been organized to perpetrate a fraud or perhaps has set up a dummy subsidiary to evade some pertinent statute, the courts may very well "pierce the corporate veil" and render the responsible stockholders individually liable.

The federal income tax laws treat corporations as separate taxable units and assess them on their income before dividends;[2] the stockholders are then taxed individually on the dividends distributed to them in cash[3] when the total value exceeds a certain amount.

Acting through their officers and agents, corporations can be guilty of at least certain torts and criminal violations. Moreover, the corporation has a name and an existence of its own quite apart from that of its stockholders, who indeed may enter into contractual arrangements with it and either sue or be sued by it.

18-5   Private Corporation's Powers in General   A corporate enterprise, as a creature of law, possesses only those powers which its certificate of incorporation (framed in light of applicable statutes) confers upon it; but it includes "implied

---

[1] Or to its creditors in case the corporation is insolvent.

[2] Various expenses, of course, such as reasonable salaries, are deductible by the corporation for income tax purposes.

[3] A special rule pertains where *stock* dividends are involved.

powers" which are necessary or incidental to the accomplishment of the purposes for which the organization was created.

What might be called typical general powers of a corporation would include the following: (1) to buy or sell real or personal property; (2) to sue or be sued in its own name; (3) to make contractual commitments; (4) to invest its funds; (5) to loan or borrow money for its business purposes;[4] (6) to make bylaws; and (7) to appoint officers and agents.

Most state corporation acts carry express limitations on the privileges and authority of corporate entities. For example, loans by the corporation to its directors or officers are commonly prohibited, as are dividend payments to stockholders when the effect of such payments would be to render the concern insolvent.

18-6 Ultra Vires Activity   A corporation may enter into only such contracts as come within the scope of its express or implied powers. Any contracts which go beyond this point are spoken of as *ultra vires.* If the contract in question runs counter to a statute or to established public policy, it will normally be judicially treated as unenforceable, although some decisions have required a benefited party to pay for what he got, perhaps on a theory of estoppel or of ratification. On the other hand, if the ultra vires contract is not "illegal" in the strict sense of the word but is merely "unauthorized,"[5] a more difficult question is presented. As a general proposition, the courts seem to resolve the matter in this fashion: (1) if the corporation has exceeded its authority in making a particular contract but there has been full performance on both sides, the courts will not interfere one way or another; (2) if the unauthorized contract is entirely executory on each side, neither party can enforce it; (3) if the contract is partly executed, the party which received the benefit cannot urge ultra vires as a defense when pressed for counterperformance.[6]

It is usually held that the directors, officers, and agents who enter into ultra vires contracts on behalf of their corporation do not thereby render themselves personally liable to the other contracting party, either on a warranty of authority basis or otherwise. The other party is chargeable with a degree of constructive notice regarding the limitations contained in the corporation's articles or imposed by statute.

The commission by a corporation of an act beyond the scope of the powers conferred upon it[7] does not, in and of itself, put an end to the corporate existence. What it does do is furnish the state's attorney general, via a "quo warranto"

---

[4] Both large and small corporations upon occasion find it necessary or desirable to borrow money, commonly issuing bonds as evidence of indebtedness. Interest on these obligations is payable to the corporation's creditors before any dividends to stockholders.

[5] Stockholders may ordinarily ratify ultra vires action by management to the extent that the action in question is not utterly void because it is in contravention of statute or public policy.

[6] The judicial treatment of ultra vires contracts, particularly those in category (3) above, varies considerably from one jurisdiction to the next. The defense of ultra vires is not generally favored.

[7] A corporation is not relieved of liability for a tort simply by reason of the fact that the activity out of which the tort arose was ultra vires. The key question relates to the individual who committed the delict: was he acting within the scope of his authority from his principal, the corporation?

proceeding, with possible justification for claiming forfeiture of the document evidencing the corporation's right to do business.

18-7   Forming the Corporation   Each state has its own laws under which private business corporations are formed and operated[8] for profit.[9] Procedures and requirements of the creating process differ with the state of domicile, but most call for prescribed fees to be paid and for information to be submitted to the proper government officials. Once the law has been satisfied in other respects and the *articles of incorporation*[10] duly filed with the appropriate state official, the latter typically certifies that the organization is officially in existence and ready to commence activity.

The articles of incorporation cover such items as the name[11] and location of the enterprise, the nature of its business, the amount and description of authorized capital stock,[12] the amount of paid-in capital with which the corporation will commence its existence, and the period (if any limitation is to be set) during which the corporation expects to remain in being.

A subject mentioned in the articles of incorporation may also be treated (but in greater detail) in the *bylaws,* which are rules adopted by the corporation under which its internal affairs are to be managed. Since the process of amending the articles of incorporation generally means fulfilling several requirements which may prove burdensome, items which are subject to frequent change are generally covered in the bylaws instead.

18-8   Stock and Stockholders   The sale of stock provides capital with which to operate the corporation, and the shareholders, collectively, are the owners of the enterprise. All capital stock must be authorized by the articles of incorporation, which fixes both the total amount of stock and the classes or types thereof.[13] If any change in the pattern is subsequently desired, the articles must first be amended.[14] In the issuance and sale of securities, federal laws administered by the

---

[8] There are some corporate instrumentalities created by the federal government. The Tennessee Valley Authority offers an example.

[9] Special kinds of corporate organizations (insurance companies for example) are subject to a separate set of statutory provisions as are nonprofit corporations.

[10] Terminology varies and, in some quarters, this document is referred to as the "certificate of incorporation" or as the "charter." Technically, the latter term has reference to a direct legislative grant to named individuals rather than to a paper evidencing qualification under a general enabling act.

[11] Partly to prevent confusion to the public and partly to protect prior rights of a corporation already operating in the state, a new corporation will not be permitted to take a name identical or substantially similar to that of such other corporation.

[12] This figure can subsequently be changed only by vote of the stockholders supporting a formal amendment. The term *capital stock* means something far different from the term *capital,* which refers to the assets owned by the corporation.

[13] If par-value shares are to be utilized, the value in question will be stipulated in the basic incorporation papers.

[14] The adoption of certain important amendments or the consummation of certain transactions (merger, for example) may invest dissenting stockholders with the right to return their stock (to the corporation) in exchange for payment of its fair, appraised value.

Securities and Exchange Commission must be checked, along with such state provisions ("blue sky laws" primarily) as are applicable.

The two basic categories of capital stock are *common* and *preferred*.[15] Generally speaking, the former carries voting rights and thus ultimate control of the enterprise, while the latter has preference in regard to asset distribution but lacks the "control" aspect. Frequently, a corporation will be authorized to issue both cumulative and noncumulative preferred stock. The distinction between the cumulative and the noncumulative varieties is discussed in the following passage from *Day v. United States Cast Iron Pipe & Foundry Co.*[16]

> While dividends (not in liquidation) upon each can, of course, only be paid out of profits or surplus. . .the dividends upon cumulative preferred stock have at all times and for all years past and present, until paid, priority in payment over any and all unpaid dividends upon common stock, whether the net earnings for any particular past or present year were or were not sufficient to pay the stipulated cumulative dividends upon preferred stock for that year; whereas the like priority of dividends upon noncumulative preferred stock (wholly or partially as the case may be) is limited to the unpaid dividends for those years when such net earnings were sufficient (wholly or in corresponding part) to pay such dividends.

The stockholders are not, in their individual capacities, agents of the corporation, and really they can act for it only by voting at meetings.[17] They do elect the directors and thus retain a measure of control in that they can eventually unseat directors whose conduct displeases them. But the directors, by virtue of their position, are expected to exercise independent judgment in furtherance of the corporation's business, and the stockholders must not seek to influence their elected representatives.[18]

Apart from voting powers,[19] stockholders, generally speaking, have a number of rights and privileges, chief among which are these:

1 A right to share ratably in whatever dividends are declared on the particular class of stock held (dividends, of course, are generated by profits and may, in most states, be declared and paid to the extent of *surplus*).

2 A "preemptive right," which means that, when the corporation becomes authorized to increase its outstanding capital stock by the issuance of new shares, its

---

[15] Usually preferred stockholders receive dividends at a fixed rate, while the dividend rights of common stockholders are limited in the long run by the earning power of the company.

[16] N.J.Eq. 736, 126A, 302, 304–305 (1924).

[17] The frequency of shareholder meetings, any relevant notice requirements, what constitutes a quorum, what proportion of the vote is necessary to carry certain propositions, and other matters pertaining to the conduct of meetings are set forth in the articles of incorporation, the bylaws, and the corporation statutes.

[18] Agreements among shareholders which directly or indirectly limit the discretion of directors are usually held invalid as contrary to public policy. A director cannot bargain away his privilege of independent judgment, which he is supposed to utilize for the benefit of the shareholders as a group.

[19] Corporation laws in the several states generally provide for and regulate proxy voting; moreover, the solicitation of proxies must, in every respect, meet Securities and Exchange Commission requirements. The proxy is deemed to bear a fiduciary relationship to the stockholder appointing same. A proxy designation is revocable unless coupled with an interest.

stockholders of record, in their respective ownership proportions, have first chance to subscribe for the additional shares.[20]

3 A right to participate in the distribution of assets should the firm be liquidated.

4 A privilege of immunity from personal liability for corporate debts.

5 A right to reasonable inspection of the concern's records. Where necessary, this right can be enforced by *mandamus* proceedings.

6 A right, under certain conditions, to bring suit in behalf of the corporation.[21]

18-9   The Directors   The business activities of a corporation and the management of its property are controlled by its board of directors, operating in accordance with its bylaws. The directors, in turn, select a president and other administrative officers who are ordinarily removable at the pleasure of the board. The method by which the members of the original board are named varies, but thereafter the directors are elected by and are ultimately responsible to the stockholders. Broadly speaking, all matters relating to the legitimate activity of the enterprise, as outlined in its articles of incorporation (or charter), may be handled by the directors without the sanction of the stockholders. For outsiders, the directors are agents of the corporation;[22] for the stockholder-owners and the corporation itself, the directors fill a role closer to that of trustees. Strictly speaking, they cannot be "trustees" in the full sense of the term, since a trustee is legal owner of the trust res and title to corporate property is not vested in the directors, but rather in the corporation itself. However, it is certainly true that the directors are fiduciaries and have a substantial duty of loyalty and diligence to the interests of the corporation.

The directors cannot act individually, and generally they may exercise their powers only as a unit brought together at a properly called board meeting with a quorum in attendance. If the full board proves difficult to assemble by reason of size or geographical complications, it may be possible to provide for the delegating of some functions to an executive committee, composed of a relatively few members of the parent board.

Compensation to directors for their services may take the form of fees for attending meetings, fixed salary, or some other arrangement. Unless the bylaws provide otherwise, a director cannot be removed before the expiration of his term except when good cause is shown.

The standard of care to which directors are held is that of "reasonable persons" in their particular positions. In other words, any particular director is required to give such care and attention to the duties of the position as are corporate directors

---

[20] The doctrine of preemptive right of shareholders is a judicial interpretation of general principles of corporation law to the effect that existing stockholders are owners of the business and are entitled to have that ownership continued in the same proportion. *Katzowitz v. Sidler*, 24 N.Y.2d 512, 301 N.Y.Supp.2d 470 (1976).

[21] The matter of stockholder litigation will be dealt with in Art. 18-11.

[22] Accordingly, notice to or pertinent knowledge of a director will be imputed to his principal.

in general.[23] Ordinarily, directors acting in good faith and within the bounds of their authority are not liable for the disastrous consequences of an honest mistake in judgment.[24]

Where the facts show fraud or gross negligence resulting in a loss to the corporation, violence has been done to their "duty of care," and the errant directors are in a vulnerable position. In fact, statutes in a number of jurisdictions impose upon directors personal responsibility under stated conditions of malfeasance, including such overt action as (1) paying dividends out of capital when there are no profits, (2) making loans out of corporate funds to a director or officer, and (3) authorizing publication of false financial reports.

Mention has been made of the director's fiduciary capacity and of his duty of loyalty to the corporate organization. This gives rise to the common-law doctrine that, if a given director's vote is needed for corporate approval of a contract in which he has a personal interest, such a contract, without regard to the fairness of its terms, is voidable at the option of the corporation. Furthermore, while a director is not precluded from selling his own property to his corporation at a profit,[25] any such self-dealing transactions are subject to careful scrutiny.

An allied problem involves the director's relationship with an individual stockholder from whom he intends to purchase shares of the corporation with which they are mutually connected. There exists a divergence of opinion about whether the director is in a fiduciary position with respect to the individual stockholder such that he must disclose facts (known to him by reason of his position but presumably unknown to the stockholder) which tend to increase the apparent value of the shares that are about to change hands. Clearly, the director must refrain from active fraud or active concealment of significant information, but beyond that his duty is unclear.

18-10   The Officers   While retaining ultimate control of the situation at the policy-making level, the board of directors will name a slate of company officers[26] (of which the principal ones are likely to be board members) to whom considerable administrative authority may be delegated. Officers may derive their power

---

[23] A director who absents himself from meetings does not thereby necessarily secure exemption from responsibility for actions taken by the board.

[24] To encourage qualified persons to serve as board members with reasonable assurance that they will not suffer financially thereby, many states have passed laws which either empower corporations to indemnify directors and officers (and sometimes certain employees) with respect to the cost of defending themselves against groundless accusations of mismanagement, or which take the form in some cases of affording the directors, etc., a statutory right to indemnification, even if there is nothing in the bylaws or the charter and no indemnification agreement as such in existence. In most instances the individual is entitled to be indemnified (with the money perhaps provided through the medium of insurance) unless he is shown to have been guilty of bad faith or to have been engaged in substantial misconduct.

[25] On the other hand, the director (or officer, for that matter), having learned of his firm's interest in a piece of realty, cannot "race" the corporation to such property, purchase it, and then sell to the firm.

[26] Some jurisdictions permit direct selection of officers by the stockholders.

from statutes, articles, bylaws,[27] or resolutions of the board. In addition to the express powers conferred upon them, the officers, especially the president, have a certain amount of "apparent authority" which stems from the nature of their position.

The principal executive official is the president, whose function may be stated broadly as that of running the everyday operations of the corporate enterprise. This officer has at least implied authority to make such contracts as are reasonably related to the ordinary activities of the corporation. There is a rebuttable presumption of authority on his part to perform any act within the scope of the company's business. Established practice of the particular corporation also enters into the picture. Thus, if the board traditionally acquiesces when the president exceeds his delegated authority in entering into a certain type of contractual commitment, the president will be deemed clothed with "apparent authority" as regards the outside world, and the corporation will be held responsible for any breach on its part of a contract so made. Whether or not a specific action taken by the president comes within some aspect of his authority is a question of fact (for a jury's determination, if need be). If there is any reasonable doubt about the president's capacity to bind the corporation to a particularly important contract in contemplation, the other party involved will do well to be on the safe side and insist upon a certified copy of a board resolution granting express authority to the president to proceed.

Ordinarily, in the typical small corporate setup and with the president on hand, the vice president has no contract-making powers. The treasurer maintains financial records, prepares reports therefrom, and (under the broad supervision of the board) receives and disburses corporate funds. More often than not, the treasurer has not express or apparent general authority to borrow money, though the board might ratify a treasurer's overreaching in this respect. The remaining official in the standard organizational arrangement is the secretary. He is largely a recording officer who keeps the minutes and various other official records, supervises the mechanics of stock transactions, and has custody of the corporate seal.

All officers, or any one of them, may by and large be discharged at any time even though they hold long-term employment contracts. If the firing, however, is not "for cause," the officer involved has available against the corporation an action seeking damages for breach. An increasing number of employment contracts written for executive officers carry for their protection a liquidated-damages clause to apply in the event of their summary dismissal (for a reason other than misconduct).

18-11 Stockholder Suits A stockholder as an individual has no right of action against third persons, including corporate directors and officers, for damages sustained by the corporation itself, and this is so even where the stockholder in question is sole owner of the business and even where the delict involved may

---

[27] Where a power is one which by generally accepted practice is in the domain of a particular officer rank, it is usually held that a bylaw purporting to eliminate that power is ineffective as against a stranger who has no actual notice of the said bylaw.

demonstrably have resulted in reducing the value of his stock holding. If the wrong is one affecting the corporate entity and its stockholders in general, the cause of action belongs to the corporation. Should the latter, however, for one reason or another (perhaps because controlled by some of the wrongdoers) fail to proceed, suit may be filed on the corporation's behalf by one or more stockholders, acting "derivatively," the term stemming from the fact that such stockholder's privilege of bringing an action derives from the right of the corporation itself. Before essaying the role of nominal plaintiffs, such stockholders must establish either (1) that they have asked the directors to sue in the corporate name and have been refused for no justifiable reason, or (2) that it would be futile to make such a request (possibly because of alleged collusion between the directors and the prospective defendant). If and when the derivative suit is brought, any benefit flowing from the action is taken by the corporation as the real party in interest and only indirectly by the stockholder-plaintiffs.

18-12   Miscellaneous Control Devices Available to Stockholders   There are several legitimate means by which an organized minority of the stockholders can secure representation on the board and a measure of control over the corporation's destiny. One of these is the system of cumulative voting, which, when authorized, works in this fashion: each share is entitled to as many votes as there are directors to be elected. Assuming three directors are to be chosen, the stockholder has three votes per share and can cast them all for one person or can distribute the votes as desired.

The "voting trust," where legal, is another useful stockholder device. The agreement involved works a separation of voting power from beneficial ownership of the stock, and any applicable statutory restrictions or limitations must be considered. The essence of the voting-trust arrangement is evident in this extract from a statute[28] on the subject:

> One or more shareholders of a corporation may by agreement in writing deposit shares with or transfer shares to a voting trustee or voting trustees for the purpose of vesting in such trustee or trustees, or a majority of such trustees, the right to vote thereon for a period not exceeding ten years, upon the terms and conditions stated in such agreement. . . .

Not to be confused with the foregoing is the so-called "voting agreement," whereby a number of stockholders undertake to vote their shares as a group in some manner provided for in the understanding among them.

Once acquired, a measure of control may be perpetuated through such mechanisms as (1) using a capital-stock structure, which includes a significant percentage of preferred and of nonvoting common stock, (2) staggering the expiration dates of directors' terms of office, (3) granting long-term employment contracts to key officials, and (4) efficiently soliciting and using proxies.

18-13   Subsidiary and Affiliated Corporations; Holding Companies; Consolidations and Mergers   Modern-day corporations in great numbers have established

---

[28] Conn. Gen. Stat., §33-338, as amended 1969.

subsidiaries[29] or worked out affiliations with independent concerns. Where the subsidiary is judicially determined to be merely the instrumentality of the parent corporation, the latter will be potentially liable for the torts and debts of its offspring. In deciding whether the subsidiary is truly a separate entity designed in all respects to stand on its own feet, the court will check for signs of (1) frequent, direct intervention by the parent in management affairs, (2) underfinancing of the subsidiary, and (3) commingling of assets as though the two ostensibly separate concerns were one and the same.

The so-called *holding company* is, in a sense, a "super-corporation"[30] created to hold such a dominant interest in one or more other companies as to be able to prescribe or, at least, to materially influence, through the medium of the voting power which attends upon ownership of all or most of the outstanding stock, the management policies the controlled companies will pursue.

Combination of several previously separate corporations can take the form of "consolidation" or of "merger," either process being subject to whatever statutes are relevant. When a consolidation occurs, the corporations involved lose their identities and a new concern emerges. The result is distinguishable from what happens in a merger, where one of the participating corporations survives and no completely new entity is created.

18-14   Dissolution and Winding Up; Reorganization   When financial reverses befall a corporation and insolvency impends, management may recommend and the stockholders approve voluntary dissolution of the business, following whatever steps are set forth in the applicable corporation act. The foregoing move would bring about lawful termination of the company's existence, as would involuntary dissolution pursuant to court order, or as would expiration of any limited time (for the corporation's existence) set by law or charter provision. The various jurisdictions commonly prescribe a "winding-up period" of several years in which the complicated corporate affairs may be drawn to a close.

Short of dissolution or being forced into bankruptcy by its creditors, a corporation which is experiencing severe financial difficulty might, in an appropriate case, seek to avail itself of the reorganization procedure provided for under Chapter XI of the Federal Bankruptcy Act.[31] Through this avenue the struggling corporation may be able to obtain a new lease on life.

QUESTIONS

1 What are (*a*) the distinctive features of the corporate form of business organization? (*b*) its advantages? (*c*) its powers?

---

[29] Subsidiaries are organized, as a rule, to handle a given phase of the parent's business, that is, its foreign activities. Thus, the *X Company,* incorporated in New York, operates retail outlets in Great Britain and Ireland through *X Company, Ltd.*

[30] Thus characterized in *Kelley, Glover & Vale v. Heitman,* 220 Ind. 625, 44 N.E.2d 981, 985 (1942).

[31] 11 U.S.C., 1101 et. seq. (1978).

2 How and by whom is a corporation (*a*) authorized? (*b*) organized? (*c*) controlled?

3 Who "owns" the corporation? Who runs it?

4 What is a corporate charter? What is its function? When may a corporation be "defective"?

5 How can the ownership of a corporation be revised? How can its management personnel be changed?

6 What is a "proxy"? How does it operate?

7 How does a public corporation differ from other business corporations? Illustrate.

8 What is a subsidiary corporation? May this type of organization prove more advantageous than having a branch of the parent organization? Explain.

9 How can a corporation be dissolved? Can it be reorganized?

10 If a corporation is chartered in New York, can it do business in New Jersey?

11 What is the difference between the "capital" and the "capital stock" of a corporation?

12 Differentiate between a de jure corporation and a de facto corporation.

13 What is meant by "foreign" corporation?

14 Explain the income tax status of a corporation and of its stockholders.

15 Can a corporation be sued (*a*) by an outsider? (*b*) by one of its stockholders?

16 How can a corporation borrow money? On what security?

17 Discuss the ultra vires concept as it relates to corporate affairs.

18 Can ultra vires actions be ratified? By whom?

19 Can a corporation increase the number of shares above that authorized by its charter? If so, how?

20 Does a corporation pay dividends out of earnings or out of surplus?

21 Define directors of a corporation. What are their duties, fees, powers, responsibilities?

22 Who elects (*a*) the directors of a corporation? (*b*) the officers?

23 What are the duties and powers of the officers of a corporation?

24 What rights does a common stockholder have?

25 May a director also be a stockholder of the corporation? Discuss.

26 What benefits may accrue from professional incorporation? What detriments?

27 Discuss the rights and privileges of bondholders of a corporation.

28 A corporation decided to go out of business. After paying off its creditors and bondholders, the firm still had assets totaling $2 million. There were 10,000 shares of preferred stock of $50 par value and 50,000 shares of common stock (no par value) outstanding. How much per share will each common stockholder receive as a liquidating payment?

29 The directors of a corporation voted to buy all its steel for a given year from the *XYZ* Steel Company. Can the stockholders do anything to reverse the decision? If the directors voted to sell the business outright, would the stockholders have any voice in the matter? Explain.

30 The *XY* Corporation went into bankruptcy. Jones owned 100 shares of the firm's common stock. Can the corporation creditors collect anything from Jones? Explain.

31 Smith owned 100 shares of General Motors Corporation common stock. What effect would Smith's death have upon (*a*) operation of the corporation and (*b*) the shares outstanding in the decedent's name?

32 Corporation *A* was chartered for the sole purpose of operating a fleet of oil tankers. Without charter amendment, can it legitimately go into the business of building tankers?

33 The president of *ABC* corporation died. Must the corporation be reorganized and, if so, what steps are necessary?

CHAPTER 19

TORTS

GENERAL DISCUSSION

19-1 Definition and Scope   The term *tort* is very similar to *wrong*; yet "tort" is not intended to include any and all wrongful acts done by one person to the detriment of another, but only those for which the victim may demand legal redress. Torts may be committed intentionally or unintentionally, and with or without force. It may be said that tortious acts consist of the unprivileged commission (or omission, as the case may be) of acts whereby another individual receives an injury to his person, property, or reputation.

A *tort* is distinguished from a *crime* in that the former is a private injury on account of which suit may be brought by the affected party, while the latter is an offense against the public for which any retribution must be sought by the appropriate governmental authority. Obviously, it is entirely possible for a single act to constitute at once a tort and a crime.

Contract liability (which is confined to the parties to the agreement) can likewise be differentiated from *tort liability*. Contract actions are afforded the innocent part as a means of protecting his legitimate interest in having whatever promises are made to him fulfilled. Tort actions, on the other hand, seek to safeguard the interest in freedom from various kinds of harm.

19-2 Classes of Torts   Torts may be quite simply divided into two broad classes. The first of these is *property torts*, and this group includes all injuries to property, be it realty or personalty. The second is *personal torts*, embracing all harms to the victim's reputation, feelings, or physical well being.

The injuries in either category can stem just as well from nonfeasance, from misfeasance, or from malfeasance on the part of the wrongdoer, who is known as the *tort-feasor*.

19-3   Bases of Tort Liability   Commonly stated are three bases of tort liability:

1 Negligence.
2 Intentional interference.
3 Strict liability for reasons of public policy.

To maintain a successful tort action, the wronged party must establish (1) that the defendant owed him a duty, (2) that the duty was breached, and (3) that damage was suffered as a result. Where the alleged tort is founded on negligence, the plaintiff must also be ready to show his own freedom from contributory negligence.

NEGLIGENCE

19-4   Definition   Causing harm through negligent conduct is a basic form of tort. An act or omission which brings injury to another or others may be redressable even though the individual so acting or omitting had absolutely no intention of occasioning such injury. It is sufficient that, in the exercise of proper diligence, the harm should have been foreseen and prevented.

The subject of negligence as a basis of responsibility in tort is a vast one. Among the various concepts involved at one time or another are these: duty, proximate cause, contributory negligence, imputed negligence, gross negligence, assumption of risk, and last clear chance.

Broadly speaking, *negligence* consists of a failure to follow such a pattern of behavior as, under the circumstances, a reasonable person would have pursued, or, contrari-wise , of doing what such reasonable ordinary person would definitely not have done. Thus negligence is definable as conduct which is abnormally likely to cause harm to others; that is, which is, all circumstances considered, unreasonably dangerous, though not intentionally so.

Usually, whether one is or is not negligent is a question of fact, but there are some instances of negligence as a matter of law. Thus, when a legislature has dictated a certain procedure to be followed in a given set of circumstances, failure to comply with the statutory edict may be held to be negligence per se.

19-5   The Elements of Negligence   In order to prevail in his negligence suit, the plaintiff must show that the alleged tort-feasor owed him a duty of care under the circumstances and breached that duty. Furthermore, he must prove that he suffered actual loss or injury and that there was a definite casual connection between the defendant's conduct and the damage complained of. Even where the plaintiff is successful in demonstrating each of the foregoing points, however, all will be lost if the adversary can substantiate a defense of contributory negligence on plaintiff's part. But when the defendant's conduct is shown to have been *grossly* negligent to the point of recklessness, a plea to contributory negligence will avail him nothing.

The burden of proof is clearly on the plaintiff's shoulders to establish the wrongful conduct of the defendant and the necessary causal connection between that and the damage to plaintiff. The defendant is called upon to defend actively *only if* the court feels the plaintiff's evidence, viewed in its most favorable light, might be sufficient to warrant a finding in the latter's favor. Otherwise, the defendant would be entitled to a directed verdict dismissing the complaint.

19-6   The Standard of Care   One of the fundamental propositions which must be sustained by the plaintiff in a successful negligence action is that the defendant, under the circumstances which pertained at the time of the alleged tort, owed a duty of care and that plaintiff is one within the class to which the duty was owed. The concept of the "normally prudent man," developed by common law over a period of many years, will be applied to the defendant's behavior to see if it conforms to a standard of reasonableness in the light of the apparent risk. There are a number of ingredients which go into this "standard-of-care" yardstick: (1) so-called "normal" intellectual capacity, memory ability, and the like; (2) such minimum knowledge, skill, and experience as is deemed common to nearly everyone; (3) whatever additional or superior knowledge, skill, and experience the particular defendant may possess;[1] (4) the alleged tort-feasor's own physical traits, handicaps, etc.[2]

Obviously, the standard of an abstract "average man" is exceedingly difficult to delineate, let alone apply on a uniform basis. The defendant may have conscientiously sought to do the best he could and yet find that the jury regards his behavior as falling short of the prescribed standard of intelligence, prudence, etc., which society anticipates in its members. Since some sort of external model seeems the only logical thing to use, there are bound to occur some marginal cases of liability without fault (as where defendants with low IQ, slow reaction time, and the like are involved). Incidentally, the standard of care is the same whether it is the defendant being tested or whether, with contributory negligence an issue, the plaintiff's own conduct is under scrutiny.

In determining the proper standard of conduct in a given situation, it is not at all unusual to have considerable testimony introduced respecting the general "custom of the trade" and defendant's adherence to, or departure from, such accepted pattern.

The jury is supposed to make the ultimate decision about what the fictional "reasonable man" would have done under the set of circumstances which con-

---

[1] Thus, in medical matters a family doctor will be held to that degree of care expected of physicians generally, while a specialist has an even higher standard to meet.

[2] The conduct of a physically disabled person will be judged in the light of the handicap and of his awareness of that impairment. The awareness is one of the circumstances which has an obvious bearing upon the reasonableness of his behavior.

fronted the defendant. Directed verdicts are rare in negligence actions, and the ordinary accident case, barring settlement during trial, eventually reaches the jury room.

19-7 Proximate Cause and Foreseeability Though the judicial approach to the matter of proximate cause is by no means uniform throughout the country, the prevailing tendency seems to be to hold the defendant responsible even for some rather extraordinary consequences of his negligence as long as the chain of direct causation remains unbroken by an intervening cause. In other words, where it is found that the defendant was indeed negligent and that his conduct was a *substantial cause*[3] in fact of the injuries under consideration, his vulnerability is clear, even though the precise damage which resulted was not to be anticipated. The question of liability is always anterior to the question of the measure of the consequences which attend the liability, but the latter remains a mighty important question, and it is indeed difficult fairly to delineate the "zone of danger" raised by defendant's conduct in a given situation. There obviously has to be a line drawn somewhere. The concept of substantial cause is supposed to preclude recovery for remote damage which defendant's conduct cannot definitely be said to have occasioned. There is always considerable pressure to restrict liability when certain unforeseen consequences are very much out of the ordinary.

The landmark case respecting proximate cause is that of *Palsgraf v. Long Island R.R.*[4] A man carrying a harmless-looking package jumped aboard a moving train and, in so doing, lost his balance and seemed about to fall. A guard on the car, sensing the danger, reached forward to help him in and another guard on the platform pushed from behind. During these maneuvers the package became dislodged, fell to the rails, and exploded. The concussion threw down some scales at the other end of the platform, many feet away. The falling scales struck and injured plaintiff Palsgraf, an intending woman passenger. In the litigation which followed, the majority decision exonerated the defendant railroad, holding that the consequences befalling the particular plaintiff were not foreseeable and that defendant consequently could not be regarded as guilty of any negligence with reference to her. The mandate was that plaintiff must sue in her own right for wrong personal to her, and not as vicarious beneficiary of a breach of duty to another. In other words, the majority opined that negligence is not actionable unless it involves the invasion of a legally protected interest, that is, the violation of a right. The conduct of defendant's guards, if tortious in relation to the holder of the package, was yet not a wrong in its relation to this plaintiff, who had been standing some distance away.

The minority view in *Palsgraf*, which appears nowadays to have become more or less the general rule, was that the railroad employee's negligence in dislodging the package from the commuter's arms was the proximate cause of the plaintiff's injuries and was therefore actionable. The dissenting judges proceeded on the

---

[3] In this connection *substantial cause* does not necessarily mean *sole clause*.
[4] 248 N.Y. 339 (1928).

theory that where there is an act which unreasonably threatens the safety of others, the doer is liable for all its proximate consequences, even where injury results to one who would generally be thought to be outside the radius of danger. Harm to some person being the foreseeable result of the negligent act, not only that one alone, but all those in fact injured may complain. In discussing proximate cause, the minority said:

> What we do mean by the word "proximate" is, that because of convenience, of public policy, of a rough sense of justice, the law arbitrarily declines to trace a series of events beyond a certain point. This is not logic. It is practical politics. . . .
>
> It is all a question of expediency. There are no fixed rules to govern our judgment. There are simply matters of which we may take account.. . . There are some hints that may help us. The proximate cause, involved as it may be with many other causes, must be, at the least, something without which the event would not happen. The court must ask itself whether there was a natural and continuous sequence between cause and effect. Was the one a substantial factor in producing the other? Was there a direct connection between them, without too many intervening causes? Is the effect of cause on result not too attentuated? Is the cause likely, in the usual judgment of mankind, to produce the result? Or by the exercise of prudent foresight could the result be foreseen? Is the result too remote from the cause?

Mention has been made of alien forces which break the chain of causation and may relieve defendant of further consequences of his negligence. Such relief is afforded when, the intervening force is one which defendant could not have foreseen or one which is patently not a normal incident of the risk initially created by defendant's tortious conduct. As an example, let us assume that the negligence of driver X brings about a collision with Y's automobile. Y steps from his car to remonstrate with X, slips on the icy pavement, and injures himself in falling. X's negligence was the proximate cause of the collision damage to Y's car, and X is clearly liable therefor, but many courts today would likely hold that the ensuing fall by Y was not X's responsibility since Y's slipping on the ice was an effective intervening cause of the personal injury to him. It is apparent, of course, that "but for" the collision there would have been no occasion for Y to alight from his car at the particular time and place, but the tort by X could be deemed the remote rather than the substantial cause of Y's bodily mishap.

While the judge will, in a negligence action, charge the jury on what constitutes proximate cause and related concepts, the jury must in every case apply a broad factual test to determine whether or not the evidence indicates that there actually exists the vital causal connection between defendant's alleged tort and plaintiff's alleged injury.

19-8 Violation of Statute Unexcused violation of a specific minimum standard of conduct dictated by legislative fiat is today almost universally treated in this country as negligence per se as long as the injury involved is of a nature which the statute was aimed at preventing and the plaintiff is within the category of persons the law sought to protect. It is open to the defendant to show that he did the best he could to obey the statute or to show by way of excuse that the violation was necessary in order to further some public policy, as, for example, deliberately crossing a solid highway line in an effort to avoid hitting a child who had suddenly run onto the road.

Relatively few jurisdictions deviate from the negligence per se concept. The minority position is that statutory violation is some evidence supporting an allegation of negligence but is by no means conclusive by itself. The jury must weigh the violation along with other factors in the picture.

Not all statutory provisions, of course, are framed in terms of precise standards of conduct. The typical reckless-diving provision is an elastic one and requires interpretation in the light of a particular fact situation to determine whether or not there has been a violation.

19-9 Gross Negligence The term *gross negligence*, which really connotes recklessness or wanton disregard for the rights of others, represents the conscious failure to exercise diligence in an effort to prevent an injury which the situation indicates is very likely to occur in the absence of special precautions. Interposing a plea of contributory negligence will avail the defendant nothing where his own negligence was of the gross variety.

19-10 Imputed Negligence As the term implies, *imputed negligence* is, in ef- fect, charging Peter with Paul's delict. Under certain circumstances, the negligent conduct of *A* will be attributed to defenseless *B* and will bind *B* in his own capacity as plaintiff or defendant, as the case may be. The most common illustra- tion is the imputing of a servant's or agent's negligence to the master or principal, respectively.

A plaintiff will be barred from recovery through having imputed to him the negligence of a third person only if their relationship is such that, as respects harm caused to others, the plaintiff, as a defendant, would have been liable for the negligence in question. Thus, the negligence of the operator of one of two vehicles in an accident will not generally be imputed to his passenger so as to prejudice the latter's action against the driver of the other car.[5]

19-11 Contributory Negligence As a general proposition, any negligence on the plaintiff's own part which contributes to the injury of which he complains will bar recovery therefor.[6] It matters not how slight such contributory negligence may be in relation to the defendant's wrongful conduct.

---

[5] If it can be shown, however, that the passenger and his driver were engaged in a so-called "joint venture," the driver's negligence may very well be imputed. Much depends, of course, upon the pertinent statutes and common law of the particular jurisdiction.

[6] A fairly commonplace example of contributory negligence would be the plaintiff's failure to take a reasonable alternate route to that of obvious danger. In one actual situation, a general contractor was erecting the steelwork of a building. As the crane operator was attempting to swing a heavy girder into place (attached by means of a sling) a workman very foolishly walked underneath. At that moment the hook holding the sling broke, and the girder fell on the workman, killing him instantly. The wife of the deceased sued the contractor for damages. In deciding the question of liability, the court considered such points as these: (1) the crane had successfully erected two similar girders before the accident occurred; (2) the load slightly exceeded the rated capacity of the crane; (3) the hook had been properly tested for a 50 percent overload by the crane manufacturer; (4) no one ordered the workman to walk under the girder, but no one tried to stop him either. The court concluded that, although the contractor did slightly overload the crane, the workman was undoubtedly guilty of contributory negligence and this fact precluded any recovery under the circumstances.

There are several instances in which ordinary contributory negligence is not a defense. One of these is the situation wherein defendant's conduct in violating a statutory standard of care is negligent per se. Another involves gross negligence of the defendant. Parallel to this latter is defendant's tortious conduct of a willful, deliberate nature. The line of demarcation between wanton and inadvertent negligence is none too clearly drawn, and it is up to plaintiff to cause the court to look at defendant's behavior with more than mild resentment in order to obviate any possible defense of contributory negligence.

For yet another limitation which serves to confuse the picture on contributory negligence, see Art. 19-14 on last clear chance.

19-12   Comparative Negligence   The doctrine of *comparative negligence* provides that an injured party will not be barred from any recovery by contributory negligence, but that there will be an apportionment of damages or responsibility in proportion to the relative fault of the parties involved. Thus, the "all or nothing" aspect of contributory negligence, which may work great hardship where the fault of one party is slight, is modified or eliminated.

Rejected for many years for largely administrative or inertial reasons, some form of comparative negligence is not the prevailing rule in the majority of American jurisdictions, by means of either judicial or legislative action.

In *Li v. Yellow Cab Company of California*,[7] the state supreme court, in abrogating the rule of contributory negligence concluded that:

> (1) The doctrine of comparative negligence is preferable to the "all-or-nothing" doctrine of contributory negligence from the point of view of logic, practical experience, and fundamental justice; . . . (3) given the possibility of judicial action, certain practical difficulties attendant upon the adoption of comparative negligence should not dissuade us from charting a new course, leaving the resolution of some of these problems to future judicial or legislative action; (4) the doctrine of comparative negligence should be applied in this state in its so-called "pure" form under which the assessment of liability in proportion to fault proceeds in spite of the fact that the plaintiff is equally at fault as or more at fault than the defendant;[8]

19-13   Assumption of Risk   Generally speaking, a person cannot voluntarily put himself in a position of manifest danger and then expect to collect from the defendant for the not-unexpected injuries which follow. The assumption of risk defense requires a showing that plaintiff had knowledge of the chance he was running and yet voluntarily ignored the danger. Some adequate warning must have been given, except where the risk is so obvious that a person of ordinary intelligence would readily sense the likelihood of impending trouble. Where assumption of risk is shown, it goes only to the dangers normally associated with the

---

[7] 13 Cal.3d 809, 119 Cal. Rptr. 858, 861 (1975).
[8] See also *Daly v. General Motors Corp.,* 20 Cal.3d 725, 144 Cal. Rptr. 380, 575 P.2d 1162 (1978), where the California Supreme Court held the principles announced in *Li* applicable to an action in strict liability.

undertaking. For example, a person contracting to paint a bridge assumes the risk of falling off, but not the danger of electrocution from negligently installed wiring.

There are many exceptions, limitations, etc., related to the assumption-of-risk defense—too many for adequate treatment here. For one thing, worker's compensation laws, as regards injuries falling within their scope, have deprived the employer of his onetime common-law defenses, including that of assumption of risk. Also, the doctrine of assumption of risk, insofar as it is but a variant of contributory negligence, is subsumed under the general process of assessing liability in proportion to fault utilized in the comparative negligence system.

Related to the discussion of assumption of risk, is the plight of the Good Samaritan. If *A* is hurt or in danger and *B* chances upon him in his predicament, *B* is under no legal obligation to tender assistance. If he accepts the moral responsibility, however, and takes steps to assist *A*, he runs the risk that his attempts to aid will prove negligent in some respect and that, despite his humanitarian intentions, he will consequently be liable for any damage caused.

19-14 Last Clear Chance The doctrine of *last clear chance*,[9] which has been called everything from "an inherent limitation in the defense of contributory negligence"[10] to a "a thinly disguised theory of comparative negligence," undeniably plays a leading role in many a negligence case. Where both plaintiff and defendant have been guilty of conduct falling short of what would be expected of the "reasonable man" in the particular situation, last clear chance focuses attention upon the time sequence of events and regards the defendant as answerable for the damage if—as between the two negligent parties—he had the last opportunity through use of the due care to avoid the mishap and its consequences.

For a not-uncommon example, assume plaintiff, having imbibed freely, stumbles in the road and falls to the pavement in a drunken stupor; the stage is now set for operation of the last-clear-chance rule, since plaintiff has, by his own negligence, placed himself in a helpless position of peril. If an approaching driver sees[11] the danger in time and is in a position to avert an accident by exercising ordinary care and prudence, he will be responsible for injuries stemming from his failure to seize the opportunity. One way of putting it is that the law prefers the individual who is negligently unaware of his peril to the individual who is negligent despite having seen the danger.

One limitation ordinarily imposed by American courts on the last-clear-chance doctrine relates to the defendant's ability to avoid the harm. If defendant has

---

[9] The doctrine of last clear chance is subsumed under the general process of assessing liability in proportion to fault where the doctrine of comparative negligence has been substituted for that of contributory negligence. *Li v. Yellow Cab Company of California,* 532 P.2d 1226, 119 Cal. Rptr. 858 (1975).

[10] If the evidence indicates that defendants indeed had the last clear chance to prevent the injury complained of and plaintiff's active negligence had ceased before the point of injury was reached, then the contributory negligence of plaintiff is not a bar to his recovery, and the entire loss must be borne by defendant.

[11] Or, in many jurisdictions, "should have seen."

noted plaintiff's helplessness and perceived the danger but is incapable by reason of some *prior* negligence of his own of averting disaster, no clear chance was his at the crucial time, and the general rule has no applicability. Thus, in the example outlined in the preceding paragraph, if defendant makes every reasonable move in an effort to avoid hitting the prostrate form but is stymied by mechanical failure his helplessness, though brought on by neglect, is as complete as the plaintiff's and means that no real last opportunity was his.

Last clear chance is clearly related to the principle of proximate cause. Despite his own antecedent negligence, plaintiff is permitted to recover because his conduct, when tested as a cause of the injury, is seen to have been rendered "remote" by the supervening negligence of defendant, which latter wrong is then left as the sole proximate cause of the damage. A jury charge on last clear chance will not be forthcoming unless the judge feels the circumstances warrant belief that reasonable persons *might* find that, notwithstanding plaintiff's own negligent behavior, defendant was the party with the vital opportunity to avoid the accident. Generally speaking, the last-clear-chance test is in order primarily where plaintiff and defendant are on different levels of activity; for example, plaintiff pedestrian is struck by defendant's automobile while plaintiff is crossing the street.[12]

To sum up the essentials of last clear chance: If the jury finds that plaintiff's own negligence materially contributed to his injuries, they can nevertheless grant him damages if they find that, (1) after the plaintiff's conduct had created for him a perilous situation, (2) defendant discovered (or should have discovered) not only this fact, but also that plaintiff apparently would not without intervention escape, (3) defendant, armed with such knowledge, had the opportunity to save plaintiff by exercising reasonable care, and (4) defendant failed to utilize the chance open to him.

19-15   Res Ipsa Loquitur   Literally translated, *res ipsa loquitur* means "the thing speaks for itself." This doctrine has the effect merely of laying the foundation for a permissible inference that defendant was indeed negligent as alleged. It has application where plaintiff is able to show three things:

1 The mishap in suit was one which would not ordinarily occur unless there was negligence on someone's part
2 The instrumentality or appliance whose careless construction or use occasioned the injury was in the exclusive possession and control of the defendant
3 There was no voluntary, inexcusable action on plaintiff's part which contributed to the damage.

Where plaintiff can successfully prove the points just listed, it is apparent that any evidence about the true cause of the injury is more than likely accessible primarily (if not, indeed, exclusive) to the defendant. This circumstance is sup-

---

[12] For an excellent discussion of the last-clear-chance principle against the factual background of the typical auto-pedestrian mishap, see *Correnti v. Catino,* 115 Conn. 213 (1932).

posed to justify res ipsa entitle the plaintiff to get to the jury even though he cannot show any direct evidence of the defendant's negligence.[13] By making out a res ipsa case, he has, purely by circumstantial evidence, created an inference of defendant's negligence, which the jury has the option of accepting or rejecting in the light of whatever the defendant is able to put in by way of explanation.[14]

Defendant's rebuttal evidence in most res ipsa cases consists of showing due care on his part and on the part of his agents and servants as respects the construction, inspection, maintenance,[15] and use of the instrumentality which is asserted to have brought about plaintiff's injury.

Two decisions well illustrate use of the res ipsa loquitur doctrine. The first of these is *J. C. Penney Co. v. Livingston.*[16] Plaintiff two-year-old injured his hand when it inexplicably became caught between two steps of a store's escalator. Plaintiff, by his father, brought an action based on the res ipsa theory; the appellate court, affirming a verdict for the plaintiff, agreed that all necessary elements were present. The closest question in the case was whether, according to common knowledge and experience, the accident would not have happened reasoned in this fashion: it is well known that children are attracted by an escalator and that the ordinary escalator is completely safe even for youngsters. Inasmuch as plaintiff's hand did become caught in the moving parts, there is a logical inference that the particular escalator was unsafe for use by children and that defendant store was consequently negligent in making the device available to such persons.

The other case was decided in Louisiana in 1973, and involved a claim of medical malpractice.[17] The plaintiff's son injured his elbow while playing in a high school football game and was taken to the hospital for treatment. Upon admission to the hospital, the injured elbow was the only source of complaint. The plaintiff's son was given a shot which put him to sleep while the defendant performed a closed reduction upon the injured elbow. Following the operation, the patient complained of severe leg pains, and a large lesion was found upon his upper left thigh. While in the defendant's care the patient also suffered injuries to his reproductive organs to the extent that his capacity to reproduce was destroyed. The plaintiff alleged that the injuries to his son's leg and reproductive organs resulted from an act of negligence by the hospital and the physician, the details of which were peculiarly within the knowledge and control of the defendants, and thus the doctrine of res ipsa loquitur was applicable against them.

The court held that the rule applies when the facts shown suggest the negligence of the defendant as the most plausible explanation of the injury. The court

---

[13] Where plaintiff is able to set out in his complaint *specific* negligent acts or omissions of defendant, *res ipsa* has no applicability.

[14] In several states, the effect of a *res ipsa* case is somewhat stronger, that is, the raising of a definite presumption which shifts the burden of proof to defendant. In such jurisdictions, the court will *require* the jury to make an inference of defendant's negligence and then see whether the evidence subsequently put in by defendant is sufficient to overcome that initial setback.

[15] In order for the maintenance to have been negligent, the defect must have been observable and must have been present long enough so that it should have been found on reasonable inspection.

[16] 271 S.W.2d 906 (Ky. App. 1954).

[17] *McCann v. Baton Rouge General Hospital,* 276 So.2d 259 (La. 1973).

observed that the injury occurred within the confines of the special service rooms at the hospital and that, since the surrounding circumstances were peculiarly within the knowledge of the defendants and were unavailable to the plaintiff, the failure to allege specific acts of the defendant producing the injury was not fatal to the complaint.

One must not succumb to the temptation to urge res ipsa loquitur in every instance, however. For example, in one case[18] a patient was experiencing some problems and difficulties with his right arm and contacted the defendant, an orthopedic surgeon. The surgeon diagnosed the problem as the result of a nonunion of an old fracture and advised the plaintiff that he could perform a compression plating procedure by which the nonunion could be repaired. Following the surgery, the plaintiff experienced difficulty moving the fingers of his hand and an infection was identified. The problems persisted, requiring a number of additional surgical procedures and the plaintiff's right arm was finally paralyzed. The plaintiff contended that the case should have been submitted to the jury and to the doctrine of res ipsa loquitur urging that the mere occurrence of the infection was sufficient to establish an issue for the jury to consider. After noting the conditions upon which the application of res ipsa loquitur is appropriate, the court stated the question to be whether an infection ordinarily does occur in the absence of negligence. The testimony showed that the hospital had a postoperative infection rate well below the national average and the plaintiff claimed that the low incidence of infection at the hospital meant that infection does not ordinarily occur. But the court, agreeing with the defendant, found that this fact did not suggest that when an infection does occur, it is the result of negligence.

19-16   Malpractice   Negligence on the part of the family doctor in the treatment of a patient is hard to prove; plaintiff must show by *expert* medical testimony that the defendant physician failed to measure up to the standard of conduct common to his colleagues practicing in the same comparable community.[19] And those who are licensed as special practitoners, as chiropractors, for instance, are entitled to have testimony about a proper standard of care applicable to them given exclusively by members of their own particular "school" within the profession.

There is, of course, a logical explanation for any extra measure of protection afforded doctors and the like in malpractice cases. Professional men are commonly insured against liability for negligence, but it is not the possible immediate money loss which hurts the most; a doctor's reputation in the community and thus his future livelihood can be ruined as the result of one malpractice recovery against him.

In an effort to control medical malpractice awards and resulting high premiums

---

[18] *Wilson v. Stilwill,* 309 N.W.2d 989 (Mich. 1981).

[19] In a medical malpractice case, the use of expert testimony is allowed to establish that as a matter of common knowledge the incident in question would not have occurred had the defendant physician adhered to proper standards. *Bucklew v. Gross Bard,* 435 At.2d 1150 (N.J. 1981).

for medical malpractice insurance, several jurisdictions are investigating the possibility of voluntary or mandatory arbitration of medical malpractice claims.

SPECIAL TOPICS

19-17 Standard of Care for Children Courts in most jurisdictions stand ready to give a child plaintiff every advantage on the issue of contributory negligence, while normally holding a child defendant to the adult standard of reasonable care. Thus, the tender years and limited experience of the child plaintiff are factors taken into consideration when evaluating his conduct.

Parents are not responsible at common law for the nondirected torts of their children, but legislatures here and there have altered the picture and made parents vicariously liable under certain circumstances.[20] One idea behind such legislation is that elders who have been placed in a vulnerable position financially are more likely to exert a restraining influence on the juveniles under their roofs.

As a practical matter, children are seldom sued unless there is insurance coverage in the background. Incidentally, minor defendants (unlike adults) are not vicariously liable for the torts of their employees or agents, ostensibly owing to the fact that the employer-employee and principal-agent relationships are contractual in nature and infants cannot generally be held on their contracts.

19-18 Tort Injury to Minors When an unemancipated child is hurt, two different causes of action arise against the negligent defendant. One of these is to be pursued in the injured child's name for his pain and suffering and his lost earning capacity for the years after majority is reached. The other is available to the parent legally responsible for the injured child's care and the suit is to recover for medical expenses incurred and for loss of the child's services during minority. Some courts say that any contributory negligence on the child's part serves as a bar to the parent's recovery.

Assume that the injury to the child stemmed from an auto accident which was attributable to the combined negligence of the child's father and of the other driver. If the child sues the operator of the other vehicle, it is commonly held that the parent's contributory negligence is not a bar. However, should the parent as plaintiff seek recovery against such third person for medical expenses in connection with the treatment of the innocent child, the defense of contributory negligence would presumably prevail.

19-19 Torts of Mentally Retarded Persons The general rule is that a mentally incompetent individual is responsible for such of his torts as do not require a finding of malice, intent, and the like.[21] As far as damage resulting from negli-

---

[20] Often on the rationale that the parents are themselves negligent in failing to control the conduct of minors resident in their households.

[21] This would exclude slander, libel, etc.

gence is concerned, the insane person or the individual of very low mental capacity is generally held to the same standard of care as would be prescribed for people of average intelligence and in good mental health. The rationale seems to be that the innocent victim should not be deprived of recovery just because the tort-feasor was mentally deficient.

Similarly, a person who is intoxicated (and who has thus temporarily impaired his mental competency) is liable for his torts. Intoxication does not excuse failure to exercise due care.

19-20   Governmental Tort Liability   The sovereign is protected at common law by a blanket exemption from liability for negligence of the agents through whom it must act.[22] There have developed, however, some substantial limitations on this immunity, since government is free to pass laws, frequently styled "tort claims acts" permitting suit against it under stipulated conditions.[23] However, such an enabling statute, since it is "in derogation of the sovereignty of the state," will be strictly construed. The doctrine of *sovereign immunity* has also faired poorly of late in the hands of the judiciary.[24]

Municipal corporations are not "sovereign" in the strict sense, but they, like school districts are deemed to be acting for the state and thus share its immunity as long as they are performing "governmental" functions, such as operating police and fire departments. Where, on the other hand, the city, town, village, etc., act in *proprietary*[25] capacities, for instance, own and operate transit systems or other public utilities, they will be responsible for torts of their agents or employees committed within the scope of their agency or employment. The line between governmental and proprietary operation is understandably difficult to draw at times. Highway defects are a fruitful source of municipal tort litigation. Maintenance of highways is generally regarded as a governmental function, but there are numerous statutes on the books waiving the common-law exemption and imposing civil liability on the government unit for negligence in connection with road upkeep. These laws normally require a showing by plaintiff of the unreasonably dangerous condition of the road and of adequate notice to the proper authorities regarding the defect.

A city or other unit of government cannot be sued, for example, for such things as failure to pass a regulatory ordinance, which would prohibit riding bicycles on sidewalks. This type of decision is regarded as a matter of discretion and beyond the reach of the individual citizen.

---

[22] There is no comparable shield where the government unit is guilty of maintaining a nuisance. See Art. 19-28 on the general subject of nuisance.

[23] New York State, for example, has set up a special court of claims to handle suits brought against it. See also Article 12, Constitution of the State of Louisiana, Section 10A, which states that "neither the state, a state agency, nor a political subdivision shall be immune from suit and liability in contract or for injury to person or property."

[24] See, for example, *Massengill v. Yuma County,* 104 Ariz., 518, 456 P.2d 376 (1969), where the court stated that the doctrine had been relegated "to the dust heap of history," and *Carpenter v. Johnson,* 231 Kan. 783, 649 P.2d 400 (1982), where the court said that liability is the rule and immunity the exception under the Kansas tort claims act.

[25] Related terms sometimes used are *ministerial* and *corporate.*

19-21 **Negligence of Government Officials** Government officials, while performing discretionary or quasi-judicial functions are not liable for negligence.[26] If officials were not granted such immunity it would be difficult to find public servants.

19-22 **Charitable Institutions** In some jurisdictions, charitable institutions have complete immunity from tort liability; in other jurisdictions, the immunity pertains only where the would-be plaintiffs are nonpaying recipients of hospital services or similar benefits.[27] Still another line of decisions has held that, as respects negligence of its "ministerial" employees (custodians, scrubwomen, etc.), a hospital is answerable to outsiders, such as the delivery boy who slips on defective flooring, but not to any of its patients, whether or not they are charity cases.

Apart from a consideration of the class of persons to whom the institution may be liable in tort, further distinctions have been based on the relationship which the actual tort-feasor bears to the charitable institution. Thus, a hospital may ordinarily be responsible for damage resulting from the fault of its ministerial employees, but nurses and doctors using the hospital facilities in the performance of their professional functions are deemed to be independent contractors and not agents of the institution. The hospital cannot be held for negligence of these independent contractors, and this assumes only that the hospital exercised reasonable care insofar as it had a voice in their selection.[28]

19-23 **Liability of Landowners** In the use and enjoyment of his premises, the landowner or occupier must exercise reasonable care for the protection of those who are lawfully in the immediate vicinity, such as persons using a nearby public highway. With respect to those who actually set foot on his property, the owner's duty of care may vary with the characterization of the visitor as trespasser, licensee, or invitee.

Trespasser One who comes upon another's property without either (1) the consent of the occupier or (2) the legal privilege to enter irrespective of the question of express or implied consent[29] is called a *trespasser*. To him, the landowner or occupier owes merely a duty to refrain from willful misconduct.[30] There is no obligation to foresee or to seek out the presence of the trespasser and no respon-

---

[26] And they are not generally held for the defaults of their subordinates, irrespective of whether such defaults pertain to ministerial or discretionary functions.

[27] There is something of an anomaly in permitting the paying patients to sue a hospital in tort while denying recovery to charity patients; the latter group is presumably more in need.

[28] However, there are isolated cases which have drawn a distinction between negligence and mistakes of judgment on the part of hospital nurses, treating the institution as liable where negligence was found to be present. In one instance, a nurse, having determined that a delirious patient would be safer with sideboards on her bed, adjusted the boards in a negligent manner, and the patient ultimately was injured in a mishap directly traceable to that negligence. The court reasoned that, had the nurse made the wrong decision in the first place about the need for boards, the hospital would nonetheless be immune from liability because the nurse had simply used faulty judgment on a matter of discretion. But, once the decision was made to affix the boards, the ensuing negligence by the nurse placed the hospital in a vulnerable position.

[29] A police official armed with search warrant would fit into this second privileged category.

[30] Thus, no traps may be set for the unwelcome visitor, nor may intentional injury be done him.

sibility to him for defects or conditions of natural origin existing on the land. A few courts have deviated from the norm and taken the position that, once a trespasser's presence is discovered, he is owed reasonable care for his safety.

The *attractive-nuisance doctrine*[31] relates solely to trespassing children and amounts to a limitation on the general rule of nonliability to trespassers unless intentional harm can be shown. The land occupier is vulnerable under the attractive-nuisance principle where he is harboring on his premises a highly dangerous instrumentality, or condition, of a nature likely to attract children of tender years (who will not appreciate the danger involved). Part of the attractive-nuisance concept is supposed to be a weighing of the degree of risk of injury to minors against the utility to the owner of maintaining the complained-of condition. If the latter factor assumes paramount importance, there should be no damage recovery by the child trespasser.

Licensee A tolerated intruder, or *licensee*, is distinguished from a trespasser in that he has the owner's permission (express or implied) to be on the premises. He is present, though, for his own purpose, convenience, or gratification and has no contractual relationship with the owner. Reasonable care must be exercised to warn the licensee of hidden dangers of which the owner is aware, but obvious defects are something else again. The owner or occupier is not required to take affirmative steps to protect a licensee but is under a duty to discover his presence if reasonably possible, to avoid active negligence toward him, and to refrain from willful misconduct toward him.

A good example of a licensee is the social guest in the home of his host. Firemen and policemen who enter upon private property in the discharge of their duty are typically treated as licensees unless they have been called by the owner, in which latter event they step up to the status of invitee and are consequently owed a higher standard of care.

Invitee An *invitee* (also known as a *business visitor*) is one who by invitation has entered upon the premises for some business purpose of mutual interest or benefit to him and to the owner. The store customer affords an everyday illustration;[32] the proprietor must assume the affirmative duty of maintaining the establishment in a safe condition so as to avoid danger to those who come within the broad orbit of the business invitation. Thus, to avoid potential liability periodic and thorough inspections of the premises would have to be made and all indicated safety precautions taken. The owner must warn the invitee of all dangers other than those of which the latter is aware or those which are readily apparent.

---

[31] The attractive-nuisance doctrine was first clearly set forth in *Keefe v. Milwaukee & St. Paul Railway Co.,* 21 Minn. 207 (1875). In this case a child was injured while playing on a railroad turntable. The turntable was not fastened or fenced or in any way protected although it could easily have been locked to prevent this type of accident. The court held for the plaintiff and had this to say: "The difference between the plaintiff's position and that of a voluntary trespasser, capable of using care, consists in this, that the plaintiff was induced to come upon the defendant's turntable by the defendant's own conduct, and that, as to him, the turntable was a hidden danger, a trap."

[32] The customer is an invitee even though he buys nothing and, indeed, has no real intention of making a purchase. It is possible for a person to be a full-fledged invitee with respect to one part of the store building and a mere licensee (or even trespasser) with respect to other portions thereof.

The trend of the law of premises liability is clearly toward the elimination of technical status positions which protect certain classes from liability. Many jurisdictions now hold that distinctions between invitees and licensees have no logical relation to the duty owed by the possessor of the premises, and the tendency is toward the substitution of the broad test of reasonable care for the status rules. The Supreme Court of California, for example, has said:[33]

> There is another fundamental objection to the approach to the question of the possessor's liability on the basis of the common-law distinctions based upon the status of the injured party as a trespasser, licensee, or invitee. Complexity can be borne and confusion remedied where the underlying principles governing liability are based upon proper considerations. Whatever may have been the historical justifications for the common-law distinctions, it is clear that those distinctions are not justified in the light of our modern society and that the complexity and confusion which has arisen is not due to difficulty in applying the original common-law rules—they are all too easy to apply in their original formulation—but is due to the attempts to apply just rules in our modern society within the ancient terminology.
>
> Without attempting to labor all of the rules relating to the possessor's liability, it is apparent that the classifications of trespasser, licensee, and invitee, the immunities from liability predicated upon those classifications, and the exceptions to those immunities, often do not reflect the major factors which should determine whether immunity should be conferred upon the possessor of land. Some of those factors, including the closeness of the connection between the injury and the defendant's conduct, the moral blame attached to the defendant's conduct, the policy of preventing future harm, and the prevalence and availability of insurance, bear little, if any, relationship to the classifications of trespasser, licensee, and invitee and the existing rules conferring immunity.

19-24 Liability of Lessors Generally speaking, there is nothing in the law which would prevent an owner from leasing a broken-down piece of property as long as he reveals to the prospective tenant defects which are unknown to the latter and which are not readily discoverable upon inspection. Most courts obligate the landlord merely to give the lessee such information as is in his possession and do not hold him responsible with regard to additional defects he *should* have known about by reason of his presumed familiarity with his own property. Thus, a lessor will not during the term of the rental agreement be liable for harm occasioned the lessee or third parties in consequence of the condition of the leased premises.

Particularly where a multiunit dwelling is involved, the landlord will retain control over a portion of the premises, as, for instance, common approaches, stairways, and halls. Where tenants or third persons[34] are permitted use of such

---

[33] *N. Rowland v. Christian,* 70 Cal. Rptr. 97, 443 P.2d 561 (1968). See also *Pridgen v. Boston Housing Authority,* 308 N.E.2d 467 (Mass. 1974), where "all lawful visitors" are grouped as a class and *Webb v. City and Burrough of Sitka,* 561 P.2d 731 (Alaska 1977), where the court abolished the classifications of trespasser, licensee, and invitee in relation to property owner liability.

[34] This would include, of course, business visitors and social guests of the tenant.

common appurtenances, it is incumbent upon the lessor to keep these parts in reasonably safe condition.

With respect to the portion of the premises exclusively demised to the tenant, landlord is under no obligation to repair or improve unless the lease or some other contractual arrangement requires him to take such action. If he promises to repair certain defects and fails to do so, the tenant has an obvious contract action against him. If, on the other hand, the landlord actually attempts repair, whether or not acting pursuant to a binding promise, and carries through in such a negligent manner that the tenant or someone else on the premises in the right of the tenant is injured, the landlord's liability in tort is clear.[35] When the lessor enters the lessee's domain to correct defects or make improvements, his contractual relationship with his tenant is not a factor, and he is treated as would be any outsider there to accomplish a particular job. That is to say, his conduct will be tested by the usual standard of reasonable care. Where the landlord would otherwise be liable for his own negligence, most courts will not permit him to absolve himself from responsibility merely by inserting an exculpatory clause in the lease agreement.

Where property is rented for a purpose known to the landlord to involve admission of the general public, a member of that public may recover against such landlord for injury traceable to defects in the property existing at the outset of the lease arrangement, whether or not the landlord was at the time actually aware of the unsafe condition. Somewhat similarly, should the lease negotiations alert the lessor that the tenant contemplated utilizing the property for obviously dangerous activities, the lessor may well be held responsible to persons in the vicinity of the premises for harm done them as the result of such activities.

19-25   Liability of Vendor of Goods   In the absence of an appropriate exclusion or disclaimer, sales by tradesmen are generally subject to implied warranties to the effect that the goods involved are merchantable and fit for the purpose which the buyer is known to have in mind.[36]

Where the items for sale are obviously of a dangerous nature, the supplier has an exposure which may assume considerable proportions. One who manufactures, supplies, sells, or repairs a chattel which is either inherently dangerous[37] or will become dangerous if defectively assembled or repaired[38] owes a duty of care not only to the immediate purchaser but also to the third parties who might reasonably be expected ultimately to utilize the article in question. Certainly the bellwether case in point is *MacPherson v. Buick Motor Co.*[39] The facts were these:

---

[35] Some courts draw a distinction where the repairs are gratuitous and say that the landlord's liability for negligence extends only to the tenant and the members of his family in permanent residence with him. Another approach where the repair undertaking is not pursuant to contractual obligation is to hold the defendant landlord liable only for *gross* negligence.

[36] *The Uniform Commercial Code* covers details of these and other facets of the sale of goods.

[37] For example, poisons, explosives, firearms.

[38] For example, an automobile.

[39] 217 N.Y. 382 (1916).

defendant, a manufacturer of automobiles, sold one of its products to a retail dealer, and the latter resold to plaintiff. While plaintiff was in the car, it suddenly collapsed, and he was thrown out and injured. One of the wheels had been made of defective wood, and its spokes crumbled into fragments. The wheel had not been made by defendent but was supplied it by another manufacturer. The evidence showed that the wheel's defects could have been ascertained by reasonable inspection and that such an inspection was omitted by the defendant. In affirming judgment for plaintiff, the New York Court of Appeals included in its opinion the following:

> If the nature of a thing is such that it is reasonably certain to place life and limb in peril when negligently made, it is then a thing of danger. Its nature gives warning of the consequences to be expected. If to the element of danger there is added knowledge that the thing will be used by persons other than the purchaser and used without new tests, then, irrespective of contract, the manufacturer of this thing of danger is under a duty to make it carefully. That is as far as we are required to go for the decision of this case. There must be knowledge of a danger, not merely possible, but probable. It is *possible* to use almost anything in a way that will make it dangerous if defective. . .
>
> Beyond all question, the nature of an automobile gives warning of probable danger if its construction is defective. This automobile was designed to go fifty miles an hour. Unless its wheels were sound and strong, injury was almost certain. It was as much a thing of danger as a defective engine for a railroad. The defendant knew the danger. It knew also that the car would be used by persons other than the buyer. This was apparent also from the fact that the buyer was a dealer in cars, who bought to resell. . .
>
> We think the defendant was not absolved from a duty of inspection because it bought the wheels from a reputable manufacturer. It was not merely a dealer in automobiles. It was a manufacturer of automobiles. It was responsible for the finished product. It was not at liberty to put the finished product on the market without subjecting the component parts to ordinary and simple tests.

The decision in *MacPherson* did not extend the liability of the seller beyond the purchaser of the product. At present, the manufacturer's liability exposure extends to other users and consumers of the product and to those foreseeably within the vicinity of the product's use.[40]

19-26 Guest Statutes Many states have laws which serve to place restrictions upon a car owner's liability to guests injured while riding gratis in his vehicle. Such guests must, as a prerequisite to recovery, prove willful misconduct or gross negligence attributable to the defendant. Typical of these "guest statutes" is that of Indiana.[41]

---

[40] The Second Restatement of Torts, Section 395, states that "a manufacturer who fails to exercise reasonable care in the manufacturer of a chattel which, unless carefully made, he should recognize as involving an unreasonable risk of causing physical harm to those who use it for a purpose for which the manufacturer should expect it to be used and those whom he should expect to be endangered by its probable use, is subject to liability for physical harm caused to them by its lawful use in a manner and for a purpose for which it is supplied."

[41] Indiana Code §9-3-3-1 (Burns, 1973).

"The owner, operator, or person responsible for the operation of a motor vehicle shall not be liable for loss or damage arising from injuries to or death of a guest while being transported without payment therefor, in or by such motor vehicle resulting from the operation thereof, unless such injuries or death are caused by the wanton or willful misconduct of such operator, owner, or person responsible for the operation of such motor vehicle."

The term "guest" as employed in the foregoing text, has been construed to mean one who accepts the invitation of the driver and takes a ride in the automobile either for his own pleasure or on his own business without conferring any benefit upon the driver other than the mere pleasure of companionship.

At least one state supreme court has held that its guest statute is unconstitutional as denying equal protection and due process of law and as impermissibly closing courts and denying remedy by due process of law to some but to all of the people of the state.[42]

Where the guest rider is injured in a collision brought about by the negligence of his driver and the negligence of a third party, most jurisdictions say the former's contributory negligence will not bar the guest's action against the third party.

19-27   Family-Purpose Doctrine   This is another principle of tort law which the automotive age brought into being. Many courts hold the car owner responsible for damage done through the negligent driving of members of his family. The underlying theory is that the owner, by permitting his immediate relatives to operate the car for their pleasure and use, has made that purpose his business and thus rendered the driver his "servant."[43] The final step, then, is to say that the "master" is vicariously liable for the torts of his "servant."[44] Quite obviously, the family-purpose principle is another instance of getting at the so-called "deep pocket" of the party most likely to be able to pay for the harm. A collateral effect of the owner's vulnerability under the family-purpose doctrine is that considerable care is likely to be exercised in selecting family members permitted to drive.

Various states have enacted statutes which seek to make certain that there will be a financially responsible defendant available in motor-vehicle cases. One type of provision, for example, requires an adult's countersignature on the license application for a driver under eighteen, such endorser to be answerable for injuries occasioned by negligence of the juvenile operator.

19-28   Nuisance   *Nuisance* is a condition, usually longstanding and of indefinite anticipated duration,[45] whose existence is directly injurious to others. A nui-

---

[42] *Primes v. Tyler,* 331 N.E.2d 723 (Ohio 1975). The constitutionality of the Indiana Guest Statute was upheld in *Sidle v. Majors,* 536 F.2d 1156 (1976).

[43] As long as general use of the vehicle is afforded by the family member, the mere fact that instructions by the owner about speed, distance, etc., were ignored (and that adherence to such instructions would have avoided the mishap) will not serve to relieve the owner of liability.

[44] Technically, of course, the servant (the car operator) is likewise vulnerable but in many instances is found to be uninsured and practically insolvent. Where such is not the case, plaintiff will probably proceed against the driver and will not need to raise the family-purpose argument.

[45] The old insistence on a showing of continuity as a condition precedent to recovery for nuisance is, however, being wittled away.

sance may or may not be lawful, and this is true whether it is of the *private* or of the *public* variety.[46] The term "nuisance" does not have reference to any particular type of conduct which results in the invasion of plaintiff's interests. The nuisance may stem from defendant's negligence, from his employment of a dangerous instrumentality, or from conduct by which he fully intends to do harm.

A plaintiff will frame his complaint on the basis of nuisance primarily when he anticipates some barrier[47] to a recovery on a standard negligence theory. Certain defenses commonly employed in negligence cases are not available when absolute nuisance is the gravamen of the complaint.

Private Nuisance   This type grew up as an adjustment of differences between adjacent landowners, permitting the aggrieved party recovery against the other for unreasonable and substantial interference with plaintiff's enjoyment of his own premises. A private nuisance is a wrong arising from the use of one's own property in such a manner as to do violence to the rights or interests of another.[48] Technically, the harm need be nothing more than discomfort, annoyance, or inconvenience in order to be actionable.

Apart from taking the law into his own hands, the party adversely affected has open to him an action at law for damages or in equity for an injunction.

Public Nuisance   Where the objectionable condition is such that the public at large is adversely affected, the nuisance is characterized as a *public* one. Obstructing a common highway and thus impeding traffic would constitute a public nuisance. Unless some person suffers a special injury distinguishable from harm done the general public, he as an individual has no damages recovery available against the perpetrator of the public nuisance.

19-29   Emotional Disturbances[49]   The case law on the issue of whether or not damages are recoverable for emotional disturbances, such as fright, with or without accompanying physical distress, and with or without a showing of an "impact" chargeable to the defendant, has undergone substantial evolution. Many jurisdictions will not allow recovery for emotional disturbance or accompanying physical distress in the absence of actual impact. Courts have shown more reluc-

---

[46] It is possible for a given condition to constitute at the same time a private and public nuisance.

[47] For example, a plea of contributory negligence or of governmental immunity.

[48] Frequently, maintenance of a nuisance on *A*'s property will materially reduce the value of *B*'s adjoining acreage.

[49] It should be noted at the outset that recovery for emotional disturbance is not normally available in the usual breach of contract situation. See *Fiore v. Sears, Roebuck and Company,* 144 N.J. Superior, 74, 364 A.2d 572 (1976), where the plaintiff alleged that he had undergone extreme mental anguish as a result of the defendant's failure to conform to its guarantee with respect to the installation of a roof and the multiple efforts made by him to have the defect corrected, and the manner in which the claim had been handled and the harassment to which he had been subjected. The court said that as a matter of basic justice when one can at the time of contracting reasonably foresee mental harm resulting from breach of contract, defendant ought to be liable in damages if his actions are either willful or reckless, but held that what was alleged in this case was nothing more or less than the normal breach-of-contract situation. The court added, however, that the allegation regarding the manner in which the claim had been handled involved a claim of tortious conduct for which an award for emotional disturbance might be proper, if the other conditions prerequisite to such liability were present.

tance to allow recovery for emotional distress not accompanied by any physical distress.

A recent case from Virginia is characteristic of the existing state of the law. There, the plaintiff was standing in the doorway of her house when an automobile crashed through the front porch of the house, causing her severe fright and agitation but resulting in no contemporaneous physical contact. The evidence indicated that the plaintiff suffered from severe nervousness, chest and arm pains, and menstrual irregularities as a result of the shock of the collision. Addressing the question of physical impact, the court said:

> Many eminent scholars have considered the rule and are virtually unanimous in condemning it as unjust and contrary to experience and logic.
> Our research reveals that a total of 35 jurisdictions have considered the rule. Of these at least 25 have either completely rejected it or abandoned it as being unsound. Since 1929 every jurisdiction which has considered the issue, except the Supreme Court of Washington. . .has either abandoned the rule or refused to adopt it.
> A rapidly increasing majority of courts have repudiated or not followed the "impact rule" for the reasons that the early difficulty in tracing a resulting injury back through fright had been minimized by the advance of medical science.. . .
> We adhere to the view that where conduct is merely negligent, not willful, wanton, or vindictive, and physical impact is lacking, there can be no recovery for emotional disturbance alone. We hold, however, that where the claim is for emotional disturbance *and* physical injury resulting therefrom, there may be recovery for negligent conduct, notwithstanding the lack of physical impact, provided the injured party properly pleads and proves by clear and convincing evidence that his physical injury was the natural result of fright or shock proximately caused by the defendant's negligence. In other words, there may be recovery in such a case if, but only if, there is shown a clear and unbroken chain of causal connection between the negligent act, the emotional disturbance, and the physical injury.[50]

Courts have been generally reluctant,however, to take the final step and remove the requirements that physical injury accompany the emotional injury. But the Supreme Judicial Court of Massachusetts, in the case of *Agis v. Howard Johnson Company*,[51] has moved in that direction.

The plaintiff in the case was a waitress in a local Howard Johnson restaurant. The manager of the restaurant informed all of the employees that there "was some stealing going on," but that, since the identity of the person responsible was not known he would, until the person responsible was discovered, begin firing all the present waitresses in alphabetical order. The plaintiff was the first to be discharged. The plaintiff did not allege any bodily injury, but became greatly upset, began to cry, and allegedly sustained emotional distress, mental anguish, and loss of wages and earnings.

---

[50] *Hughes v. Moore*, 214 Va. 27, 197 S.E.2d 214 (1973). See also *Culbert v. Samsons Super Markets, Inc.*, 444 A.2d 433 (Me. 1982), where the court allowed a suit for emotional distress without a requirement of impact and without the plaintiff being in the zone of danger.

[51] 355 N.E.2d 315 (Mass. 1976).

Upholding the right to recovery, the court said:

The most often cited argument for refusing to extend the cause of action for inten-tional or reckless infliction of emotional distress to cases where there has been no physical injury is the difficulty of proof and the danger of fraudulent or frivolous claims. There has been a concern that "mental anguish, standing alone, is too subtle and specu-lative to be measured by any known legal standard," that "mental anguish and its consequences are so intangible and peculiar and vary so much with the individual that they cannot reasonably be anticipated," that a wide door might "be opened not only to fictitious claims but to litigation over trivialities and mere bad manners as well," and that there can be no objective measurement of the extent or the existence of emotional distress.

While we are not unconcerned with these problems, we believe that "the problems presented are not. . .insuperable" and that "administrative difficulties do not justify the denial of relief for serious invasions of mental and emotional tranquility. . . ."

In light of what we have said, we hold that one who, by extreme and outrageous conduct and without privilege, causes severe emotional distress to another is subject to liability for such emotional distress even though no bodily harm may result. However, in order for a plaintiff to prevail in a case for liability under this tort, four elements must be established. It must be shown (1) that the actor intended to inflict emotional distress or that he knew or should have known that emotional distress was the likely result of his conduct, (2) that the conduct was "extreme and outrageous" and was "utterly intoler-able in a civilized community," (3) that the actions of the defendant were the cause of the plaintiff's distress, and (4) that the emotional distress sustained by the plaintiff was "severe" and of a nature "that no reasonable man could be expected to endure it."

In a recent emotional distress case from California, a doctor negligently diag-nosed the plaintiff's wife as a syphilitic and instructed her to advise her husband of the diagnosis, causing him serious emotional distress. Finding foreseeability on the part of the doctor of the reaction on the husband to be critical, the court held that the husband had stated a cause of action for negligent infliction of emotional distress even though the doctor had never spoken to the husband.[52]

19-30 Survival Statutes and Wrongful-Death Acts At common law no civil action would lie against the tort-feasor for an injury resulting in the death of the victim. The injured party, while he lived, had available a suit for damages, but the right of action abated at his demise and could not be maintained by his personal representatives. The anomaly of the common-law situation was that a fatal injury left the victim's estate and beneficiaries without recourse, while lesser hurts could be mended with damage money recoverable. Even in the latter circumstance, however, plaintiff would be thwarted by death of the wrongdoer prior to entry of judgment against him.

The common-law picture has been radically altered by statutory provisions in most jurisdictions, though the applicable laws vary widely. What is probably the prevailing pattern is to have a "survival-of-action" provision and a so-called "wrongful-death statute" complementing each other so as to afford to the parties damaged a thoroughgoing remedy for the entire loss sustained through defen-

---

[52] *Molien v. Kaiser Foundation Hospitals*, 616 P.2d 813 (Cal. 1980).

dant's malfeasance. Defenses such as contributory negligence, which would have been at hand for use against the deceased victim had he lived, generally continue to be available after he has left the scene.[53]

The survival statutes typically provide that a right of action for negligence belonging to the injured person does not terminate at his death, but an action to recover damages may be instituted or taken over by his estate.[54] Such an action would seek recompense for pain and suffering endured by the deceased for expenses incurred and for loss of earnings up to the time of death.

Unlike a true survival act, the wrongful-death statute[55] creates a new cause of action and does not simply transfer to his successor in interest a right to sue possessed by the injured person before his demise.[56] In some states the new cause of action goes to the administrator of the deceased, while in others specific relatives are named as beneficiaries. In either case, recovery is limited to the monetary loss sustained by the parties for whose benefit the action is brought, that is, the loss suffered through being deprived of what they figured to receive of the victim's earnings from the date of his death on through the rest of what would have been his anticipated lifespan. In other words, the wrongful-death action is essentially a suit for injury to the property rights of certain beneficiaries favored by the statute, and the measure of damages is the pecuniary value of decedent's life to his next of kin. Under the majority rule, wrongful-death actions do not survive the tortfeasor.

Survival acts and wrongful-death statutes perform separate functions in providing recourse for the varied damage done. The United States Supreme Court, in discussing the role of these complementary statutes, has put it this way.[57]

> Although originating in the same wrongful act or neglect, the two claims are quite distinct, no part of either being embraced in the other. One is for the wrong to the injured person, and is confined to his personal loss and suffering before he died, while the other is for the wrong to the beneficiaries, and is confined to their pecuniary loss through his death. One begins where the other ends, and a recovery upon both in the same action is not a double recovery for a single wrong, but a single recovery for a double wrong.

19-31   Intrafamily Torts   Briefly stated, parent and unemancipated child can sue each other for torts affecting property interests but not, as a general rule, for personal wrongs, intentional or otherwise. A parallel situation pertains where the

---

[53] See, by way of example, New York Estates, Powers and Trusts Law, §5-4.2 (McKinney 1967).

[54] See, for example, *id.* §11-3.2. Incidentally, the same statutory provision deals also with the other side of the story and stipulates that no cause of action for injury to person or property shall be lost because of the death of the person liable for the injury.

[55] Lord Campbell's Act (otherwise known as the Fatal Accidents Act), 9 and 10 Vict. c. 93 (1846), was the forerunner of the modern-day wrongful-death acts.

[56] Thus, New York Estates, Powers and Trusts Law §5-4.1 (McKinney 1967) provides: "The personal representative. . .may maintain an action to recover damages for a wrongful act, neglect or default which caused the decedent's death against a person who would have been liable to the decedent by reason of such wrongful conduct if death had not ensued. . . ."

[57] *St. Louis, I.M. & Southern R. Co. v. Craft,* 237 U.S. 648, 658 (1914).

husband-wife relationship is involved. Before the advent of what are commonly termed "married women's acts,"[58] not even property torts were actionable between spouses. The theory seems to have been that the ideal of domestic bliss could be advanced by preventing the marital partners from bringing each other into court on tort claims.

This doctrine known as interspousal immunity is dying a slow death. The modern approach is well illustrated in a case in which a wife slipped and fell on a snow-covered driveway where the husband was in control of the premises and responsible for sanding, salting, or shoveling after a snow storm. The court termed the reasons that had supported the common-law rule of interspousal immunity "antediluvian assumptions" and negated the doctrine in the circumstances.[59]

TORTS WHICH ARE USUALLY INTENTIONAL

19-32 Introduction One who purposely commits a tortious act is answerable for any and all damage flowing from the wrong. There are, of course, a great many different types of torts normally intentional in nature.

19-33 Fraudulent Misrepresentation To prevail when the allegation is one of fraud, plaintiff must prove (1) that defendant made a false, material statement of fact, (2) the defendant either showed a reckless disregard for the truth or knew that what he advances as true was actually false,[60] (3) that defendant intended to induce reliance by plaintiff, (4) that plaintiff did with justification rely upon the accuracy of the statement, and (5) that plaintiff was damaged in consequence of such reliance. Note particularly (4) in the above list; a seller, for example, is entitled to "puff" his wares to a degree, and the buyer cannot safely take all statements at their face value.[61]

19-34 Defamation *In general,* a statement is defamatory if its natural tendency is to subject the victim to public ridicule, contempt, hatred, or the like. The court decides whether an odious interpretation could reasonably be drawn from the questioned statement, and the jury's function is to determine whether the statement was actually undertook in the harmful way alleged by the complainant. The defamatory meaning of the wording used may be obvious or may arise only in the light of the surrounding circumstances, which the plaintiff would have to show.[62] It must be proved that the alleged wrongful remarks were communicated

---

[58] These statutes gave the wife control over her separate property, rendered her individually responsible for her own torts, and accomplished other similar reforms.

[59] *Brown v. Brown,* 409 N.E.2d 717 (Mass. 1980).

[60] This prerequisite is called *scienter.*

[61] The doctrine of *caveat emptor* ("let the buyer beware") in mercantile transactions has the earmarks of a contributory-negligence theory.

[62] Thus, a statement that twins were born to Mr. and Mrs. Jones appears innocent enough, even if false, until it is shown that the declarant knew the Joneses were very recently married.

by defendant[63] to some third person[64] and that the statement in question clearly related to plaintiff and not merely to some large indeterminate group of which he happened to be a part.[65] Under the prevailing common-law policy, truth[66] and privilege are complete defenses in defamation suits.

Slander and Libel   Defamation is traditionally divided into slander and libel. The yardstick which measures statements for traces of slander or libel varies with the times, with the geographical location, and with other, less definite factors. *Slander* is oral defamation published without legal excuse, whereas *libel* takes the form of written statements, pictures, images, etc.[67] All true libel is actionable per se; that is, the law will infer that third persons have read the objectionable remarks and that damage has been done to plaintiff's reputation. In most slander cases, on the other hand, plaintiff must make a showing of actual damage. In many jurisdictions slander is actionable per se in the following four instances:

1 When defendant has imputed to plaintiff the commission of an indictable crime or of an act involving moral turpitude
2 When defendant has asserted that plaintiff is suffering from some loathsome disease
3 When defendant has imputed unchastity to a female plaintiff
4 When defendant has made an assertion detrimental to plaintiff's business; for example, by claiming that he is unable to perform his job properly

Measure of Damages   Recovery has been limited to those damages reasonably foreseeable by the person defaming and has not been extended to cover further injury resulting to plaintiff from totally unexpected repetition of the statements arousing adverse feelings toward him. If very offensive words are spoken in a public place before a large gathering of people, punitive damages may be recovrable.

Liability Varies   Much depends upon the status of the individual defending the defamation suit. For instance, the publisher of a newspaper is strictly liable if defamatory material appears in her paper, since she presumably has control over the printed contents. The corner news vendor is responsible only if he should have known the defamatory character of the publication he sells.

Defenses   Apart from *truth, privilege* is the key to successful defamation defense. Certain positions carry with them an absolute privilege to defame, without regard for motive or for reasonableness of conduct. The judge in his courtroom (or

---

[63] Publication by plaintiff himself will not suffice.

[64] The outsider may be plaintiff's wife, but communication by defendant merely to his own wife is not the requisite "publication." Some courts say that an "appreciable minority" of the people in a community must have their opinions of plaintiff lowered before he has the basis for a successful suit.

[65] The test is whether a fair percentage of the people who know plaintiff believe the material refers to him.

[66] When truth is the defense, it matters not for what evil motive declarant made the statement sued upon.

[67] Hence, improper remarks about *X* dictated to one's secretary constitute slander. When she transcribes the dictation, libel has entered the scene.

the witness on the stand) epitomizes this category. Legislative proceedings,[68] communications between husband and wife, and publications made with the consent of plaintiff are other situations in which *absolute privilege* applies.

Qualified privilege likewise affords an adequate defamation defense. This type is conditioned upon defendant's reasonable behavior and lack of malice. Newspaper columnists, music critics, radio commentators and various others enjoy a qualified privilege protecting their reports, comments, and criticisms. A plea of qualified privilege will not avail defendant if plaintiff can show that the privilege was abused, for example by excessive publication.

The Supreme Court decision in *New York Times Company v. Sullivan*[69] and related subsequent opinions preclude recovery for defamation by public officials in the absence of actual malice. The "public official" concept has been extended to include a wide variety of public figures other than officials and a wide variety of well-known individuals have been denied recovery in defamation cases because of an absence of a showing of express malice.

Mitigating Damages   What might be called *partial defenses* in mitigation of damages include such points as (1) the belief of defendant in the essential truth of his statements, (2) the bad reputation of plaintiff, and (3) prompt retraction by defendant of the defamatory remarks.

19-35   Right of Privacy   A person's life history, name, and likeness constitute the physical indicia of his individual existence; these things are in a sense, "property rights" and will be protected against unprivileged invasion, particularly when the violator is actuated by commercial considerations. One is entitled to a certain amount of seclusion, far from the glare of unwanted publicity. Thus, the seeds of numerous invasion-of-privacy suits are sown when pictures of individuals, particularly in embarrassing situations, are taken and published without their consent. In suing for invasion of privacy, plaintiff need show no special damage of a pecuniary nature, recovery does not rest on defamation, and truth is no defense.

If the individual whose privacy has been violated is or was formerly a public figure, his chances of recovering damages through suit are slim. People in public life are deemed to have waived their rights of privacy,[70] and anything of news value about them becomes a matter of legitimate interest to the man in the street and thus a proper subject for publication. For example, a government official's likeness is considered common property, and his consent to its general use presumed.

19-36   Malicious Prosecution   By means of an action for malicious prosecution, a person may recover for harm done his reputation, business, etc., as the

---

[68] Verbatim reporting by a magazine of a senator's defamatory speech is as privileged as the address itself.

[69] 376 U.S. 254 (1964).

[70] Similarly, a private citizen may participate unwillingly in some event of such public concern that he cannot complain of his involvement in the attendant publicity.

result of an unsuccessful criminal proceeding brought against him maliciously and without probable cause.[71] This type of suit is not favored in the law, and plaintiff faces a heavy burden of proof. The essential things he must show are:

1 The party he now sues was responsible for the bringing of the criminal charges.

2 The criminal proceedings terminated in the tort plaintiff's favor.[72]

3 There was an absence of "probable cause" in connection with the criminal charges.

4 The tort defendant was guilty of malice[73] or was spurred by a primary motive other than that of bringing an offender to justice.

Probable cause is always an effective defense in a malicious-prosecution action and is usually evidenced by showing that a grand jury indictment was returned or that the magistrate bound over the criminal defendant at a preliminary hearing. Advice of counsel, acted upon in good faith, is likewise a defense to an allegation of malicious prosecution.

19-37 False Arrest and Imprisonment Unless an unjustified arrest is followed by some sort of judicial proceedings, the victim cannot bring an action for malicious prosecution and will be restricted to a suit alleging false imprisonment.[74] An arrest usually involves confinement, and a person who brings about the (undeserved) detention of another through the auspices of a third party (policeman[75]) is as fully responsible as though he had physically taken matters into his own hands.[76] When an arrest is made without a warrant, the moving party has gambled that a crime has indeed been committed and that an arrest is therefore proper; and even when a crime undeniably has occurred, such party must be in a position to demonstrate that he had reasonable grounds for believing the arrested individual was responsible.[77] The rules are far different when the moving party swears out a complaint and has a valid warrant issued. Under such circumstances the element of probable cause has been passed upon by responsible authority, and the complainant is consequently insulated from any possible false-arrest suit (though, if he acted maliciously, there might eventually prove to be a basis for a malicious-prosecution action).

---

[71] There is a companion tort commonly called *vexatious suit* and based on the wrongful institution of civil proceedings.

[72] An acquittal is not necessary; abandonment of the prosecution will suffice. The theory underlying requirement (2) above is that disposition of the prosecution in a manner favorable to the tort plaintiff tends to indicate freedom from guilt. This factor, plus a showing that the accuser acted maliciously and without probable cause, establishes the tort, that is, the malicious and unfounded charge of crime against an innocent person.

[73] Lack of probable cause may imply malice.

[74] Technically, the tort itself if "false imprisonment," and one way of committing the tort if through "false arrest."

[75] The policeman's own liability will depend upon the reasonableness of his action under the circumstances.

[76] Incidentally, the right to make an arrest in a given situation carries with it the privilege of using *reasonable* force.

[77] Honesty of belief is alone insufficient, as are lofty motives.

In the broad sense of the term, *false imprisonment* may be defined as (1) any confinement of another within boundaries fixed by the tort-feasor (2) for any measurable time whatever and (3) irrespective of whether actual harm is caused, if (4) the action taken is intended to imprison the other person, (5) if he is aware of being confined, and (6) if the imprisonment is neither consented to nor otherwise privileged. The unlawful restraint must be total; that is to say, defendant must have closed to his victim every reasonable means of escape. On the question of whether the restraint involved is sufficient to support a false-imprisonment cause of action, much depends upon the physical circumstances. Attempted confinement of a male athlete, aged twenty, in a first-floor room with an open window in view is far different from imprisonment in the same quarters of an elderly woman of considerable girth. In general, if there is an obvious means of egress, free of danger, there may be inconvenience involved for the victim but there is no true imprisonment.

19-38   Interference with Business Relations   As a general proposition, one who induces X to breach a contract with Y is liable in tort to the latter, since the law imposes upon strangers to a contractual arrangement the duty to refrain from willful interference with its performance.[78] A person is privileged to invade an existing contract interest when he acts in protection of an equal or superior interest, and there is a somewhat broader privilege when the interference is indirect and incidental. Some types of contracts that are "terminable at will" may be invaded (under the guise of competition) with relative safety.

Prospective as well as existing contracts may be intentionally interfered with in such way as to give the injured party a right of action. Deliberate interference with a person's privilege to enter into advantageous economic relations is tortious, and plaintiff may recover probable damages unless defendant can show justification.[79]

19-39   Slander of Title   When plaintiff brings an action for disparagement of his goods or property or an action for slander of title, he must show the following:

1 That the defendant communicated to third persons statements, purporting to be factual, vilifying plaintiff's goods or casting a cloud of doubt on his title.
2 That the statements were untrue.
3 That in consequence of defendant's conduct plaintiff suffered actual damage, perhaps because others, believing defendant's allegations, refused to deal further with plaintiff. The damage must be real and not just speculative or potential.

There is some conflict of authority about whether defendant must also have acted

---

[78] Put another way, purposeful inducement to breach is a prima facie tort, actionable in the absence of privilege.
[79] Such as that he was merely engaging in bona fide competition for business.

maliciously[80] and with knowledge that the statements he made were false. If defendant's words pertain to plaintiff's goods but necessarily imply his personal dishonesty,[81] plaintiff's best vehicle for recovery is perhaps an action for defamation.

19-40 Assault; Battery *Assault* is a wrong to the person of the victim and may be defined as an intentional[82] act which creates a reasonable apprehension that offensive bodily touching or injury is at hand. Mere words do not alone constitute an assault; there must be some physical maneuver by the tort defendant, coupled with the apparent present ability to carry out the immediate threat posed. The phrase "apparent present ability" poses a subjective test in that the important consideration is whether the plaintiff actually feels the defendant can and will follow through. The tort of assault has been committed when defendant threatens plaintiff with a revolver which the latter thinks is loaded.[83] Such a wrong as this occasionally results in real physical damage, as where the frightened plaintiff is a pregnant woman, or an elderly person with heart trouble. A threat of *future* violence is not actionable as an assault. A warning by defendant that he will shoot plaintiff the following week falls into this category.

Battery is the intentional and offensive touching of another, even to the slightest extent. In most instances, of course, assault and battery go together, though there are a fair number of civil suits involving assault alone.

A showing of provocation will mitigate damages in an assault and battery action. In addition, there are several typical defenses to such a suit:

1 Self-defense or defense of a third person. The degree of force used must be reasonable in the particular circumstances.

2 Defense of one's property. Again, the means employed must be reasonable.

3 Enforcement of discipline, as by a person *in loco parentis* (acting in the place of a parent), a schoolteacher, for example, subject to statutory limitations.

19-41 Conversion This tort is committed through the unauthorized assumption and exercise of ownership rights with respect to the property of another. Conversion takes various forms, such as stealing, intentionally destroying or altering property, using property without authority, buying from a known thief or a "fence," selling goods belonging to someone else, willfully refusing to surrender a person's property upon demand, and misdelivering goods (as by a bailee).

The defendant in these cases is normally answerable for damages measured by the market value of the chattel at the time and place of its conversion, together with interest on this sum up to the time of trial. Where the property involved is something like used household furniture or books, recovery may be limited to the original cost less depreciation. On the other hand, when the misappropriated article is one of highly fluctuating value, such as a stock certificate, plaintiff may

---

[80] Most courts seem to require some showing of malice or improper motive.
[81] For example, an allegation that plaintiff butcher knowingly sold diseased meat.
[82] The requisite intent may be inferred from the nature of the act.
[83] It matters not that the gun actually was without cartridges.

be able to recoup a sum equivalent to the highest market price reached by the converted property in a reasonable period after discovery of the tort.

When someone has wrongfully made off with his goods and plaintiff sues in conversion, he is seeking monetary recompense for the wrong. He may, though, elect to pursue a cause of action in *replevin,* which means he wants the actual chattel itself returned rather than its cash equivalent.

19-42 Trespass Anyone who interferes with possessory rights in property or enters upon land without permission or privilege commits a trespass and is potentially liable for at least nominal damages even though the motive may have been beyond reproach. There are, of course, various instances of privileged invasion. Thus, the innocent owner of escaped goods has the right to go upon the land of another to reclaim them, though he is responsible for any removal damage done and for possible injury caused by leaving the goods on the land of such other person for an unwarranted period of time.

LIABILITY WITHOUT FAULT

19-43 Respondent Superior[84] Under the doctrine of *respondeat superior,*[85] one person is held vicariously liable under certain conditions for the tortious conduct of another. The principle is applicable only when the relationship between the person sought to be charged for the damage done and the actual wrongdoer is shown to be that of (1) master and servant[86] or (2) principal and agent, and only to the extent that the tort-feasor on the occasion of the wrongful deed was acting within the scope of his employment or agency. While in fact accomplishing his ends through remote control, the master or principal is, in the eyes of the law, himself acting. The idea underlying vicarious liability is, of course, to permit the injured third party access to the "deep pocket."[87] Technically, both master *and* servant are liable to *X* for the torts of the servant, and the latter, in turn, is answerable to his master.

Persons engaged in a true *joint enterprise*[88] are, without any wrongful conduct of their own, each responsible to injured outsiders for the fault of other participants in the undertaking as long as the tortious behavior occurred within the purview of the joint enterprise. To constitute such an enterprise, wherein the negligence of one participant is imputable to the others, there must be a community of interest in the purpose of the undertaking and an equal right for each affiliated individual to govern the conduct therof.

---

[84] For applications of this doctrine see Chap. 16, Agency, and Chap. 17, Partnerships.
[85] Let the principal answer.
[86] That is, employer and employee, etc.
[87] Respondeat superior is normally inappropriate with respect to minors; they cannot be held for the wrongs of their servants.
[88] This term is essentially synonymous with *joint adventure* or *common enterprise* in legal contemplation.

19-44 Nondelegable Duties As a rule, when *A* (perhaps a building owner or a prime contractor) hires *B* to perform certain tasks in the capacity of an *independent contractor,*[89] *A* cannot be held for *B*'s torts unless there is proof of clear negligence in the selection of *B* or a showing that the work in prospect was of such a nature as to be inherently dangerous to others.

Certain tasks and duties are of such import or nature that they cannot be delegated to the extent necessary to relieve *A* of his potential tort responsibility. Thus, the obligation of a property owner to keep his premises safe for invitees cannot be transferred to an independent contractor employed to effect repairs; under normal circumstances, negligence of the latter will render the owner, though personally blameless, liable to injured business visitors.

19-45 Worker's Compensation Acts The worker's compensation acts[90] and similar laws permit recovery for personal injuries and certain diseases arising out of and in the course of employment. Several occupational groups, such as domestic and agricultural workers, are presently excluded from coverage under most worker's compensation acts.

The basic test of liability under the statutes is work connection rather than fault; if the requisite work connection is found, the employer pays[91] despite the absence of any wrongful conduct on his part. In return for the readily available compensation-act benefits, the employee and his dependents surrender (as regards any injury covered by the act) their common-law right to sue the employer for damages.

While worker's compensation legislation varies from one jurisdiction to the next, the typical pattern expressly excludes from coverage injuries which are self-inflicted or which stem from willful misconduct on the part of the employee claimant.

19-46 Extrahazardous Activity Broadly speaking, a person who engages in an activity which poses a real menace to others is held strictly accountable for the natural consequences. Such person is under strict or absolute liability for the mischief he causes, and the fact that he took all reasonable precautions will not relieve him of ultimate responsibility. He is, to all intents and purposes, regarded as an insurer of the safety of outsiders, and this is true even though the activity involved may be entirely lawful, proper, and even necessary.

Perhaps the example most pertinent for engineers[92] is afforded by the innumerable damage-from-blasting cases. Some decisions have drawn a distinction be-

---

[89] As the term "independent contractor" is used here, *A* (the owner) controls only the final result and not the means by which it is accomplished; *A* is not privileged to interfere with the details of performance.

[90] All states now have such acts.

[91] The employer can, of course, insure his burden of compensation liability.

[92] The application of the *strict liability* doctrine has been greatly extended in recent years, but not to engineers in general. See *Swett v. Gribaldo, Jones & Assoc.,* 40 Cal. App.3d 573, 115 Cal. Rptr. 99 (1974). See also, Chap. 20, Professional Liability.

tween (1) damage done by rocks or other substances propelled as the result of the dynamiting and (2) damage occasioned purely by the concussion, imposing liability without negligence in the first instance but not in the second. The trend today, though, appears to be toward repudiating any differentiation of the above sort and holding the defendant responsible in every such blasting case regardless of the fault element. An Oregon decision[93] is illustrative of what seems to be the general pattern. The following passages are of particular significance.

> This is a case of first impression in this court. It presents the question whether damages may be recovered for injury to real property caused by concussion or vibration from blasting operations where it is neither pleaded nor proven that the defendant was negligent in the conduct of such operations. . . . We are satisfied that. . .the rule of liability should be the same whether the blaster injures his neighbor's land by casting rocks, debris or other material on it or by setting in motion concussions or vibrations of the earth or air. In neither case is it necessary for a plaintiff to allege and prove negligence. . . . Our recent decision in *Brown v. Gessler,* 191 Or. 503, 512, 230 P.2d 541, 545, 23 A.L.R.2d 815, definitely places this court on the side of the courts of this country which have adopted as law the rule of *Rylands v. Fletcher.* Justice Blackburn in 1 Exch. 278 stated the rule as follows: "We think that the true rule of law is, that the person who for his own purposes brings on his lands and collects and keeps there anything likely to do mischief if it escapes, must keep it in at his peril, and, if he does not do so, is prima facie answerable for all the damage which is the natural consequence of its escape. He can excuse himself by showing that the escape was owing to the plaintiff's default; or perhaps that the escape was the consequence of vis major, or the act of God; but as nothing of this sort exists here, it is unnecessary to inquire what excuse would be sufficient."
>
> It requires very little extension of this doctrine to apply it to the facts of the present case. And there is slight difficulty in holding that one who engages in blasting operations which set forth in motion vibrations and concussions of the earth and air which reach to another's land, no matter how far distant, and shatter his dwelling, commits a trespass no less than one who accomplishes the same result by the propulsion of rocks or other material. . . .

19-47 Dangerous Instrumentality   One who harbors, maintains, or utilizes anything commonly regarded as a *dangerous instrumentality*[94] will be held to the very highest standard of care, amounting really to strict liability for whatever injurious consequences befall those within the circle of danger. If Smith keeps a wild animal, dangerous by nature, on her place, and there ensues injury to neighbor Jones, Smith's negligence is presumed, and she is under absolute liability. The same situation pertains if the animal is a domestic one but its owner knows that it has vicious propensities.[95] Ordinary contributory negligence of the harmed person, as in all instances of "strict liability," is no defense.[96]

---

[93] *Bedell et ux v. Goulter et al.,* 199 Ore. 344, 261 P.2d 842 (1953).

[94] For example, firearms or explosives.

[95] Some jurisdictions have done away with this requirement of scienter and have imposed absolute liability even where a dog owner, for instance, is not aware of the animal's shortcomings.

[96] However, there is a real question when plaintiff, knowing the dog was inclined to fits of anger, deliberately provoked him.

19-48 No-Fault Insurance A number of jurisdictions, dissatisfied with the application of the negligence system to automobile accident cases, have adopted plans which to some degree provide for reparation of auto accident losses without regard to fault.[97]

No-fault plans vary considerably in detail but generally have in common two features: (1) regardless of fault an injured insured motorist receives recompense from his own insurer; (2) tort liability is abolished at least to the extent of benefits received. A few programs eliminate or curtail the payment of benefits for pain and suffering but most retain conventional tort liability as the basis for recovery in cases of serious injury.

The constitutionality of no-fault schemes has been upheld.

QUESTIONS

1 Define "tort"; differentiate between tort and crime.
2 Distinguish contract liability from tort liability.
3 What elements are necessary to support a tort action?
4 Define and illustrate "negligence."
5 Define and illustrate "contributory negligence."
6 Define and illustrate "comparative negligence."
7 What is the meaning of (a) the "standard of care"? (b) the "normally prudent man"?
8 Define and illustrate "proximate cause."
9 Distinguish among "negligence," "gross negligence," "assumption of risk," and "last clear chance."
10 Explain and illustrate the doctrine of *res ipsa loquitur.*
11 What is malpractice? Illustrate.
12 What might be the advantage of arbitration in malpractice cases?
13 In general, are minors and mentally retarded persons legally responsible for their negligent actions? Elaborate.
14 What is the liability status of governmental officials in respect to their negligence? Of government itself? What is the trend of the law in this regard?
15 Name one circumstance which might result in tort liability for (a) a landowner, (b) a lessor of land, (c) a lessee of land.
16 Differentiate among "trespasser," "licensee," and "invitee." What degree of care does the landowner owe to each?
17 Has a vendor of goods any special tort liability? Explain.
18 What is (a) a "public nuisance"? (b) "private nuisance"?
19 What is the difference between "libel" and "slander"? What elements would support a suit for defamation? What defense is often presented? What if the plaintiff is a public official or well-known personality?
20 What is the tort of "conversion"?
21 Explain the doctrine of respondeat superior.
22 Can torts stem on occasion from inaction as well as from improper action?
23 Can there be several wrongdoers involved in a single tort? Can plaintiff bring action against one, or must he include all?

---

[97] See, for example, 39 NJSA 6A-1 et seq.; Annotated Laws of Massachusetts, Chap. 90, Section 34A et seq. Massachusetts was the first state to enact a no-fault scheme (in 1970).

24 In what circumstances may recovery be had for emotional distress?

25 What duties may there be which a property owner cannot with impunity delegate to an independent contractor who is doing some work for the former?

26 What is "worker's compensation"? How does the system operate?

27 Hart had a dog that was known to be vicious, and Hart was ordered to keep him on a leash. One day the dog broke away and bit the milk carrier, who had come to make her customary delivery. Can the latter expect to collect damages from Hart?

28 Varney operated one of two 5-ton cranes in a factory. Both cranes were to act simultaneously in lifting a 9-ton machine. The other operator was busy elsewhere, so Varney, who was in a hurry, tried to lift the machine with his crane alone. The cable broke after the load was 10 feet above the floor, and a worker who happened to be walking under the crane but who had nothing to do with the operation was injured. Has the worker a valid cause of action against Varney?

29 Perkins was about to make a contract with Green. Murray persuaded him not to do so, but to engage Black. Green brought action against Murray. What do you think of Green's prospects in court?

30 *X* was shingling the roof of *Y*'s house under a contract. *X* accidentally dropped his hammer. It rolled down the roof and injured a neighbor's child who was playing nearby. Is *X* legally responsible? How about *Y*?

31 Horton's manager, Garnett, was guilty of negligence in connection with certain behavior which endangered Horton's property. Fortunately, the trouble was detected before Horton suffered actual injury. Do you think that Horton could sue and expect to collect anything? Explain.

32 In the preceding case, assume that Garnett's negligent act caused a fire which damaged $5,000 worth of Horton's property, and that the latter brought suit. Upon whom does the burden of proof fall?

33 Jackson was driving a loaded truck, gross weight 15 tons. He was on his way from his sand pit in the country to the neighboring city. He started to cross an old bridge which was posed "10 tons maximum," and he had been over this structure many times with excess loads. However, this time the bridge collapsed. The town authorities sued Jackson for damages. What will be the probably result?

34 *A* rented a suburban house and lot from *B*. Having no garage, *A* customarily parked her automobile under a large tree (on *B*'s property) bordering the driveway. During a severe ice storm, a large branch broke off the tree and fell on *A*'s car, causing $300 worth of damages. *A* tried to collect from *B*. What do you think of *A*'s case?

35 Vines, a plumbing contractor, sold Wilson a 40-gallon hot-water tank and installed it in Wilson's basement. Four days later the tank burst, flooding the basement and ruining some goods which Wilson had stored there. Can Wilson collect for the cost of replacing the goods as well as of replacing the tank?

26 Murray had invited Carson to spend a week with her at at her cottage at Indian Lake. One morning, Carson proposed that they go fishing in Murray's canoe. Murray hooked a large fish, and, during the excitement which followed, the canoe capsized. Both women lost all their fishing tackle, much of it expensive. Carson claimed that Murray should compensate her. Has she a just claim? Why?

37 On Tuesday, farmer Jones bought three of Holley's cows, and delivery was made that afternoon. A week later, Jones discovered that one cow had a serious growth which later caused its death. Jones claimed that Holley had defrauded him. What would plaintiff Jones have to prove in order to substantiate his charge?

CHAPTER 20

PROFESSIONAL LIABILITY
OF ARCHITECTS
AND ENGINEERS

20-1 Scope of the Problem The works of architects or engineers affect countless people, some directly and others indirectly. They may function as designers, supervisors, employers, or as employees; the extent of liability exposure may depend upon their status. Where negligence has come into the picture and damage has ensued, courts have been confronted with the necessity of determining the precise nature and limits of the obligations assumed by architects and engineers.

To what standard of care and skill is an architect or engineer to be held? To whom is a duty owed and under what circumstances is that duty breached so that compensatory damages should be recoverable? May an architect or engineer be held liable without a showing of fault on his part? How long after a project is completed may an architect or engineer be held liable for damages? Such questions have proved to be a fruitful source of litigation over the years, and to this day it can hardly be said that the central issues are thoroughly resolved and concise rules of law established. Decisions continue to go off in diverse directions and often turn upon the peculiar facts of a case, but some general guidelines can be set forth with reasonable assurance. It should be kept in mind that many of the acts or omissions which may give rise to professional liability on the part of architect and engineer are classified as torts, and the rules set forth in Chap. 19 are applicable to those situations.

20-2 Definitions and Distinctions For purposes of this chapter the term "architect" will be treated as interchangeable with the term "engineer." Purists will be quick to point out that there are discernible differences in functions of the respective professions, but the fact remains that the architect on a job often doubles as his own engineer, and vice versa. Thus, where only one party serves the owner as plan-designer and supervisor of construction, it does not matter in a court of law whether such party be called the "architect" or the "engineer." The situation becomes more complex when both the architect and an engineer are on the scene, with separate areas of responsibility, or where one is employed by the other.

An architect, the practice of whose profession is the subject of various state laws, is one whose occupation entails the utilization of expertise in the formulation of designs, detailed plans, and specifications for use by a contractor in the erection or alteration of a building. The practice of architecture may likewise include, as an adjunct to the designing function, the supervision of construction under such plans from excavation to project completion.[1]

It should not be assumed, however, that because an architect is licensed in one state means that the laws of professional liability of that state govern his conduct in every instance. It is more likely that the law of the state where the incident occurred will be applied.

As a planner, the architect is generally an independent contractor but as an inspector, his role is more that of agent for the owner. In his supervisory capacity, the architect or engineer is supposed to guard against substandard workmanship and to prevent material deviations from the plans and specifications; in short, he assumes the duty (within limitations)[2] of seeing to it that the owner receives the building for which he bargained.

It is not at all unusual for the architect to employ engineers of one type or another. Their responsibilities may include the preparation of specialized plans,

---

[1] By way of comparison, one definition of a "professional engineer" is the following, appearing in §6701 of the California Business and Professions Code (West 1975): ". . . a person engaged in the professional practice of rendering service or creative work requiring education, training and experience in engineering sciences and the application of special knowledge of the mathematical, physical and engineering sciences in such professional or creative work as consultation, investigation, evaluation, planning or design of public or private utilities, structures, machines, processes, circuits, buildings, equipment or projects, and supervision of construction for the purpose of securing compliance with specifications and design for any such work."

[2] The American Institute of Architects in its 1967 form of owner-architect agreement has incorporated these provisions, clarifying areas of responsibility during the construction phase:

"The Architect shall make periodic visits to the site to familiarize himself generally with the progress and quality of the Work and to determine in general if the Work is proceeding in accordance with the Contract Documents. On the basis of his on-site observations as an Architect, he shall endeavor to guard the Owner against defects and deficiencies in the Work of the Contractor. The Architect shall not be required to make exhaustive or continuous on-site inspections to check the quality or quantity of the Work. The Architect shall not be responsible for construction means, methods, techniques, sequences or procedures, or for safety precautions and programs in connection with the Work, and he shall not be responsible for the Contractor's failure to carry out the Work in accordance with the Contract Documents. . . .

"The Architect shall not be responsible for the acts or omissions of the Contractor, or any Subcontractors, or any of the Contractor's or Subcontractors' agents or employees, or any other persons performing any of the Work."

destined to become part of the architect's composite scheme, and also supervision of certain portions of the work so that construction proceeds according to plans and specifications.

20-3   The Architect's Undertaking   The employment of an architect, and the setting of his fee (often expressed as a percentage of the cost of construction), are matters of contract, and it is that document which sets forth his rights, duties, and obligations. In a typical situation the architect might undertake to render the following professional services involved in the erection of a proposed building:

1 Participating in necessary conferences and preliminary studies
2 Interpreting physical restrictions as to the use of the land
3 Examining the site of the construction
4 Preparing and/or interpreting soil, subsoil, and hydrologic data
5 Preparing drawings or verifying and interpreting existing drawings of existing facilities or construction
6 Assisting in procuring of financing for the project
7 Assisting in presentation of a project before organizations possessing approval-disapproval power
8 Preparing drawings and specifications for architectural, structural, plumbing, heating, electrical, and other mechanical work
9 Assisting in the drafting of forms of proposals and contracts
10 Preparing cost estimates and controls
11 Obtaining bids from contractors
12 Letting contracts with owner's written approval
13 Inspecting the contractor's work on a regular basis, including the checking of shop drawings[3] (but without dictating the method or means by which the contractor seeks to accomplish the desired results)
14 Interpreting for the contractor the meaning of the drawings and specifications
15 Ordering the correction or removal of all work and materials not in strict conformity with specifications
16 Keeping accurate books and records
17 Preparing as-built drawings which show construction changes and final locations of mechanical and electric lines
18 Issuing certificates of payment

20-4   The Requisite Degree of Skill   In antiquity, the architect-engineer was a combination of designer and builder and was exposed to liability which was most severe. For example, the Code of Hammurabi required: "If a builder builds a house for a man and does not make its construction firm, and the house which he

---

[3] Be sure that the contract states that the checking of the shop drawings (and the approval of any samples of materials) by the design professional is only to see that they conform to the intent of the design. The contract drawings are to show the final desired result, whereas the shop drawings are to show how specific parts are to be fabricated and installed—matters for which the contractor should be responsible.

has built collapses and causes the death of the owner, that builder shall be put to death." Similar liabilities attached to lesser forms of damage.

Today, the services of experts are sought because of their special competence, and an individual who practices a given profession represents to those who deal with him that he possesses and will utilize the knowledge and skill commonly attributed to those engaged in that calling. Thus, an architect or engineer, whether in designing a building or in supervising the work or in issuing a payment certificate, owes to the person retaining him a duty to use his best judgment and to exercise without neglect the particular skill normally employed by architects or engineers of good standing working on comparable projects in the same locality.

On the other hand, those who hire architects or engineers are not justified in expecting infallibility; they purchase service, not insurance. The design professional, whether technically an architect or an engineer, does not in the normal case guarantee perfect plans, flawless judgment, or satisfactory results. In the absence of special agreement, the prevailing view is that the designer's undertaking is merely to exercise ordinary professional skill and diligence and to conform to accepted architectural standards. A breach of this obligation, as by negligently furnishing defective plans, will render him liable to the owner for any damage or expense which the latter consequently incurs and will also militate against any right to full compensation for services rendered.

Negligence may also result from the breach of a statute or ordinance, without the usual necessity of proving a want of reasonable care. For example, a design engineer who violates a state building code may be held negligent per se if his violation results in an injury, even if the violation did not result from any negligence.

20-5 Duty to the Owner-Employer   Perhaps the paramount duty an architect or engineer owes to the person who engages his services is that of good faith and loyalty. Thus, fraudulent issuance of a payment certificate will subject the wrongdoer to liability for all *resulting* damages.

Apart from fraud, actionable transgressions lie in the area of negligence. An architect renders himself liable to his employer where he breaches his duty to exercise the requisite care and expertise. Claims against him may be grounded upon such diverse shortcomings as (1) preparing defective plans, (2) delaying construction by tardy completion of plans, (3) significantly underestimating costs, or (4) specifying inferior materials. An action by the owner to recover damages may follow upon one or more of these errors, or litigation may stem from the reluctance of the owner to honor the architect's fee. In either event the owner, in order to prevail, must demonstrate that the architect fell below the appropriate standard of care in, for instance, preparation of plans—and that the faulty plans were substantially followed: hence the defective structure came about by reason of the plans in suit and not as the result of poor work on the contractor's part. In other words, the owner must show a causal connection between the alleged negligence of the architect and the alleged injury sustained, such as a defective structure or delayed completion.

20-6 Measure of Damages for Defective Plans or Supervision Where the architect or engineer is answerable to the owner for negligence in preparation of plans or in supervision of construction, there are two distinct rules by which damages may be measured. Where the effect is of such a nature that it lends itself to correction without undue expense, difficulty, or delay, the cost of repair is regarded as the proper damages indicator. On the other hand, where the defect does not fit the foregoing pattern and a remedy would entail tearing down and rebuilding, the appropriate measure of damages is regarded as the difference between the value of the impaired building as it stands and the value it would have enjoyed if it had been correctly designed and constructed. By way of illustrating the latter situation, assume that a warehouse was supposed to have 100,000 square feet of usable floor space, but somehow the plans were drawn and a structurally safe warehouse was built with only 95,000 square feet. If it is not feasible to add to the building, the owner is left with property of lesser worth than was intended. In the circumstances he would probably be awarded damages in an amount equivalent to the difference between the value of the building as turned over to him and the value of the building for which he had contracted.

The application of this rule may work a hardship on the owner if the building as constructed is completely unsuitable for his purposes and he cannot sell it, or where the building is unsuitable for his purposes and the cost of constructing a suitable building has risen substantially during the original construction or during the litigation. In some jurisdictions, damages for the delay or loss of business may be awarded.

20-7 Liability to Third Parties In the past, courts usually took the position that an architect's exposure in damages for negligent performance of any of his duties would not extend to strangers to the contractual arrangement between himself and the owner-employer; in other words, no one could recover damages unless he could demonstrate some connection ("privity") with the underlying contract.

The strictures of the Code of Hammurabi, for example, applied only when the owner or a member of his family was killed or injured. Courts today generally reject that rationale which was earlier in vogue and hold architects, engineers, and contractors liable to outsiders lawfully on the premises for injuries the "proximate cause" of which is the defendant's negligence in preparing plans that result in malfunctioning of one sort or another in connection with the erection of a structure constituting a hazard to the public at large or some segment thereof.

Put another way, an architect's failure to exercise the ordinary skill of his profession may expose him to damage claims brought by third persons where there is a direct causal connection between the negligent performance or nonperformance of duty, be it planning or supervisory in nature, and the forseeable harm suffered by the plaintiff. Third parties may include business visitors to the property, passers-by, or employees of the contractor, although in the last instance, some jurisdictions require the contractor's employee to prove the architect committed an affirmative act of negligence as opposed to an omission.

The liability to third persons may also be extended to adjoining property owners. For example, an architect was held liable when an adjoining landowner suffered damages as a result of improper drainage of sewage water from a construction site, the result of a design defect. The architect's argument that the designs
were formulated to provide adequate drainage upon completion of the project, not
during the construction stage, was rejected.[4]

Architects have also been held liable for damage to property which was not
adjoining. In one case, an engineer was held liable where he prepared plans which
failed to reveal a water main which was broken by the contractor, causing the
plaintiff's home to be damaged when the water rushed into it under abnormal
pressure.[5]

Lawsuits against architects by employees of the contractor usually arise as a
result of the failure of worker's compensation benefits to cover the injured employee's damages fully. Recovery under worker's compensation plans are usually
limited to medical expenses and to fixed amounts for disabling injuries and lost
wages. Also, the employee relinquishes his common-law rights to sue his employer
for the injury, although the right to sue third persons whose negligence caused the
injury is retained.

Typical of the modern view, the broad rule on an architect's liability to third
persons for negligent performance of professional duties is shown clearly in *Mon-
tijo v. Swift.*[6] The plaintiff was seeking damages for injuries suffered when she fell
while descending a stairway in a bus depot. She alleged that the fall was caused by
the negligent construction of the stairway in that the handrails attached to the
adjoining tile walls did not, contrary to accepted practice, extend to the bottom of
the stairs, and the wall tiles were set at such an angle as to create the illusion that
the bottom of the stairs was at a place which, in fact, was a step above the bottom.

The defendant architect had, as an "independent contractor," both planned
and supervised the construction work. Given that state of affairs, the court said,
he was under a duty to exercise ordinary care "for the protection of any person
who foreseeably and with reasonable certainty may be injured by his failure to do
so, even though such injury may occur after his work has been accepted by the
person engaging his services."[7]

A more recent case has held that the duty of the architect is to use the standard
of care ordinarily exercised by members of that profession, but placed more emphasis on the foreseeable use of the building and persons or property that might
be injured thereby.[8] The building involved in this case was designed as an apart-

---

[4] *McCarthy v. J. P. Cullens & Son, Corp.,* 199 N.W.2d 362 (Iowa 1972).

[5] *Crockett v. Crothers,* 285 A.2d 612 (Md. 1972).

[6] 219 Cal.App.2d 351, 33 Cal. Rptr. 133 (1963).

[7] Notice how the liability extends into the "future."

[8] *Karna v. Bryon Reed Syndicate,* 374 F.Supp. 687 (1974), applying Nebraska law. See also *Parliament Towers Condominium v. Parliament House Realty, Inc.,* 377 So.2d 976 (Fla. App. 1980) where the
court said that privity of contract between an architect who designs and supervises construction of a
condominium project and a subsequent purchase of the condominium unit is not an essential element
for the purchaser to recover damages in a negligence action against the architect.

ment house but had been converted into a hotel; the plaintiff, a guest of the hotel, was injured when he walked through an inner glass door next to which a registration desk had been placed without the knowledge of the original architect. The conversion of the building to a new use was found not to be forseeable, and no liability for the plaintiff's injuries attached to the architect.

20-8   The Trend toward Liability without Fault   There can be no doubt that the trend in personal injury law is toward the imposition of liability without a showing of negligence or fault.

The strict liability format has probably had its greatest impact in the field of products liability, under which a manufacturer can be held strictly liable for tort. Under this theory, a plaintiff can recover if he can prove that his injury or damage resulted from a condition of the product, that the condition was an unreasonably dangerous one, and that the condition existed at the time the product left the manufacturer's control.

It seems a simple step to extend the products liability theory to ill-designed structures. Indeed, at least one case found a designer-developer-builder of mass-produced houses strictly liable where an injury resulted when a defective plumbing system discharged scalding water onto the purchaser. However, the court focused on the defendant's role as the developer-builder rather than as the designer of the structure, so the case is less than definitive.[9]

Implied warranty, a doctrine from the law of sales, has also been applied in an attempt to broaden the design professional's liability. A few courts have held that an engineer or an architect who furnishes specifications and plans for a contractor thereby warrants or guarantees their sufficiency for their intended purpose. The majority view, however, is that the architect does not imply or warrant a satisfactory result. Thus, a California court said:

> ... the well settled rule in California is that where the primary objective of a transaction is to obtain services, the doctrines of implied warranty and strict liability do not apply. Those who sell their services for the guidance of others in their economic, financial, and personal affairs are not liable in the absence of negligence or intentional misconduct.
>
> This rule has been consistently followed in this state with respect to professional services [citations omitted].
>
> Those who hire such services are not justified in expecting infallibility, but can expect only reasonable care and confidence. They purchase service, not insurance.[10]

Even if an architect is found to be subject to an implied warranty of expertness, such a warranty would not run to the general public but only to the architect's employer. Thus, an action by a hotel guest against an architect when the guest fell down a spiral staircase was unsuccessful, the court stating that the architect is only charged with the duty to exercise reasonable care, technical skill, and ability in the performance of the contract.[11]

---

[9] *Shipper v. Levitt and Sons, Inc.,* 44 N.J. 70, 207 A.2d 314 (1965).
[10] *Allied Properties v. John Blume & Assoc., Engineers,* 25 Cal.App.3d 848, 102 Cal Rptr. 259 (1972).
[11] *Gravely v. Providence Partnership,* 549 F.2d 958, (4th Cir. 1977), applying Virginia law.

Attempts to apply a pure version of strict liability to engineers have also met with failure. For example, in *LaRossa v. Scientific Design Company,*[12] the court declined to apply strict liability to an engineer who designed a chemical plant where a workman was exposed, during the plant's startup, to a carcinogenic dust from a chemical catalyst. In that case, the impossibility of proving a negligent act on the part of the defendant was found to be absent and was said to be an element which normally gives rise to the document of strict liability.

Where an innocent plaintiff is injured and a well-off or highly insured design professional is the defendant, an unofficial doctrine known to lawyers as the "deep pocket theory" may be applied, thereby resulting in essentially the imposition of strict liability, though phrased in terms of a relaxation of the strict negligence standard. Thus, courts occasionally find a violation of the negligence standard, that is, reasonable care, even where the breach of duty has been small, in an effort to compensate a completely innocent plaintiff.

Whether or not the design professional is insured against personal liability is irrelevant in the determination of his liability for the plaintiff's injuries, but it can hardly be doubted that widespread knowledge that professional people are usually well insured has contributed to the increasing frequency of court judgments against them.

An Oregon case[13] points up several of the principal issues which play a part in many lawsuits brought against architects or engineers by injured third parties. A pedestrian, hurt when high winds caused a masonry wall to collapse upon him, sued the architect, the owner, and the contractor.[14] The trial court rendered judgment against the architect alone, and he appealed. In affirming, the Supreme Court of Oregon held that the architect, who was paid for assuming overall responsibility for designing the building (of which the wall in question was a part), had a nondelegable duty to meet building-code design provisions, including structural engineering requirements. Although the consulting engineer retained by the architect was the party who actually performed the faulty design work which failed to comply with a city building ordinance, the architect was held to be responsible in damages to the pedestrian for injuries sustained when the wall failed.

The appellate court dealt with three basic questions. The first of these was whether the plaintiff has proved negligence on the part of the designers of the building; the trial court has correctly instructed the jury that, if the wall was not designed so as to comply with the city code, such a violation would represent statutory negligence. The second question was whether there existed a causal connection between the alleged design defect and the injuries plaintiff received. The defendant had argued that the hurricane-force wind gusts on the critical day were of such unprecedented velocity as to amount to an "act of God." Despite an

---

[12] 402 F.2d 937 (3d Cir. 1968), applying New Jersey law.

[13] *Johnson v. Salem Title Co.,* 425 P.2d 519 (1967).

[14] This is a common trial technique. By bringing an action against several defendants, the plaintiff can often get each of them to blame the others for the injury, thus making negligence seem widespread and acute.

instruction to that effect, the jury chose to believe that a proper wall would have stood up to the unusual winds. The supreme court found that there was evidentiary support for the verdict on this issue.

The final question was whether an architect can avoid liability for faulty design by delegating the planning function to an independent consulting engineer, incorporating the latter's work into the comprehensive set of drawings, and putting his signature upon the package. Recognizing the general rule to be that the employer (here the architect) of an "independent contractor" (here the consulting engineer) is not liable for such contractor's negligence, the opinion notes that there are significant exceptions to the rule in that the employer remains liable when the work involved is "inherently dangerous" or when as a matter of law an employer's duty cannot be delegated. Plaintiff in this case contended, and persuasively, that the architect was under a nondelegable duty to design the wall in conformity with the building code. The following extract is taken from the supreme court opinion:

> This exception to nonliability has been adopted by the American Law Institute:
>
> "One who by statute or by administrative regulations under a duty to provide specified safeguards or precautions for the safety of others is subject to liability to the others for whose protection the duty is imposed for harm caused by the failure of a contractor employed by him to provide such safeguards or precautions. . . ."[15]
>
> The question, therefore, is whether an architect is required to comply with the building-code provisions concerning the structural engineering of walls in designing a building. If he is, then he cannot avoid responsibility by subcontracting that part of his work to others. . . .
>
> The evidence below showed that the architect assumed and was paid for assuming the over-all responsibility for designing the building. He contracted with the owner to provide all the drawings and specifications necessary for the construction of the building. The engineers were strangers to the owner. The building permit was issued upon the architect's plans, and the building was constructed according to those plans. Any engineering work incorporated in the plans became a part of the architect's design. Since the defendant assumed the benefits and burdens of designing the building, he assumed the responsibility of meeting the building-code design provisions, including the structural engineering requirements. His duty to meet the minimum safety standards of the building code was, therefore, nondelegable.
>
> The defendant argues that, as a professional man, the standard of care applicable to him is measured by the practices of the reasonably prudent architect in the same locality. Since it is the custom of architects in Salem to refer engineering work to consulting engineers, he argues, his only obligation was to select a reliable engineering firm. This he did, thereby fulfilling, according to his view, his duty of exercising due care.
>
> Defendant's reliance on local customs among his fellow professionals is misplaced. Selecting a competent engineering firm only indicates that the defendant himself was not negligent. It begs the question of vicarious liability. . . .

A similar case from different jurisdiction[16] involved as defendants licensed mechanical and professional engineers who had designed a supermarket pylon which fell during a windstorm, struck plaintiff, and rendered her a paraplegic. She was a

---

[15] Restatement (Second), Torts §424 (1965).
[16] *Laukkanen v. Jewel Tea Co.,* 78 Ill.App.2d 153, 222 N.E.2d 584 (1966).

member of the public and was endeavoring to enter the store during a severe thunderstorm accompanied by hail and high winds.

The complaint alleged that (1) defendants prepared architectural drawings and specifications for a building which they knew was to be erected for public use as a retail grocery store; (2) construction was performed in substantial conformity with such plans; (3) defendants knew that the concrete-block pylon would be located over the main entrance and that any defects would not be discoverable by the public; (4) defendants failed to exercise the care which should be employed by reasonably well-qualified engineers planning a similar building, in that defendants failed to design a proper pylon and they knew or should have known that (a) the pylon as planned was structurally unsound, (b) it was incapable of withstanding the wind-load force reasonably to be expected at the building site, and (c) it was of an aerodynamic shape which rendered it "eminently dangerous" in high winds; (5) the collapse of the pylon and the injury to plaintiff were the direct and proximate result of the defendants' negligence.

Specifically denied in the defendant's answer were the allegations of negligence and of proximate cause, and affirmatively argued was the proposition that the cause of the collapse was an act of God for which defendants were not responsible. They further contended that their sole duty was to prepare the plans and specifications and check them for structural adequacy, and that they had no supervisory obligations in connection with the actual construction.

At the trial an expert witness testified to his opinion that the pylon would not withstand winds in excess of 75 miles per hour because of the use of lightweight concrete block. Another expert stated that the wind experienced on the day the pylon fell was unusually intense but was still within the range of winds to be expected at the location involved.

In light of the evidence, it was held that the jury was justified in its findings that the failure of defendants to specify with particularity that, in constructing the pylon, a standard heavy concrete block should be utilized, instead of the light-weight-aggregate concrete block designated for use in other portions of the building, resulted in a pylon with strength inadequate to withstand severe winds. By reason of which failure it was found that defendants did not exercise the degree of skill ordinarily and customarily used under similar circumstances by members of their profession. The opinion proceeds as follows:

> The defendants can escape liability for this occurrence as a matter of law only if the duty which they had to furnish adequate plans and specifications for the construction of this building did not extend to plaintiff as a member of the general public. Defendants contend that the duty extended only to the party who contracted for their services, and since the plaintiff was neither a party to the contract nor in privity with someone who was she cannot recover directly from them. Defendants point out that they were not hired to supervise the construction of the building they designed. They say light-weight block was not the building material they intended for the pylon. They desired standard, heavy-duty blocks to be used and they say their lack of opportunity to observe the construction or object to the use of light-weight blocks in the pylon insulates them from direct liability to the plaintiff.
>
> We do not believe the privity of contract is a prerequisite to liability. . . . The

building in question was certified as structurally adequate. It was to be occupied and used by the public for business purposes, and the defendants knew it. . . . We conceive that the defendants owe a duty to respond in damages to those members of the general public who can be reasonably anticipated to be present in the structure they designed when negligence in design is a causal factor in injuries sustained through collapse of the building.

The parties agree that the legal duty of the defendants here is the same as the duty imposed upon an architect under the same circumstances. That duty is described in 5 Am. Jur. 2d, Architects sec. 23, at 686-87, as follows:

"Liability rest only on unskillfulness or negligence, and not upon mere errors of judgment, and the question of the architect's negligence in the preparation of plans is one of fact and within the province of the jury. . . ."

While these defendants did not supervise construction of the building, the jury here found it to be their plan that was defective, not the *construction.* . . .

We believe that the evidence in this case created a question of fact for the jury on the issues of negligence and proximate cause, and that the jury was justified in concluding that the defendants were negligent in designing a building containing the pylon at the location chosen by defendants to be constructed of light-weight concrete block, and that the injury to the plaintiff was forseeable and was proximately caused by the defendant's breach of duty to members of the public. . . .

20-9   Defensive Posture   An architect or engineer is not helpless when an injured party seeks to hold him legally responsible for damage occasioned by professional negligence. Frequently asserted defenses include contributory negligence and assumption of risk on plaintiff's part. A decision of the Louisiana Supreme Court[17] illustrates other defensive contentions which proved compelling. In that case the architects had approved a subcontractor's shop drawing which failed to specify a pressure-relief valve for a hot-water boiler. Ultimately an explosion killed plaintiff's husband, an employee of the subcontractor in question. The defendant architects argued that the shop drawing was not intended to include all equipment required for installation of the system; furthermore, the subcontractor did not rely on the shop drawing while installing the boiler, so approval of such drawing by the architects could hardly be regarded as the "proximate cause" of the explosion and attendant tragedy. In the process of finding for the defendants the court pointed out that the architects could not be charged under the supervision phase of their contract with a duty to inspect the *methods* employed by subcontractors; it follows that, without a duty, there could be no actionable breach.

A slight different twist to the "proximate cause" issue is to be found in a leading case decided in 1957 by the New York Court of Appeals[18] and standing for the proposition that neither the architect nor the contractor was liable to outsiders in connection with faulty designs, once the structures were accepted by the owner, where the danger was obvious or readily discoverable upon inspection.

---

[17] *Day v. National Radiator Corp.,* 241 La. 288, 128 So.2d 660 (1961).
[18] *Inman v. Banghamton Housing Authority,* 3 N.Y.2d 137, 164 N.Y.S.2d 699.

The basic theory was that the owner's acceptance and subsequent maintenance of the defective property breaks the causation chain and relieves the architect of responsibility in injured third parties.

Some decisions have been handed down to the effect that an architect's liability to third parties for his supervisory oversights is limited to situations of active wrongdoing as distinguished from a mere failure upon inspection to detect a contractor's error. Knowingly permitting material deviation from the established plans or intentional overlooking of improper workmanship will, as might be imagined, pave the way to a plaintiff's verdict.

Contributory negligence has occasionally been successful as a defense in actions against architects and contractors. For example, in one case a construction worker engaged in the construction of a school building was killed when he fell while climbing in a "dead space" in the ceiling of the school gymnasium in order to reconnect a sprinkler pipe. There was testimony by coworkers to the effect that work on suspended pipes was customarily accomplished from below and that the deceased walked across a 15-inch beam that was 36 feet from the floor without a safety rope. On this evidence, the court reversed a trial judge's decision that the deceased was not contributorily negligent as a matter of law.[19]

A statute of limitations may also be a defense available to the architect or engineer. For example, in the case of *Bormin's Incorporated v. Lake State Development,*[20] the court, after rejecting a contention that an architect is a warrantor of his plans, found that the statute of limitations begins to run from the moment of the discovery of the defect. Many states have statutes which bar suits after a specified period, often ten years, has elapsed since the "omission complained of."

The Supreme Court of Oregon has held that a statute of limitations begins to run at the moment the defendant's negligent act occurs,[21] thus possibly decreasing the period in which a suit might be brought against the architect.

In response to the waning of the privity requirement many states have now enacted statutes of limitations applicable to the third party liability of architects and engineers, and the constitutionality of these statutes has been upheld in several instances. In one case,[22] a statute barred actions against architects and engineers for injuries that occurred more than six months after the time of occupancy of the completed structure. This statute was upheld, the court saying that the power of the legislature to determine under what conditions a right may accrue and the period within which it must be exercised cannot be doubted and that if the legislature can abrogate a common-law right it can surely provide that a particular right be barred if not brought within a specific period of time.

Defenses based upon statutes of limitations are of substantially less use in cases where the suit is brought by a third person who has suffered physical injury. In such cases the statute of limitations does not ordinarily begin to run until the

---

[19] *Bartak v. Bell, Galyardt and Wells,* 629 F.2d 523 (4th Cir. 1980) applying South Dakota law.
[20] 220 N.W.2d 363 (Mich. App. 1975).
[21] *Joseph's Inc. v. Burns and Co., Inc.,* 491 P.2d 203 (Ore. 1971).
[22] *O'Brian v. Hazelet and Erdal,* 410 Mich. 1, 299 N.W.2d 336 (1982).

injury has been suffered, even though the negligence of the architect or engineer might have taken place many years before. On the other hand, the period in which a lawsuit may be brought in tort cases is typically much shorter than that which applies in contract cases, and many states require that court actions be brought within one year of the date of the injury.

As a defensive measure, the design professional should take positive steps to make certain that the contract specifications and drawings are not to be used by the owner or anyone else for any purpose outside of the original project without written permission of the design professional to do so. The reader should remember that the drawings are stamped with the seal of the architect or engineer, or both. That means that these persons are legally responsible for the safety of their design. Use of such material on any other project may not yeild a safe and satisfactory result.[23] If such use is desired by the owner, he should be required to engage and pay the original design professional for all additional services and for use of the original papers, or he should secure the latter's written permission for such use. Claims for liability resulting from any such subsequent use of the plans and specifications can follow the line back to the original design professional.

20-10   Measure of Damages for Personal Injury   When plaintiff suffers a personal injury, the courts begin with the assumption that damages compensate for pecuniary losses caused by the defendant's wrong. Such damages are not necessarily restricted to actual loss in time or money but may also include awards from pain and suffering, disfigurement, medical expenses, decreased life expectancy, and the like. Except in the case of damages for pain and suffering, however, compensatory damages must be capable of measurement in dollars and be capable of estimation with a pecuniary standard. Although compensation in tort cases is intended to put the plaintiff in the same financial position he was in prior to the tort, such theories would seem to have limited applicability where damages are awarded for pain and suffering. In an action to recover for personal injuries, the plaintiff must recover in one suit for both past and prospective injuries, including future pain and suffering, value of medical servics, and impaired earning capacity.

20-11   Miller v. DeWitt   In this chapter there has been at least some discussion of each of a number of concepts and crosscurrents which relate to the professional liability of architects and engineers for damages stemming from imperfect work with which they have had some connection. Recognizing the courts have adopted varying points of view and that it is not entirely appropriate to pinpoint any single decision as "representative," it would seem, nonetheless, that the case in the subheading,[24] could be cited as affording a useful recital of the various inherent problems and several legal principles which pertain.

The negligence action was brought by three employees of the contractor against

---

[23] There are likely to be various circumstances or site conditions which make reuse of the original material unsuitable for the proposed project unless various modifications are made.

[24] 59 Ill. App.2d 38, 208 N.E.2d 249, modified at 37 Ill.2d 273, 226 N.E.2d 630 (1967).

the architects and against the owner for injuries sustained as the result of the collapse of the roof of a school gymnasium building during remodeling operations. The defendant architects filed a third-party complaint against the contractor. The state supreme court determined that the evidence justified the finding that the architects, who had the right to interfere and stop the work if the contractor began to shore the roof in an unsafe and hazardous manner, were negligent in failing to watch over the shoring operation, which was a major part of the remodeling job; however, the third-party complaint should not have been dismissed before the presentation of evidence, since there was the possibility of valid jury findings to the effect that the personal injuries resulted directly from improper construction methods and techniques of the contractor and the architects were liable to plaintiff employees only by reason of the "passive" fault of failing to stop the work—in the events of which finding the architects should be able to recover over against the actively negligent party.

The central issue in the case concerned the obligations of design professionals who undertake both the planning and supervision functions.[25] The defendant architects took the position that their responsibility consisted of drafting plans and specifications and of performing such inspection of the actual construction as would assure the owner that the contractor had compiled with the drawings and specifications; the finished product would accordingly be architecturally pleasing, constructively sound, and functionally useful. It was maintained that supervising architects have neither the right nor duty to direct or in any manner control the methods or means by which the contractor accomplishes the desired results. Plaintiff insisted that under the facts of the case the architects had a clear duty to prevent the builder from carrying out his work in a faulty manner; the power of architects to stop is tantamount to a power of economic life or death over the contractor, and that authority, exercised in such a relationship, carries commensurate legal responsibility.

The owner-architect contract had this to say regarding supervision of the work:

> The Architect will endeavor to guard the Owner against defects and deficiencies in the work of contractors, but he does not guarantee the performance of their contracts. The supervision of an Architect is to be distinguished from the continuous personal superintendence to be obtained by the employment of a clerk-of-the-works.

The owner-contractor agreement, in its turn, contained a number of provisions, including several on supervision, which proved important in the lawsuit:

> The Contractor shall take all necessary precautions for the safety of employees on the work. . . .
>   The Architect and his representatives shall at all times have access to the work

---

[25] If an architect or engineer is not expected as part of his duties and responsibilities to perform continual inspection of field operations, that fact should be made evident in the contractual arrangement with the owner. On the other hand, if it is understood that the architect or engineer is indeed to personally supervise the conduct of fieldwork, his compensation from the owner should reflect this function and cover the cost of any additional personnel whom the architect or engineer may have to engage to assist him.

wherever it is in preparation or progress and the Contractor shall provide proper facilities for such access and for inspection. . . .

The Contractor shall keep on his work, during its progress, a competent superintendent and any necessary assistants, all satisfactory to the Architect. . . .

The Contractor shall promptly remove from the premises all work condemned by the Architect as failing to conform to the Contract. . . .

The Architect shall have general supervision and direction of the work. He is the agent of the Owner only to the extent provided in the Contract Documents. . . . He has authority to stop the work whenever such stoppage may be necessary to ensure the proper execution of the Contract. . . .

The Contractor shall provide all bracing, shoring and sheeting as required for safety and for the proper execution of the work. . . .

Existing structural steel shall be carefully shored and braced as required for installation of new connecting steel. Existing truss reused shall be carefully removed, revised and reerected as called for. . . .

The Appellate Court affirmed the judgment entered below on a jury verdict for plaintiffs against defendant architects. In the process of arriving at its decision to affirm, such court noted these points:

1 The terms of architects' employment are governed by contract provisions, and in this instance their duties were not limited to preparation of plans and specifications;

2 In contracting for their services, the architects implied that they possessed sufficient skill to enable them to perform the required services at least ordinarily and reasonably well, and that they would exercise such skill without neglect;

3 They held themselves out as experts in their line of work and were retained because they were believed to be such;

4 The efficiency demonstrated by particular architects in their planning function is tested by the yardstick of the skill usually attributed to one in that profession;

5 Architects must guard against defects in the plans as to design, materials and construction, and must keep abreast of the improvements of the times;

6 Absent special agreement, architects do not undertake perfection in plan or satisfaction in results, and they are liable only for failure to employ reasonable care and expertise;

7 Liability rests on unskillfulness or negligence, not upon mere errors of judgment, and the question of the architects' negligence is one of fact and within the province of the jury;

8 Architects' liability or negligence resulting in the erection of an unsafe structure, and consequently in personal injuries to individuals lawfully and foreseeably on the premises, may be based upon their supervisory activities or upon faulty plans or both;

9 Their exposure is not limited to the owner who employed them, and privity of contract is not a prerequisite to liability;

10 There can be no recovery against the architects unless it be established that they were negligent and that this way was the proximate cause of the injuries in suit;

11 Under these documents and factual circumstances, the position and author-ity of the architects were such that they necessarily labored under a duty to supervise the project with due care. With respect to the general contractor, too much control rested with the architects for them not to be placed under a duty imposed by law to perform without negligence their several functions, including supervision and direction of the work;

12 The architects were to be paid, in part, to use their best efforts to guard against the contractor's doing such things as erecting inadequate and unsafe tem-porary supports and shoring.

13 Both the contractor and the architects by virtue of their positions and their commitments to the owner were required to exercise care and skill for the protec-tion of employees on the job.

When the case reached the Supreme Court of Illinois, that tribunal upheld the finding of actionable negligence on the part of defendant architects. It was clear from the evidence, the court noted, that the architects did not prepare detailed specifications for the temporary shoring of the gymnasium roof, nor did they compute on the plans the load that would be placed upon the shores, nor did they provide the contractor with a safety factor to be used in the shoring, nor did they oversee and inspect the shoring as used. The court reasoned that any finding of negligence must be based upon one or more of these omissions. Despite their argument that the shoring here was method or technique of construction over which they had no control, the architects were held to have had the contractual right to interfere with and halt the work if the contractor began shoring operations in an unsafe and hazardous manner in violation of its written agreement with the owner. While conceding that the defendant architects were under no duty to specify the precise method the contractor should use in shoring, the supreme court concluded this would not obviate the architects' right and obligation under the contracts to insist that whatever method was selected be a safe and adequate one:

> From a careful examination of the record we conclude that, if the architects knew or in the exercise of reasonable care should have known that the shoring was inadequate and unsafe, they had the right and corresponding duty to stop the work until the unsafe condition had been remedied. . . . If the architects breached such a duty they would be liable to these plaintiffs who could forseeably be injured by the breach.
>
> Here it appears that the shoring and removal of part of the old gymnasium roof was a major part of the entire remodeling operation and one that involved obvious hazards. We think that the shoring operation was of such importance that the jury could find from the evidence that the architects were guilty of negligence in failing to inspect and watch over the shoring operation.

20-12   Small Things May Prove Important   An error or miscalculation need not be of dramatic size in order to lead to significant damage; even a seemingly small and unimportant detail may prove to be the cause of serious trouble. An illustration follows:

The architect on a store-building job engaged an engineer to design the struc-tural work and to check the shop drawings of structural steel and reinforcement, but it was understood that he was not to have the function of inspecting the project during construction. Fig. 20-1a shows a marquee or canopy over the entrance to the large store building *ABC*. The marquee *BD* was held at *B* by a connection to the steelwork of the building at structure. At *D*, it was supported to two tie rods *CD* fastened to the building at *C*. This last connection failed suddenly during a snowstorm shortly after completion of the structure. In the ensuing collapse of the marquee, a pedestrian rightfully on the premises was severely injured, and the question naturally arose as to who (among architect, engineer, and builder) was to blame.

The contract drawings for the structural design showed an eyebolt at *C*, which was to be connected to the structural steel frame of the building somewhat as shown in Fig. 20-1b. Here it is obvious that the pull *P* of a tie rod will cause a horizontal force P_h which tries to pull the eyebolt horizontally. *P* also causes a vertical force P_v acting downward so to produce bending in the eyebolt. The vertical force must be resisted by a reaction *R* in the wall of the building. Furthermore, the tie rod was to be connected to the eyebolt by means of a bolt through a clevis as illustrated in sketch (*c*). This clevis would cause one-half of the pull *P* to be applied equally on both sides of the eyebolt.

When the structure was built, a rod with a long loop at the end, similar to that shown in sketch (*d*), was substituted for the eyebolt in (*b*). It is obvious here that the force *P* now acts at a point much farther from the wall, causing greater bending in the rod where it goes into the wall. Furthermore, there was a similar loop used at the top end of each tie rod. The two loops were connected side by side by means of a bolt as shown in (*e*), resulting in a tendency for the force *P* to twist the horizontal loop about the reaction *R*. With the magnitudes of all forces remaining practically unchanged, the bending moment applied now in the rod near point *C* is much larger than before and, in addition, there is a large twisting moment. Furthermore, the looped rod was, for some obscure reason, threaded clear up to point *E,* about 1 inch inside the outer face of the masonry wall. The threads automatically reduced the cross section of the steel rod at this point. These rods failed suddenly at point *E* in sketch (*d*), and the masonry of the wall spalled under the rods at the outer face of the brickwork.

An investigation after the accident revealed that:

1 The engineer knew nothing about the substitution of the looped rods for the eyebolts called for in his design and was unaware of the omission of the clevis. Apparently the shop drawings had not been revised for his approval. As stated previously, the engineer was given no opportunity for field inspection, which might have enabled him to make sure that his design was executed as he intended.

2 Moreover, it developed that the architect was not consulted about the substitution of the looped rods for the eyebolts and clevises. Although he had kept watch over the project as a whole and may even have seen the looped rods, he was unaware that a significant substitution had been effected.

3 The contractor for the steelwork either had the looped rods on hand or found he could more readily procure them than have the eyebolts and clevises manufactured. He was apparently not conscious of any danger in the circumstances, and

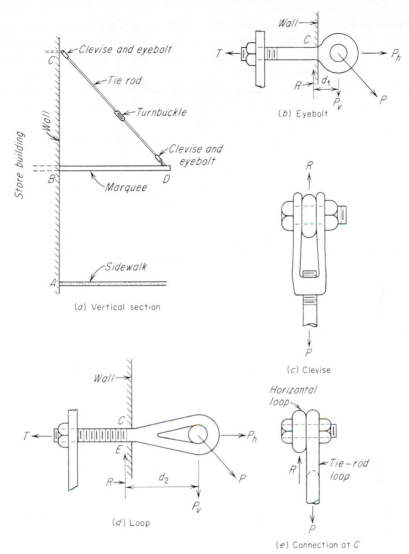

FIGURE 20-1
Example of structural failure.

seemingly did not bother to consult anyone before making the unfortunate substitution.

The foregoing situation points up several principles to bear in mind:

1 An architect who is not also an engineer should engage a competent engineer to assist in complicated technical work. He should hold the latter responsible for engineering matters and give him every chance to follow things through adequately.

2 The architect should have the engineer examine important engineering features during construction. In the case above outlined, if there had been an inspection opportunity, the engineer was the logical one to have noticed the substitution and realized its significance. The architect should not be expected to cope with such engineering technicalities.

3 A contractor should not change structural features without consulting the architect or engineer, as the case may be. He should obtain approval of any such changes before following through with them.

4 Liability insurance carried by the architect, the engineer, and the contractor serves to protect each in his respective phase of the work.

5 It is important for all concerned with structural safety to be careful and conservative, even in regard to what appear to be minor details. Economy is desirable, but only as tempered by the exercise of good judgment.

20-13   Indemnification Clauses   Indemnity provisions of one type or another have found their way with increasing frequency into contracts for both public and private work. Where fault occurs during the performance of the construction work and is not brought about by design errors or by architect's failure to transmit important instructions, there appears to be considerable logic in the argument that the contractor, in the role of the active tort-feasor, should indemnify and hold harmless both the owner and the architect or engineer in any litigation which may take place.

20-14   A Possible Case to Think About   An unfortunate accident occurred during the construction of a large shopping center. The structure was a building having over 75,000 square feet of floor area. The contractor was building one of the cinder block masonry walls which, at the time of the accident, was over 100 feet long and 15 feet high. It was apparently standing free without any lateral bracing and without any roof supports bearing on it.[26] A strong wind of about 35 miles per hour arose. As a result of lateral pressure caused by this wind, the wall suddenly toppled over, killing two men and injuring a third one who happened to be working alongside the wall. Obviously, this was a serious accident. Someone was to be held responsible.

Here are some pertinent questions relating to this case:

1 Did the owner tell the architect exactly what he wanted the construction to be? For example, he might have said that he wanted his building to be the same construction as a certain other one which he cited as an example. He may have specified the overall dimensions, inside clearances, and the use of masonry bearing walls. He may also have stated that he wanted a structure which would cause a minimum of expenditure. If the owner made any of these requirements, is he in any way to blame for the accident? Did the owner engage a clerk-of-the-works to handle the field inspection for him?

2 Did the architect (or his engineer) raise any questions about the safety of the

---

[26] The weight of such roof members might have helped to stabilize the wall.

structure? Even if the owner did indicate in considerable detail what he wanted, does this relieve the architect of his responsibility for the safety of the structure? Has he any excuse for not investigating all features of his building to see that they are safe? Is his responsibility confined to the safety of the completed structure only? Should he have seen (or foreseen) that such a freestanding wall would be endangered by a strong wind? Did he fail to see what was being done in the field by the contractor? Should he have "sensed" that a wall over 100 feet long and 15 feet high was dangerous if a bad storm occurred? Did he see the situation, realize that the stability of the wall was weak, but do or say nothing about remedying the situation? Can he take shelter by citing other walls that are similar and were built safely? If the wall as designed did not violate the local building code, is this a real defense for the architect?

3 Did the contractor build the wall without any question regarding its stability? Was the omission of lateral bracing his own doing? Did he thoughtlessly neglect to raise any question with the architect concerning the construction? In other words, is the contractor alone responsible for the accident? Were winds of 30 to 40 miles per hour an "act of God" when winds of such velocity frequently occur in this region? Will worker's compensation insurance protect the contractor from any legal action brought against him?

What is likely to be the legal action taken by the families of the deceased and by the injured man? It is probable that any suit for damages brought by them will take the form of an action against all three parties—the contractor, the architect, and the owner—hoping to collect from *somebody*. We can see that all three parties will be in trouble and must defend themselves. By thinking about these questions, readers will appreciate the possible seriousness of any legal actions which may be brought against them. They will also recognize the need to exercise care with designs and operations so as to preclude such action.

QUESTIONS

1 Explain why both architects and engineers are concerned with the matter of professional liability.

2 What are the customary rights, duties, and obligations of an architect or engineer?

3 List the customary duties of an architect or engineer in connection with a construction project.

4 Discuss the normal inspection duties of architects and engineers in connection with construction operations in the field.

5 Discuss the normal responsibility of architects and engineers in connection with their designs.

6 What is meant by "requisite degree of skill"? Is the architect or engineer to be infallible?

7 What duty does the architect or engineer owe to the owner or employer?

8 Upon what grounds might an owner or employer make claims against an architect or engineer?

9 Discuss two customary measures of damages in cases of negligence associated with structural failures or deficiencies.

10 May architects or engineers be liable for damages to third persons in case of negligence? Illustrate.

11 Does liability for negligence necessarily cease after completion of a contract?

12 In case of injury claimed to be due to negligence, why may the injured party sue the owner, architect, engineer, and contractor? What can the architect and engineer do? Who must prove the negligence?

13 Can an architect or engineer shift responsibility for his design if he hires someone else to help and uses the latter's designs in his own drawings?

14 What is meant by "nondelegable" duty?

15 If a designer adheres to an applicable building code, has he done all that is essential to avoid liability for his design?

16 Discuss the meaning of a "reasonable prudent" architect or engineer.

17 Discuss the idea of claiming an "act of God" in defending against claims of negligence relating to a structural failure.

18 Discuss possible claim that "privity of contract" is a prerequisite to liability.

19 What may be the result of a breach of a statute or ordinance by a design professional?

20 What are the elements of damage for personal injury?

21 Discuss the trend toward liability without fault.

22 What is the applicability of implied warranty to construction defects?

23 From when does a statute of limitations begin to run?

24 Discuss "defective plans" versus "defective construction."

25 Discuss "contributory negligence" and "assumption of risk" as defenses for the architect and engineer against claims by third parties.

26 Should the designer be responsible for methods of construction used by the contractor? Explain.

27 Discuss "active wrongdoing" versus "failure upon inspection to detect the contractor's error."

28 Discuss "passive fault" versus "active negligence."

29 If the designer is to supervise field operations, (*a*) what duties may this involve? (*b*) what authority? (*c*) what liability?

30 If the designer is to be responsible for safety of fieldwork, should he or she be paid extra for such responsibility? Discuss this matter.

31 Referring to the case described in Art. 20-11, discuss the situation of the architect and the possible expenses to be met by him.

32 Discuss the possible use of indemnification clauses. Illustrate.

CHAPTER 21

INSURANCE AND BONDS

21-1 Introduction   The manifold hazards to which business enterprises and personal affairs are subject have led to the development of a great many different types of insurance and bonds, each with its own purpose and characteristics. New forms—exemplified by insurance against atomic perils—are constantly being developed to meet emerging demand.

It would be well to distinguish between insurance contracts and bonds. The former involves two parties, the policyholder and his carrier; if the exposure insured against materializes, the insurance company must, within policy limits, make good the loss. Bonds, however, are tripartite contracts among the insured (called "the obligee"), the individual (called "the principal") who is under obligation to the insured, and the bonding company (called "the surety"). If a loss impends *and* the principal defaults, then the surety pays, and, under terms of the bond, looks to the principal for reimbursement.

21-2 Nature of Insurance   Broadly speaking, an insurance contract is an agreement whereby one party (the insurer) undertakes for a stipulated consideration (the premium) to indemnify the other party (the insured) against pecuniary loss arising from the particular peril described in the policy and to pay to the insured or his nominee a stated or ascertainable sum of money or its equivalent on the occurrence of the untoward event against which the insurance was afforded. Insurance represents contractual security against anticipated potential damage or liability. It is a device for shifting the exposure or "risk" from the insured to his carrier. By accepting from many policyholders, similarly situated, premiums paid into a general fund, the insurance company is able to underwrite each separate risk at a charge which is a small fraction of the exposure being assumed on that particular risk. The technique of insurance is to utilize the statistical law of large numbers; by assuming a moderate amount of liability upon each of a great number of risks, ample room is allowed for the operation of the law of averages, so that the element of chance is minimized and the degree of uncertainty reduced. Predicting future experience figures is much more of an accurate proposition when many separate risks are involved in the calculation.

21-3 Insurable Interest A valid contract of insurance presupposes a so-called "insurable interest" on the part of the applicant. The term refers to an interest of such a nature that, should the subject matter be destroyed or damaged, it would mean actual money loss for the individual involved. In life insurance, the insurable interest must exist at the time the policy is issued and not necessarily thereafter. Every person is recognized as having an unlimited insurable interest in his own life. When insuring the life of another, however, there must be indication that the applicant would have reason to desire the continuation of the would-be insured life.

Where property coverages are concerned, the picture is somewhat different. Here the policyholder must be able to demonstrate an interest in the subject matter of the insurance such that he would suffer financial loss at the time property damage occurs.

21-4 The Policy Is a Contract The insurance policy is the document which sets forth the agreed-upon limits of the liability which the issuing company has contracted to assume. Agreement between insured and insurance company is subject to various statutory provisions and to the general principles of contract law, noteworthy among the latter being that rule of interpretation which holds that ambiguities in the language of an agreement are to be construed strictly against the drafter (the insurer in this instance).

21-5 Governmental Regulation In a celebrated case in 1944,[1] the Supreme Court of the United States reexamined its previous position[2] and held that the business of insurance, when conducted across state lines, is interstate commerce and, as such, is subject to whatever degree of regulation the Congress shall deem it advisable to exercise for the national welfare. In an effort to resolve the resulting confusion, the McCarran-Ferguson Act[3] was passed by Congress in 1945. It declared the continued regulation and taxation of the insurance business by the several states to be in the public interest. With certain provisos regarding applicability of federal antitrust laws and several other identified statutes, McCarran-

---

[1] *United States v. South-Eastern Underwriters Association,* 322 U.S. 533.

[2] Many years ago a Virginia statute regulating foreign insurance companies was found not to offend the commerce clause of the federal constitution because "issuing a policy of insurance is not a transaction of commerce." *Paul v. Virginia,* 8 Wall 168 (1869).

[3] 15 U.S.C. §§1011–1015 (1976).

Ferguson stipulated that no act of Congress shall be construed to invalidate or supersede any state law regulating or taxing the business of insurance unless such federal enactment specifically related to insurance.

Each of the states has an insurance supervisory official, commonly with the title of superintendent (or commissioner or director), whose department administers the insurance code, a comprehensive set of statutory provisions dealing with all phases of a rather complicated business. Among others, there are sections imposing restrictions with respect to investments and dealing with the setting up of reserves for both known and contingent liabilities. Insurance companies are required to file detailed annual statements showing assets, liabilities, premium and claim figures, and other financial data, and they are subject to periodic examination by insurance departments of the states in which they operate. A substantial portion of governmental regulation is directed toward assuring carrier solvency.

Some portions of the insurance laws of a given state will relate to domestic companies only, while others pertain as well to foreign companies admitted to do business in the jurisdiction. Similarly, some provisions have pertinence to a particular type of insurance organization (for example, a stock company or a fraternal society) and not to others, or perhaps exclusively to a stated line of insurance (such as accident and sickness coverage). Suffice it to say that an insurance company entered in all or even a majority of the states is faced with an imposing array of different laws and regulations with which it must comply.

21-6 Some Types of Insurance Coverage There is no shortage of different lines or types of insurance. Representative of these are:

1 *Life Insurance.* This coverage may be afforded on either an individual or a group basis and further subdivides itself into a variety of special formats. For instance, it may be written as *declining term,* as *paid up,* as *participating,* etc. Actuarial studies and mortality tables, which form the cornerstone of the life insurance business, are founded on the law of averages and thus on probabilities.

2 *Annuities.* Traditionally, annuities involved repetitive benefit payments of a fixed amount each, but in recent years the *variable* annuity has entered the scene and the concept has achieved considerable popularity. As with policies of life insurance, annuities come in a variety of forms. One common type is the *annuity certain,* which calls for payments during the annuitant's lifetime, but with the further guarantee that, notwithstanding his early demise, payments will continue (to his named beneficiary) during the unexpired portion of the "certain" period. Quite different is the *life annuity,* under which payments cease upon the death of the annuitant, irrespective of how soon that may occur.

3 *Health Insurance.* Included in the category are hospital, surgical, and medical coverages, disability loss-of-income protection, and others. The policies involved may or may not include a noncancellable or a guaranteed renewable feature and may be written in the individual, group, or blanket form.

4 *Property Insurance.* Damage to one's own property from such diverse causes as fire and theft is commonly insured against. The two most familiar forms are automobile and homeowner policies. In the first of these protection for the insured with respect to loss of or damage to his own car by collision or assorted other perils is normally combined with protection against liability to others for causing them bodily injury or property damage. The typical homeowner's policy, which includes comprehensive personal liability coverage along with insurance against loss of the owner's house and contents from a variety of possible misfortunes, illustrates the modern trend toward broad, multiperil, "package" policies.

5 *Title Insurance.* Here the carrier undertakes to indemnify to a stated extent against loss arising as the result of defects in title to real estate in which the policyholder has an insurable interest.

6 *Insurance against Legal Liability.* Anyone may, in the daily course of his private life or of his business or professional activities, inadvertently injure the person or property of another. Various types of liability insurance have been designed to protect the covered individual from loss resulting from his exposure to damage recoveries stemming from accidents attributable to him. To put it another way, the insurance company providing liability protection promises to pay on behalf of the policyholder damages for which such policyholder becomes legally obligated by reason of personal injuries to other persons or loss occasioned to their property. Previously mentioned was one well-known example of liability insurance—the usual automobile policy which, with stated limits, indemnifies the insured against loss which he would otherwise sustain when through his negligent operation of a motor vehicle others are injured or killed and their property is harmed or destroyed.[4] There are any number of specialized liability insurance contracts, such as those created for (*a*) physicians and (*b*) architects and engineers.

7 *Worker's Compensation Insurance.* Generally, accidental personal injuries and certain occupational diseases arising out of and in the course of employment are compensable to the affected employee under worker's compensation laws enacted in the several states. Public and private employees subject to such laws,[5] some of which are "compulsory" and others "elective," may insure that stated obligations through a government-operated fund (where such is available), or through a carrier admitted in the particular state to write worker's compensation insurance, or they may furnish evidence of adequate financial ability of their own; only employers with rather extensive resources are positioned to self-insure, and many of these prefer to avoid that alternative. In practice the vast majority of those employers subject to worker's compensation laws purchase insurance with respect to their liability thereunder.

8 *Reinsurance.* This involves a contract under which an originating carrier (called "the ceding insurer") procures coverage for itself in a second carrier (vari-

---

[4] A liability insurance policy is not identical to an indemnity policy. Unlike a policy of liability insurance an insurer's obligation to pay an indemnity policy usually arises only when the insured sustains a loss by reason of payment to injured third party. *White v. Goodville Mutual Casualty Company,* 596 P.2d 1229 (Kan. 1979).

[5] Excepted employments vary with the jurisdictions but normally include farmworkers, domestics, and "casual" employees.

ously called "the reinsurer" or "the assuming insurer") with respect to part or all of an insurance risk of the originating carrier.[6]

21-7   Professional Liability Insurance for Architects and Engineers   It is not surprising that appropriate insurance protection, despite its cost, has come to be regarded by a great many practicing architects and engineers as a virtual necessity,[7] particularly in the light of the tendency of plaintiffs to "sue everybody in the hope of collecting from somebody." Premium rates are adversely affected to the extent that frequent and liberal verdicts against architects and engineers are encountered.

Typical policy "exclusions" would embrace such claims as those arising out of (1) failure to complete drawings or specifications in timely fashion, (2) liability of others contractually assumed by the insured, and (3) losses intentionally caused by the insured. To avoid an inordinately high premium, a deductible[8] feature is customarily included. Above such deductible, and not to exceed a figure shown as the "aggregate amount payable" during a given policy period, the insurer undertakes to make certain liability payments on behalf of the insured and to defend any suit (groundless or otherwise) against the insured complaining of negligence in the performance of professional services and seeking compensation on account thereof.

By way of a concrete example, here is the insuring clause and a portion of the exclusions from the professional liability policy involved in *Runyan v. Continental Casualty Co.:*[9]

1. Coverage

To pay on behalf of the insured all sums which the insured shall become obligated to pay be reason of the liability imposed upon the insured by law for damages resulting from any claim made against the insured arising out of the performance of professional services for others in the insured's capacity as Architects and/or Engineers, and caused by any error, omission or act of the insured or of any person employed by the insured, or of any others for whose acts the insured is legally liable. . . .

V. Exclusions

The company shall not be liable with respect to any claim made against the insured:. . .

(f) for injury to, or destruction of property resulting from any error, omission or act of the Named insured, his agents or employees, not arising out of the customary and

---

[6] So defined in §731.126, Oregon Insurance Code.

[7] In addition to the prospective financial outlay, an architect or engineer as defendant in what amounts to a "malpractice" suit finds his "stock in trade," his professional reputation, in jeopardy.

[8] A deductible clause provides for a specified reduction from any loss payment due.

[9] 233 F.Supp. 214, 217 (D.Ore. 1964). See also *United States v. United States Fidelity and Guaranty Company,* 601 F.2d 1136 (10th Cir. 1979), which contains an appendix providing examples of coverages and exclusions which specifically refer to architects and engineers.

usual performance of professional services for others in the insured's capacity as Architect and/or Engineer and including the failure or omission on the part of the insured to effect or maintain insurance, or any required bonds. . . .

Litigation between insured and insurer does materialize upon occasion, normally triggered by the insurer's declination of a claim as outside the scope of the coverage afforded. One such case[10] involved a policy insuring the architect against liability "because of injury to or destruction of tangible property, including the loss of use thereof, all in direct consequence of any negligent act, error or omission of the Insured resulting in *accident*" [emphasis added]. The decision ultimately turned on the interpretation of the word[11] "accident." It seems that the contractor on a school building project had, with the approval of the architect, used inferior or defective concrete (which failed to meet the specifications of the building contract) in constructing certain bents. It proved necessary to repair them and the contractor sought to recoup his out-of-pocket cost from the architect. The latter, confronted with his insurer's denial of responsibility, compromised the outstanding claim and took legal action against the carrier, which interposed a defense that plaintiff's loss was not within the coverage of the policy because no "accident" had resulted from any error or omission of the insured. The appellate court refused to accept the defendant's argument and reversed the lower court's holding, which had been adverse to the plaintiff architect. An "accident" is an occurrence which is unforeseen and unexpected and undesigned, reasoned the appellate court, and that test was met by the use of inferior concrete in constructing bents and their incorporation into the building—all of which happening was caused in part by the fault of the architect in approving defective material:

> Whether a building has been injured by the use in its construction of defective material is ordinarily a question of fact, but it is a question of law if the fault is unquestionably so great as to materially depreciate the value of the building or create a condition which demands correction. And we cannot doubt that in the present case the incorporation into a school building of the defective bents caused injury and damage to the structure through the creation of an intolerable condition.

In another case,[12] architects insured under a professional liability policy brought a successful suit to recover from the carrier for amounts expended to settle a claim by a hospital district for alleged malpractice which caused a heating system to be constructed in faulty fashion, with the result that serious corrosion occurred. The defendant insurer contended that there had been no architect neg-

---

[10] *Gogerty v. General Accident, Fire and Life Insurance Corp.*, 238 Cal.App.2d 574, 48 Cal. Rptr. 37 (1966).

[11] A number of words may give rise to dispute. See, for example, *Johnson v. A.I.E. Insurance of Des Moines*, 287 N.W.2d 663 (Iowa 1980), "occurrence"; *Travelers South Insurance Co. v. C. J. Gayfers and Company Insurance*, 366 So.2d 1199 (1979), "completed operations"; *C. Raymond Davis & Sons, Inc. v. Liberty Mutual Insurance Co.*, 467 F.Supp. 17 (1979), "expected"; *Alterman v. Hanover Insurance Corp.*, 363 A.2d 1102 (Conn. 1975), "control" and "structure"; *Cargill, Inc. v. Liberty Mutual Insurance Company*, 488 F.Supp. 49, affirmed 621 F.2d 275 (D.C. Minn. 1979), "business risk," "property damage."

[12] *Scott v. Potomac Insurance Co. of the District of Columbia*, 217 Ore. 323, 341 P.2d 1083 (1959).

ligence or omission within policy terms. The testimony disclosed that the precise failure which was a material causal factor in the subsequent damage was called to the direct attention of plaintiff architects and an opportunity offered to correct the deficiency within the policy period. The trier of fact, said the Supreme Court of Oregon, could well have determined that this failure by the architects, after direct warning, was an act of negligence for which they were liable. Judgment for architects, as against insurance company, affirmed.

Mention was made of the fact that the architects knew nothing of the particular tubing material employed or of its proper use and relied on the representations of the manufacturer and of the heating engineer employed by the architects to design the heating system. The Supreme Court of Oregon had some rather sharp things to say regarding that situation and quoted with approval the following from an ancient New York case:[13]

> The architect who undertakes to construct a house that is to be heated by steam is groping in the dark unless he knows how large a chimney is required. It is as necessary that the architect should know what is needed to make the steam-heating apparatus serviceable, as it is that he should know how sewer gas is to be kept out of the house. No one would contend that at this day an architect could shelter himself behind the plumber, and excuse his ignorance of the ordinary appliances for sanitary ventilation by saying that he was an expert in the trade of plumbing. He is an expert in carpentry, in cements, in mortar, in the strength of materials, in the art of constructing the walls, the floors, the staircases, the roofs, and is in duty bound to possess reasonable skill and knowledge as to all these things; and when, in the progress of civilization, new conveniences are introduced into our homes, and become, not curious novelties, but the customary means of securing the comfort of the unpretentious citizen, why should not the architect be expected to possess the technical learning respecting them that is exacted of him with respect to other and older branches of his professional studies? It is not asking too much of the man who assumes that he is competent to build a house at a cost of more than $100,000, and to arrange that it shall be heated by steam, to insist that he shall know how to proportion his chimney to the boiler. It is not enough for him to say, "I asked the steamfitter," and then throw the consequences of any error that may be made upon the employer who engages him, relying upon his skill. Responsibility cannot be shifted in that way.

21-8 Owner's Protective Liability Insurance Like the architect and/or engineer and like the contractor the owner too can take out insurance. While he ordinarily is not liable for the negligence of a contractor selected with care, there are instances where responsibility may nevertheless be ascribed to the owner, as where an accident occurs through omission of a duty that the law does not allow to be delegated. An example might be injury resulting from faulty excavation under a street, where the city (as owner) has the duty of keeping that street in safe condition for traffic, a duty which cannot successfully be delegated to a contractor.

The owner's protective liability policy affords insurance against liability for

---

[13] *Hubert v. Aitken,* 1 N.Y.S.711, 712 (Common Pleas 1888).

damage arising out of operations performed for the insured by independent contractors or attributable to omissions or negligent supervision by the insured owner in connection with such operations.

21-9 Contractor's Liability Insurance Policies are commonly issued to contractors[14] insuring them against liability to members of the public for damages or injuries stemming from the performance of contracts. The insuring provisions vary with the issuing carrier. One type of policy in general use undertakes, subject to specified exclusions, to pay on behalf of the insured contractor all sums which he shall become obligated to pay by reason of the liability imposed upon him for damages because of injury to or destruction of property caused during the policy period by accident and arising out of defined hazards.[15]

The outcome of lawsuits involving the proper interpretation of contractor's liability policies generally rides on the word "accident" and whether or not the facts surrounding a given loss render it accidental and thus within policy coverage. Three examples will demonstrate the point. In the first of these, a building occupant brought an action against a subcontractor participating in remodeling operations, basing the complaint on allegations of the subcontractor's negligence in permitting quantities of dust and plaster to be deposited on such occupant's merchandise. Ultimately the insured subcontractor sued his carrier for wrongful refusal to defend on his behalf the prior action. The subcontractor held a policy covering liability for property damage caused by "accident." The decision was that the insurer had no duty to defend; where harmful conduct is reasonably calculated to cause substantial damage of the very sort that materializes, the damage is not the result of "accident."[16]

A similar problem arose where an insured removed several water hydrants from property it claimed under a quitclaim deed. The property was also claimed by someone else whose deed predated the insured. After a finding that the property was not his, the insured paid damages to the real owner and then sued its accident insurer to recover the damages paid. In this case, the court held that where the injury resulted from a mistaken belief on the part of the insured there was no accident within the meaning of the policy since the insured acted intentionally.[17]

The same issue, but opposite result, can be found in *O'Rourke v. New Amsterdam Casualty Co.*[18] A sudden and unpredicted rain in Albuquerque in October, a normally

---

[14] Subcontractors also are very likely to be policyholders. It is certainly not unusual to find liability insurance protecting the subcontractor against its contractual obligation to indemnify the general contractor "against liability imposed upon the contractor by law for damages because of injury to or destruction of property caused by accident due to any act or omission of the subcontractor, his agents or employees, arising out of and during the prosecution of the work of the type subcontractor as contemplated under this agreement."

[15] For some actual policy language in point, see *Leebov v. United States Fidelity & Guaranty Co.,* 401 Pa. 477, 165 A.2d 82 (1960).

[16] *Casper v. American Guarantee & Liability Insurance Co.,* 408 Pa. 426, 184 A.2d 184 A.2d 247 (1962).

[17] *Foxley & Co. v. U.S. Fidelity & Guaranty Co.,* 277 N.W.2d 686 (Nebraska 1979).

[18] 68 N.M. 409, 362 P.2d 790 (1961).

dry month, was deemed an "accident" within a roofing contractor's liability policy, and the insurer was therefore liable to pay for rain damage to a house, the roof of which had not been completed.

Similarly, where an insured mismanufactured precast concrete walls and they had to be removed, refabricated and reerected at the manufacturer's expense it has been held that the definition "accident" is "an unexpected, unforeseen, or unde- signed happening or consequence from either a known or unknown cause, and that under that definition the manufacturers unintentional and unexpected error was an accident entitling them to recover."[19]

Another case arose under a ". . . manufacturers and contractors schedule liabil- ity policy" issued to a painting contractor.[20] It seems that, during the course of painting the wood gables on a large number of houses in a building development, paint accidentally (and almost inevitably) splashed on the shingle siding of some fifty-two houses, and extensive repainting of such siding became necessary at considerable cost to the painting contractor. In entering judgment for the insurer, the appellate court found that the damage was not caused by "accident" and hence was outside the purview of the policy. The opinion is almost entirely de- voted to a discussion of the proper meaning of the word "accident" in the context of this fact situation, and the court quotes with approval from various earlier decisions. Here are some extracts which put the problem well in focus:

> The word "accident" is not defined in the policy, and the term must therefore be inter- preted in its usual, ordinary and popular sense. Webster has defined it as "an event that takes place without one's foresight of expectation; an undesigned, sudden, and unex- pected event; chance, contingency." . . . It seems clear that an accident is the antithesis of something likely to occur, forseeable in due course. If an occurrence is the ordinary and expected result of the performance of an operation, then it cannot be termed an accident. To constitute an accident, the occurrence must be "an unusual or unexpected result attending the operation or performance of a usual or necessary act or event. . . .
>
> Schnoll's evidence clearly established that the dripping of paint was an expected occurrence in the course of the operation. In fact, the secretary-treasurer of the firm testified that "it is almost impossible to keep the drippings from falling." . . . To hold that the resulting damage was caused by accident within the meaning of the policy would be, in effect, to constitute appellant a guarantor of perfect performance. Such liability was never intended. . . .
>
> It is true that courts construing insurance contracts have differed in judgment whether the fact that the responsible cause is an intentional act suffices in itself to preclude the resulting injury from being "accidental" damage. . . . But if, in addition to being intentional, the harmful conduct is reasonably calculated to cause substantial damage of the very sort that occurs, all courts seem to recognize that the damage is outside of the insured risk. . . .

21-10 Builders' Risk Policies   Builders' risk policies are available to contrac- tors or owners to protect against pecuniary damage to their separate property interests during the construction or renovation of a building. From the time mate-

---

[19] *Yakiam Cement Products Co. v. Great American Insurance,* 608 P.2 254 (Wash. 1980).
[20] *Schnoll & Son v. Standard Accident Insurance Co.,* 190 Pa. Super 360, 154 A.2d 431 (1959).

rials or equipment arrive at the construction site, the owner and contractor are vulnerable to loss from a number of exposures, such as vandalism and the elements. This type of policy may be limited so as to cover loss resulting from certain stipulated causes only, such as fire, or it may be much broader in scope. Sometimes a single policy covering a building in process of construction will be issued in the names of both the owner and the contractor. In that event, and assuming that no explicit policy provision covers the point in some other fashion, the rights of the respective parties to any claim proceeds recoverable would seem to depend upon the relative interests of each at the time of the loss.

Where separate policies cover owner and contractor, difficult questions arise. In many instances the construction agreement obligates the contractor to assume the risk of damage occurring to the structure or its contents during the building process. Under these circumstances, court decisions are not uniform on whether or not the owner may recover from *his* "builders' risk" carrier for loss which manifests itself while construction is yet continuing. In at least the large majority of those cases where this matter has been argued, the contractor had not (at the time the owner-insurer litigation commenced) fulfilled his commitment to bear the loss. If the contractor, by reason of his obligation to repair damage arising before the structure is finished and turned over to the owner, actually does assume the cost (presumably through his insurance company), it is normally contended that the owner has suffered no damage and has no valid claim against his insurer.

Reference is made to *Lititz Mutual Insurance Co. v. Lengacher*[21] for an illustration of a lawsuit involving builders' risk insurance. Such a policy protected the contractor's interest in a dwelling, during its construction, against loss through fire. Meanwhile, the owner, obliged by a provision of the building contract to do so, independently procured a "fire and extended coverage insurance policy in an amount sufficient to fully protect these improvements at all times during construction." After work had reached an advanced stage and the contractor had been paid about four-fifths of the agreed price, the threat of fire did materialize, and substantial damage resulted. The contractor effected the necessary repairs, as he was legally obligated to do, and sought to recoup his cost from his insurance carrier. The owner, who had sustained no loss, did not file a claim under his policy. It was held that the contractor's insurance company was primarily liable and was not entitled to have the loss prorated between it and the owner's insurance company, notwithstanding the owner's aforementioned duty to procure insurance protection. The two companies insured separate and distinct interests in the same property.

St. Paul Fire and Marine v. Murray Plumbing and Heating Corporation[22] also

---

[21] 248 F.2d 850 (7th Cir. 1957).

[22] 135 Cal. Rptr. 120 (Cal. App. 1976). See also *J. F. Shea Co., Inc. v. Hydes Plumbing and Heating,* 619 P.2d 862 (Nev. 1980), where an employee of the subcontractor caused a fire while welding. The court there found that the contractor could recover under his builder's risk insurance for his losses as well as the subcontractor and his employee. The contractor's insurer was not entitled to subrogation against the subcontractor's insurer as the liability for this type of risk fell under the builder's risk policy of the contractor and not under the public liability or property damage insurance of the subcontractor.

presents a situation involving a builder's risk policy. A building under construction had suffered severe water damage when a 4-inch cast iron pipe broke, causing extensive flooding. The alleged cause was negligence on the part of the subcontractor. The insurance company reimbursed the contractor and then sought to recover against the subcontractor through its right of subrogation. However, the subcontractors had been listed as additional insureds under the policy. The insurance company argued that the policy did not extend coverage to them for the loss so as to prevent subrogation against them and that they were contractually obligated to indemnify the general contractor under the provisions of the subcontract. Finding in favor of the subcontractors, the court rejected the contention that the policy provided only property coverage and not liability coverage, thus creating a right of subrogation against the subcontractors. The policy was said to have been procured by the contractor for the mutual benefit of himself and the subcontractors; the court noted that, since subrogation exists only with respect to rights of the insurer against third persons to whom the insurer owes no duty, no right of subrogation could arise in favor of the insurance company against the subcontractor.

21-11   Bonds in General   There are hundreds of different kinds of bonds grouped into two broad categories: fidelity and surety. The risk of dishonesty is an element common to all bonds, but those in the surety class normally involve as well the risk of nonperformance, whether because if incapacity, inefficiency, or some other cause.

21-12   Fidelity Bonds   These are instruments executed by the principal as well as the surety, or *underwriter,* and designed to indemnify the named third party, *obligee* or *insured,* against loss occasioned by a dishonest act commited by the principal. The usual fidelity bond protects an employer against employee wrongdoing, such as larceny, embezzlement, or forgery; typically, an employer will pay the premium on such bond.

Fidelity bonds are written for commercial concerns, schools, clubs, and a whole host of other entities.

21-13   Surety Bonds   Under such a bond, which again is executed by both principal and underwriter, the latter guarantees performance by the principal and is responsible for making good to the insured any pecuniary loss sustained through the principal's failure to fulfill his contractual obligation. It matters not whether the difficulty is due to incompetence, dishonesty, or just plain bad luck. In other words, the surety has assumed responsibility for the principal's ability, trustworthiness, and financial strength. Breach by the principal gives rise to a remedy for the obligee against both principal and surety.[23]

[23] Obviously, the quality and financial standing of the surety are matters of considerable importance to the obligee (owners), who may well require the bidding contractor to submit for approval the name of his intended surety. If the owner does not feel that he can accept that particular surety, he may demand the substitution of another one or may reject the bid outright. After a designated surety has been approved, the contractor should not be allowed to effect a change without the owner's consent.

It may be wise for the engineer to insert in the contract documents a provision that the contractor shall secure a satisfactory replacement surety if the original one gives indication of financial difficulties before the building project is completed. In very large contracts it is not unusual to specify that multiple sureties shall be involved in the bonding arrangements.

Surety bonds are classified as either common law or statutory. The former type is an obligation "voluntarily" undertaken and the underwriter is free to insert such limitations and restrictions on its liability as it chooses. Statutory bonds, on the other hand, are those required by law in certain circumstances—as, for example, where contractors are engaged in government projects. Generally such bonds must comply strictly with the terms of the pertinent statutory provision and are conditioned to protect the public body as well as suppliers of material and labor.

21-14   Bonds in Public Work   A further word is in order regarding distinctions in the bonding picture between public and private work. In the case of the latter, any requirement for bonding is a matter for owner-contractor negotiation. It is not uncommon these days for an owner to insist upon surety bonds to protect himself against the contractor's breach and against claims of laborers and materialmen.[24] However, a substantial percentage of private construction work is still not bonded.[25]

Generally speaking, statutes now mandate some form of surety bond or bonds in connection with any kind of public works contract. At the federal level the Miller Act[26] requires that, before any contract in excess of $2500 and involving construction, alteration, or repair of any public building or public work of the United States is awarded to any person, such person shall furnish to the United States:

1 A performance bond in such amount as the officer awarding the contract shall deem adequate for the protection of the United States.
2 A payment bond for the protection of all persons supplying labor and materials in the prosecution of the work.

Appropriate bond forms may be secured from the General Services Administration of the federal government.

Similar legislation, sometimes called "little Miller Acts," has been adopted

---

[24] Except where some specialized instrument is stipulated by law, the so-called "AIA forms" are in general use for both public and private construction work. It is understood that these bond forms were the product of cooperation between surety companies and the American Institute of Architects.

[25] It is possible to allow the contractor to substitute (for the performance bond) cash or approved securities, depositing them with the owner or to his credit until the contractor has completed the work and the owner has accepted the end result. This alternative procedure, however, is seldom desirable and is rarely used. For one thing, the possible loss or misplacement of money or securities presents a danger from the owner's standpoint, since he will be responsible for the safekeeping. Also, stocks and bonds may decline in value during the life of the contract, and if trouble develops, the securities may have to be sold under adverse market conditions. The owner would then lose part of the protection afforded by the collateral. It is easy to see that, if the contract is breached, the handling of the deposit may cause difficulties and arguments. Furthermore, not every contractor is in a position to deposit collateral without impairing his fincancial ability to complete the contract.

[26] 40 U.S.C., starting with §270(a) (1969).

rather uniformly in the several states, causing surety bonds to be furnished by contractors engaging in the works projects for the states. The purpose is the dual one of assuring completion of the construction and remuneration to those who furnish labor and materials. The standard pattern seems to be to require all builders seeking public contracts to file bid bonds—with the successful bidder having then to execute a performance bond and a payment bond (occasionally termed a "labor and materials bond"). Claimants' rights and remedies under these bonds vary according to applicable local law and the conditions of the surety's obligation assumed in line with such law.

While it is the general contractor who must furnish bonds to the public body, a good portion of the actual work is let out on subcontracts. Accordingly, the general contractor, in his turn, may find it expedient and prudent to require performance and payment bonds from his various subcontractors; these bonds, of course, are meant to protect the general contractor against loss in the event a "sub" fails to fulfill his commitments.

21-15   Contract Bonds   Contract bonds are essentially credit risks; the underwriter will do well to consider in advance the contractor's financial status as well as his technical competence, experience, current work schedule, and the adequacy of his bid.

This group of suretyship obligations subdivides itself into bid bonds, performance bonds, payment bonds, and maintenance bonds.

Bid Bonds   A surety company issuing a *bid bond* thereby guarantees that the named contractor (as "principal") will, if his bid prevails, sign the contract document and furnish any requisite performance and payment bonds; its limited function does not include assurance of satisfactory completion of the project itself. In other words, the bid bond could be said to insure the good faith of the contractor in entering his bid. Should he fail to accept the contract as awarded and supply the required further bonds at that juncture, the surety must be prepared to answer in damages. These could include (1) the expense to the obligee of again putting the job out to bids, and (2) the out-of-pocket loss of the difference between the original low bid and the price at which a contract is ultimately signed. The typical bid-bond penalty is perhaps 10 percent of the contract amount.

This particular bond terminates when the principal signs a contract and supplies further bonding protection or when the contract is awarded to some bidder other than the designated principal.

Instead of a bid bond to accompany each proposal, a deposit (perhaps of a certified check) is normally acceptable. The engineer, the owner, will call for a deposit of such size as would seem to afford adequate protection for the owner. It is not unusual for the sum to be calculated in terms of a stated percentage of the bid price.

Performance Bonds   This bond makes certain that, should the principal fail to perform any of the terms and conditions of this contractual undertaking, the owner[27] is protected by the surety against loss up to the amount of bond penalty.

---

[27] Variously referred to also as "insured" or "obligee."

Often required in an amount equal to 100 percent of the contract price, the performance bond shifts from the owner much of the risk of damage arising from default by the contractor.

The manner in which the owner is made whole depends on the precise terms of the surety's commitment,[28] but the latter must in some fashion (1) promptly remedy the principal's failure, or (2) itself complete the contract, or (3) assist the owner to get the contract relet and fulfilled and then pay to the owner the extra cost to him of the substituted completion.[29]

Notice that the bond is not intended to provide a means for the owner to make a profit. It is designed to prevent, or at least to reduce, extra cost to him in case the contractor fails to perform.

Assume that a contractor secured a lump-sum contract for the building of a store. The contract price was $1,000,000; the bond was $500,000. After a little over one-half of the work was completed, the contractor was forced into bankruptcy. The owner had paid him a total of $450,000. The surety and the owner then agreed to engage another contractor to finish the job on a cost-plus-fixed fee basis. The cost of completing the store was $600,000, making a total out-of-pocket cost to the owner of $1,050,000. That is, of course, $50,000 more than the original contract figure. Under these circumstances, the surety was obligated to pay only the extra $50,000 and not the entire face amount of the bond.

On the other hand, assume that the amount of the bond had been but $100,000 and that the cost of completing the structure in the preceding case was $700,000 bringing the total cost to $1,150,000 (that is, $450,000 plus $700,000). The cost above the contract price was thus $150,000. The surety was therefore liable for the full $100,000 of the bond, and even this did not make the owner "whole"—all of which shows that it is unwise to call for too small a performance bond.

If the surety had agreed to take over the foregoing store job and to have the work done under whatever plan he thought best, the owner would be expected to pay the original contract price, and the surety would still make good on any excess up to the limit of the bond.

If the payment to the owner under the bond does not fully reimburse him for the extra expense occasioned by the contractor's breach, the owner may try to

----

[28] An example is afforded by the following quotation from "Document No. A-311 (Formerly 107) 1963 Edition," as approved by the American Institute of Architects: "Whenever Contractor shall be, and declared by Owner to be in default under the Contract, the Owner having performed Owner's obligations thereunder, the Surety may promptly remedy the default or shall promptly

(1) Complete the Contract in accordance with its terms and conditions, or

(2) Obtain a bid or bids for submission to Owner for completing the Contract in accordance with its terms and conditions, and upon determination by Owner and Surety of the lowest responsible bidder, arrange for a contract between such bidder and Owner and make available as work progresses (even though there should be a default or a succession of defaults under the contract or contracts of completion arranged under this paragraph) sufficient funds to pay the cost of completion less the balance of the contract price; but not exceeding, including other costs and damages for which the Surety may be liable hereunder, the amount set forth in the first paragraph hereof. The term 'balance of the contract price,' as used in this paragraph, shall mean the total amount payable by Owner to Contractor under the Contract and any amendments thereto, less the amount properly paid by Owner to Contractor."

[29] See *Continental Realty Corp. v. Andrew J. Crevolin Co.,* 380 F.Supp. 246 (S.D. W.Va. 1974).

recover from the defaulting contractor. But the chances of success are slim because the latter is probably in financial straits. Breach by the contractor generally creates a bad situation and leaves the whole construction job in difficulty. For one thing, a contractor who is about to default will seldom exert himself to make things easy for his successor. It is apparent that the owner should be extremely careful at the outset to see that the contractor he selects is capable, reliable, and solvent. Even though he holds a performance bond as some protection, one of the things the owner dreads is the financial failure of his contractor.

Release of Surety There are circumstances in which the surety will be relieved of its liability, as where the obligee, without the consent of the surety, violates a condition of the bond; for example, the obligee might ignore a requirement that he advance no sums to the principal (contractor) except in exchange for sworn statements from all persons who had done work on the job or furnished material, such statements to evidence the fact that these persons had been duly paid.

A case illustrating an obligee act which served to release the surety is *Mauney v. Hartford Accident and Indemnity Co.*[30] The bond in suit provided that:

> The obligee shall, at the times and in the manner specified in the contract, fully comply with all the terms thereof, and if the obligee default in the performance of any matter or thing agreed or required in this bond, or in the contract, the surety shall thereupon be relieved of any liability hereunder.

It seems that the obligee, without obtaining the written approval of the surety as called for in the conditions of the bond, changed the terms of payment provided for in the building contract. It was held that this course of conduct on the part of plaintiff-obligee sufficed to discharge the surety from any obligation. In the process of so holding, the court rejected a contention by plaintiff that the surety had prior knowledge of the irregular payments and had waived its right to claim a release.

Payment Bonds To guard against difficulties arising from the contractor's failure to pay for labor and materials, it is desirable to require that the contractor furnish a *payment bond* (also called *labor and materials bond*) to protect the owner against loss if the contractor should indeed fail to pay subcontractors and other suppliers. There are statutes providing that contractors engaged in public jobs shall give not only a bond to "secure" performance but also one to protect materialmen and laborers, suit upon the bond by these third persons being expressly or implicitly authorized.[31] Like the performance bond and unlike the bid bond, the payment bond is most typically written in the same amount as the full contract price.

---

[30] 68 Ga. App. 515, 23 S.E.2d 490 (1942). See also *Airtol Eng. Co., Inc., v. U.S. Fidelity & Guaranty Co.,* 345 So.2d 1271 (La. App. 1977) where the plaintiff made payments in derogation of the contract which called for payments to be made after materials were in place. Payment has been made for some shop drawings and some materials that the evidence indicated had not yet been installed. The court found that since these payments were in derogation of the job-site-installed clause that they were not covered by the bond and thus the surety was not liable.

[31] *Fort Smith Structural Steel Co., v. Western Surety Co.,* 247 F.Supp. 674 (W.D. Ark. 1965).

A bond given pursuant to a statute must be read with the statute and extended as well as limited in its scope by the statute, unless violence would be done to the language of the bond by such a construction.[32]

Here is an example of how unfortunate the failure to secure a payment bond can be. An owner made plans for a new house and thought he would economize by making the drawings and preparing the contract himself instead of engaging an architect to do so. He let the construction contract to a "friend." When the work was about 90 percent complete and when the builder had been paid that portion of the contract price, he quit. During the next few days several concerns which had furnished materials, several subcontractors who had installed plumbing and heating fixtures, and others who had done miscellaneous work appeared on the scene with claims against the owner for what was due them from the contractor. The owner was shocked when he learned that he was required under the circumstances to pay these bills. Before he was through with the construction, the job had cost approximately 40 percent more than the original contract price. His only recourse was to take action against his former friend, the builder.

A case presenting a different twist is *Parliament Insurance Co. v. L. B. Foster Co.*[33] wherein a materialman brought suit against a surety to recover on a subcontractor's performance-payment bond executed in favor of the original contractor. Noting that all doubts in the construction of such bonds must be resolved against the surety, the court held that the subcontractor's bond insured to the benefit of the materialman and that it was immaterial if some of the materials were furnished to the contractor prior to the execution of the bond, so long as they were used in the work which was covered by the bond.

Maintenance Bonds Faulty workmanship and poor materials may lead to problems which do not manifest themselves until well after the completed building has been accepted by the owner. The function of the *maintenance bond* is to cover the owner by binding the surety and its principal to correct defects which develop belatedly. An exclusion is customarily provided for defects (1) which cannot legitimately be blamed on the contractor or (2) which are not discovered before a certain period of time elapses after contract completion.

QUESTIONS

1 Differentiate between insurance contracts and bond.
2 Explain the general nature of insurance.
3 How does the law of averages enter the picture in the case of insurance?
4 Define "insurable interest."
5 Explain how an insurance policy is a contract.
6 Discuss regulation of insurance by government.
7 Explain the basic difference between life insurance and property insurance.
8 Define and illustrate annuities (fixed and variable).

---

[32] *Nelson Roofing & Contracting Inc. v. C.W. Moore Co.*, 245 N.W.2d 866 (Minn. 1976).
[33] 533 S.W.2d 43 (Tex. Civ. App. 1975). See also *General Insurance of America v. Century Indemnity Co.*, 384 So.2d 1305 (Fla. App. 1980).

9 Define and illustrate accident and sickness insurance.

10 Discuss insurance against legal liability.

11 What is subrogation?

12 What is worker's compensation insurance and how does it function?

13 What is meant by "reinsurance"? When may it be utilized?

14 Discuss professional liability insurance for an architect or engineer. What may be protected? What might be excluded?

15 What might be the claimed effect of lack of conservatism in a design for a structure?

16 In liability insurance, what is meant by "accident"?

17 Discuss accident to a person versus damage to a structure as far as insurance is concerned.

18 Describe "owners' protective liability insurance" and what it might cover.

19 Describe "contractors' liability insurance" and what it might cover.

20 Illustrate damages to a structure which might happen but not be an "accident."

21 Define and illustrate "builders' risk policy."

22 Define and illustrate "fidelity bonds."

23 Define and illustrate "surety bonds."

24 What is the difference between common-law and statutory bonds?

25 In connection with bonds, define "obligee," "principal," and "surety."

26 Define and illustrate "contract bonds."

27 What types of surety bonds are in common use in the construction industry?

28 Explain the function of a bid bond. Illustrate.

29 How is the size of a bid bond determined? When does it cease to be effective?

30 A contract required submission of a bid bond for $10,000 for each proposal. The lowest bid was $500,000 and was made by Morse. The next lowest bid was $520,000 and was made by Whipple. Morse later refused to sign the contract. The owner then accepted Whipple's proposal and awarded the contract to the latter. (*a*) What is the obligation of Morse's surety? (*b*) What does the owner gain or lose? Explain.

31 Explain the function of a performance bond. Whom is it designed to protect?

32 Who usually determines the size of a performance bond? On what do they base this determination?

33 Does a performance bond guarantee personal performance by the contractor? Explain.

34 Why should the surety have the right to object if major revisions of the contract are made?

35 What happens to the performance bond when the contract has been completed satisfactorily?

36 Should a performance bond be oral or written? Why?

37 Should a performance bond be made to cover 100 percent of the contract price? Can it do so?

38 A clause stating that the contractor is to furnish a performance bond for $50,000 specifies the required surety by name. Is this customary? Why? Should the owner have the right to approve or reject the surety?

39 A contractor furnished a performance bond having a face value of $100,000. The contract price was a lump sum of $1 million. The contractor had completed over 60 percent of the work and had been paid $540,000 by the owner. Then the contractor defaulted, leaving unpaid bills of $100,000. The owner had withheld $60,000 of partial payments due the contractor for work done prior to the default. The owner engaged another contractor to finish the job for $475,000. Does the surety for the performance and payment bonds have to pay? If so, how much? Explain.

40 When and why might the engineer require multiple sureties?

41 Who ultimately pays the fee for a surety bond? Explain.

42 Kiley performed part of the work required by his contract with Daniels, then defaulted. Kiley's work was, in part, very unsatisfactory. By agreement with the surety, Daniels had another contractor finish the job for $15,000 additional cost, and she also had this contractor replace Kiley's defective work for $8000. The face value of the performance bond was $25,000, and it promised "satisfactory performance by the principal." Daniels presented the surety with a bill for $23,000. Can Daniels collect this sum? Explain.

43 Is the owner expected to obtain a profit by resort to a performance bond? Why?

44 If the contract calls for multiple sureties, can one surety take over the complete obligation?

45 Does the surety take over when the contractors get into difficulties or only when they quit?

46 A surety posted a performance bond for $30,000 for a contractor who was to strengthen and repave an old bridge. A hurricane and flood carried away the bridge a week after the contractor started work. What do you think is the obligation of the surety? Explain.

47 If the lowest responsible bidder refuses to sign the contract and if she furnished a bid bond of $10,000 with her bid of $102,000 on the job, and if the lowest bid was $105,000, what must the first bidder's surety pay?

48 What happens to the performance bond if the owner cancels the contract?

49 Under what conditions may a surety be released?

50 What is a "labor and materials bond," and how does it function in a construction contract?

51 Define and illustrate a "maintenance bond."

CHAPTER 22

REAL PROPERTY

22-1 Nature of Real Property *Realty* as distinguished from *personalty,*[1] consists of the following:

1 The land itself
2 All buildings, trees, and other fixtures of any kind thereon
3 All rights and privileges incidental or "appurtenant" to the land, that is, used with the land for its benefit, as in the case of a watercourse or of a passage across the property of another.

Assume that you agree to buy a farm from *A*. Barring an agreement to the contrary, you will acquire the land, the structures existing thereon, and all trees, shrubs, and other long-lived vegetation growing on the premises. However, you are not automatically entitled to crops grown by *A* and ready for harvest. The dam and the 5-acre pond on *A's* farm are part of the realty as of the time of your purchase and thus become yours, but the same is not true of *A's* boat, which is regarded as personalty and which does not, therefore, pass with the land. The farm machinery, stock, poultry, etc., owned by *A* are his personal property, and title to same does not pass with the farm unless you buy these items or *A* donates them to you. On the other hand, after agreeing to sell the farm to you, *A* has no right to tell a landscape gardener that he can take 200 cubic yards of topsoil for $3 per cubic yard.

22-2 Ownership To own property of any sort entails the rights of possession,

---

[1] In a broad sense personal property includes everything which is the subject of ownership and which is not properly described as real estate.

reasonable use, and disposition. There are various kinds and degrees of ownership of real property.

Sole ownership involves only one person. *Joint tenancy* is a relationship whereby two or more people together own an interest in land, and each individual concerned has precisely the same rights in respect of that interest as have his cotenants. *Tenancy in common* is another multiple-ownership from where two or more persons have undivided but nonetheless distinct shares in a property interest.

The generic designation *real property* is usually said to be sufficiently comprehensive to embrace all estates in land that is, estates in fee simple, estates for life, estates for years, etc. A *fee simple* is the largest estate known to the law. It is a freehold estate[2] of inheritance carrying an unlimited power of alienation. Where land is acquired by *A* and his heirs absolutely and without any end or limitation being imposed upon the estate, *A* is said to hold title in fee simple. He can, if he desires, dispose of such fee during his lifetime. Otherwise it will pass under his will or descend to his heirs if he dies intestate. A *life estate,* on the other hand, is property interest whose duration is limited to the life of the person holding same or to the life of some other named individual. During the continuance of the life estate, the tenant has the right to full enjoyment and use of the land. However, existence of a life estate necessarily means that the fee exists elsewhere than in the life tenant, and the latter has no power to destroy the remainderman's interest in the property.

Ownership of realty varies also in respect to the existence and extent of any encumbrances. Holding title free and clear is a pleasant situation but no more common these days than to have the property burdened with a substantial mortgage, with unpaid assessments for streets or sewers, or with any number of other items.

In connection with the general subject of ownership, mention should be made here of the necessity of writing where interests in realty are concerned. Section 5-707 of the New York General Obligations Law (McKinney 1978) is typical of statutes patterned after the famous English Statute of Frauds (29 Car II). Section 5-507 reads:

> **1** An estate or interest in real property, other than a lease for a term not exceeding one year, or any trust or power, over or concerning real property, or in any manner relating thereto, cannot be created, granted, assigned, surrendered or declared, unless by act or operation of law, or by a deed or conveyance in writing, subscribed by the person creating, granting, assigning, surrendering or declaring the same, or by his lawful agent, thereunto authorized by writing. But this subdivision does not affect the power of a testator in the disposition of his real property by will: nor prevent any trust from arising or being extinguished by implication or operation of law, nor any declaration of trust from being proved by a writing subscribed by the person declaring the same.
>
> **2** A contract for the leasing for a longer period than one year, or for the sale, of any real property, or an interest therein, is void unless the contract or some note or memo-

---

[2] At common law, *freehold estates* in real property included only estates of inheritance and estates for life.

randum thereof, expressing the consideration, is in writing, subscribed by the party to be charged, or by his lawful agent thereunto authorized by writing.

3 A contract to devise real property or establish a trust of real property, or any interest therein or right with reference thereto, is void unless the contract or some note or memorandum thereof is in writing and subscribed by the party to be charged therewith, or by his lawfully authorized agent.

4 Nothing contained in this section abridges the powers of courts of equity to compel the specific performance of agreements in cases of part performance.

22-3 Condominiums Lack of land in congested areas, the advantages of common maintenance, tax benefits of ownership as opposed to rental, and the high cost of single-family residences have all contributed to the growth in popularity of condominiums and cooperative apartments. Every state now has a statute allowing for or regulating condominium construction and ownership.[3]

There are significant legal differences between condominium and cooperative forms of ownership. A Condominium is typically a multiunit dwelling, in many cities a former apartment house, each of whose residents has exclusive ownership of an individual unit and an undivided interest in the common areas, such as land, halls, foundation, lobbies, and installations for common services such as heating, utilities, and air conditioning. Apartment maintenance is generally the province of the individual owner, but the owner's association maintains common areas and facilities. Funds for common maintenance are derived from dues or assessments on unit owners.

A cooperative apartment is a multifamily dwelling in which each resident has an interest in the entity owning the building, as opposed to the building itself, and a lease entitling him or her to exclusive use of a particular unit. The cooperative owner is thus a stockholder rather than a property owner.

22-4 Encumbrances An *encumbrance,* broadly speaking, is any burden upon land depreciative of its value. An existing encumbrance is adverse to the interest of the landowner, though it does not preclude his conveying the property in fee simple. If he has contracted to sell the land free and clear of all encumbrances, he must, obviously, pay up back taxes and take other steps of like nature before the closing date arrives.

The list of encumbrances is lengthy and varied. Thus, a California statute[4] stipulates that encumbrance is a term which "includes taxes, assessments and all liens upon real property."

This provision has been interpreted[5] to include any right to or interest in land which may subsist in another to the diminution of its value, but consistent with the passage of the fee. An encumbrance may include whatever charges, burdens, obstructs, or impairs the use of land or impedes its transfer. Examples include a

---

[3] For example, see McKinney, New York Con. Laws, Real Property Law §339(d)(ii).
[4] Calif. Civil Code §1114 (West 1982).
[5] *American Title Company v. Anderson,* 52 Cal.App.3d 255 (1975); *Evans v. Faught,* 231 Cal.App.2d 698 (1965).

covenant running with the land, limiting the use of property; building restrictions; a reservation of a right-of-way; an easement; an encroachment; a lease; a deed of trust; or the pendency of a condemnation proceeding.

22-5 Mortgages A mortgage contract between debtor and creditor affords the latter by way of security against the outstanding debt a charge upon the land of the former.[6] It is an encumbrance created to secure a debt and takes the form of a defeasible conveyance of the realty described. The defeasance clause declares the conveyance void if the debt in question is discharged by the stipulated day. The mortgage transaction, under what seems to be the general rule, gives to the grantee-creditor (the "mortgagee") a lien against the property[7] but leaves with the grantor legal title (technically, an *equity of redemption*) and all incidents of ownership. In some jurisdictions the so-called *common-law* or *title* theory obtains, by which the mortgagee becomes the owner of the legal estate and would be entitled to possession unless this last right is expressly reserved by the mortgagor. The mortgagor retains, in any event, an equity of redemption conditioned upon satisfaction of the debt.

Typically, the debtor ("mortgagor") signs a bond or a promissory note along with the mortgage. This arrangement adds the debtor's personal liability to the security which the mortgage represents.

When the mortgagor fails to fulfill his mortgage commitments, the mortgagee's usual recourse[8] is to invoke the judicial proceeding known as *foreclosure,* by which the property covered by the mortgage is subjected to sale and all rights of the delinquent mortgagor, including his equity of redemption, are ultimately cut off. If the foreclosure sale does not provide sufficient funds with which to liquidate the mortgage debt, the mortgagee-creditor may sue on the note and secure a *deficiency judgment* for the balance due. On the other hand, any surplus realized from the sale belongs to the mortgagor.

A mortgage and the debt for which it stands may be assigned by the mortgagee to an outsider. The latter "stands in the shoes of the mortgagee" and finds his interest subject to all equities which may have existed between the parties to the mortgage. Up to the time he is actually notified of the assignment, the mortgagor may safely pay to the mortgagee any sums in reduction of the outstanding obligation.

When the debt is paid and the mortgage released, whatever paper the mortgagor receives evidencing the discharge should be recorded as was the mortgage itself, so that any subsequent title search will reveal that the burden represented by the mortgage no longer attaches to the land.

---

[6] To be effective as against subsequent parties in interest, the real property mortgage should be recorded, in the same manner as a deed, in the appropriate office of the official in the county which is the situs of the realty.

[7] Note that, since the mortgagee acquires some interest in the realty, the statute of frauds applicable in the particular jurisdiction must be satisfied.

[8] He can, if he desires, sue on the note and ignore the mortgage. However, this would be an unusual procedure.

22-6 Servitudes *Servitude* designates some privilege in reference to the land and held by one other than the landowner. A servitude amounts to a charge or encumbrance on the real estate. The three categories of servitude are easement, license, and profit a prendre; they will be considered separately in the following articles.

22-7 Easements An *easement* is a liberty or right which one person has to use the land of another for a specific purpose. It exists distinct from the ownership of the soil itself and can be created by grant, by implication,[9] by eminent domain, or by prescription.[10] A deed conveying an easement not only must describe the land to which the use granted attaches but also must detail the particular use permitted.

Other characteristics of the easement are as follows: (1) it is assignable; (2) it is inheritable; (3) it may be transferred independently of the land to which appurtenant; (4) a person cannot have an easement in his own lands.

The existence of an easement presupposes two distinct tenements:[11] the *dominant estate,* to which the right belong, and the *servient estate,* upon which the easement or servitude is imposed.

Here is an example of the functioning of easements. An industrial corporation desired to build a high-tension power line from its new generating station on the shores of Lake Superior to its taconite plant some 70 miles inland. The contemplated power line would cross in its journey a number of pieces of private property, and easements were therefore obtained from the serveral owners. The latter can continue to utilize the involved property for any purpose not inimical to the existence and operation of the power line. They cannot, however, change their minds and compel the corporation to remove its towers and cables. And, when the owners eventually sell the affected lands, the new property holders are bound by the easements previously granted.

An easement is extinguished by agreement of the several interested parties or upon cessation of the need therefor. Thus, in the above example, should the

---

[9] Three conditions are essential for an easement by implication. They are the ownership of the dominant and servient estate by a common grantor followed by a separation of title; use of easement prior to the separation in an apparent, obvious, continuous, and manifestly permanently manner; and necessity that the easement is essential to the beneficial enjoyment of the dominant estate. *Miller v. Schmitz,* 400 N.E.2d 488 (Illinois 1980).

See *Johnson v. Robinson.* 26 Md.App. 568, 338 A.2d 88 (1975), where a seller of real estate conveyed a parcel of land which was accessible from a public road only through land retained by the seller or through land of a stranger.

Implied easements may, depending upon the circumstances, run either way, in favor of grantor or grantee. In *Blanchet v. Ottawa Hills Co.,* 63 Ohio App. 177, 25 N.E.2d 861 (1939), the owner of a house, equipped with sanitary facilities which were served by a private underground drain, sold the lot on which the house was located and retained title to lands through which the drain ran. The court held that the purchaser of the house acquired an "implied easement" for the continued use of the drain to its outlet.

[10] That is, by adverse possession. See Art. 22-18.

[11] The word "tenement" comprehends everything of a permanent nature which may be held. It signifies an estate in land or some estate or interest connected with, pertaining to, or growing out of the realty.

industrial corporation tear down its original high-tension line in area *A* and replace it with a new one through area *B,* the easements acquired to effect passage of the line though its now-abandoned route would terminate, and new easements from property owners in area *B* should be secured.

22-8   Licenses   A *license* amounts to an excuse for what would otherwise be a trespass. A mere license to use the land of another in a given way or for a stated purpose differs from an easement in that the latter confers an interest in the land and may not be terminated at the pleasure of the servient owner. A license is simply a personal, unassignable, revocable privilege, which is generally granted orally.

By way of illustrating the concept of license, suppose a railroad grants a trucking concern permission to use an existing private roadway crossing over the tracks, on condition that the crossing be removed upon notice from the railroad and that the trucking outfit promise to indemnify the railroad against all claims arising from use of the crossing by the trucking or its invitees. Such an arrangement constitutes a license rather than an easement.

22-9   Profit a Prendre   *Profit a prendre* is taken from the French and means, literally, "profit to take," the words "from land" being implied in supplement. A profit a prendre, popularly shortened to the single word "profit," is an interest in realty and (under some statutes, at least) must be created by a properly executed writing. In this respect it differs from a license. In a comparison of profit and easement, both of which create interests in real property, the distinguishing feature of a profit is the right to take from the land of another part of the soil or a product of same in which there is supposed value, while the distinguishing feature of an easement is the absence of any right in the privilege holder to participate in the profits of the soil.

22-10   The Doctrine of Support   A landowner enjoys the common-law right to lateral support,[12] which is the right to have his land in its natural state[13] supported by the adjoining land. The abutting proprietor may excavate his land up to the very boundary line, and he may use the soil as he chooses provided he replaces by artificial means the support thus removed. If such abutting owner, without

---

[12] Of importance particularly in connection with mining operations is the companion right to *subjacent support,* this being the right of land to support from land which underlies it.

[13] It has been held that, before the right can be extended to buildings or other improvements imposed upon land, negligence or malice of the defendant (adjoining landowner) must be shown, with the exception of those instances where the building was erected on the premises shown, with the exception of those instances where the building was erected on the premises before a part of the land contiguous thereto is sold. *Commonwealth v. Solley,* 384 Pa. 404, 121 A.2d 169, 171 (1956). It might be pointed out that this also works the other way. If a landowner builds improvements and then sells the improved land, retaining a contiguous tract, his duty of lateral support is increased with respect to the improvements on the land he sold. *Bay v. Heine,* 9 Wash. App. 774, 515 P.2d 536 (1973).

Incidentally, the negligence referred to in *Solley* can be predicated upon a departure from normal methods of excavation established by accepted engineering practices.

making provision for substitute support, as by a retaining wall, removes the soil from his own land so near his neighbor's property that the latter's soil will crumble away under its own weight, the actor is responsible for damages so caused. Put another way, he who withdraws the necessary lateral support of another's land must respond in damages for any substantial subsidence of such land, absent an intervening cause or some other reason occasioning relief from liability. The actionable wrong is not the excavating itself, but the allowing of the neighbor's land to fall.[14]

A clear explanation of the highly technical and rather complicated concept of the right to lateral support is the following, taken from *Sime v. Jensen:*[15]

The right of lateral support from adjoining land consists in having the soil in its natural condition remain in its natural position without being caused to fall away by reason of excavations or improvements made on adjacent land. . . . The right to lateral support is said to be a natural one of property, arising from the fact that in a state of nature all land is held together and supported by adjacent lands by operation of the forces of nature. Ownership of land is acquired and held subject to the rights and burdens arising from that situation. Supported land has a right of lateral support from that which naturally affords it support. Supporting land is burdened with affording such support to land which it naturally supports. . . .

As a necessary corollary, it follows that there is no right of lateral support for land where the natural condition thereof has been altered through man's activity so as to create need for lateral support where none existed in a state of nature, as where an owner raises his land above his neighbor's by filling. The right of lateral support "does not include the support needed because of. . .artificial alterations in the supported land. . . . The measure of this right of the other and of this duty of the actor is the natural dependence of land upon land, and the right and duty are not enlarged by alterations of the natural condition." (Restatement. Torts §817c.) An owner who by filling raises his land above the level of adjoining land has no right of lateral support from the latter and is under the duty to keep soil used in so raising his land off adjoining land. . . . The owner who by filling raises the level of his land above that of his neighbor's is bound to build a retaining wall or other structure if necessary to keep such soil within his own line. . . .

A retaining wall or other structure to keep raised land from falling cannot be erected on adjoining land without invasion of the right of the owner thereof to its exclusive use and enjoyment. An owner who raises the grade of his lot must build such a wall or

---

[14] Since a landowner has an absolute right to have his land in its natural state laterally supported by the lands of adjoining landowners, the adjoining landowners are absolutely liable for damages caused by removal of support even though they are free from any negligence. The rule is different, however, where the case involves injury to buildings caused by the withdrawal of lateral support, and liability must be based on a negligence theory since there is no absolute right to lateral support of buildings. *Spall v. Ganota,* 406 N.E.2d 378 (Ind. App. 1980). See also *Urosevic v. Hayes,* 590 S.W.2d 77 (Ark. 1979).

There is a contrary view. In *Gladin v. Vonengehn,* 575 Pac.2d 418 (Col. 1978), the court said that the application of strict liability should not be based upon whether the thing damaged is natural or artificial, but rather the distinction must hinge on whether the artificial condition created on the plaintiff's land contributed to the injury, or whether the subsidence would have occurred even if the land had remained in its natural state.

[15] 213 Minn. 476, 7 N.W.2d 325 (1942).

structure entirely on his own land. . . . The adjoining owner is entitled to a mandatory injunction to compel the removal of a wall encroaching on his land. . . . The adjoining owner cannot be compelled to pay any part of the cost of a structure necessary to hold the raised land of his neighbor. . . .

There are statutes which deal with the doctrine of support. One example is California Civil Code Section 832 which reads:

Each coterminous owner is entitled to the lateral and subjacent support which his land receives from the adjoining land, subject to the right of the owner of the adjoining land to make proper and usual excavations on the same for purposes of construction or improvement, under the following conditions:

1 Any owner of land or his lessee intending to make or to permit an excavation shall give reasonable notice to the owner or owners of adjoining lands and of buildings or other structures, stating the depth to which such excavation is intended to be made, and when the excavating will begin.

2 In making any excavation, ordinary care and skill shall be used, and reasonable precautions taken to sustain the adjoining land as such, without regard to any building or other structure which may be thereon, and there shall be no liability for damage done to any such building or other structure by reason of the excavation, except as otherwise provided or allowed by law.

3 If at any time it appears that the excavation is to be of a greater depth than are the walls or foundations of any adjoining building or other structure, and is to be so close as to endanger the building or other structure in any way, then the owner of the building or other structure must be allowed at least 30 days, if he so desires, in which to take measures to protect the same from any damage, or in which to extend the foundations thereof, and he must be given for the same purposes reasonable license to enter on the land on which the excavation is to be or is being made.

4 If the excavation is intended to be or is deeper than the standard depth of foundations, which depth is defined to be a depth of nine feet below the adjacent curb level, at the point where the joint property line intersects the curb, and if on the land of the coterminous owner there is any building or other structure the wall or foundation of which goes to standard depth or deeper then the owner of the land on which the excavation is being made shall, if given the necessary license to enter on the adjoining land, protect the said adjoining land and any such building or other structure thereon without cost to the owner thereof, from any damage by reason of the excavation, and shall be liable to the owner of such property for any such damage, excepting only for minor settlement cracks in buildings or other structures.

22-11 Rental of Real Property *Lease or rental* of realty relates to a contractual arrangement between two parties designated in common parlance as *landlord* and *tenant,* whereby the owner divests himself for a period of time of the possession and use of his property in exchange for a stipulated compensation in the form of *rent.* The lease may be granted for life, for years, "at will," etc., but always for some period short of the time interest which the lessor has in the premises.[16] At the

---

[16] That is to say, a lease creates a lesser estate from the greater. If the *entire* interest were being conveyed, the transaction would more properly be designated an "assignment" than a "lease."

close of the lessee's term, the owner (landlord) has the absolute right to reenter and make full use of his property.

A lease is a contract and is governed by the same rules as other contracts. Ordinarily, the landlord assumes the expenses occasioned by taxes, insurance, and reasonable upkeep on the rented property. The use to which leased premises are put must be in line with that agreed upon between the parties. Thus, renting a structure for living purposes does not entitle one to utilize it as a store[17] or other business establishment.

Certain covenants[18] are implied in every lease. These include a covenant by the tenant to pay rent, and covenants by the landlord (1) that he has title to the property and (2) that the tenant will have "quiet enjoyment" of the premises. The latter covenant is breached by eviction, actual or constructive.

If the terms of the lease do not expressly preclude such, the tenant can assign[19] or can sublet. An *assignment* transfers the tenant's entire interest in the remaining portion of the term. The assignee is deemed to have accepted such obligations as the lease imposes and is responsible for the rent. Without a release from the landlord, the assignor retains a secondary liability with respect to the rent; he becomes, in a sense, a surety. A *sublease* results when the tenant transfers a portion of the unexpired term but reserves some to himself. There is no relationship between landlord and subtenant, and no rent can be recovered by the landlord from the subtenant.

An oral lease for more than one year is usually unenforceable under the statute of frauds.[20] A lease of appreciable duration should be recorded so as to give any prospective purchasers of the property due notice of the true situation.

22-12   Disposing of Real Property   Real property may pass by intestate succession,[21] or may be devised under the will of the owner. During his lifetime the owner may convey[22] the realty by way of gift or sale. Relatively infrequently, transfer of real property from one owner to another is effected on an involuntary basis. Cases of this nature include (1) forced sale for back taxes; (2) sale attendant upon foreclosure of a mortgage; (3) bankruptcy (in which event any nonexempt

[17] Such an attempted use might well run into the additional obstacle of zoning restrictions limiting the area to residences.

[18] A *covenant* is defined as an undertaking to do, or, in the case of a negative covenant, *not* to do, a stated act. The remedy for breach of an affirmative covenant is by damages or, sometimes, by a bill for specific performance. When a negative covenant is violated, the appropriate relief is an injunction. Some covenants are personal in nature, while others "run with the land" and hence pass from one beneficiary to his successor in interest.

[19] Even where a prohibitory clause exists, the landlord may waive his right to object by accepting rent from the assignee.

[20] See the New York provision quoted in Art. 22-2.

[21] Statutory provisions govern disposition of a decedent's property where no will is to be found. Typically, the state prescribes that the surviving spouse receive the use for life of something like one-third (of what is left after payment of debts and charges against the estate): the remaining two-thirds would go to the children.

[22] *Conveyance* is a word of broad meaning which comprehends the several modes of transferring title to real estate from one person to another.

real property of the bankrupt passes to the trustee for the benefit of the creditors); (4) acquisition by adverse possession, a concept to be considered in some detail in Art. 22-18; and (5) eminent domain, regarding which see Art. 22-19.

22-13   Mechanics of the Sale Transaction   One very common means of disposing of real estate is by sale. The ordinary sequence of events is (1) the signing of a contract of sale, which is an agreement to buy, on the one hand, and to sell, on the other, a given piece of property upon terms stated; (2) title search and mortgage arrangements by the prospective purchase;[23] (3) the closing, wherein the vendor ("seller") delivers a deed to the property in exchange for payment of the agreed-upon consideration; (4) recording of the deed and other documents in the official land records at the situs of the property.

22-14   Contract of Sale   A *contract of sale* is an executory contract for the sale of realty. It is not a transfer of title, and it is only after the would-be purchaser has fulfilled the several conditions[24] of the contract that he has a right to demand delivery of the *deed of conveyance,* vesting in him title to the property.[25]

The contract of sale ordinarily covers a number of points, among them the following:

1 Full description of the property involved
2 Selling price and terms of the sale
3 Determination of which party will pay for such items as the water bill from the date of the closing to the end of the current period for which a bill is rendered
4 Statement about existing encumbrances or restrictions of any sort affecting title
5 Statement about whether purchaser will assume any mortgage which may be outstanding or about what other disposition will be made of such obligation. Perhaps the most common arrangement is for the seller to discharge his mortgage, using a portion of the sale proceeds, and for the purchaser to bring his own bank into the picture as a new mortgagee on the property
6 Type of deed the seller will furnish
7 Date and place for the official clossing

---

[23] The purchaser may desire title insurance, which would indemnify him in a stated amount against loss arising through defects or encumbrances (affecting the title) which may be in existence when the policy is issued.

[24] Primarily, payment of the indicated price. A portion of the selling price will be paid at the time the bond for deed is executed, and this down payment is recited in the instrument as part of the consideration for the seller's promise to deliver a deed. The balance of the agreed-upon price is to be forthcoming at the closing.

[25] If the seller should renege on his promise to transfer the property, the buyer presumably can secure a remedy of *specific performance,* forcing the seller to go through with his bargain. Damages would be an inadequate remedy in many cases, since each parcel of real property is regarded as unique, and the buyer is entitled to the precise land he bargained for, and not simply to its supposed monetary equivalent.

When the title search[26] has been completed[27] and ambiguities of whatever sort resolved to the satisfaction of both parties, the actual closing is the next item of business. At the closing, any mortgage transactions are consummated, transfer taxes are paid, the selling price (adjusted in minor respects for amounts due purchaser or seller) is turned over, and the deed is delivered. Then the documents will be recorded, and third parties are thus put on notice that new ownership of the property is now an accomplished fact.

22-15   The Deed; Function and Types   A *deed* is a written instrument by which the grantor conveys to his grantee some interest, right, or title in or to certain real estate described in the document. The purpose of the deed is to declare the fact of conveyance and to stand as evidence of the transfer of title. It represents the act of but one of the parties, being signed only by the grantor, and, unless a gift is intended, is drawn in fulfillment of the grantor's commitment under a previous agreement with the grantee.

The statutes prescribing the requisites of deeds vary with the jurisdiction. Every deed should contain, among other things, the names of the parties concerned, apt words of conveyance, and a full description of the property in question,[28] indicating what buildings, appurtenances, and privileges are included with the land. The instrument will also recite the consideration for the transfer, the quantity of inter-

---

[26] An "abstract of title" may have been supplied. Such abstract (the object of which is to enable the buyer and his attorney to pass readily upon the validity of the title) is a concise and orderly statement covering all matters, such as conveyances and encumbrances, which appear on the public records and which affect title to the reality in question.

[27] If the buyer has indeed arranged for new financing of his own, his mortgagee, if a bank, will normally have its own lawyer conduct a title search in protection of the bank's interest, and probably charge this to its customer, the buyer, who usually will not then bother to have still another search made.

[28] It is not unusual to find that, with reference to a particular piece of reality, a detailed map has been drawn as the outgrowth of a survey and the establishment of precise boundary lines; a sample map is shown in Fig. 22-1. Among other ways, property descriptions may be expressed in terms of lot numbers and maps, or in terms of who occupies adjoining premises, or in terms of *metes and bounds*. The last-named form is common in certain areas of the country, particularly in the northeast. The phrase "metes and bounds" refers to the boundary lines or limits of a tract, with their terminal points and angles; such boundaries may be identified by relating them to fixed objects, either natural or artificial. Here is the substance of an actual description using metes and bounds; obviously, the impermanence of certain markers, and the probable difficulty in locating them after many years have passed, are among the frailties inherent in such a description:

"Commencing at the east end of a stone wall on the southerly side of Hill Road at a point on the west line of premises of *A,* said point being the northeast corner of premises herein conveyed; thence westerly along the southerly side of said Road 1000 feet to an iron fence post on the southerly side of the Road near telephone pole marked 127/34; thence southeasterly along line of iron pins marking northeast line of other lands of Grantors, approximately 125 yards to center line of old lumber road near pile of large stones; thence southerly along other lands of Grantors, following center line of road marked by iron pins to south bank of a brook; thence westerly along south bank of brook to east line of premises of *B*; thence southerly along *B*'s east line to hilltop and north line of premises of *C*; thence easterly along *C*'s north line to stone wall on west line of premises of *D*; thence northerly along said stone wall, easterly along a fence and northerly along a fence, all along premises of *D,* to stone wall on south line of premises of *A*; thence westerly along said stone wall and northerly along fence marking *B*'s west line to beginning, 100 acres more or less."

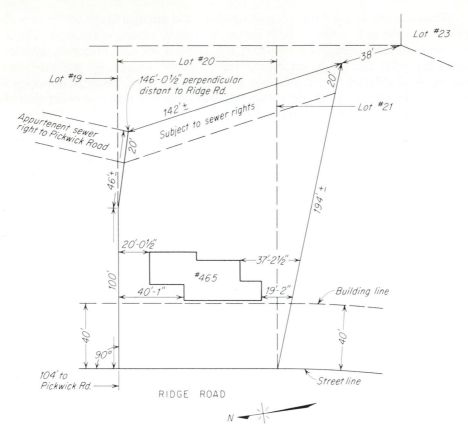

FIGURE 22-1
Plot map accompanying deed.

est conveyed, and any conditions, reservations, and the like. The deed must be properly executed in the presence of whatever number of witnesses the applicable statute requires and must actually be delivered to the grantee. Unlike a will, a deed is irrevocable and takes effect upon delivery, thus passing to the recipient an immediate or "present" interest.

There are two principal types of deeds in general use today:

Warranty Deed   This is the kind the buyer invariably wants but may have difficulty getting in a "seller's market." The *warranty deed* contains five express covenants which run with the land: seisin,[29] quiet enjoyment, freedom from encumbrances, further assurance, and title. Reproduced below is the form of such deed which appears as a sample in Section 258 of the New York Real Property Law.[30] Note particularly the above-listed five covenants:

---

[29] *Seisin* means the possession, by right, of an estate of freehold.
[30] McKinney (1968).

DEED WITH FULL COVENANTS

Statutory Form A   (Individual)

This indenture, made the —————— day of —————— nineteen hundred and ——
————, between —————— (insert residence) party of the first part, and ——————
(insert residence) party of the second part:

Witnesseth, that the party of the first part, in consideration of —————— dollars,
lawful money of the United States, paid by the party of the second part, does hereby
grant and release unto the party of the second part, —————— and assigns forever, all
—————— (description), together with the appurtenances and all the estate and rights
of the party of the first part in and to said premises.
   To have and to hold the premises herein granted unto the party of the second part —
—————— and assigns forever. And said —————— covenants as follows:
   First. That said —————— is seized of said premises in fee simple, and has good
right to convey the same;
   Second. That the party of the second part shall quietly enjoy the said premises;
   Third. That the said premises are free from incumbrances;
   Fourth. That the party of the first part will execute or procure any further necessary
assurance of the title to said premises;
   Fifth. That said —————— will forever warrant the title to said premises.
   In witness whereof, the party of the first part has hereunto set his hand and seal the
day and year first above written.
   In presence of:

Quitclaim Deed   By this form the grantor turns over such title and interest as
he may have in the property but makes no promise whatever in the nature of a
covenant. A quitclaim deed covering a piece of real estate does not purport to
convey anything more than the interest of the grantor as it existed at the time the
instrument was executed. As a rule, such a deed is ineffectual in transferring to the
grantee any title subsequently acquired by the grantor. This type of deed is employed when a person wants to sell an interest of which he is not sufficiently sure
to merit chancing a warranty deed.
   A sample version of a quitclaim deed, as contained in Section 258 of the New
York Real Property Law, follows:

QUITCLAIM DEED

Statutory Form D   (Individual)

This indenture, made the —————— day of —————, nineteen hundred and ——
————, between —————, (insert residence), party of the first part, and ——————,
(insert residence), party of the second part:
   Witnesseth, that the party of the first party, in consideration of —————— dollars,
lawful money of the United States, paid by the party of the second part, does hereby
remise, release, and quitclaim unto the party of the second part, —————— and assigns
forever, all (description), together with the appurtenances and all the estate and rights of
the party of the first part in and to said premises.

To have and to hold the premises herein granted unto the party of the second part,
————— and assigns forever.

In witness whereof, the party of the first part has hereunto set his hand and seal the day and year first above written.

In presence of:

22-16   Recordation   Generally speaking all conveyances of real property must be recorded[31] in order to be effective as against subsequent parties in interest. In this connection, Section 291 of the New York Real Property Law reads in part:

> Every such conveyance not so recorded is void as against any person who subsequently purchases or acquires by exchange or contracts to purchase or acquire by exchange, the same real property or any portion thereof, or acquires by assignment the rent to accrue therefrom. . ., in good faith and for a valuable consideration, from the same vendor or assignor, his distributees or devisees, and whose conveyance, contract or assignment is first duly recorded.

This means that a buyer who fails to record his deed has no claim to the land if his unscrupulous vendor subsequently executes another deed to the same property, and the second buyer (having made a good faith purchase without knowledge of the prior transaction) does record his deed. The initial purchaser is left with a damages action against the double-dealing vendor.

22-17   Dedication   *Dedication* is any process, formal or otherwise, by which a landowner intentionally sets his property apart for public use. Property dedicated for a particular purpose generally cannot be used for another purpose,[32] and the use of property for purposes other than those for which it was dedicated cannot be authorized by ordinance.[33]

But, dedication is usually regarded as carrying with it the right of the public to utilize the land in a way which accords with changing times and conditions. Thus, real estate dedicated for public use as a highway may continue to be so employed, not only in the fashion in which streets or highways were ordinarily used as of the time of dedication, but also to accommodate the most recently invented vehicles and modern methods of travel.[34]

Effective dedication requires the presence of an unequivocal offer by a competent person to commit his property and a clear acceptance on the part of the intended recipient. Acceptance is occasionally evidenced by passage of a resolution of some sort, but is much more likely to take the form of action or conduct which is unmistakable in reliance upon the outstanding offer and which is of such a nature as to bespeak acceptance. An illustration might be the start of public school construction on a site dedicated for the purpose.

---

[31] Typically, in the office of the clerk of the political subdivision which is the situs of the realty concerned.

[32] *Big Sur Properties v. Mott,* 62 Cal.App.3d 99, 132 Cal. Rptr. 835 (1976).

[33] *Zumbrotta v. Strafford Western Emigration Company,* 290 N.W.2d 621 (Minn. 1980).

[34] *McCraw v. City of Dallas,* 420 S.W.2d 793, (Ct. of Civ. App. of Tex. 1967).

A number of jurisdictions have statutory provisions stipulating how land may be committed to public use; *statutory dedications* are grants made in a manner conforming to such provisions. These laws are not exclusive, however, and dedications effected outside their purview are referred to as *common-law dedications.* It has been said that, under a common law dedication, the public acquires simply an easement, or such an interest in the land as is necessary for its use.[35]

22-18   Acquiring Title to Real Property by Adverse Possession   Upon occasion, title to real property, or a right in respect of such property, is acquired by a process known as *adverse possession.* If *A* occupies a parcel of land belonging to *B* without the license or consent of the latter and such possession and utilization of the premises continues uninterruptedly for a period of years,[36] the claimant's dominion over the property is said to ripen into title.[37] To reach that state of affairs, however, the occupancy by the adverse possessor must have been open, notorious,[38] continuous,[39] actual, exclusive, and hostile, that is, under a claim of right which runs counter to the true owner's interest. Acquisition of title through operation of the doctrine under discussion is not an easy matter, since the adverse possessor must overcome the presumption that his occupancy of another's land is in subordination to the title of the real owner. Thus, even a token payment of rent destroys the requisite pattern, since such a payment by the occupant acknowledges that ownership and control belong elsewhere.

Quite commonly a "right-of-way" comes about through the medium of adverse possession. Occasionally a property owner will take positive steps to see to it that that sort of thing does not happen in connection with his holdings. For example, a country club owns a suspension footbridge over a body of water in Connecticut. This bridge is often used by nonmembers as part of a shortcut across the lake and the golf course. Once a year the club exercises its prerogative and closes the bridge, posting a no-trespassing sign to warn outsiders to stay away. Such action is taken to destroy any continuity of use on the part of nonmembers and thus to preclude any possible contention that a right-of-way has been acquired through adverse possession.

22-19   Eminent Domain   The right of *eminent domain* enables the federal government or the state or those to whom the power has been lawfully delegated to take private property and appropriate it to public use. The owner's interest is safeguarded in two ways: (1) his property can only be taken by due process of law

---

[35] *North Spokane Irrigation District No. 8 v. County of Spokane,* 86 Wash.2d 599, 547 P.2d 859 (1976).

[36] The precise number depends upon the statutes of the state wherein the real estate is situated.

[37] Title achieved through adverse possession is as perfect a title as one transferred by deed from the owner.

[38] The requirement that a would-be adverse-possession claim be manifest from the nature or circumstances of the possession is imposed so that the real owner may be informed of the possessor's apparent intent and not misled into acquiescence in what the owner might otherwise reasonably suppose to be a mere trespass.

[39] Occasional entries upon someone else's property for such a purpose as cutting timber is not adverse possession within the meaning of the law.

and (2) he is entitled to payment of reasonable compensation. Statutes of the particular jurisdiction will set forth the exact procedure which must be followed in condemning the land and in settling any dispute over the amount the owner should be paid.[40]

The right of eminent domain is an inherent and necessary attribute of sovereignty. Were it not for the existence of such a power, the government would find it difficult or impossible, short of paying truly exorbitant prices, to secure, for instance, whatever land it needs for construction of a state highway or for the site of a new school building.[41]

22-20   Zoning   The essential purpose of zoning is to rationally coordinate land-use planning to promote orderly development and preservation of property values.[42] Another court defines zoning to mean "division of a city by legislative regulations having to do with structural and architectural designs of buildings and of regulations prescribing use to which buildings within designated districts may be put."[43] Through this process a municipality is mapped out as a grouping of separate areas, in each of which only certain designated uses of land are permitted, to the end that the community may develop in an orderly manner in accordance with a comprehensive scheme and for the good of its residents as a whole. As one example, social welfare is promoted by establishing residential areas free from the offensive smoke, odors, noise, and commotion which normally are native to industrial operations. The assurance that other parcels in a given residential area will not be devoted to objectionable uses can be thought of as compensating a particular owner for his inability (because of zoning restrictions) to utilize his own lot for other than residential purposes.

While they are related concepts, zoning and planning are readily distinguishable one from the other. The former is concerned primarily with the use and regulation of realty and structures, while the latter term is of broader scope and significance, contemplating the evolvement of an overall program or design covering the present and future physical development of the total community and all aspects thereof (parks, streets, etc.).

Zoning regulations derive from the "police power" of the duly constituted governmental unit. Property rights are held subject to such power, and the legality of zoning has long since been established beyond doubt. Perhaps the statutory scheme in Connecticut will serve as a reasonably typical example of the conferring of zoning powers upon local communities. Section 8-1 of the Connecticut General Statutes provides that any municipality may, by vote of its legislative body, "adopt the provisions of this chapter and exercise through a zoning commission

---

[40] For an example of laws dealing with eminent domain, see Chap. 835, Title 48, Connecticut General Statutes (1978).

[41] The proceeding enforcing the right of eminent domain is sometimes referred to as condemnation.

[42] *Fargo Cass County v. Harwood,* 256 N.W.2d 694 (N.D. 1977).

[43] *City of Moline Acres v. Heidbreder,* 367 S.W.2d 568, 572 (Mo. 1963), quoting *Black's Law Dictionary,* 4th ed., and cases therein mentioned.

the powers granted hereunder." Section 8-2 indicates the purpose of zoning, and tells *what* can be regulated and *how*:

The zoning commission of each city, town or borough is authorized to regulate, within the limits of such municipality, the height, number of stories and size of buildings and other structures; the percentage of the area of the lot that may be occupied; the size of yards, courts and other open spaces; the density of population and the location and use of buildings, structures and land for trade, industry, residence or other purposes, and the height, size and location of advertising signs and billboards. Such zoning commission may divide the municipality into districts of such number, shape and area as may be best suited to carry out the purposes of this chapter; and, within such districts, it may regulate the erection, construction, reconstruction, alteration or use of buildings or structures and the use of land. All such regulations shall be uniform for each class or kind of buildings, structures or use of land throughout each district, but the regulations in one district may differ from those in another district, and may provide that certain classes or kinds of buildings, structures or use of land are permitted only after obtaining a special permit or special exception from a zoning commission, planning commission, combined planning and zoning commission or zoning board of appeals, whichever commission or board the regulations may, notwithstanding any special act to the contrary, designate, subject to standards set forth in the regulations and to conditions necessary to protect the public health, safety, convenience and property values. Such regulations shall be made in accordance with a comprehensive plan and shall be designed to lessen congestion in the streets; to secure safety from fire, panic, flood and other dangers; to promote adequate light and air; to prevent the overcrowding of land; to avoid undue concentration of population and to facilitate the adequate provision for transportation, water, sewerage, schools, parks and other public requirements. Such regulations shall be made with reasonable consideration as to the character of the district and its peculiar suitability for particular uses and with a view to conserving the value of buildings and encouraging the most appropriate use of land throughout such municipality.

Zoning regulations may be made with reasonable consideration for the protection of existing and potential public surface and ground drinking water supplies and may provide that proper provision be made for sedimentation control, and the control of erosion caused by wind or water. Such regulations may also encourage energy-efficient patterns of development, the use of solar and other renewable forms of energy, and energy conservation. The regulations may also provide for incentives for developers who use passive solar energy techniques in planning a residential subdivision development. The incentives may include, but not be limited to, cluster development, higher density development and performance standards for roads, sidewalks and underground facilities in the subdivision. Such regulations shall not prohibit the continuance of any nonconforming use, building or structure existing at the time of the adoption of such regulations. Any city, town or borough which adopts the provisions of this chapter may, by vote of its legislative body, exempt municipal property from the regulations prescribed by the zoning commission of such city, town or borough; but unless it is so voted municipal property shall be subject to such regulations.

The courts will insist that zoning restrictions on the rights of landowners not be arbitrary, unreasonable, or discriminatory, and that they reveal a genuine concern for the public welfare. However, a presumption of validity attaches to decisions reached by appropriate administrative bodies, and determinations by zoning authorities will not lightly be overturned.

QUESTIONS

1 Define real property; differentiate it from personal property.
2 What is a quitclaim deed? A warranty deed?
3 If a man buys some land and the deed he receives states "in fee simple," what does this mean for the purchaser?
4 What are the essentials in a deed?
5 What is a title search? Why is one made?
6 Explain the function and operation of eminent domain.
7 How are boundaries generally recorded?
8 What is an easement? By what means is one obtained? Can the easement be (a) sold, or (b) relinquished?
9 What general limitations are there upon one's use of his own land?
10 Do water and mineral rights usually accompany the transfer of title to land?
11 What aspects of the property should one investigate before buying a given piece of real estate? Explain fully.
12 What is meant by an "encumbrance" in regard to real property? Give serveral illustrations.
13 What is a lease? What restrictions are normally imposed upon the use of leased property?
14 What is a license for the use of land? How is it obtained? How is it terminated?
15 What is a condominium? A cooperative?
16 Distinguish between lease and license. Illustrate.
17 What is the meaning of "profit a prendre"?
18 What is a mortgage on real property?
19 Can a mortgage be assigned? Can a lease be assigned?
20 Explain the proper steps to take if one is (a) to mortgage his real property, (b) to pay off a mortgage.
21 If a piece of real estate is sold, what of the existing mortgage thereon? Does the mortgagee have any effective objection to the sale or to the identity of the purchaser?
22 Why does a mortgagee generally require the mortgagor to sign a bond or a promissory note as well as the mortgage itself?
23 How may realty be conveyed?
24 Explain the steps to be taken in selling a piece of realty.
25 What is an "abstract of title"?
26 What are metes and bounds?
27 Where should the instrument evidencing the conveyance of real property be recorded? Why the recording?
28 If a man buys some land, what rights, broadly speaking, does he acquire regarding such land?
29 Can two or more persons own a given piece of land? If so, on what basis?
30 If two persons own a given piece of land, can one sell it despite objections by the other? Explain.
31 What is a "life estate" in real property? If a man has a life estate in land, can he sell the property itself? Why?

32 Can ownership of real property be conveyed orally? Why?

33 What is a "covenant" as the term is employed in the law of real property? Illustrate.

34 What remedies are available for breach of a covenant?

35 As regards realty, distinguish between sale and lease.

36 Under what conditions is real property transferred without the "approval" of the owner, and very likely against his wishes?

37 What points are generally covered in a bond for deed?

38 Explain what steps are involved at the "closing" of a real estate sale.

39 On what basis is the state able to take land (needed, for example, for a new highway) regardless of the wishes of the property owners involved? What is the justification?

40 What is an "implied easement"?

41 What is the "right of lateral support"? Illustrate.

42 Distinguish between "statutory" and "common-law" dedication.

43 What is the purpose of zoning?

44 Distinguish between zoning and planning as related to land use.

46 Brown owned 25 acres of woodland situated off the highway behind Sullivan's property, and she had an easement across the latter's land for access to the woodland. Thompson offered Sullivan a handsome price for his property, intending to construct a housing development thereon. Can Sullivan (or Thompson, after the sale is consummated) terminate Brown's easement without her consent? Elaborate.

46 What is the mortgagee's situation if the mortgagor fails to pay the taxes, a sewer assessment, and insurance on the property? Explain. What steps can the mortgagee take to enforce his rights?

CHAPTER 23

WATER RIGHTS

23-1 Introduction This chapter explores the rights of real property owners to the use and benefit of water which is either upon, adjacent to, or under their lands. Case law and statutory provisions respecting various aspects of water rights differ with the jurisdiction, and the legislatures in those states plagued with inadequate rainfall have enacted comprehensive control measures regarding the use of water.

A "water right" has no physical form, but is a legal interest akin to real property and may be bought and sold like any other property. The holder of the interest is privileged to make reasonable use, beneficial to him, of the water involved.

23-2 Definitions and Distinctions In contradistinction to those which seep or percolate through the soil or which flow in subterranean streams, *surface waters* are an accumulation of natural precipitation (rain or melting snow). They appear upon the surface of the ground in a diffused state and follow no well-defined course. Ultimately any identifiable segment of surface water will either be absorbed by the land, will evaporate, or will reach some lake[1] or channelized stream, whereupon it ceases to be categorized as "surface water." Conversely, flood waters becoming detached from a stream swollen by excessive rainfall and spreading over

---

[1] The term "lake" is taken to mean a body of water whose characteristics are (1) permanency, (2) location in a depression in the earth's surface, and (3) no perceptible flow (that is, the water is basically at rest). The distinction between a lake or pond and a pool is that the last named is generally a relatively small enlargement of a stream frequently above or below a rapid or waterfall and revealing an appreciable current as water flows through it.

the land, never to return to the main current, are generally though not universally regarded as "surface waters."

A *natural watercourse,* normally of great benefit to the land through which it flows, is described as a stream established when surface water begins to move in a definite direction and forms a regular channel or bed with discernible banks and sides;[2] such a stream, which ultimately empties into some other body of water, enjoys a measurable flow and a regular, though occasionally diminished or interrupted, source of supply. These concepts are deeply rooted in American law and have not changed in over 100 years. "To prove the existence of a watercourse it must be made to appear that the water usually flows in a certain direction, by a regular channel with banks or sides. It need not be shown to flow continually; it may be dry at times but it must have a well defined and substantial existence."[3] The flow of water need not be continuous but, to constitute a true watercourse, the stream must be distinguishable from mere surface drainage occasioned by freshets or other extraordinary causes.[4] Whether or not the stream qualifies as *navigable* depends upon whether it is susceptible of being utilized in its natural condition as a commercial highway upon which travel or trade may readily be conducted. The terms *public waters*[5] and *navigable waters* are not infrequently treated as synonymous, with the words *private waters* being employed to denote "nonnavigable" streams and the like.[6] The other basic grouping of waters, besides those which, in one form or another, overlie the land, is referred to as *groundwater.* This grouping in turn is divided into (1) *percolating water* and (2) water flowing in *channelized subterranean streams,*[7] the existence and location of which are ascertainable from surface indications such as to belie the need for exploratory excavation. To put the distinction another way, all subsurface water which does not travel in known channels is said to be percolating water. Percolating water, deemed a part of the soil which contains it, either flows in some unknown course or simply seeps or oozes[8] its movement through the ground being perpendicular or horizontal, or a little of each.

In the absence of actual knowledge, it is presumed that underground water is

---

[2] *Lowe v. Lowe Realty Co.,* 138 Ind.App. 434, 214, N.E.2d 400 (1966). *Johnson v. Whitten,* 384 A.2d 698 (Me. 1978).

[3] *Ullian v. Cullen,* 325 N.I.2d 593 (Mass. App. 1975), quoting *Ashley v. Wolcott,* 11 Cush. 192 (1853).

[4] *Reed v. Jacobson,* 169 Neb. 245, 69 N.W.2d 881 (1955).

[5] These are often of considerable size and commonly used for such public purposes as bathing, fishing, and navigation.

[6] It has been held that title to the beds of nonnavigable rivers lies with the riparian owners rivers rather than with the state *Prazzek v. Drainage District,* 237 P. 1059 (Kan. 1925).

A transfer of title of riparian land bounded by a nonnavigable waterbody also transfers title to the threadline of the bed, unless such is specifically excluded. *Snake River v. United States,* 395 F.Supp. 886 (D.C. Wyo. 1975): see also Oregon Revised Statutes 9303 (1975).

[7] Some authorities do not agree that water moving in underground streams can properly be denominated as "groundwater." In any event, regardless of the accuracy of the terminology employed, rights and liabilities respecting subterranean streams, as distinguished from percolating water, are generally governed by the rules of law pertaining to streams on the surface of the ground.

[8] *Jones v. Home Building & Loan Assn. of Thomasville,* 252 N.C. 626, 114 S.E.2d 638 (1960).

percolating and not a stream.[9] Percolating water may be precipitation[10] which has slowly filtered through the soil, or it may represent moisture which has escaped through the bed or banks of a stream and thence worked its way through the ground.

Water law is a highly statutory subject and varies greatly from state to state. Principal rules discussed in the following sections provide a general conceptual framework, but any water-right problem requires specific reference to the law of the jurisdiction in which the watercourse is found.

23-3   Riparian Rights   Perhaps the term riparian rights is more familiar than any other in the overall subject of water rights. *Webster's Third New International Dictionary* defines riparian right as the right of one owning land having access to and use of the shore and water.[11]

At the common law a person owning land bordering a nonnavigable stream owns the bed of the stream to the *filum acquae,* or thread of the stream, and may make reasonable use of its waters. A riparian owner is one who owns land on the bank of a river, or one who is the owner of land along, bordering upon, bounded by, fronting upon, abutting or adjacent and contiguous to and in contact with a river.[12] It follows that *riparian land* is a parcel which is bounded by or includes a portion of a natural watercourse. Technically, only one whose realty abuts a *river* is a "riparian owner," and the purists would say that one whose land touches upon a *lake* is more accurately referred to as a "littoral owner"; but it appears that "riparian" is generally regarded as an acceptable word when the land involved borders upon either river or lake.[13]

Unless the circumstances or the express terms of the transaction indicated a different intent, at common law a grant of land hugging a nonnavigable river conveyed property rights in the stream bed to the center or "thread," which is an

---

[9] *Woodsum v. Pemberton,* 412 A.2d 1064 (N.J. 1980). Artesian water, in contrast, exists under pressure in seams of rock or in porous strata of soil underlying impervious soils which prevent its escape. When a drilled hole provides an outlet channel, the water will rise in the hole until the pressures are equalized, or it may actually flow out at the surface if the pressure behind it is sufficient.

[10] When a storm occurs some of the precipitation falling in the area may run off promptly, some may evaporate, some may be taken up by vegetation, and some may penetrate into the ground or even into seams in the underlying rock.

[11] *Webster's Third New International Dictionary,* G. & C. Merriam Co., Springfield, Mass., p. 1960 (1981).

[12] The riparian proprietor must own land in actual contact with water; proximity alone does not suffice and carries no benefits relating to use of the water. To illustrate this point, assume that a man owns a strip of land 500 feet wide and extending 1000 feet along the east bank of a stream. He is a riparian owner with respect to that stream, whereas the owner of the property "inland from" this man's holdings is not a riparian owner because his land does not actually border the stream. It is possible, of course, that a nonriparian owner might acquire special water privileges through grant or otherwise, and it is also possible that the riparian proprietor's use of "his water" might be so negligently handled or might take such an ultrahazardous form as to occasion liability "running to" his neighbor, the nonriparian owner.

[13] If a person is contemplating the purchase of property adjacent to an artificial pond (or lake), he should carefully investigate the question of just what rights he will have to the water area and its use. It may be that the one who made the pond still owns part or all of the submerged land, and can empty the pond or even remove the dam if he so desires.

imaginary line at the middle of the main channel,[14] with the water being measured at its lowest stage. Riparian rights can be thought of as vested property rights, which are "attached to" the land and which may not legally be invaded or taken from the owner without just compensation. Such riparian proprietor is entitled to protect his interest by seeking to enjoin any injurious interference with the stream.

To the extent that a rule of broad applicability exists in this country on the subject of the legal rights of a riparian proprietor, every riparian owner is entitled to the natural flow in its accustomed channel of the content of a stream running through or along his land,[15] subject to the reasonable use of the water by those similarly entitled, for the ordinary purposes of life. In other words, a riparian right is reciprocal in character as to other riparian rights in the same body of water. Each riparian proprietor in turn has the right (which some sources say is "neither gained by use nor lost by disuse") to make reasonable utilization of the watercourse which abuts his property.[16] The right of any given owner is subject, however, to the equal right of all other "similarly-situated" riparian proprietors likewise to make reasonable use of the water during its passage through or along their respective parcels of land. What is perhaps a corollary of the foregoing proposition is the well-settled rule that any unwarranted obstruction of the channel, or diversion of the water, or comparable move by other owners which works injury to a given riparian proprietor entitles the latter to compensatory damages,[17] or an injunction in appropriate circumstances.[18]

The concept of riparian rights seems to be an inherently difficult one, and it

---

[14] Where a meandering river is involved, such as a portion of the mighty Mississippi (whose main channel in some places is slowly being altered by erosion and sedimentation), even state boundaries are likely to be altered somewhat with the passage of time. While on the subject of dividing lines, it might be noted that boundaries of land bordering ponds and lakes also give rise to uncertainty upon occasion. In the case of large lakes, boundaries may be set in the same terms as are usual with tidal waters, that is, the "normal highwater mark"; or they may be established by survey or by prior appropriation, etc. At any rate, the boundary lines should be carefully ascertained, rather than merely assumed, when one intends to take any action involving lakeside properties.

[15] Such flow to be undiminished in quantity (except as affected by an act of God) and unimpaired in quality.

[16] And he has the right to any accretions (deposits of soil) which the passing stream leaves on his property.

[17] In a case where the defendant had initiated a stream channelization project in order to reclaim 20 acres of land for farming purposes the plaintiff alleged that the result of the straightening of the stream was an increase in the velocity of the water causing scouring and abrasion to creek banks as well as increased flooding, thereby damaging the plaintiff through deposits of sand, silt, and extensive corrosion. Finding for the plaintiff, the court said that "although we have never specifically held a riparian owner may be liable for accelerating or hastening the flow of a water course to the injury of another, we believe permitting recovery for this type of damage is consistent with our prior decisions requiring a land owner to exercise due care for his neighbors when he avails himself of a water course." *Oakleaf Country Club, Inc. v. Wilson,* 257 N.W.2d 739 (Iowa 1977).

[18] *Calvaresi v. Brannan Sand & Gravel Co.,* 534 P.2d 652 (Colo. App. 1975); *Roberts v. Brewer,* 276 So.2d 574 (Ala. 1973).

An injunction is appropriate where the evidence establishes that the injury will be either continuous or repetitive. For example, the issuance of an injunction was held proper where the defendant had constructed several pits on his property to collect surplus water to enable him to irrigate his farmland and where on several occasions the pit wall failed to hold and discharged water and silt onto the plaintiff's land. *Delp v. Laier,* 288 N.W.2d 265 (Neb. 1980).

might be helpful to quote briefly from each of the several court decisions which discusses aspects of the subject in succinct fashion:

1 We. . .reassert the riparian right to be a vested property right inhering in and a part and parcel of the abutting lands, a right not gained by use or lost by disuse, a right (qualified only by the correlative rights of other riparian proprietors) to use the entire ordinary and natural flow of the stream for all lawful riparian uses and also to have all such flow come down to the land undiminished other than by the lawful uses by upper riparian proprietors or by the rights of those who have otherwise obtained a superior claim to the use of a portion of the stream. This right to use the water of the stream we hold to be entitled to the same respect and protection at the hands of the law as any other vested property right (*Fall River Valley Irr. Dist. v. Mt. Shasta Power Corp.*).[19]

2 An abutting or riparian owner of a lake, suitable for fishing, boating, hunting, swimming, and other domestic or recreational uses. . .has a right to make such use of the lake over its entire surface, in common with all other abutting owners, provided such use is reasonable and does not unduly interfere with the exercise of similar rights on the part of other abutting owners, regardless of the navigable or public character of the lake and regardless also of the ownership of the bed thereof.[20,21]

3 In Wisconsin riparian rights vary in accordance with the nature of the body of water. With respect to the ownership of the bed of the stream, a riparian owner owns to the thread (the geographical center) of the stream. . . . The title of the riparian owner is, however, a qualified one, subject to the paramount interest of the state. . . . However, the owner of the land abutting a natural lake or pond owns to the waterline only, since title to the submerged lands beneath a permanent body of natural water belongs to the state. An abutting property owner on a natural lake, except for the right of access, has no more rights as a riparian than any other member of the public. . . . Riparian rights include the right to use a body of water "for bathing, swimming and boating purposes." . . . An artificial lake located wholly on the property of a single owner is his to use as he sees fit, provided, of course, that the use is lawful. He may if he wishes reserve to himself or his assigns the exclusive use of the lake or water rights.[22]

4 In Washington, riparian owners of land on nonnavigable lakes hold in common and may exercise, subject to the rule of reasonableness, such rights as access, swimming, fishing, bathing, and boating. . . . Thus, the rights possessed by each of the parties are (1) the ownership of land abutting on the lake, (2) the conditions and covenant contained in the deeds, and (3) the specific undivided interest in the bottom of the lake associated with each of their parcels. What must be determined on remand is whether the portion of the dock which extends into the lake is legal considering the riparian rights of the parties. The test for deciding whether a given structure is legal is twofold; it must (1) constitute a riparian use and (2) be reasonable.[23]

"Reasonable use" is often mentioned with the consideration of how a riparian owner properly may utilize the water which flows past his acreage. The rather nebulous test to be applied is that, considering all the facts and circumstances of

[19] 202 Cal. 56, 259 P. 444, 448 (1977).
[20] Where injury to a riparian owner from use of lake water is incidental to reasonable enjoyment of the common right of all the riparian owners, such injury, it has been said, can demand no redress.
[21] *Flynn v. Beisel,* 257 Minn. 531, 102 N.W.2d 284, 290–91 (1960).
[22] *Mayer v. Grueber,* 29 Wis. 2d 168, 138 N.W.2d 197, 202–204 (1965).
[23] *Hefferline v. Langkow,* 15 Wash. App. 896, 552 P.2d 1079, 1081 (1976).

a particular situation,[24] a use which works substantial injury to the common right as between riparian proprietors is an "unreasonable" use. In one case[25] a small creek skirted the property of riparian owner *A* and subsequently flowed past owner *B*'s land. *A* pumped a great deal of water from the creek for irrigation of his field crops and, in the process, virtually stopped the flow of water onto *B*'s land and deprived the latter owner of a supply adequate for domestic uses and for livestock. It was held that *B* had a clear right to receive his proportionate share of the water for lawful purposes and that *A*'s monopolistic employment of the limited contents of the stream for extensive irrigation constituted an "unreasonable" use under the circumstances and an invasion of *B*'s legitimate interest in utilizing some of the water.

One type of difficulty arises where a riparian owner substantially diverts or blocks a natural watercourse, thus damaging downstream property by depriving it of much-needed flow.[26] By the same token, a lower owner cannot with impunity obstruct a surface stream, thereby flooding the lands and buildings "above" him by preventing the water from moving along as it otherwise naturally would. One is not privileged to flood the property of another, absent prescriptive right to do so or a grant of privilege by such other person.

Common law riparian rights may be legislatively altered.[27] Thus, statutes may

---

[24] By way of illustration, what is regarded as "reasonable use" in agricultural territory might well differ from the standard employed where a stream flows through a suburban community. No matter what the geographical location, however, no purpose is "reasonable" which is not lawful and beneficial to the taker—or where it can be shown that his utilization of the quantity of water in question was motivated solely by spite against his neighbors.

[25] *Great American Development Co. v. Smith,* 303 S.W.2d 861 (Ct. of Civ. App. of Tex. 1957).

[26] One often-repeated maxim in the field of water rights is "let it flow as it has been accustomed to flow." Thus, if a land owner has in the past enjoyed the benefit of a stream flowing over or adjacent to his land, such owner cannot have his reasonable use of this water prevented or markedly curtailed without his consent or without due process of law. Similarly, he in his turn would incur liability were he to interfere with the privileges of other rightful users.

Here are two situations in which the aforementioned "flow" maxim has pertinence:

1 The water from a spring flowed across *A*'s property, through a ditch, to a pit dug by him; thereafter it flowed underground into *B*'s cellar, causing damage thereto. The court awarded damages to *B* as plaintiff, reasoning that, if *A* (the owner of the dominant estate) had done nothing to change the natural flow of water over or from his property, he would not be liable for any water damage to *B*'s property, regardless of whether the water came from springs, streams, or the surface of the land. If, however, the proprietor of the upland had diverted quantities of surface water and discharged it upon the lower land without consent of the owner of the latter property, the upland owner, *A*, would be liable. The important question to answer is whether or not the ditch and pit were natural in origin or were artificial and created by *A*; only in the latter circumstance would *A* be responsible for the consequences. Inspection by the court, and the testimony of *B*'s witnesses, showed that the pit, at least, was artificial and designed to receive the flow of surface water. Consequently, *A* was required to pay *B* for the cellar damage done by the escaping spring water.

2 Jones built a dam to make a small fishpond on his property but thereby caused water to flow on or across his neighbor's land, and the latter became entitled to seek compensation for any resulting damage. Similarly, if the pond had caused harm to the neighbor because of a change it occasioned in the groundwater conditions, or if the pond diverted an appreciable volume of water away from the neighbor's land so that he no longer had the use of such water as a surface stream, a spring, or a source of supply for his well, the neighbor could again institute an action for damages.

[27] For a graphic illustration, see *In Re Waters of Long Valley Creek Stream System,* 549 P.2d 656 (Cal. 1979), where the state of California sought to statutorily adjudicate the rights of all riparian owners within the Long Valley system. There the court held that the state water resources control

require riparian landowners to acquire irrigation permits before using water, a requirement that in effect abrogates the common law riparian right of irrigation.[28]

23-4 Appropriative and Prescriptive Rights A number of jurisdictions in the arid and semiarid western portions of the country have adopted the doctrine of appropriation and in some of these states the body of common law surrounding riparian rights coexists; that is, such states recognize both appropriative and riparian water rights.

The term *appropriation* means diverting or taking water from some stream (or other natural source of supply) in accordance with local law and with the intention of applying the appropriated water to some beneficial use.[29] Such intention must be fulfilled within a reasonable time by actual application to the designed purpose of all the water taken. In other words, appropriation of water is the intent to take, accompanied by some physical manifestation of the intent, and for a valuable purpose.[30] Thus, a Colorado decision[31] recognized that the use of water in springs and ponds for the maintenance of livestock and for other domestic needs constituted appropriation of water to beneficial use.

Loosely speaking, appropriation includes any taking of water for other than riparian or overlying uses, and unappropriated water flowing in any natural channel is subject to appropriation. In California, to give one example, the one-time practice of appropriating water in any fashion one desired by self-help was terminated when, many years ago, the statutory method of appropriation was made exclusive.[32] A decision[33] from that jurisdiction describes the essential nature of appropriative and prescriptive[34] rights:

board was authorized to decide the priority of unexercised riparian rights but that it could not extinguish them altogether.

[28] *Omernick v. Department of Natural Resources,* 71 Wis.2d 370, 238 N.W.2d 114 (1976), interpreting Wisconsin Statutes Annotated 3018 (1973).

[29] Beneficial uses may include domestic consumption, stock watering, industrial purposes, and a variety of others.

[30] Since one need not "appropriate" what he already owns, an "appropriator" in the legal sense is one who takes water, either by surface diversion or from underground by wells, without possessing a property right entitling him to such taking.

[31] *Town of Genoa v. Westfall,* 349 P.2d 370 (Colo. 1960); see also *McClellan v. Jantzen,* 26 Ariz.App. 223, 547 P.2d 494 (1976).

[32] See California Water Code §1225 (West 1974).

[33] *San Bernardino Valley Mun. Water Distr. v. Meeks and Daley Water Co.,* 226 Cal.App.2d 216, 38 Cal. Rptr. 51 (1964). This was a proceeding in eminent domain to condemn the prescriptive and appropriative water rights and diversion facilities of defendant mutual water companies.

[34] Under certain conditions continuous exercise of a claimed right to divert and use water can ripen into what is known as a *prescriptive right* against a riparian or overlying owner, assuming either that such owner knows of and acquiesces in such adverse use or that knowledge and acquiescence will be presumed because the adverse use was so obvious and yet had gone unchallenged. Additional essential elements in prescription have been variously expressed, but they would appear to call for actual use under a claim of right by the taking party, which use is hostile and adverse (as to the property owner's interests), and continuous and uninterrupted for a considerable period of time, the length of which varies with the jurisdiction. For typical decisions discussing the acquisition of water rights by prescription, see *Sibbett v. Babcock,* 124 Cal.App.2d 567, 269 P.2d 42 (1954); *Moore v. California Oregon Power Co.,* 22 Cal.2d 725, 140 P.2d 798 (1943); and *Smelcer v. Rippetoe,* 24 Tenn.App. 516, 147 S.W.2d 109 (1940). The Moore case quotes this passage from prior opinions in the same jurisdiction:

Perhaps the most persuasive factor in our consideration of the character of the water rights here involved is that as appropriative and prescriptive rights, they are usufructuary. That is to say, there is no right to any particular water flowing in the stream, only a right to take from the stream a certain amount of flowing water; a right that does not come into being until both the means of diverting and the means of using water have been completed.

A statutory scheme may preclude the acquisition of water rights against the state by prescription. It has been held that the California water codes are a comprehensive scheme for granting appropriative rights and prescriptive rights have been described as parasites because the only way to obtain such rights is to take water rights from someone else.[35]

A dominant distinction between riparian rights and the doctrine of appropriation is the principle of equality among riparian owners, which contrasts with the concept of time priority ("first come, first served") among appropriators.[36] The two types of rights are further distinguishable in that a riparian owner's use of water must only be measured in terms of "reasonableness," while the appropriator's use measure is predetermined, at least in its maximum. A right of appropriation is a right to a definite quantity of water;[37] a riparian owner, by comparison, is not entitled to a specific volume of water but rather to have the stream flow to and over his land as it is naturally wont to do, and for his reasonable use.

States which have adopted appropriation principles differ as to what waters are subject to appropriation but, in one way or another, they all recognize that private rights must be subject to the important concerns of government with respect to navigable waterways. As to just how and under what circumstances an appropriative right can be acquired[38] and taking of the water properly accomplished, statutes may prove very precise. By way of example Arizona has a number of statutory provisions on the subject of which the following[39] are illustrative:

§45-141. Right of Appropriation; Permitted Uses

A. Any person or the state of Arizona or a political subdivision thereof may appropriate unappropriated water for domestic, municipal, irrigation, stock watering, water power, recreation, wildlife, including fish, mining uses, for his personal use or for

---

"Any person may obtain exclusive rights to water flowing in a stream or river by grant or prescription as against either riparian owners on the stream or the prior appropriation of the water by other parties. But the right acquired by prescription is only commensurate with the right enjoyed. The extent of the enjoyment measures the right." (page 805).

[35] *People v. Shirocow,* 605 P.2d 859 (Cal. 1980).

[36] *Cappaert v. United States,* 426 U.S. 428 (1976); *Colorado River Conservation District v. United States,* 424 U.S. 800 (1976).

[37] The top limit can be thought of as set by the requirement that the appropriator's use be "beneficial." Subject to that limitation, the holder of an appropriative right enjoys a priority claim to the available water.

[38] At the other end of the line, the appropriative right may terminate under various circumstances, abandonment being one.

[39] Arizona Revised Statutes, Title 45, Chap. 1, Art. 3 (1956).

delivery to consumers. The person or the state of Arizona or a political subdivision thereof first appropriating the water shall have the better right.

B. To effect the beneficial use, the person or the state of Arizona or a political subdivision thereof appropriating the water may construct and maintain reservoirs, dams, canals, ditches, flumes and other necessary waterways.

§45-142. Application for Permit to Appropriate Water

A. Any person, including the United States, the state or a municipality, intending to acquire the right to the beneficial use of water, shall make an application to the director of water resources for a permit to make an appropriation of the water. The application shall state:
1. The name and address of the applicant.
2. The water supply from which the appropriation is applied for.
3. The nature and amount of the proposed use.
4. The location, point of diversion and description of the proposed works by which the water is to be put to beneficial use.
5. The time within which it is proposed to begin construction of such works and the time required for completion of the construction and the application of the water to the proposed use.
B. The application also shall set forth:
1. If for agricultural purposes, the legal subdivisions of the land and the acreage to be irrigated.
2. If for power purposes, the nature of the works by which power is to be developed, the pressure, head and amount of water to be utilized, the points of diversion and release of the water and the uses to which the power is to be applied.
3. If for the construction of a reservoir, the dimensions and description of the dam, the capacity of the reservoir for each foot in depth, the description of the land to be submerged and the uses to be made of the impounded waters.
4. If for municipal uses, the population to be served, and an estimate of the future population requirements.
5. If for mining purposes, the location and character of the mines to be served and the methods of supplying and utilizing the waters.
6. If for recreation or wildlife, including fish, the location and the character of the area to be used and the specific purposes for which such area shall be used.
C. The application shall be accompanied by maps, drawings, and data prescribed by the director.

§45-143. Criteria for Approval or Rejection of Applications; Restrictions on Approval; Municipal Use

A. The director shall approve applications made in proper form for the appropriation of water for a beneficial use, but when the application or the proposed use conflicts with vested rights, is a menace to public safety, or is against the interests and welfare of the public, the application shall be rejected.
B. An application may be approved for less water than applied for if substantial reasons exist therefor, but shall not be approved for more water than may be put to a

beneficial use. Applications for municipal uses may be approved to the exclusion of all subsequent appropriations if the estimated needs of the municipality so demand after consideration thereof and upon order of the director.

§45-147. Relative Value of Uses

A. As between two or more pending conflicting applications for the use of water from a given water supply, when the capacity of the supply is not sufficient for all applications, preference shall be given by the department according to the relative values to the public of the proposed use.

B. The relative values to the public for the purposes of this section shall be:
1. Domestic and municipal uses. Domestic uses shall include gardens not exceeding one-half acre to each family.
2. Irrigation and stock watering.
3. Power and mining uses.
4. Recreation and wildlife, including fish.

23-5 Overlying Rights; Groundwater   All ground or subterranean water, unless qualifying as underground streams or lakes whose boundaries and course are well defined and readily ascertainable, is referred to as "percolating" water.

Rights and liabilities to subterranean streams are generally governed, by the rules of law applicable to surface streams.[40] The rule in regard to percolating water seems to be slowly changing over in this country from one of substantially unrestricted privilege on the part of the overlying owner to one of "reasonable use." The latter concept is that:

Each owner of land overlying the same general underground supply of water may take such water on his own land for any beneficial use theron, so long as such taking works no unreasonable injury to other land overlying such waters; that, if the natural supply is not sufficient for all such owners, each is entitled only to his reasonable proportion of the whole, and that each may apply to the courts to restrain an injurious and unreasonable taking by another, and to have the respective rights adjudicated and the use regulated so as to prevent unnecessary injury and restrict each to his reasonable share.[41]

In *Rank v. (Krug) U.S.*[42] the court said:

An overlying owner. . .has the right to take water, regardless of its source, from the ground underneath his land for use on his land which lies within the watershed or basin. The overlying right is. . .based on ownership of the land and is inseparably annexed to the soil. The overlying right, though the manner of its exercise may be different, is

---

[40] *Tehachapi-Cummings County Water Dist. v. Armstrong,* 49 Cal.App.3d 992, 122 Cal. Rptr. 918 (1975). At least one court, however, has held the rules applicable to percolating waters to be different from those applicable to either subterranean streams or surface streams. Thus, the law in Texas is that the owner of overlying land owns percolating water and can do with it as he pleases and may use it for any purpose. *Bartley v. Sone,* 527 S.W.2d 754 (Tex. App. 1975).

[41] *City of San Bernardino v. City of Riverside,* 186 Cal. 7, 198 P. 788 (1921).

[42] 142 F.Supp. 1 (1956), reversed in part and affirmed in part *Cal. v. Rank,* 293 F.2d 340 (9th Cir 1961); injunction modified on rehearing 307 F.2d 96 (1962), affirmed *Fresno v. Cal.,* 372 U.S. 627 (1963).

identical in law with the riparian right. These rights are absolute property rights against the world, except that as between riparian owners similarly situated, and as between overlying owners similarly situated, their rights are correlative and common with those in a like situation.[43]

Each proprietor should be considerate of the needs and rights of others. It is obvious that a spring or a farmer's well which serves as its owner's source of water is very important to him.[44] Generally speaking, it is improper for another person in some fashion to knowingly intercept and exhaust the supply, causing the spring or well to go dry. On the other hand, when one digs or drives a well he cannot foretell exactly where the water will come from. Nor can he know, as a rule, whether trouble will ensue for someone else. In many instances, it is difficult to draw the line between the rights of the respective parties in a situation of this kind.[45]

The trend of the law, especially in the eastern half of the United States, is toward treating rights to percolating water as limited by reasonable use. In a case from Wisconsin,[46] the court said:

It makes little sense to make an arbitrary distinction between rules to be applied to water on the basis of where it happens to be found. There is little justification for property rights in groundwater to be considered absolute while rights in surface streams are subject to a doctrine of reasonable use.

The other side of the picture is that, if one takes some action which results in *raising* the water table to the detriment of others, he may be subject to claims for damages. Thus, one man owned about two acres of land extending upstream along a brook (which ran through his property) and behind the house of a neighbor. He built a small dam so as to create a little pond entirely on his own property. However, the pond occasioned a rise in the level of the groundwater and caused

---

[43] Some of the western states which apply the doctrine of appropriation with regard to watercourses are similarly disposed where groundwaters are concerned.

[44] Groundwater is understandably important with respect to agriculture. Operations which lower the water table unduly may deprive the land of the moisture needed for the growing of crops, while activity which raises the water level too much may flood the land or render it so swampy as to be unfit for farming.

[45] In a typical case, *Canada et al. v. City of Shawnee,* 179 Okla. 53, 64 P.2d 694 (1937), the city constructed wells on its own property from which to obtain water for sale to its residents. As a result of the extensive pumping from city lands and the lands of neighboring farmers, the wells of the latter went dry, and the court granted plaintiff farmers a remedy. As another example, a metallurgical plant in the arid region of northern Mexico had to sink wells about 2000 feet deep into a subterranean valley to secure needed water. Before being allowed to do this the plant owners had to agree to supply water to residents in the area in the event the deep drilling operations dried up the shallow wells of these persons.

The doctrine of reasonable use does not ordinarily permit percolating waters to be used off the lands from which they are pumped if thereby others whose lands overlie the common supply are injured or damaged. For example, injunctive relief was held proper where the plaintiff sued to enjoin various mining companies from transporting water off their land overlying the common water supply to other tracts of land for use in mining operations. *Farmers Investment Co. v. Bettwy,* 558 P.2d 14 (Ariz. 1976).

[46] *State of Wis. v. Michels Pipeline Constr., Inc.,* 63 Wis.2d 278, 217 N.W.2d 339, 345 (1974), rehearing denied, 219 N.W.2d 308 (1974).

moisture to enter the neighbor's basement. The first property owner found it advisable to discontinue the pond in order to avert a suit for damages.

It is very difficult to evaluate potential damage which might be caused by a disturbance of the previously existing conditions regarding percolating water. All sorts of trouble can follow substantial interference with the groundwater level. The engineer should endeavor to anticipate what will happen as the result of his operations and take any indicated precautions.

23-6 Surface Water; the Common-Enemy Doctrine Surface water has been defined earlier herein as "a natural accumulation of precipitation that appears on the surface of the ground in a defused state following no well-defined course." The owner of land upon which surface water is found is normally free to seize and use it as he wishes, even though in so doing he may completely exhaust the supply at hand.

In the generally accepted pattern, *hostile* surface water (normally occasioned by long and heavy rains) is considered a "common enemy" and the consensus view seems to be that each landowner is entitled to defend himself against such enemy as best he may by operations conducted on his own land and that, if he does not exercise such right wantonly, carelessly, or unnecessarily, any resulting damage to his neighbor's property is damnum absque injuria.[47]

What would appear to be the harshness inherent in the common-enemy doctrine, which is in vogue in a number of states, is ameliorated somewhat by such commonly accepted modifications as that a landowner cannot collect the offending surface water into an artificial channel and pour unnatural quantities of it upon the land of another to the latter's severe detriment. In other words, the rights afforded by the common-enemy theory must be exercised within reasonable limits so as not to do needless violence to the privileges of others. A landowner incurs liability only when his harmful interference with and alteration of the natural movement of surface water is "unreasonable." The quoted word poses a question of fact to be resolved in any given case upon a reading of all the circumstances, including the amount and foreseeability of harm caused.

Construction in flood-prone areas is dangerous, and the Federal Insurance Administration of the Department of Housing and Urban Development has taken steps to issue restrictions on construction in flood-prone areas along coastlines and rivers. The regulations are to force property owners in such dangerous areas to secure federal flood insurance, to flood-proof any new structures and utilities, to flood-proof existing structures and utilities, or not to build there at all. Municipalities are to make appropriate zoning regulations to control these matters in areas below the 100-year flood levels as determined by flood-hazard boundary maps to be issued by F.I.A.

23-7 Navigable Waterways There are various instances in the United States of major rivers serving to mark the junction point between one state and another.

---

[47] Freely translated, loss which does not give rise to an action for damages against the party causing it.

The Mississippi, the Colorado, the Ohio, the Connecticut, the Hudson, and others perform this function along at least a portion of their routes to the sea. The precise location of the boundary line "within" the stream may be the subject of a dispute such as persisted for some years between the states of New York and New Jersey.[48]

In the early days of the United States there was uncertainty about whether the federal government or the states were to be responsible for the improvement and control of navigable waterways. The former ended up with the major job and has charge of such works as:[49]

1 Dredging of harbors and channels
2 Protection of shores against erosion
3 Establishment and maintenance of various types of signals, buoys, and the like for the aid and protection of ships
4 Establishment of pierhead and bulkhead lines[50]
5 Determination of locations, depths, and widths of navigable channels
6 Determination of horizontal and vertical clearances for bridges

When planning any work involving one or more of the items in the preceding list, the engineer must be sure to ascertain and comply with all applicable regulations and governmental requirements.

23-8   Reservoirs   A municipality desiring to construct a reservoir for water supply cannot simply buy the needed land and proceed as it wishes with the building of a dam. Owners for many miles downstream have rights to the use of the water flowing from the watershed involved, and, in many cases at least, their approval of the project will have to be obtained in advance of construction of the reservoir project.

A prime example is afforded by New York City's wish to divert to its own supply system vast quantities of water from the Delaware River. Even though this stream rises in New York State, the watershed covers parts of Pennsylvania and New Jersey as well, and the river waters serve these states over a distance of many miles. New York City was not entitled on its own to take whatever water it wished to obtain and, in the process, ignore the interests of Pennsylvania and New Jersey. The issue became one for discussion among all parties involved in an effort to

---

[48] For an instance of how some perplexing questions can arise regarding exact boundary lines, consider this situation. It is asserted that governmental land grants in the eighteenth century had given Kentucky the rights to that segment of the Ohio River which flowed past its territory with the limit of Kentucky's jurisdiction being established as the low-water mark on the *north* side of the river. When Ohio subsequently joined the Union, the state's Constitution of 1802 (Article VII, Section 6) recited that Ohio was "bounded. . .on the south by the Ohio River to the mouth of the Great Miami River."

[49] Through its control of navigable waterways the federal government probably can have something to say also about what activity (especially regarding projects which may affect navigation and interstate commerce) occurs in connection with those nonnavigable streams which are tributaries of these waterways.

[50] Pierhead lines are boundaries beyond which no pier construction should project toward navigable water; bulkhead lines set the riverward limit of any bulkhead or other fixed shore development. Between the two boundaries is the area for piers and other facilities for docking ships.

arrive at a sort of treaty or compact spelling out an equitable arrangement for the allocation of the available supply.

Another illustration concerns the possible use of the water of the Connecticut River by the city of Boston. The river flows between the states of Vermont and Connecticut. Just because it is located in Massachusetts, and might thus be in a position to appropriate the available water before same reaches Connecticut, does not entitle Boston to take unilateral action to the detriment of others with rights worthy of protection.

In some instances similar to those outlined above, it may be decided that a specified minimum flow will be maintained for the benefit of those located downstream from a dam, or there may be a guarantee that water from the system will be supplied to them. Failure to fulfill this obligation would then position the victimized persons to claim damages for loss of the use of water.

The objections of a few riparian owners cannot always be allowed to prevent the construction of works which are clearly in the public interest. The exercise of the government's right of eminent domain in the taking of property, court proceedings to determine the extent of damage resulting from the taking of the water in question, or arrangements to supply those proprietors in need with water from the reservoir are some of the steps which may be resorted to in order to settle disputes standing in the way of a contemplated reservoir project.

23-9   Rights to Water Power   The right to develop a major hydroelectric power site may be a matter of vast importance, and it is quite possible that there will be laws affecting the specific location. There are often complicated questions as to whether the right to develop the available power belongs to the federal government, to the state in which the site lies, to contiguous states jointly, or to private interests.

Assume that a medium-sized stream runs through private property and affords the potential for the development of a minimum of 3000 kilowatts of electric power, the generation of this power being possible even under ordinary low-water conditions. However, the property owner has not yet made any attempt to develop this natural resource. Assume further that a neighboring city proposes to build upstream a municipal reservoir which will greatly reduce the flow of the stream below the dam in dry weather. This will mean that the downstream owner cannot rely upon generating adequate power continuously at his site if he should in the future desire to exploit the resource which he had previously neglected. Has this property owner the right to claim damages for the loss to him of this potential power development? This question cannot be answered without a detailed study of the situation, but the following four points, among others, would require consideration:

1 Is the private owner's potential power development a sound project from an engineering standpoint?
2 Would it cost more to develop this power than it is worth?

3 Even at present (with no city reservoir in being) is the minimum stream flow reliable?

4 Is there a market for the sale of the prospective power, or could the private owner himself use it profitably?

If the private owner's potential project is feasible and offers a good investment, its worth would be measured primarily by the capitalized value of its estimated earnings, and this sum very possibly might be claimed by the owner as compensation for the loss of his chance to undertake future development.

In one case in Connecticut, an industrial plant had built a low dam across the adjacent river to impound water for use in its bleachery. The plant owners also had studied the possibility of building a small hydroelectric power development at this dam. Meanwhile, a power company secured the right to construct a large dam downstream from this plant and was authorized to acquire land that would be flooded below elevation 200. After completion of the power company's dam the plant owner's land was inundated to the extent that no power could be developed at that site. The bleachery claimed damages for the loss of this potential power development. In that case, however, the court found that the plaintiff's hydroelectric power-generating potential was of little value because the development of the available power would have cost more than the expense of purchasing the equivalent energy from the power company.

23-10 Major Irrigation Projects Major irrigation projects may become advisable for the general welfare in certain areas continually troubled with inadequate rainfall. Such projects inevitably mean diverting vast quantities of water from some stream or other, whose flow beyond the point of diversion is thereupon drastically reduced. In such circumstances some downstream proprietors may suffer loss. Nevertheless, it would be difficult to justify permitting such proprietors unreasonably to prevent the constructoon of irrigation works which can be expected to prove of vital importance to the region as a whole.

23-11 Mining Water plays an important role in some mining operations. Various rules and customs have grown up, particularly in the western states, regarding the use of water in that connection.

In the early days many mining claims, as in the so-called "gold fields," were established on public property by prospectors who subsequently secured land grants issued by the government. These prospectors, in working their claims, were well aware of the great importance of an adequate water supply. The general rule for appropriation and use of water in mining operations seems to have been, at the outset, that the prior appropriator was entitled to protect himself against subsequent appropriators. In time this was modified somewhat by the introduction of the principle of "reasonable use." Ultimately, customs prevalent in the mining community regarding the utilization of water came to be recognized by the courts and even found their way into some statutory law.

Mining has occasioned its share of "water rights" litigation over the years. For one thing, it is obvious that mining shafts, tunnels, and pits are likely to disturb the previously existing condition of groundwater (and therefore of the streams) in the vicinity of the mining operations. In one case[51] a farmer brought a successful action against a gold-mining company because a drainage tunnel excavated by the latter dried up the spring-fed creek upon which the plaintiff was dependent for irrigation, domestic, and household uses.

23-12 Flood Control *Flood Control* is the management of the flow in a stream in such a way that, even in times of maximum runoff from the land in the watershed, serious flooding of the land downstream can be prevented.

One recently developed means for flood control is the construction of a dam with outlets of limited capacity which are continuously open but will allow only a moderate amount of water to flow through, the effect being the temporary impounding of excessive floodwaters. Under ordinary conditions, the river water can flow as usual but, when a big storm occurs, part of the resulting water is retained behind the dam because it cannot quickly escape through the gates. After the storm is over, the impounded water will gradually flow through the gates until the reservoir is again low. This arrangement evens out the flow below the dam and inhibits flooding of any consequence.

Private work contemplated along major rivers may have to be coordinated with plans and requirements of governmental agencies and the engineer should investigate the situation before proceeding very far.[52] As time progressed, the federal government took over more and more of the work connected with flood control on the large rivers. This was done partly because of the great cost involved and partly because such projects for the public good usually concerned more than one state.

Even flood control, seemingly an innocuous subject on the desirability of which most people can readily agree, has occasionally led to litigation. In one situation[53] a railway company had a bridge across a Snohomish River in Washington. The flood-control district, a municipal corporation, brought an action of condemnation, asking for an "order of public use and necessity" for an easement to tie its dikes into the railroad's embankment; at the same time the district sought a judicial determination as to whether it would incur any liability for damages to the railroad bridge caused by the increased water height and velocity resulting from construction of dikes along the bank, upstream from the bridge. The Supreme Court of Washington held that the district could indeed condemn what it needed of the railroad's right-of-way (while in the process compensating the latter

---

[51] *Eckel v. Springfield Tunnel & Development Co.,* 87 Cal.App. 617, 262 P. 425 (1927).

[52] Thus, there may be certain legal restrictions regarding just what can be constructed along a given stream. For example, under the authority of the general statutes, the Water Resources Commission of Connecticut has established on several streams in the state "lines beyond which in the direction of the waterway, no obstruction or encroachment shall be placed by any person, firm or corporation, public or private, unless specifically authorized by the Commission." Of course, such lines are intended to prevent interference with the channel of a stream, and the restriction may be needed in areas of congested developments in order to maintain enough discharge capacity of the stream to prevent flooding of adjacent lands.

[53] *Marshland Flood Control Dist. v. Great Northern Ry. Co.,* 71 Wash.2d 365, 428 P.2d 531 (1967).

for the invasion of its property). However, since liability for damages is not limited to recompense for the property appropriated but also includes injury to the remainder of the property not actually taken, the district must stand ready to pay the railroad the cost of rendering its damaged bridge once again operational. The dikes were designed to keep the river water within the confines of the stream bed and thus to prevent flooding. Despite this salutary purpose, the unfortunate and unavoidable side effect of harm to the railroad's embankment and bridge entitled the injured property owner to a remedy in damages.

An interesting situation was presented in a recent flood control case[54] in which a landowner was allowed recovery for flood damage. The defendant, the power company, owned flood easements but had taken steps to prevent extensive flooding. The court noted that since the landowners had relied on the defendant's flood-control practices, the power company had a duty not to unnecessarily flood their land.

23-13   State Water Codes   Customs and regulations regarding riparian rights and the use of water have changed as the nation developed. Throughout the country the various governmental authorities are endeavoring to enact laws and compose executive regulations or codes which will serve to meet the problems of their particular areas. Some of these laws and codes relate to water conservation, irrigation, pollution, obstructing the flow of streams, and industrial usage of water; some to the establishment of regulatory bodies; and some to the construction of dams and other public works destined to aid in water supply and control problems. A number of jurisdictions have provisions bearing upon the use of water which is in the soil as well as water which is above the stream or lake bed. It is probable that the increasing need for water, and for its purification and wise utilization, will bring about still more laws and regulations affecting water rights.

In arid states, such as some in the southwest, water is the heartbeat of the economy. It is not surprising, therefore, to find that such states have enacted rather elaborate sets of laws[55] dealing with almost every aspect of the subject of

---

[54] *Kinz v. Utah Power & Light Co.,* 526 F.2d 500 (9th Cir. 1975).

[55] Even constitutional provisions can be found which are in point. Thus, Art. 14 §3 of the California Constitution reads in part:

"It is hereby declared that because of the conditions prevailing in this State the general welfare requires that the water resources of the State be put to beneficial use to the fullest extent of which they are capable, and that the waste or unreasonable use or unreasonable method of use of water be prevented, and that the conservation of such waters is to be exercised with a view to the reasonable and beneficial use thereof in the interest of the people and for the public welfare. The right to water or to the use or flow of water in or from any natural stream or water course in this State is and shall be limited to such water as shall be reasonably required for the beneficial use to be served, and such right does not and shall not extend to the waste or unreasonable use or unreasonable method of use or unreasonable method of diversion of water. Riparian rights in a stream or water course attach to, but to no more than so much of the flow thereof as may be required or used consistently with this section, for the purposes for which such lands are, or may be made adaptable, in view of such reasonable and beneficial uses; provided, however, that nothing herein contained shall be construed as depriving any riparian owner of the reasonable use of water of the stream to which his land is riparian under reasonable methods of diversion and use, or of depriving any appropriator of water to which he is lawfully entitled."

waters. By way of showing how comprehensive are these statutory treatments, it might be noted that Title 45 of Arizona Revised Statutes (1956) lists these chapter and article headings, among others:

Chapter

A. State Water Code
  1. Public Nature and Use of Water
  2. Arizona Resources Board
  3. Appropriation of Water
  4. Rights to Water
  5. Reservoirs and Canals
  6. Determination of Conflicting Rights
  7. Ground Waters
B. Interstate Streams
  1. Interstate Stream Commission
  2. Lake Mead Contract
  3. Colorado River Compact
  4. Upper Colorado River Basin Compact
  5. Distribution of Federal Funds Derived From Colorado River
C. Dams and Reservoirs
  1. Supervision of Dams, Reservoirs, and Projects
  2. Construction of Livestock Storage Dams
  3. Abatement of Menace Created by Dam
D. Agricultural Improvement Districts
  1. Formation
  2. Administration
E. Drainage Districts
F. Irrigation Districts
G. Irrigation Water Delivery Districts
H. Soil Conservation Districts
 I. Provisions Applicable to More Than One District
J. Flood Control
K. Weather Control and Cloud Modification
L. State Water and Power Plan

QUESTIONS

1 Define (*a*) "riparian rights," (*b*) "riparian lands," (*c*) "riparian owner (or proprietor)."
2 Define (*a*) "watercourse," (*b*) "surface waters," (*c*) "percolating waters."
3 What is meant by "reasonable use" of water as applied in matters relating to riparian (or water) rights?
4 How may the riparian rights of neighboring landowners vary? Explain.
5 What are the common-law principles of riparian rights? What uses of water do these common-law principles allow?
6 What is "reasonable" use as respects riparian rights?
7 What is meant by the right of use of water through "prescription"; through "prior appropriation"?

8 List (in order of probable priority) the various uses of water which may be involved in questions of riparian rights.

9 Who is likely to own the land constituting the bed (*a*) of a small creek? (*b*) of a navigable stream?

10 Is there any "legal" limit to the amount of water which a farmer may take from a watercourse adjoining his land for use in the irrigation of his land? Explain.

11 Does the owner of a hydroelectric power site have exclusive right to develop the water power at that site as he pleases?

12 What is the difference between percolating water and artesian water?

13 Who ordinarily has the right to the use of percolating waters? Within what limits?

14 What are the general principles of the "miners' law" governing water use in the west?

15 Who may set up and enforce (*a*) a water code, (*b*) regulations affecting navigable waterways, (*c*) rules regarding prevention of pollution of streams?

16 What is meant by "floodwater"? Who can use it or guard against it?

17 What is meant by "flood control"? Who has charge of this (*a*) in major rivers? (*b*) in small streams?

18 What is the difference between a reservoir and an ordinary lake? What should the planners of the former consider regarding persons living downstream?

19 Discuss the problem of pollution of water and possible remedies therefor.

20 In a given state, who may have charge of matters relating to the use and control of water?

21 If a riparian owner possesses a site which is suitable for the development of hydroelectric power, can she sell the "power rights" without selling the land, or vice versa?

22 Carter bought some low and partially swampy land for a housing development. She filled in the land to raise the elevation of the surface approximately 5 feet. This caused the level of the groundwater to rise upgrade from Carter's land, and this rise in the water level turned Smith's meadow into a swamp. What can Smith do about the matter?

23 Turner, a farmer, cut drainage ditches to drain a swampy pond in his pasture. Bradley, a neighboring farmer, then found that his spring for watering his stock dried up. What can Bradley do about his former spring?

24 The *X* Silver Company has discharged chemical wastes into *Y* Creek for 25 years. Assume that the company now plans to expand its plant 100 percent. (*a*) Can the *Y* Creek Fish and Game Club, which recently bought 100 acres of land downstream on *Y* Creek, compel the company to stop polluting the stream? (*b*) Can the club do anything about the pending additional pollution?

25 Assume that the city of Southford plans to build a reservoir on Oak Creek, using most of the water available and designing the spillway so as to divert the excess water to Roaring Brook. (*a*) Has the city the right to do this? (*b*) What steps should be taken by the city in clearing the way for construction of the reservoir? (*c*) What steps may be taken by riparian owners downstream along Oak Creek to protect their rights?

26 Clark built a dam in Rice Brook (entirely on his own land) and, during the months of July and August, piped all the water to his new irrigation system. Dawson, who owned a farm downstream, tried to get an injunction to require Clark to permit part of the water to reach Dawson's property. Do you think that Dawson will be successful? Explain.

27 Alton built a small dam on his own property to create a small fishpond. The overflow from the pond continued downstream in the same channel as before, crossing Braxton's land. Has Alton damaged Braxton? Explain.

28 The *X-Y* Power Co. built a power dam on Hoton River. The company agreed to limit the level of the impounded water to elevation +520 feet. During an unusual spring thaw and rain, the spillway jammed with ice and the water rose to elevation +529 feet, flooding several cottages. The owners of these cottages sued the company for damages caused by this flooding. What do you think of their rights in this situation?

29 A brass-manufacturing company plans to build a sheet-and-tube mill on the *X-Y* River at a site which it has purchased. There will be considerable acid waste from pickling tanks. Can the company discharge this waste into the river as it sees fit? What steps can others take to prevent this?

30 Assume that the riparian owners downstream from the plant described in the preceding case did not realize, until after the plant was in operation, that acids would be discharged into the river. What can they do about the pollution problem at that time?

CHAPTER 24

ENVIRONMENTAL LAW

24-1 **Introduction** The growth in importance of federal environmental law is well illustrated by the fact that the Environmental Protection Agency has expanded since its creation in 1970 to be the largest federal regulatory agency. Indeed, it is now a rare industrial or commercial activity which is not undertaken subject to some manner of environmental constraint.

The general trend seems to be a shift from federal assistance to states to either direct federal regulation or mandatory federal guidelines for state implementation. This trend would seem to reflect both a dissatisfaction with state achievements and an awareness that many pollution problems are national or regional in scale.

While the subject is largely statutory it should be kept in mind that many actions resulting in pollution damage may be tortious and subject to private abatement or damages under the principles discussed in Chapter 19.

24-2 **National Environmental Policy Act, Generally** Perhaps the single most significant example of federal environmental legislation has been the National Environmental Policy Act (NEPA) of 1969.[1]

NEPA established the encouragement of a productive harmony between man and his environment as a national policy, established the Council on Environmental Quality, and required federal agencies to include in every recommendation or report on proposals for legislation and other major federal actions significantly affecting the quaility of the human environment an environmental impact statement.

---

[1] 42 U.S.C. 4321.

The watershed NEPA case of *Calvert Cliff's Coordinating Committee v. U. S. Atomic Energy Commission*[2] set the tone for the voluminous NEPA litigation to follow by observing that the sweep of NEPA is extraordinarily broad, compelling consideration of any and all types of environmental impact. The court in *Calvert Cliff's* noted the requirement that all agencies must use a systematic, interdisciplinary approach to environmental planning and evaluation in decision making which may have an impact on human environment, and required federal agencies to consider environmental issues, just as they consider other matters within their mandates, to the fullest extent possible. Congress did not, however, establish environmental protection as an exclusive goal; rather it desired a reordering of priorities, so that environmental costs and benefits will assume their proper place along with other considerations.

24-3   Council on Environmental Quality   The National Environmental Policy Act established the Council on Environmental Quality in the executive office of the President to assist and advise the President in the preparation of the environmental quality report required annually by NEPA, to analyze and interpret environmental information and communicate it to the President, and to review and appraise the programs and activities of the federal government in light of the environmental policy set forth in the act. As well, CEQ is to develop and recommend to the President national policies to promote the improvement of the environmental quality, to conduct investigations relating to ecological systems and environmental quality, to document and define changes in the natural environment, and to furnish such studies and reports with respect to matters of policy and legislation as the President may request.

By executive order, the Council on Environmental Quality is also authorized to prepare guidelines to aid in the preparation of environmental impact statements.

24-4   Environmental Impact Statements   The key requirement of the National Environmental Policy Act, and the one which has resulted in the most litigation, is that which requires federal agencies to prepare environmental impact statements on all major federal actions significantly affecting the quality of the human environment. The purpose of the environmental impact statement requirement is not to produce an objection-free document, but rather to give Congress, federal agencies, and the public a useful decision-making tool. Subsequent environmental rationalizations of decisions already fully and finally made do not meet the requirement that federal agency actions be examined and environmental consequences considered "to the fullest extent possible." NEPA does not impose a permit scheme of any sort but agency failures to comply with NEPA's procedural requirements have resulted in litigation delaying or killing many major construction projects.

Environmental impact statements are required for all major federal actions,

[2] 449 F.2d 1109 (D.C. Cir. 1971).

and this requirement includes not only federal construction projects and proposals for legislation but also all activities which are undertaken pursuant to a federal license, permit, or financing. Federal officials are entitled to "dream out loud" without filing an environmental impact statement, but a statement is required when a proposal moves beyond the dream stage into some tangible form. Environmental impact statements must be prepared at the earliest time possible prior to the implementation of the proposed major federal action so that alternative courses of action with less severe environmental consequences can be considered. As well, it is clear that the environmental impact statement should be prepared before irretrievable commitments are made or options precluded by agency action. It is the federal agency itself, not environmental groups or others, that is required to produce environmental impact statements. However, the collection and analysis of environmental data are often contracted to universities or other research organizations.

The environmental impact statement must contain an analysis of the environmental costs and benefits of the proposed projects, including all significant primary environmental affects and all substantial secondary environmental effects. Aesthetic considerations are part of the "quality of the human environment" which must be considered in an environmental impact statement, as are future potential circumstances, mitigation methods, and reasoned analyses in response to conflicting data or opinions on environmental issues. An environmental impact statement must not be merely a catalogue of environmental facts, but also must explain its course of inquiry, analysis, and reasoning. Responsible opposing views must also be included within the environmental impact statement, but highly remote or conjectural consequences need not be discussed.

The language of the National Environmental Policy Act calls for a detailed statement, and this requirement mandates that the environmental consequences of a particular project be sufficiently spelled out to allow an executive to arrive at a reasonably accurate decision regarding the environmental benefits and detriments to be expected from the project. It must also provide a record upon which a decision maker can arrive at an informed decision.

The environmental impact statement must contain a thorough discussion of alternatives to the proposed project, including the elimination of the project altogether. Also required is a presentation of the environmental risks incident to all reasonable alternative courses of action. The alternatives to be considered naturally depend upon the type of project involved. In connection with a federally funded freeway project, for example, the environmental impact statement should consider all possible alternatives to the proposed freeway including changes in design, different systems of transportation, changes in the route, and abandonment of the project entirely.

In an environmental impact statement dealing with a proposed lease of oil and gas lands off the coast of Louisiana, alternatives considered included the elimination of oil import quotas, development of oil shale, desulphurization of coal, coal liquefaction and gasification, tar sands, and geothermal resources. Obviously, the distinction between a true alternative and a supplement is not always easy to draw.

Several hundred lawsuits challenging the propriety of federal action have been brought utilizing the National Environmental Policy Act, and long delays have often resulted. Such was the case with the construction of the Trans-Alaskan pipeline, and these delays, combined with energy shortages and the petroleum embargo of 1973, resulted in the passage of the Trans-Alaskan Pipeline Authorization Act,[3] which specifically authorized construction without further action under the National Environmental Policy Act of 1969.

24-5   Air Pollution Control   Federal law directed at the problems of air pollution has existed since the passage of the Air Pollution Control Act in 1955. Numerous other statutes with the same aim have been enacted thereafter, but until the passage of the Air Quality Act of 1967 the emphasis was upon federal assistance to state and local governments, technical assistance, and programs of research.

In 1963 the Clean Air Act provided explicit authority for federal regulatory action to abate interstate air pollution problems. Amendments in 1965 provided for greater control over automotive exhaust emissions. The Air Quality Act of 1967 established a regional framework for the enactment and enforcement of federal-state air quality standards, but it was not until the passage of the Clean Air Act Amendments of 1970 that a comprehensive attack on the problems of air pollution was made.

Requirements under the Clean Air Act are often industrywide and may have substantial economic repercussions; therefore, the requirements are subject to intense political pressure. Delays and amendments are inevitable as political and economic factors make their weight felt. The reader's attention is called to the Energy Supply and Environmental Coordination Act of 1974,[4] which amended the Clean Air Act to facilitate the burning of coal and to grant delays in compliance with motor vehicle emission requirements. Motor vehicle emission requirements were amended again in 1977, and the state of the law in that field is dynamic indeed.

24-6   Clean Air Act; Air Quality Standards   The Clean Air Act Amendments of 1970 directed the Secretary of Health, Education, and Welfare to establish nationally applicable ambient air quality standards for any pollutant or combination of pollutants for which air quality criteria had been issued. Each state is designated as an air quality control region, but interstate air quality control regions may be established where necessary to deal with interstate air pollution problems.

Along with the establishment of air quality regions, primary and secondary

---

[3] Sec. 28, Mineral Leasing Act of 1920, 30 U.S.C. 185.
[4] 15 U.S.C. 791, 792.

ambient air quality standards were translated into numbers by the promulgation of National Primary and Secondary Air Quality Standards[5] and the states were required to submit, by January 31, 1972, implementation plans which would achieve the level of air quality established by the primary and secondary standards within specific time limitations. Problems with respect to the maintenance of ambient air quality, transportation contracts, and the regulation of complex sources all made state compliance difficult. The Environmental Protection Agency exercises its authority to substitute specific provisions where state plans are deficient.

The Clean Air Act also grants almost unlimited powers of abatement to the Administrator of the Environmental Protection Agency if he determines there is "an imminent and substantial endangerment to the health of persons and that appropriate state and local authorities have not acted to abate such sources." The administrator may also establish emission standards for any type of source which emits any substance which he determines to be hazardous.

24-7   New Stationary Sources   The problem of major new air-polluting sources is addressed by the Clean Air Act Amendments of 1970, which authorize

---

[5] Following are the National Primary and Secondary Air Quality Standards: 40 C.F.R. Part 50 (1976).

FEDERAL STANDARD

Substance	Primary	Secondary
SO_2	80 μg/m³ (0.03 ppm) annual arithmetic mean	1.300 μg/m³ (0.5 ppm) maximum in 3 h
	365 μg/m³ (0.14 ppm) maximum in 24 h	
Particulates	75 μg/m³ annual geometric mean	60 μg/m³ annual geometric mean
	260 μg/m³ maximum in 24 h	150 μg/m³ maximum in 24 h
CO	10 mg/m³ (9 ppm) maximum in 8 h	10 mg/m³ (9 ppm) maximum in 8 h
	40 mg/m³ (35 ppm) maximum in 1 h	40 mg/m³ (35 ppm) maximum in 1 h
Photochemical oxidants	160 μg/m³ (0.08 ppm) maximum 1 h	160 μg/m³ (0.08 ppm) maximum 1 h
HC (corrected for methane)	160 μg/m³ (0.24 ppm) maximum in 3 h 6 A.M.–9 A.M.	160 μg/m³ (0.24 ppm) maximum in 3 h 6 A.M.–9 A.M.
NO_2	100 μg/m³ (0.05 ppm) annual arithmetic mean	100 μg/m³ (0.05 ppm) annual arithmetic mean

Maximum concentrations are not to be exceeded more than once per year.
40 C.F.R. Part 50 (1976).

the administrator of the Environmental Protection Agency to establish nationally applicable standards of performance based upon "the application of best system emission reduction which (taking into account the cost of achieving such reduction) has been adequately demonstrated."[6] State implementation plans must provide for a review of every new source, prior to construction, to determine the effects in all air quality regions of the state on the overall pollution control program. State implementation plans must also provide for an analysis to determine whether a plant will allow maintenance of the primary, secondary, and nondeterioration standards which are applicable.

As the supply of fossil fuel declines, the need for major centralized power plants may increase because of their probable greater efficiency compared to that of host of small, individual means of producing electricity and heat. Thus it may be that serious conflicts will develop between what is desirable and what is possible.

24-8 Mobile Sources The Administrator of the Environmental Protection Agency is authorized by the Clean Air Act Amendments of 1970 to set emission standards for all new vehicles and engines for use on the streets and highways and for aircraft. The act called for the establishment of standards for vehicles likely to cause or contribute to air pollution endangering public health or welfare within the constraints of available technology. It also authorized the testing of vehicles or engines or those being manufactured in order to determine whether they in fact comply with emission standards. Authority is also granted to authorize the establishment of standards respecting the composition of the chemical and physical properties of any fuel or fuel additives except in the case of aviation fuels.[7]

Litigation between the automobile manufacturers and the environmental Protection Agency continues. In one recent case[8] Ford Motor Company sought re-

---

[6] Among the possible new sources identified by the Senate Committee on Public Works were:

Cement manufacturing	Pulp and paper mill operations
Coal Cleaning operations	Rendering plans (animal matter)
Coke by-product manufacturing	Sulfuric acid manufacturing
Cotton ginning	Soap and detergent manufacturing
Ferroalloy plants	Municipal incinerators
Grain milling and handling operations	Steam electric power plants
Gray iron foundries	Petroleum refining
Iron and steel operations	Phosphate manufacturing
Nitric acid manufacturing	Phosphoric acid manufacturing
Nonferrous metallurgical operations (e.g., aluminum reduction, copper, lead, and zinc smelting)	

S.Rep. No. 91-1196, 91st Cong., 1st Sess., at 16 (1970).

[7] The reader's attention is directed to the case of *International Harvester Co. v. Ruckelshaus,* 478 F.2d 615 (D.C. Cir. 1973) in which the issue was a decision by the Administrator of the EPA denying applications of International Harvester, Ford General Motors, and Chrysler for a suspension of the 1975 emission standards. The case presents a thorough discussion of the relationship between economics, technology, and administrative power and discretion in the context of air pollution.

[8] *Ford Motor Company v. EPA,* 604 F.2d 685 (D.C. Cir. 1979).

view of an EPA determination that section 202A of the Clean Air Act should include methane hydrocarbons. Ford took the position that methane hydrocarbons do not contribute to air pollution, but the court found that Congress' intent was to include methane hydrocarbons in the total allowable amount of hydrocarbons emitted from new cars. Therefore, the standards set for 1980 and 1981 cars included a reduction of the permissible amount of methane hydrocarbons.

In another case[9] the Motor and Equipment Manufacturers Association challenged a decision by the administrator of the Environmental Protection Agency which waived federal preemption of maintenance standards for manufacturers of automobiles which allowed the state of California to impose stricter standards. The court found that section 209 of the Clean Air Act permitted the EPA to waive preemption where it was found that the state acted reasonably in concluding that the regulations were as protective of the public health and welfare as the federal regulations.

24-9 Noise Pollution The Noise Control Act of 1972[10] empowers the Environmental Protection Agency to control the noise characteristics of products distributed in commerce and to control noise produced by interstate railroads and motor carriers. Unlike many federal air and water quality laws, the noise law generally endeavors to control only noise emissions at the source, rather than ambient conditions. This is true in part because noise, unlike air and water pollution, does not accumulate in the environment but dissipates within a reasonably short distance and time.

The Environmental Protection Agency is required to propose emission standards for products identified as major sources of noise if "noise emission standards are feasible" in situations where the product is either construction equipment, transportation equipment, a motor or engine, or electrical or electronic equipment. Products not within one of these standards are feasible and requisite to protect public health and welfare. Imported products are subjects to the same standards as products made in the United States, but products made solely for export are excluded.

The control of aircraft noise is complicated by the fact that the regulation of air safety is the responsibility of the Federal Aviation Administration and the administration of the Noise Control Act of 1972 is vested with the Environmental Protection Agency. The FAA is empowered to adopt rules prescribing flight paths and procedures, traffic patterns, and preferential runway systems for specific airports, all of which are as important in the control of aircraft noise as is the actual amount of noise emanating from the vehicle. The Noise Control Act requires that the FAA consult with the Environmental Protection Agency before adopting noise rules and before any exemptions are granted under the rules. The FAA is empowered by the Noise Control Act of 1972, section 611, to prescribe whatever

[9] *Motor and Equipment Manufacturers Association, Inc. v. EPA,* 627 F.2d 1095 (D.C. Cir. 1979).
[10] 42 U.S.C. 401 et. seq.

sort of rules it finds necessary to control aircraft noise, including the authority to set noise emission standards as a condition upon aircraft-type certificates.

Standards with respect to the control of noise emission by jet aircraft are under continual revision and are complicated by the fact that the "retrofit" of older equipment is an extremely expensive proposition, and many of the noisier aircraft are to be due for replacement by quieter wide-bodied planes.

24-10   Nondegradation   One of the most intriguing and important issues to have arisen under the Clean Air Act is that of nondegradation; that is, whether a state implementation plan may allow significant deterioration of existing clean air areas, areas with levels of pollution lower than the secondary standard.

In the case of *Sierra Club v. Ruckelshaus*[11] it was held that the language of the Clean Air Act stating its purpose to be "to protect and enhance the quality of the nation's air resources so as to promote public health and welfare and the productive capacity of its population" declared Congress' intent to improve the quality of the nation's air and to prevent the deterioration of that air quality, no matter how presently pure that quality in some sections of the country happened to be. Thus, an EPA regulation permitting states to submit plans which would allow a clean air area to be degraded so long as the plans were "adequate to prevent such ambient pollution levels from exceeding the secondary standard" was upset

24-11   Water Pollution   The federal law of water pollution control has been evolving for more than thirty years, but its present stage takes the form of the Clean Water Act of 1977.[12]

The Federal Water Pollution Control Act, by its terms, applies to the navigable waters of the United States, but defines navigable waters to be "waters of the United States, including the territorial seas." The act therefore applies to waters of the United States without regard to traditional definitions involving commerce and navigability. The various provisions of the act authorize the establishment of standards and issuance of permits for discharges from "point sources," which include any discernible and discrete conveyance including pipes, ditches, containers, vessels and floating facilities, and other conduits from which pollutants are or may be discharged.

24-12   Direct Discharges   Direct discharges of pollutants are subject to both effluent standards and water quality standards under the regulatory provisions of the Federal Water Pollution Control Act and must comply with the stricter of the two.

Effluent Standards   Effluent standards are limitations on the amount of pollutants that may be discharged by particular classifications of dischargers and are based on the availability of pollution-control technology. Effluent standards are measured in terms of the amount of pollutant which may be discharged per period

---

[11] 344 F.Supp 253, *affirmed* 412 U.S. 541 (1973).
[12] 33 U.S.C. 1251 et seq.

of time or unit of production. Polluters are required to utilize the "best available technology economically achievable" by 1983, and the levels of technology are specified in regulations of the Environmental Protection Agency. The FWPCA requires the Environmental Protection Agency to establish effluents standards for new sources of pollution, and each standard must reflect the "greatest degree of effluent reduction achievable through the application of the best available demonstrated control technology, processes, operating methods and other alteratives, including a standard permitting no discharge of pollutants where practicable."

Applicants for federal permits must first obtain state certification, which requirement essentially gives the states a veto power over federal discharge permits. A state may set its standards higher than the EPA and may also establish its own permit scheme which, if approved by EPA, will supplant the EPA scheme.

Water Quality Standards   Water quality standards are rules defining required water quality for ambient water. Such standards are based on technical information regarding minimum requirements necessary to sustain the uses to which water is put. Standards for different waters may vary, depending on uses to which the water is put and the preferences of the individual states which adopt water quality standards subject to EPA approval. Water quality standards may be required for suspended solids, dissolved solids, hydrogen ion concentration, radioactive substances, plant nutrients, dissolved oxygen, temperature, fecal coliform, and the like.[13]

24-13   Nonpoint Sources   The permit scheme of the Federal Water Pollution Control Act applies only to point-source discharges, but the control of nonpoint sources of water pollution presents a perplexing problem which is addressed in section 208. Although the line is difficult to draw with precision, nonpoint sources of pollution may be agricultural, mine-related, construction activity runoff, and saltwater intrusion into rivers and estuaries, resulting from a reduction in freshwater flow from causes such as irrigation and groundwater extraction.

The complexity of the nonpoint problem is illustrated by the fact that the amount of runoff from a construction site may depend on the type of construction, the amount of rainfall, the topography of the site, the type of soil, the amount of vegetation left undisturbed, the amount of earth moving involved, and the type of control measure utilized. The Federal Water Pollution Control Act seems to recognize that effective federal control over nonpoint sources of water pollution may be impossible and thus establishes a requirement under which state or regional agencies must establish regulatory programs to control nonpoint

---

[13] R. 323.1062. Fecal coliform:

"Rule 1062.

(1) Waters of the state protected for total body contact recreation shall contain not more than 200 fecal coliforms per 100 milliliters; and all other waters of the state shall contain not more than 1,000 fecal coliforms per 100 milliliters. These concentrations may be exceeded if due to uncontrollable nonpoint sources.

(2) Compliance with the fecal coliform standards prescribed by subrule (1) shall be determined on the basis of the geometric average of any series of 5 or more consecutive samples taken over not more than a 30-day periods."

source pollution, pursuant to federal guidelines. To date, relatively little success has been achieved in controlling nonpoint sources of water pollution.

24-14 Thermal Pollution   Section 316[14] of the Federal Water Pollution Control Act reverses the general pattern of the FWPCA, which prescribes technologically based effluent standards that apply to individual discharges in the uniform manner. Dischargers of waste heat, most often from fossil fuel or nuclear power plants used in the generation of electricity, are allowed to show that any proposed effluent limitation for the thermal component of their discharge is "more stringent than necessary to assure the protection and propagation of a balanced, indigenous population of shellfish, fish, and wildlife in and on the body of water into which the discharge is to be made."

24-15 Dredged Material   The disposal of dredged or filled material is regulated under a separate permit system established by Section 404[15] of the Federal Water Pollution Control Act. The traditional role of the Army Corps of Engineers in the dredging of navigable channels led to the establishment of a scheme by which permits are issued by the secretary of the army acting through the chief of engineers subject to the application of Environmental Protection Agency guidelines. Recent legislation authorizes the states to take over Section 404 permitting in other than traditional navigable waters following EPA approval of their procedures and standards.

24-16 Oil and Hazardous Substances   A separate regulatory scheme[16] has been established to control spills of oil and hazardous substances. The Environmental Protection Agency has promulgated a list of hazardous substances and has defined what constitutes harmful discharge of each substance and of oil. An elaborate regulatory structure provides for penalties for oil spills and for failure to notify of a spill, for assessment of cleanup costs, and for federal authority to prescribe spill prevention equipment for oil-handling facilities and for vessels.

The administrator of the Environmental Protection Agency is also required, by Section 307a of the Federal Water Pollution Control Act, to set effluent standards for toxic pollutants which are defined to be any pollutant which is toxic to any organism and which may cause death, disease, or behavioral or physical abnormalities. Since almost any pollutant could be toxic to some organism if applied in sufficient quantities, EPA is not required to set a standard for every toxic pollutant but rather is required to publish a list of toxic pollutants for which effluent standards will be established. The Environmental Protection Agency must set the standards "at a level which the administrator determines provides an ample margin of safety," and the standard may be in the form of a prohibition.

---

[14] 33 U.S.C. 1326.
[15] 33 U.S.C. 1344.
[16] 33 U.S.C. 1317.

24-17   Ocean Dumping   The disposal of wastes at sea is regulated by the Marine Protection, Research, and Sanctuaries Act of 1972,[17] which is also known as the Ocean Dumping Act. This legislation establishes a permit system under which an Environmental Protection Agency permit is required for the transportation from the United States of virtually all materials for dumping in ocean water, the actual dumping of materials transported from outside the United States into the territorial sea or contiguous zone, and the transportation of materials by a United States agency or official from outside the United States for the purpose of dumping in any ocean waters. No permits may be issued for radioactive, chemical, or biological warfare agents or for any highly radioactive substance. The act applies to all ocean waters seaward of the land or internal waters, and the states are preempted from adopting or enforcing any regulations relating to activities covered by the act.

24-18   Pesticides   The negative impacts of the extensive use of pesticides—brought to public notice primarily through the cancellation by the Environmental Protection Agency of most uses of DDT—are controlled primarily by the Federal Insecticide, Fungicide, and Rodenticide Act, ad amended by the 1972 Federal Environmental Pesticide Control Act.[18]

This federal statute requires the registration of pesticides. It also requires adequate labels and directions for safe use. Amendments in 1972 added a system of classifications of pesticides using two categories: general and restricted-use products. Products placed in a restricted-use category are subject to special controls and may be applied only by trained operators. Virtually all pesticides used in the United States require federal registration by the Environmental Protection Agency.

Experimental-use permits may be issued by the EPA for some types of newly developed pesticides—as yet inadequately tested—to apply for registration. The EPA administrator is authorized, at his discretion, to "exempt any federal or state agency from any provision of this Act if he determines that emergency conditions exist which require such exemption."

The Federal Insecticide, Fungicide, and Rodenticide Act authorizes the administrator of the Environmental Protection Agency to cancel registrations of insecticides where there is an unreasonable risk to people or the environment, taking into account the economic, social, and environmental costs and benefits of the use of any pesticides.[19]

In 1979, pursuant to its emergency powers, EPA ordered an emergency ban of the herbicides 2-4-5-T and Silvex because there was a probability that spontaneous human abortions were related to the spraying of these herbicides. The ban was upheld when the court found that the EPA had considered all relevant factors and

---

[17] 33 U.S.C. 1401 et seq.

[18] 17 U.S.C. 135 et seq.

[19] See *Environmental Defense Fund Inc. v. Environmental Protection Agency,* 489 F.2d 1247 (1973), where the Court of Appeals for the District of Columbia upheld the cancellation of almost all registrations for the use of DDT except for limited public health and agriculture pest quarantine purposes.

did not make a clear error of judgment in deciding to order the suspension of the herbicides.[20]

24-19 Solid Wastes The Resource Conservation and Recovery Act of 1976,[21] as amended by the Solid Waste Disposal Act Amendments of 1980, provides federal regulation of hazardous waste, financial assistance to the states for solid-waste management planning, and funding for development, research, and demonstration of new technology. The Environmental Protection Agency is required to list specific substances deemed hazardous and to promulgate regulations regarding their disposal. Permits are required of owners and operators of hazardous waste treatment, storage, or disposal facilities, but the act applies only to wastes discharged on land, and air and water discharges are covered under the Clean Air Act and the Federal Water Pollution Control Act.

Major responsibility for regulation of solid wastes, other than those deemed hazardous, remains with the states. The EPA is directed, however, to establish guidelines for states' solid-waste management plans, which guidelines must be complied with if the states hope to obtain funding for the development of such plans.

The storage of nuclear waste is a problem of immense and technical and legal complexity. The three major areas of concern are temporary storage of spent fuel rods, reprocessing, and permanent disposal of high-level wastes. The approximately 30 metric tons of spent fuel rods produced each year are presently stored in onsite storage pools and must be isolated from the biosphere for as much as 250,000 years. The nuclear industry intended that this spent fuel would be reprocessed and used in the now defunct breeder reactor program.

In 1980 President Carter established a radioactive-waste management program which called upon the nuclear regulatory commission to determine whether and how nuclear waste could be disposed of safely. However, bureaucratic overlap between the EPA, the Department of Transportaton, and units of state and local government with authorities over transportation and land use have so far precluded the establishment of a comprehensive nuclear-waste disposal policy.

QUESTIONS

1 Discuss the factors that have led to the federalization of environmental law.
2 What were the goals of the National Environmental Policy Act?
3 Discuss the functions of the Council on Environmental Quality.
4 What is the role of the environmental impact statement in the decision-making process?
5 What is meant by the requirement that environmental consequences be considered "to the fullest extent possible"?
6 Who is required by the National Environmental Policy Act to prepare environmental impact statements?

---

[20] *Dow Chemical v. Blum,* 469 F.Supp. 892 (D.C. Mich. 1979).
[21] 42 U.S.C. 6901–6987.

7 Discuss the elements required to be included in an environmental impact statement.

8 Discuss the problems encountered in the effort to meet National Primary and Secondary Air Quality Standards.

9 What abatement powers exist in the Environmental Protection Agency in situations where health is endangered?

10 Discuss Environmental Protection Agency authority to regulate emissions from motor vehicles.

11 Discuss the problems inherent in the control of noise pollution.

12 Are there any special noise pollution problems in the case of aircraft?

13 What arguments might be made in favor of allowing air quality to decline until the point of safety is reached.

14 To what waters does the Federal Water Pollution Control Act apply?

15 What is a "point source" of water pollution? A nonpoint source?

16 Differentiate effluent standards from water quality standards.

17 Discuss the state's role in the control of water pollution under the Federal Water Pollution Control Act.

18 Discuss the problems inherent in the control of nonpoints sources of water pollution.

19 In what respect does the control of thermal pollution differ from the normal scheme of the Federal Water Pollution Control Act?

20 Which agency is responsible for the issuance of permits for the disposal of dredged or filled material?

21 Ocean dumping permits may not be obtained for what substances? To what waters does the Ocean Dumping Act apply?

22 In general, how does federal law control the use of dangerous pesticides?

CHAPTER 25

PATENTS

25-1 Definitions and Purposes A *patent* is a right granted by proper governmental authority to a person or persons entitling them, for a limited period of time, to exclude everyone else from making, selling, and using the process or article which they have patented.

Article I, Section 8, Clause 8 of the Constitution of the United States recognizes the need for stimulation and protection of inventions as a matter of public interest.[1] Accordingly, the patent system of the United States has been developed.

The patent system was inaugurated in 1790. It is administered by the Patent Office and Trademark Office (hereinafter Patent Office), which is in the charge of a commissioner of Patents and Trademarks (hereinafter Commissioner of Patents) under the secretary of commerce in Washington. Protection of inventions under the patent law is effective throughout the United States and its territories.

A patent is essentially a contract between an inventor and the government.[2] The inventor is obliged to describe his invention fully through the medium of the issued patent: the government undertakes to protect the right created by the patent. Of course, a patent on a particular article or process will be granted to the first inventor only.

The "life" of a patent in the United States is seventeen years from date of issue;[3] the inventor is afforded by the government the right to exclusive use of the

---

[1] "[The Congress shall have Power] To promote the Progress of Science and useful arts by securing for limited Times to Authors and Inventors the exclusive Right to their respective Writings and Discoveries."

[2] Courts and legislatures will treat patents as they do other contracts.

[3] The life (or term) of a "design" patent (such as for a piece of jewelry) is fourteen years. The specified term of any patent can be extended only by special act of Congress.

patented device for this period. However, this privilege is conditioned upon the *complete disclosure* of all details of the invention so that, after the expiration of the patent, anyone with the necessary know-how and finances will be in a position to make, sell, and use the device. In this sense, "patent" means "opened," or "disclosed."

Issuance of a patent does not necessarily mean that the patentee has a right to manufacture the article involved; additional governmental (or other) authorization[4] may be needed before he can do so. The patent itself merely grants to the holder the privilege of temporarily preventing *others* from using the process or making the article in question.

Placing a patent label[5] on an article is a way of serving notice that the maker relies upon his rights under the patent and presumably will prosecute anyone who violates them.[6] The words "patent applied for" have no particular efficacy at law and merely tend to deter another person from belatedly trying to get a patent on the same idea, since the first party's application would obviously be ahead of his.[7] Similarly, "patent pending" means relatively little—only that an application for a patent is being prepared, has actually been submitted to the Patent Office, or is under consideration by the Patent Office. "Application pending" means almost nothing: the so-called "inventor" may try to use this label as a means of keeping his invention for himself, believing that the use of the quoted phrase will inhibit manufacture of the article by others.[8]

If there is some secret material or process involved, the "inventor," by using an "applied for" or "pending" label, will be risking early discovery of the idea by others and may end up losing all right to patent the invention later on. The patent law will not protect an invention until the patent is actually granted.

A patent will accomplish nothing unless it is put to use. That use is the purpose of the system—profit for the inventor is more or less incidental. The patent system is supposed to make inventions available to the public sooner than they might be if the inventors and manufacturers endeavored to keep all their new developments and products strictly secret for many years. By offering an opportunity for inventors to secure protection, the patent system serves to encourage investment in research and to foster competition in invention, each imaginatively inclined person trying to outdo others in the creation of better products and processes.

Infringement of a patent is the premature use of the protected invention by another person without the permission of the inventor. The infringement may be unintentional or deliberate. It is necessary for the holder of the rights to take

---

[4] Perhaps someone else holds a patent on certain important features incorporated in and dominating the invention.

[5] The patent number should be included.

[6] The courts will not allow a patentee who has failed to affix such label and number damages for any period prior to formal notice to another party that the latter is infringing the patent.

[7] The second party might also be deterred by the expense of "tooling up" for production which may well never materialize because the article to be produced is the subject of another (and prior) patent application.

[8] "Application pending" labels might prove helpful in this way: would-be manufacturers who may wish to become licensees might approach the inventor with some satisfactory arrangement for production mutually beneficial to the inventor (ultimate patentee) and the licensee.

action in order to stop the infringer's activities; mere possession of the patent will not alone do the job. If a product or process does infringe, or seems to infringe, upon a patent, the dispute between the parties is not for the Patent Office to resolve; it is then within the jurisdiction of the federal courts.

Prior to receipt of formal patent application in respect of a given device, the Patent Office will not try to determine whether or not an alleged "invention" will infringe upon an existing patent. Neither will the Patent Office give advance counsel to the applicants about whether or not their articles or processes are worth patenting. Inventors may always apply for a patent if they wish; then the Patent Office will consider the application.

25-2 Patent Records The Patent Office, in order to promote the progress of the arts and sciences, maintains an excellent library of data on issued patents for the use of all who wish to learn what has already been developed. The information about patents is arranged in accordance with a predetermined classification, and it includes material about foreign as well as domestic patents. Persons who think they have developed a new process or product can readily ascertain whether or not someone else had preceded them. Such checking frequently avoids the waste of time and money entailed in a hopeless attempt to secure a patent in connection with an article or process which, it develops, is not original.

A copy of any given patent may be secured by written request to the Commissioner of Patents, Washington, D.C. 20231, and by payment of 50 cents per copy or 20 cents for design patents. A photocopy of a plant in color or patents exceeding twenty-five pages of drawings and specifications are available for $1.00.

There are several regular publications sponsored by the Patent Office: (1) a general information folder furnished without charge; (2) *Rules of the United States Patent Office* (50 cents); and (3) *Official Gazette of the United States Patent Office.* The last-named item is published weekly and it lists all patents and trademarks granted during the week. The patent listing is in the form of an "abstract." It is also cross-indexed by patentee, assignee, and number.

25-3 Patentability By no means is every new process or article patentable. Some of the fundamental principles regarding patentability[9] are the following:

---

[9] "Conditions for patentability" are spelled out in 35 U.S.C. §102 (1982), as follows:

A person shall be entitled to a patent unless—

"(*a*) the invention was known or used by others in this country, or patented or described in a printed publication in this or a foreign country, before the invention thereof by the applicant for a

0"(*b*) the invention was patented or described in a printed publication in this or a foreign country or in public use on sale in this country, more than one year prior to the date the application for patent in the United States, or,

"(*c*) he has abandoned the invention, or

"(*d* the invention was first patented or caused to be patented, or was the subject of an inventor's certificate, by the applicant or his legal representatives or assigns in a foreign country prior to the date of the application for patent in this country on an application for patent or inventor's certificate filed more than twelve months before filing of the application in the United States, or

"(*e*) the invention was described in a patent granted on an application for patent by another filed

1 The new development must have the element of invention. It is not to be merely a routine improvement[10] of some article or process. An improvement, to be patentable, must reveal the exercise of unusual ingenuity—it should represent an advance that would not ordinarily be expected. Expressed the other way around, the patent laws[11] state that an invention is not patentable "if the differences between the subject matter sought to be patented and the prior art are such that the subject matter as a whole have been obvious at the time the invention was made to a person having ordinary skill in the art to which said subject matter pertains."

2 The invention must not be frivolous, and it must not appear that it would have an injurious effect upon the morals and health of people.

3 The invention must be useful. This requirement may, however, be met by something (for example, a design) that merely caters to a whim of the public.

4 Broadly speaking, a person is not entitled to a patent on someone else's invention.[12]

A patent may cover such subject matter as the following:

1 A process or way of making something—often referred to as an "art"—such as a new method of manufacturing sulfuric acid

2 An article of manufacture for sale or other use, such as a new type of outboard motor

3 A piece of apparatus, such as a machine for extruding aluminum tubes

4 A composition of matter, such as a new plastic compound

5 An agricultural development, such as a new variety of plant

6 A new design whose value comes from its appearance, such as a new pattern for table silverware.

25-4 Application for a Patent The preparation of an application for a patent is a very serious and exacting task. An engineer who plans to apply might very well secure the services of an experienced agent or expert who is registered at the Patent Office. Otherwise, errors of commission or omission may be made that will prove very costly in the future, perhaps even to the extent of defeating the application.

One of the first steps to be taken is a search of the files maintained at the Patent

---

in the United States before the invention thereof by the applicant for patent, or on an international application by another who has fulfilled the requirements of paragraphs (1), (2), and (4) of section 371 (c) of this title before the invention thereof by the applicant for patent, or

"(f) he did not himself invent the subject matter sought to be patented or

"(g) before the applicant's invention thereof the invention was made in this country by another who had not abandoned, suppressed, or concealed it. In determining priority of invention there shall be considered not only the respective dates of conception and reduction to practice of the invention, but also the reasonable diligence of one who was first to conceive and last to reduce to practice, from a time prior to conception by the other."

[10] That is to say, an improvement which one might reasonably expect would be made in the normal course of events by someone who is "skilled in the art."

[11] 35 U.S.C. §103 (1964).

[12] But see the discussion of "assignment" in Art. 25–14.

Office in Washington to see whether the idea or principle of the device involved has previously been patented. The search may also reveal that someone has come so close to the same invention that prospects of the newcomer being able to secure a patent are not bright. Of course, the inventor (or some other person, on his behalf) may make this search personally, but it is preferable for him to engage an experienced investigator if the inventor is not familiar with such searching.

A patent will be granted only when there is an application properly filed with the Patent Office. The application must be complete in every respect and accompanied by payment of the fee mentioned below. The application must contain complete disclosure of the invention and satisfy the requirement that the article, composition of matter, or process be new and useful. The mere submission of an idea or suggestion for an invention is insufficient.

The application for a patent is to include a petition, addressed to the Commissioner of Patents, requesting the grant of a patent, giving a designation to the device, and stating the inventor's name, residence, and post office address. The signed application must contain a statement to the effect that, so far as the applicant knows, he[13] is truly the originator of the device which he seeks to patent. The petition is also supposed to include specifications (description), drawings (when practicable), and "claims" regarding exactly what the invention is intended to accomplish and what the patent is to cover. Proper preparation and wording of the foregoing material is extremely important: the information given must be both comprehensive and specific. The application for letters patent must be legibly printed or written in ink, and it must be in English.

There is an application filing fee and a patent issue fee.[14] These have increased over the years. At the present writing, the filing fee (except in design cases) is $300 plus "$30.00 for each claim in independent form which is in excess of three and $10 for each claim (whether independent or dependent) which is in excess of 20." For issuing each original or reissue patent, except in design or plant cases, the statute prescribes a charge of $500 plus "$10.00 for each page (or portion thereof) of specification as printed, and $2.00 for each sheet of drawing." The pattern in design cases is this: "On filing each design application, $125.00. On issuing each design patent, $125.00. For three years and six months, $10.00, for seven years, $20.00; and for fourteen years, $30.00."[15] In plant cases the filing fee is $200; issuance patent $250.

Applications—each one of which can cover no more than one invention—are given numbers in the sequence in which they are received by the Patent Office. They are examined in the order of filing and are classified according to the subject matter of the invention. They are kept confidential unless written authority is given to the Patent Office (by the applicant) to disclose certain information to

---

[13] In rare cases, subsequently to be discussed, someone else would be applying in place of the actual inventor.

[14] 35 U.S.C. §41 (1982).

[15] The total cost to the inventor of obtaining a patent may be at least $800; it could amount to far more, depending upon the cost of the development of the invention and preparation of the application (drawings, patent attorney's fees, etc.).

interested persons. If the subject of the invention is a matter that should be kept secret in the interest of national defense, the Commissioner of Patents, on his own initiative, may label the application "secret" and withhold any revelation whatever.

The application is to be made by the inventor personally. If two persons really worked together in the development of the process or article, they must jointly apply for a patent. When there are known to be joint inventors, neither one is allowed to patent the invention exclusively. However, under one of the practice rules of the Patent Office if one of the joint inventors refuses to sign the application papers or cannot be reached after diligent effort, the application may be made by the other inventor on behalf of both.

One who merely supplies money to enable the inventor to proceed with his endeavor cannot qualify as a joint inventor. Working independently and without knowledge of the other's efforts, developers of separate improvements in the same article or machine cannot jointly obtain a patent.

If an inventor dies before any filing can be accomplished, an application for a patent may nonetheless be perfected by his executor or administrator and the patent will be granted if the application meets the usual requirements. If an inventor dies after filing his application, the patent may issue to the executor or administrator if such fiduciary files the proper papers to evidence his appointment as the representative of the decedent,s estate.

Should a certain patent application conflict with someone else's somewhat similar application pending in the Patent Office, such office will conduct what is called an *interference proceeding.* An appointed board of examiners will determine from evidence submitted who is the real inventor and thus entitled to the patent. In such a situation, one can readily see how important are adequate records kept by the respective parties in interest.

25-5 Specifications In connection with a patent application, the term "specifications" denotes a written description of the invention.[16] The specifications should include illustrative drawings when these are needed for clarification. The description should cover how the device is made and how it is to be used. The data must be clear and complete, yet concise. The invention should be described in such detail that anyone skilled in the art can understand what is being said and, by following the specifications, put together the device in question. Otherwise, a patent will not be granted.

In making the disclosure, the applicant should exercise extreme care to ensure that the description of the invention is correct. He should state the suggested method for constructing the device but should be careful not to limit the ways in which it can be made.[17] In this connection, the application might use alternatives. For example, he could say that a certain portion of his machine may be a casting

---

[16] The applicant should be very careful in stating what his invention is designed to accomplish; this is often called the *statement of invention.* The inventor ought to avoid expressing aims which are either too meager or unduly comprehensive.

[17] Or the "process" accomplished, if it is an article or device which is the subject of the invention.

or may be fabricated by means of welding pieces of steel together. However, the listing of alternatives can, in certain cases, prove confusing or even harmful.

Making the *drawings* for a patent application is almost an art in itself. These drawings should be clear, neat, and so prepared as to bring out the special points which the inventor intends to emphasize. For example, the drawings depicting a mechanical device should generally show a label or number for each pertinent piece or element, so that the specifications can, through cross-reference, clearly describe each item and its function.

The requirements regarding drawings are very specific and must be complied with in every detail.[18] Models, in lieu of drawings, will not be accepted unless the substitution has been requested by the Patent Office.

The following suggested arrangement of items is one which may be observed in framing the application:[19]

1 Preamble, stating the name and residence of the applicant and the title or designation given to the invention
2 Abstract, or general statement of the nature of the invention
3 Objects
4 Brief description of the several views shown on the drawings (if the invention lends itself to such illustration)
5 Details of the invention
6 Claim or claims
7 Petition
8 Signature of applicant

25-6 Claims The term *claims* as used in connection with an application for letters patent denotes the statements describing what the invention is expected to accomplish. The claims must convey clearly the scope and character of the invention. Usually, the first portion of the claims material constitutes a general outline of what the patent is to cover. Subsequent portions will be more specific and will relate to details of the invention. The claims must be worded very carefully and accurately. If they are too broad in scope, the prior art may prevent securing a patent; if too narrow, it may be easy for another party to circumvent the patent as issued. Preparation of these claims is such an important matter that it should be handled by an expert, preferably one who is a specialist in the particular field to which the subject matter of the invention relates—for instance, chemistry or mechanical engineering.

The following example illustrates a typical set of claims. All four items are shown as they might appear in the application for one patent on the device, a card table. The drawings accompanying the claims are grouped in Fig. 25-1 and are so numbered as to accord with the corresponding claims.

Claim 1 A table comprising a platform, at least one support member, and mounting means for mounting said support member to said platform.

[18] See U.S. Patent Office, *General Information Concerning Patents,* page 5.
[19] The application, per 35 U.S.C. §111 (1982), consists of a specification, a drawing, and an oath.

Claim 2   In the article of claim 1, said mounting means comprises a hinge and locking means to maintain said support member in a desired position in relation to said platform.

Claim 3   In the article of claim 2, said locking means comprises a brace member having one end engaging said support member and an opposite end engaging said platform.

Claim 4   In the article of claim 3, said brace member comprises a first portion engaging said support member, a second portion engaging said platform, and said first second portions movably connected to each other.

In claim 1 (an "independent" claim), the claimed elements of the structure appear after the word "comprising," which elements are (*a*) a platform, (*b*) a support member, and (*c*) means for mounting the support member to the platform. An unauthorized person building a table with one leg or a plurality of legs might impinge upon this claim.

Claim 2, known as a "dependent" claim, contains the three elements of claim 1 with the added limitation that the mounting means is a hinge. If the first inventor did not have claim 2, another inventor might get a patent on a table with a hinge; in that case, the first inventor could not use a hinge without permission from the second inventor, and the second inventor could not make a certain type of table without permission of the first inventor. It is desirable, of course, for the first inventor to provide for as many modifications and improvements as possible in order to protect his position.

Claim 3 includes (by implication) all the elements of claim 2, but the locking means is identified as a brace. It is conceivable that another person might choose to hold the support member (leg) by means of a wing nut on the axis of the

FIGURE 25-1
Drawing to accompany claims for a patent.

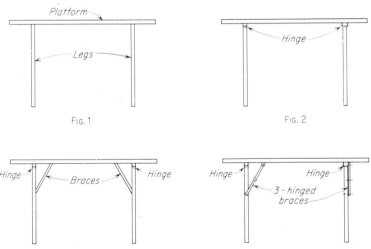

hinge—and this arrangement would not constitute a brace. However, in the process he would run afoul of claim 2, which specifies "locking means." Claim 3 is said to be "narrower," or "limited," as compared to claim 2.

Claim 4 includes all the elements of claim 3 and is even narrower than claim 3 because it calls for the brace to have *two-part* member portions. This particular claim (number 4) could be avoided by using a one-part brace which can be hinged at one end and slidable at the other end, but any such expedient would be unable to contend with claim 3. A three-part brace member would similarly encounter difficulty because the claim would "read on" the suggested three-part brace, which obviously has at least *two* parts. The added part (that is, added to the two) does not successfully circumvent claim 4.

From the foregoing it will be obvious that, in drafting claims, a great deal of thought must be given not only to the selection and arrangement of the words which will make up the claims but also to the various ways someone else could avoid the claims—which ways could be detected from the disclosure (both specifications and drawings).[20]

In general, the more elements the claim has, the more specific it becomes. One should try to make the claims broad and simple. The Patent Office will decide whether or not the claims are too broad, but it will not try to determine whether they are too narrow.[21]

Calvert[22] says

A claim covers all that it does not exclude. It excludes those structures that omit one or more essential features ("elements") recited in the claim. The claim does not exclude but, to the contrary, covers a structure containing all elements of the claim and an additional part or element.

If the invention is a method of doing something, the claims should state in detail the various steps to be taken to accomplish the desired objective. Someone else who can achieve the same result in fewer steps or in different ones can assert that his invention is a new or different method and that it therefore does not infringe upon the patent of the first inventor.

One cannot add a new material to his patent application once it has been filed and at the same time preserve the original filing date. Claims can be expanded during the course of their presentation to the Patent Office *if* the specifications mention the point involved in the attempted expansion. Of course, a new, enlarged application can be filed to replace the old, but the substituted one will necessarily bear a later date than that of the original.

If a patent is found after issue to have been based upon claims one (or more) of which is invalid, it may be possible for the patentee to disavow the invalid claim

---

[20] In the case illustrated, notice that the first claim listed *dominates* (for example, includes) the second, third, and fourth claims. It may be that a patent issued to cover several claims will be treated as though each claim were covered (or represented) by a separate patent.

[21] Notice again that, if the claims are unduly narrow, the scope of the patent may be so restricted that competitors can easily come up with a slightly different device which will not infringe.

[22] Calvert, *Patent Practice and Management* (Reinhold Publishing Corporation, New York, 1950).

(or claims), thus paving the way for the remaining, valid claims to retain patent protection.

25-7   Validity   Claims in an application for a patent are invalid if they do not meet the requirements for patentability.

It is entirely possible that an applicant has been utterly unaware of prior publication or prior art concerning the substance of his "invention." Nevertheless, proof of such prior activity will cause his application to be rejected or will invalidate[23] the patent after it has been granted.

The patent must be taken out in the name of the real inventor. If the invention is the joint product of two or more persons, proof of this circumstance will invalidate a patent issued in the name of one of the parties alone. There may be considerable disagreement about whether an invention is *sole* or *joint*. If one inventor conceives the idea for the device and then in detail instructs other persons in making of a drawing or working or model or in the conducting of a successful experiment, the resulting "invention" is really the product of the first individual alone. On the other hand, if the first person had merely a general conception of the prospective invention but had to depend upon the *unusual* skill and imagination of someone else to develop the article or process, then the same will probably be treated as a joint invention—the product of the creative efforts of both parties. If one person develops a patentable device but an assistant proposes an ingenious additional feature to add to the original scheme, then the two inventions should be separately patentable; the total product is not really a joint invention because the item attributed to the first inventor is autonomous and completely separable from that produced by the second inventor.

One may have to prove the priority of his invention in order to overcome interference[24] from the other persons interested in patenting virtually the same product, and he must be able to defend his own assertion that others have not (within a certain time period) used or "published" the invention on which he seeks a patent. Any pertinent publication or use by others more than one year before an inventor submits his application for a patent will void the latter's application.

25-8   Date of Conception of Idea   If a person develops in his mind a scheme which he thinks is potentially patentable and he makes a dated record of the new proposition, such action alone will probably be insufficient to establish the *idea*

---

[23] Assuming that the unlikely occurs and that there has been some oversight by, or lack of information in the hands of, the Patent Office, causing a patent to issue when it should not have.

[24] In case of an "interference" proceeding (conducted under the auspices of the Patent Office to determine which of serveral claimants is entitled to the patent), the date of conception of an invention will be judged by such criteria as the following:

1 The date of the first verbal disclosure to someone else
2 The date of the first written description of the invention
3 The date of the first drawing showing the invention
4 The date of the first reduction of the invention to practice

date in connection with his subsequent petition for a patent. The inventor should do something beyond simply marking his records. He might, for instance, write to an outsider, giving the general outline of his idea—and do so promptly upon conceiving it. The date of the conception in the eyes of the law for the purposes of securing a patent is likely to be the date of the first *disclosure* of the idea to someone who is competent to understand it and later to verify the fact of the disclosure.[25] If such observer makes a written record of the disclosure on behalf of the inventor and if the former signs and dates this record forthwith, the document may well be a valuable one for the inventor to have available to him.

It is apparent that, when one is developing what he hopes will be an invention, he should try to obtain at once excellent proof of the date of conception of the idea so that he will have an "airtight case" on the point if and when needed. Letters, sketches, and tentative descriptions that have been dated, witnessed, and signed—preferably by more than one competent observer—are very useful instruments in this connection. The observers should have some understanding of the invention against the possibility that they are asked to appear in support of the inventor's claims.

25-9 Reduction to Practice   It is important to prove that one's invention really works in practice, not merely under ideal laboratory conditions. An inventor should have at least one witness to the fact that the invention indeed does what it is supposed to do. This testing in operation is called *reduction to practice.* Assume that *A* invents a garage door that will automatically open and roll up overhead when a car approaches it. He should build a sample of such door, install it in his garage or that of a friend, and demonstrate it to a competent observer showing how well the mechanism operates.

Naturally, an inventor will want to maintain secrecy so that no one except his trusted friend will learn of his invention before the patent application can be submitted. This security problem is one for the inventor to work out with the aid of the friend to whom he disclosed the invention.

If the final perfection of an article or process to be patented will cost so much that the inventor cannot immediately finance the necessary work of development, he may nonetheless file his application for a patent in order to protect his rights. This course of action is known as *constructive reduction to practice*—which means that the device described in the petition does not yet exist. The inventor then may make arrangements with someone who will finance the perfection of the invention in return for a share of the benefits, on whatever terms may be mutually agreeable.

25-10 Diligence   After conceiving his idea, the inventor must be diligent in implementing it. Exactly what will be regarded as diligence in a particular set of

---

[25] This "observer" should be someone other than a person working with the would-be applicant on the invention.

perfect his invention within a few months of the conception of the idea, he probably will be deemed to have been dilatory.[26]

Let us suppose that *A* thought up the idea for an invention in January 1984,
and that *B,* who knew nothing about *A* or his activities, came up with the same
thoughts independently in June 1984. Assume also that *A* did little to perfect his
idea, whereas *B* developed the invention promptly and applied for a patent in
October 1984. Shortly thereafter, *A* heard about *B*'s application and tried to secure the patent for himself by applying in December 1984, claiming that he had
the idea first. It is probable that *B* would be granted the patent because *A* had
been comparatively dilatory. In one case,[27] the court made the diligence point in
these words:

> It is obvious from the foregoing that the man who first reduces an invention to practice
> is prima facie the first and true inventor, but that a man who first conceives, and, in a
> mental sense, first invents, a machine, art, or composition of matter, may date his
> patentable invention back to the time of its conception, if he connects the conception
> with its reduction to practice by reasonable diligence on his part, so that they are
> substantially one continuous act. The burden is on the second reducer to practice to
> show the prior conception, and to establish the connection between that conception and
> his reduction to practice by proof of due diligence.

Although it is advisable to proceed as quickly as possible to develop an invention, it is equally evident that the inventor's work should not be so hasty and
incomplete that his claims will be disallowed because of errors, imperfections, or
even failure of his invention to function as advertised.

It is well for an inventor not only to have a record of the disclosure of his
invention idea to at least one competent person (and preferably two or more)[28] but
also, from time to time, to have outsiders witness and verify his procedures and his
progress in developing the invention, always being sure that any pertinent records
kept are properly dated and signed. The witnesses ought also to observe the
functioning of any model of the invention, its performance under tests, and even
the preparation of descriptive matter and drawings.

25-11 Keeping Notes on the Invention It is very important to maintain a
running record of activities once work on the invention has commenced. This
record should be kept in a *bound* notebook so that others would find it difficult to
claim that the inventor inserted information at a date later than was actually the
case or that he deleted some pages, as would be possible with a looseleaf notebook. It is also desirable for the inventor to date and sign each page as he records
his notes. Furthermore, at intervals he might well have these notes read, dated,

---

[26] Some inventions may require years for their perfection; others may lend themselves to complete
development within a few months. The reasonableness of the time period between conception of the
idea and perfection of the invention to application for a patent thereon will be judged in accordance
with the factual circumstances which pertain in a given case.

[27] *Christie v. Seybold,* 55 F. 69, 76 (6th Cir. 1893).

[28] Involving *several individuals* will serve to reduce the possibility that—for one reason or another—
when the need for testimony arises, no one will be available.

and signed by a trusted friend (or, better still, by several friends), who will subsequently be able to verify the accuracy of the record. To avoid suspicion, such notes should not show erasures or alterations. If changes have to be made, it is best to add a supplementary statement correcting what was recorded previously. The aforementioned record of the step-by-step development of the device or process should contain all relevant data, including such information as the following:

1 When the inventor first conceived the idea
2 When and to whom he disclosed his idea
3 What he did at each step of development
4 What results he obtained along the way and whether such were good or unsatisfactory
5 When he reduced the invention to practice
6 Who has seen it work or has verified the results
7 When the witnesses saw it work
8 When and through whom the inventor took steps to prepare the application for letters patent
9 Whatever else may assist in securing the patent and in establishing the inventor's precedence over some other applicant

25-12 Ownership of the Patent Sometimes there are serious misunderstandings about who should own a patent. Here are a few principles which may serve as a guide:

1 An individual who develops his invention entirely on his own and who meets all the requirements for securing a patent is personally entitled.
2 An individual who develops his invention independently but who uses his employer's time and facilities to do so probably can secure the patent, but the employer may have "ship rights" to manufacture the article in question.
3 An individual who is part of a research organization is generally hired for the very purpose of inventing. As a rule, the employer of such an individual requires him to agree in advance to *assign* to the employer any patent which the employee receives in line with his regular work.
4 A member of a research organization may be individually entitled to retain a patent on an invention which he develops entirely outside his employer's field of interest and perhaps largely on his own time.

Before an engineer undertakes work, for or with others, which may result in an invention, there should be a definite and recorded agreement regarding the ownership of any patent or copyright for which such work might pave the way.

25-13 Secret Information When attempting to sell an idea, one must be careful about divulging details of his unpatented invention. By the same token, whoever receives information about such an invention should keep the data to himself. Ordinarily, a firm or individual is reluctant to accept confidential information from an inventor unless the latter—before disclosing any secrets—has patented the invention to which the information relates, has applied for a patent, states in writing that he will rely entirely upon any patent rights available to him, or agrees to a stated compensation should his invention be accepted by the information recipient. One can easily understand the receiving party's reluctance. He would not fancy being subjected ultimately to unreasonable claims by the inventor about the value of the latter's device. It is unwise for one to accept broad or vague ideas, the transmittal of which might position their author to claim compensation should the recipient of the disclosures later develop anything which appears to have been even remotely suggested by the volunteered information.

Negotiations regarding "secret" information may take place even when a discovery, though apparently useful, is not patentable. For example, the winner of a national cake-baking contest may sell her recipe to the sponsoring corporation in spite of the fact that the formula is not a patentable item. A different situation is created when a man with an excellent idea finds himself unable to finance the development of it; he may wish to join forces with someone who can provide the necessary money and equipment to perfect the invention. The respective rights and obligations of the several parties participating in such an arrangement should be clearly defined and the agreement reduced to writing.

Suppose a manufacturer receives from a self-styled "inventor" what appears to be an offer to disclose the idea for a new gadget of some sort. The manufacturer would do well to proceed cautiously. He ought to notify the sender that he has not examined the data and that he will not do so unless and until a proper agreement is reached between the parties, setting forth what the first party is to receive (if his idea proves worthwhile and of interest to the manufacturer) and what the latter is to pay for the disclosure. The arrangement may be such that the manufacturer shall not be obligated to treat such information as confidential if the invention proves to be (1) unworkable, (2) nonpatentable, (3) previously disclosed to others, (4) previously known to the manufacturer, or (5) valueless for some other reason.

The following form might be used by a corporation in handling receipt from outsiders of information relating to new ideas or alleged inventions:

DISCLOSURE AGREEMENT

L-M-N Corporation:
I request that *L-M-N* Corporation consider my idea relating to

[subject matter of idea very briefly summarized here]

which idea is described in the material herewith and listed below. All additional disclosures relating to the idea submitted herewith shall be subject to the provisions of this Agreement.

I understand that no confidential relationship is established by or is to be implied from this submission or from consideration by *L-M-N* Corporation of the material, and that such material is not submitted "in confidence."

By this request and submission I do not grant *L-M-N* Corporation, or any of its subsidiaries, any right under any patents on the idea submitted. I agree that, at least unless this Agreement be superseded by a different agreement in writing, I will make no claim against *L-M-N* Corporation, or any of its subsidiaries, with respect to the idea here submitted, other than possible for patent infringement.

List of Documents and Samples [follows]:
Agreement acknowledged:
L-M-N Corporation   Signed
By   Address
Dates

It appears that, in settling disputes stemming from the disclosure of inventions, the courts give considerable weight to whether the idea was completely new to the recipient and to just how the latter made use of it. In case no advance agreement had been reached regarding the measure of compensation that is to be paid the inventor for utilization of his idea, the courts will try to determine what is reasonable—perhaps an amount comparable to what might have been expected in the way of royalties for use of the process or article if it had been patented.

25-14 Licenses, Assignments, and Grants An inventor may assign his patent and all rights pertaining to it, or he may permit another person a license or grant[29] to "make, use, and vend" whatever is covered by *any* (or, perhaps, *all*) of the rights under the patent (or under the application for patent). A license may or may not be treated as exclusive, and the agreement between the parties should clarify this point. Licensing, and possibly sublicensing, arrangements may cover manufacture of the article, its use, its sale, or some combination of these.

The licensee may make payment in the form of the royalties, the total varying with the volume of sales by such licensee. On the other hand, he may pay a lump sum for the rights, or he may make any other legal arrangements agreed to by the parties involved.

Unless the applicable agreement provides otherwise, the license carries with it the implied right for the licensee either to sell the patented article or to make use of it himself.

The agreement on licensing arrangements will cover such matters as powers of the respective parties, limitations on these powers, royalties or other payments, anticipated efforts to market the device, actions which would constitute infringement by outsiders,[30] time or territorial limits affecting the license, reports due the patentee, labeling of the article, type of use authorized, and steps designed to prevent misuse. If payment is to be on a royalty basis, the licensor will want to be sure that the licensee is going to make a sincere and adequate effort to procure and market the product involved. On the other hand, the licensee will wish to be protected in the event something occurs which may cause (1) the patent to be declared invalid or (2) use of the patented item to be declared illegal.

---

[29] It seems that a "grant" generally affords permission on a more exclusive basis than does a "license." Both types of permit should be written or printed, properly signed, and recorded in the Patent Office.

[30] A patentee may remedy through civil action an infringement of his patent; there are statutory provisions covering both injunction and damages.

An assignment transfers to the assignee the inventor's interest in the patent or an agreed portion of the interest, and the assignment is effective throughout the United States. If the entire interest in an as-yet-unpatented invention is assigned, the patent will normally issue to the assignee (upon application by the inventor[31]). If such assignee is to hold an undivided part interest in the invention, the patent will normally issue to the inventor and assignee jointly.[32] If it is desired that the patent so issue, the assignment in either of the foregoing cases must have been recorded with the Patent Office not later than the date of payment of the final fee for the patent.

A certificate of acknowledgment, under hand and seal of a person authorized to administer oaths, is prima facie evidence of the execution of an assignment of a patent or applicantion for patent. Once recorded at the Patent Office,[33] an assignment, even though bearing various conditions (regarding such things as payment to the assignor), is considered "absolute" and will remain effective unless and until abrogated by action of the parties or on an appropriate tribunal.[34]

Patent pooling is an arrangment among patent owners for the interchange of patent rights so that they each have the right to use any of such patents or to license[35] others so to do.

Package licensing denotes an arrangement under which a patentee licenses others under some of his patents—perhaps under all of them. If all the patents are included in a single package, this may make for simplified accounting.

Under an agreement for *cross-licensing,* patentee *A* agrees to license party *B* on the understanding that *B* will reciprocate with respect to patents *he* holds.

One should not accept an assignment or seek a license privilege without first checking the pertinent patent records; additionally, competent advice regarding exactly what is covered by the patent in question should be elicited. Furthermore, it is desirable to effect licensing arrangments and the like directly with the original inventor, not through some third party.

25-15 Copyrights and Trademarks A *copyright,* in fashion similar to a patent, is provided for in Article I, Section 8, Clause 8 of the United States Constitution. The copyright differs from the patent right in subject matter, in the protection afforded, and in other ways. Patents are granted for machines, for processes, for compositions of matter, and for certain types of plants—but only after the Patent Office has searched the prior art and has determined that the invention is novel and useful. Copyright law relates to literary and artistic property and is essentially inherited from English law. Copyrights are under the general supervision of the Library of Congress.

---

[31] If the inventor declines to execute a patent application, or if he cannot be located, the assignee may apply—on behalf of and as agent for the inventor—on a showing that such action is necessary to prevent irreparable damage.

[32] Absent a contrary agreement, each joint owner of a patent may utilize the patented invention without accounting to the other owner or securing the latter's consent.

[33] This should be accomplished within three months after the assignment is executed.

[34] In some instances, there may be a time limit set on the duration of an assignment.

[35] It should be noted that there are monopoly and unfair trade practice questions which may arise under certain circumstances in connection with the licensing of patents.

Copyrights involve no search and can be secured by the author upon complying with simple formalities and paying a nominal fee. The author's work must be *original*, and the copyright term will consist of the life of the author and extend fifty years after the author's death.[36]

The owner of copyright has the exclusive rights to do and authorize the reproduction of the copyrighted work or to prepare derivative works based on the copyrighted item. Nevertheless, the fair use of a copyrighted work, including reproductions, is permissible for purposes such as criticism, news reporting, teaching (including multiple copies for classroom use), scholarship, or research. These activities do not constitute an infringement of copyright. In determining whether the use made of the copyrighted work in a particular case is a fair use, the factors to be considered include the following:

1 The purpose and character of the use, including whether such use is of a commercial nature or is for nonprofit educational purposes

2 The nature of the copyrighted work

3 The amount and substantiality of the portion used in relation to the copyrighted work as a whole, and

4 The effect of the use upon the potential market for or value of the copyrighted work[37]

Trademarks come within the jurisdiction of the Commissioner of Patents. A trademark is any word, name, symbol, or the like, adopted and used by a manufacturer or merchant to distinguish his goods from a comparable product manufactured by others. Trademarks may also be used to identify services and, in that connection, are generally referred to as "service marks."

Congress has enacted precise statutes covering the use of trademarks in interstate commerce, and the Commissioner of Patents has formulated specific rules for obtaining registration of such trademarks. A registration lasts for twenty years and may be renewed as many times as the owner can show continuation of active use in interstate commerce.[38]

A person should seek to register a trademark only on such materials as he intends to use in the immediate future. He should not attempt to obtain trademarks on each of a series of names or slogans which he thinks he may want to use at some time in the future—such a scatter-gun approach being designed to prevent someone else from using such names or slogans meanwhile. Generic words or terms cannot properly be registered as trademarks.

QUESTIONS

1 What is the principal objective of the U.S. patent system?

2 On what authority is the U.S. patent system based? When was this system inaugurated?

3 What is a patent? What is its "life"?

---

[36] For works copyrighted before January 1, 1978, the copyright term is twenty-eight years, with the privilege of one twenty-eight year renewal. See 17 U.S.C. §708 (1977).

[37] 17 U.S.C. 107, as amended on Oct. 19, 1976.

[38] States have their own laws regarding trademarks, as well; see, for example, New York General Business Law, Art, 24 (McKinney 1968).

4 Is a patent a contract? If so, what are the rights and obligations of the parties?

5 Who administers the patent laws of the United States?

6 What are the limits of the areas in which the U.S. patent laws are effective?

7 Does the Patent Office try to advise the patentee whether or not a patent will be a financial success?

8 List the elements which are generally necessary for patentability of an invention.

9 Describe the procedure to be followed in applying for a patent.

10 What is the meaning of the word "specifications" when referring to patent matters?

11 What is the meaning of the phrase "statement of invention" when referring to patent matters?

12 Referring to patent matters, what is (*a*) an "infringement"? (*b*) an "interference" proceeding?

13 When and why may drawings be required in an application for a patent?

14 What are the "claims" made in connection with an application for a patent? Why are they important?

15 What is meant by "complete disclosure"?

16 Differentiate between (*a*) a "sole" and (*b*) a "joint" invention.

17 Define the words "copyright" and "trademark." Illustrate how they are applied.

18 What is "fair use" of copyrighted material?

19 Of what significance is the "date of conception" of an invention?

20 What are the meaning and importance of the "date of disclosure" referring to inventions?

21 What is meant by "diligence" in matters of invention?

22 Referring to inventions, what is meant by "reduction to practice"?

23 What procedure should be followed by an inventor when developing an invention?

24 What precautions should be taken by both parties when dealing with the possible disclosure of secret information?

25 Explain and illustrate the licensing of a party to use a patent.

26 What is (*a*) a "grant" of patent rights to a party? (*b*) an "assignment"?

27 What is meant (*a*) by "patent pooling"? (*b*) by "cross-licensing" of patents?

28 What time limit is there on the life of a patent? Can this life be extended?

29 Can the Patent Office prevent the disclosure of patent data? When?

30 Why may a manufacturer stamp on an article (produced by him) the words "Patent applied for"?

31 In case of a grant of patent rights on a royalty basis, has the grantee any implied obligations with respect to the grantor?

32 If claims in an application for a patent are made extremely narrow, what may happen? If claims are extremely broad, what may happen?

33 If a patent application contains six claims and the Patent Office finds that two are invalid before issue, must the application be thrown out? What can be done?

34 Kimball was a mechanical engineer with an electric power company. Assume that she devised and built an apparatus for the disposal of coal ashes from the boilers of one of the company's plants. Two years afterward, she applied for a patent on this equipment. Do you think she could get the patent? Explain.

35 *X* was trying to develop and patent a new device for dust-collecting equipment. He had *Y*, who was in his office, assist him in perfecting the invention, and the latter showed that certain improvements, which he suggested, were vital to the successful operation of the equipment. *X* patented the invention using *Y*'s suggestions, but *Y* claimed that it was a joint invention. Is *Y* justified in his claim? Explain.

36 Jones, a machinist, made and used a clever improvement in an automatic lathe, but he did not think that the device was patentable. Two years after the device was installed at the shop where Jones was employed, a visiting sales engineer saw it, had the apparatus installed on the lathes which his company manufactured, and applied for a patent on the device. Jones heard about the patent application. Could Jones attack the patent which might be issued on the device? Explain.

37 Long developed a special formula for a cough syrup, and she kept it a secret even though she manufactured and sold the medicine. Short discovered the secret formula and produced and marketed the same medicine under a different name. (*a*) Can Short patent the formula? (*b*) Can Long stop Short from making the competitive product? Explain.

38 *X* patented a special fountain pen. *Y* copied it and started manufacturing and selling the same pen. *X* appealed to the Patent Office to stop *Y*'s activities. Will the Patent Office do this? Why?

39 What money damages can *X*, in the preceding case, collect from *Y* for goods produced before *X* learned about *Y*,s activities? How can *X* stop *Y*,s production?

40 Brown invented and patented a toy airplane. He later made a certain improvement in the toy and continued manufacture of the modified toy under the original patent. Is his improved device properly protected? Explain.

41 If a chemical product is made by a certain patented process, can someone else produce and sell the same material if he can manufacture the chemical by a different process?

CHAPTER 26

EVIDENCE

26-1 Introduction The rules of evidence form an integral part of our system of justice. Innumerable cases before the courts have turned upon the admissibility of a single bit of evidence offered by a litigant. The present chapter will attempt to highlight a few of the basic evidentiary principles.

Despite their great importance in the overall scheme of things, the rules of evidence (unlike, for example, the law of contracts or that of torts) can be classified as *procedural,* rather than as *substantive,* law. While many of the underlying ideas are universally accepted, application of some of the evidence rules will vary from one jurisdiction to another.

26-2 Definitions The *relevancy* of evidence deals with its logical connection to the fact at issue, for example, whether the existence of the evidence renders probable the existence of the fact to be proved. The *materiality* of evidence relates to the probative value of the evidence. *Material* evidence is *relevant* evidence which is of sufficient value in the particular case to justify its consideration by the judge or jury. Thus, though evidence may be logically related to proof of a fact, that relationship may be so remote as to fail to justify its consideration. *Competent* evidence is that which is fit for its intended purpose, for example, it tends to establish the fact at issue without relying on speculation. Thus, the source of the information and form of the evidence are two factors which often render evidence incompetent (for example, hearsay evidence and oral testimony as to the written terms of a contract are normally incompetent evidence).

The two categories of probative material which may be produced in court are *testimonial evidence* and *real evidence.* The former can be readily subdivided into

(1) *direct* and (2) *circumstantial. Direct evidence* is that which tends to prove a fact at issue without the necessity of proving any collateral facts. When, as regards a fact in dispute, a witness can impart knowledge which he has acquired by means of his senses, the evidence he gives is direct. Thus, a witness who personally saw or heard something relevant and material to a point in issue is in a position to give direct evidence. Very often, however, it becomes necessary for litigants to rely heavily upon a somewhat less satisfactory form of evidence, known as *circumstantial.* This type aims at establishing some collateral fact or set of facts from which the jury may reasonably infer the existence or nonexistence of the critical fact in contention. Examples of circumstantial evidence would be the fact of an accused's flight and related conduct being admissible as possibly pointing to his consciousness of guilt.

Real evidence involves observation by the tribunal itself. Thus, when a question arises whether a certain object is green or blue, the ideal proof of the true color would be the exhibit of the object to the jury. Real evidence is resorted to in an infinite variety of instances, one of the most familiar being the demonstration by plaintiff of the injury of which he complains. Where the approximate age of some individual is a material factor in the trial, personal inspection by the court or jury is often employed.

Documentary evidence comprises (1) *public* or *official records* and (2) *private documents.* The former includes all writings made by government officials setting forth data which their duties require them to record. Public documents are evidence of the facts therein stated and, if properly authenticated (by seal or otherwise), are normally admitted without the testimony of the officer who prepared them. Admission on this basis constitutes an exception to the *hearsay rule.*

Any writings which do not qualify as public records are private documents, and before these can be received in evidence, the offeror must show that the paper was indeed duly executed by the person who is alleged to have executed it.[1]

26-3 Function of Judge and of Jury In cases heard without a jury, the judge handles all phases of the trial. When there is a jury, however, a problem arises about the respective prerogatives of judge and jury. Broadly speaking, it may be said that questions of a fact are for the jury, and questions of law are for the court. As usual, there are exceptions, the primary one being that the court will decide certain "preliminary" questions of fact, such as those necessary to its determination of the admissibility of evidence or the competency of witnesses. By way of illustration, the qualifications of a would-be "expert', witness constitute a preliminary factual question for the court. Another sizable group of facts, which are treated as preliminary in nature and thus within the province of the judge, are those which are subject to *judicial notice. Judicial notice* is a convenient doctrine by which the judge will take official cognizance of a well-known fact without requiring the presentation of evidence. A proposition is indisputable when it has

---

[1] There are a few exceptions under which proof will be excused; for example, the adverse party may concede due execution of the private document in question.

been judicially noticed. The range of things of which judicial notice will be taken is broad and includes such diverse items as the provisions of the Constitution of the United States, the existence of foreign governments, the sequence of the seasons of the year, standard weights and measures, and the kicking propensity of the mule.

The jury for its part, decides upon the credibility of the witnesses it has heard examined and cross-examined and weighs the evidence presented. There should be submitted to the jury only matters in dispute about which reasonable men could differ. If the evidence offered by one side so heavily overbalances that presented by the other that a jury verdict for the latter would have to be set aside as against the weight of evidence, the judge will direct the appropriate verdict. If, however, submission of the case to the jury for determination appears warranted, the court will *charge* the jury. Along with instructions regarding the substantive law applicable to the controversy, the charge usually includes a reiteration of the issues of fact which have been submitted to proof, a statement about which party has the burden of proof, and a summary of the evidence on the record. The judge may not comment upon the evidence so as to give an indication of his personal opinion.

26-4   Burden of Proof; Presumptions   The burden of proof is the task of proving a proposition either by a preponderance of the evidence if the case is civil in nature or "beyond a reasonable doubt" if the action is a criminal one. There is no question about which party has the burden of proof in criminal cases. Regardless of the nature of the defense, the prosecution carries the burden throughout the trial, and from this circumstance stems the well-known saying that "every man is presumed innocent until proved guilty." Though it is somewhat dangerous to generalize on the rule in civil cases, usually the plaintiff must take the initiative and persuade the tribunal of the truth of his allegations. Occasionally, the defendant's answer to the complaint will admit the several points raised by plaintiff but will set up additional facts by way of defense. In such circumstances, defendant has brought up the only matters in issue, and the burden of proof is upon him. The burden of proof does not shift from one party to the other during the course of the trial. What does vary is the duty of going forward with the evidence at any particular stage.

It is imperative that the jury be correctly charged as to which side has the burden of proof, since the verdict must be against that side if the evidence presented by the several parties is in equilibrium. Having the burden of proof is generally regarded as carrying with it the right to open and close the presentation of evidence and argument.

Some matters involved in a lawsuit are not subject to proof, for example, admissions, either as contained in the pleadings or as made in open court. Thus, if defendant's answer admits the truth of an allegation in plaintiff's complaint, the latter need not introduce evidence in support of that allegation. Similarly, evidence is unnecessary to prove what is presumed, at least until the party adversely affected has taken steps to rebut the presumption. A good example of this latter proposition is afforded by the presumption of the innocence of an accused criminal. He may choose to rely upon the presumption and decline to introduce evidence in his own behalf, at least until the prosecution has assembled a substantial case against him. There are several categories of presumptions, but perhaps the basic distinction is between those which are *conclusive* (or "absolute") and those which are *rebuttable* (or "disputable"). No evidence, however strong, will be permitted to overturn a conclusive presumption, such as that grounded upon a statute which provides that a child under a given age "is deemed incapable of committing a crime."[2] Rebuttable presumptions are numerous; commonplace illustrations are (1) the presumption against suicide and (2) the aforementioned presumption of innocence, which aids persons brought to trial on criminal charges and which necessitates acquittal unless the evidence as presented establishes guilt beyond a reasonable doubt.

26-5 Competency of Witnesses; Privileged Communications It is part of the judge's function, in instances where a question has arisen, to pass upon the competency of prospective witnesses. Lack of mental capacity is probably the most common basis of disqualification, though the law affords other grounds as well.

For reasons of public policy, the law recognizes several relationships as confidential in nature and places in a *privileged* category certain information passing between the parties to such relationships. In the typical jurisdiction, the list would include confidential communications involving husband and wife, physician and patient, clergyman and penitent, and attorney and client. If we take the last-named relationship as an example, we find that communications which the client has had in confidence with his legal adviser, and which pertain to the lawyer's professional services in behalf of the client, have what amounts to a seal of silence upon them, which can be broken only by consent of the client—who is fully entitled to stand upon his privilege and thus to prevent the information involved from being used in evidence against him.

26-6 Hearsay Evidence; Exceptions The *hearsay rule* is designed to keep out all statements not tested by oath and cross-examination. Subject to a number of exceptions, hearsay evidence (whether oral or written) will be excluded upon timely objection.[3] Where testimony about the existence of a fact is not derived from the personal knowledge or observation of the witness but is based rather on a statement someone else has made, the evidence in question is *hearsay*.[4] As an illustration, suppose Smith sues dentist *A* for malpractice. Testimony by the plaintiff about what dentist *B* had told him as to the condition of his mouth is hearsay. The second dentist should be called to the stand for direct questioning about what he had said, and the adverse party thus has an opportunity for cross-examination.

The hearsay rule serves in the average case to bar considerable testimony per-

---

[2] For an example, Illinois Revised Statutes 38:6-1 (1972).

[3] Only in rare instances will a trial judge invoke on his own initiative a rule of exclusion. where they are part of the res gestae.

[4] Sometimes a point in issue will be whether or not a given remark by *X* was or was not made. Under such conditions, anyone who heard the statement made can testify to the fact. Such testimony is not hearsay, since no effort is being made to establish the truth or falsity of *X*'s statement, but simply whether it was uttered.

tinent to the issues at hand. It would exclude a great deal more were it not for a rather lengthy list of exceptions to the rule. Illustrative exceptions include:

1 Dying declarations
2 Spontaneous utterances
3 Certain declarations of pain and suffering
4 Declarations against interest
5 Book entries made in the regular course of business

In criminal cases involving homicide, *dying declarations* of the late victim will be admitted despite the hearsay rule if the following prerequisites are met: (1) the declarant, at the time the statements were made, must have been at the very point of death (that is, *in extremis*) and must have entertained no hope of recovery; (2) the declarant, if living, would have been a competent witness; (3) the statement relates to the cause of the injury. The justification for permitting dying declarations to come in is twofold: (1) the fact of impending death affords some guarantee that the statements are trustworthy; and (2) the declarant is unavailable as a witness, and evidence other than what he would have presented is difficult, if not impossible, to obtain.

Spontaneous utterances and *involuntary expressions of pain and suffering* represent two out of a number of parts of the res gestae[5] doctrine which constitutes a broad exception to the hearsay rule. Just what the res gestae doctrine encompasses is apparent from the following language, contained in a recent court decision:[6]

> The res gestae or excited utterance exception applies to statements relating to a startling act or event made spontaneously and without reflection while the declarant was under the stress of excitement and offered to prove the truth of the matter asserted. In order for statements to be admitted as part of res gestae, the statements must be reasonably contemporaneous with the event to which they relate: i.e., they must be such as to have been proximately caused by the exciting influence of the event without opportunity for deliberation or influence.

By way of an example in line with the above quotation, the general rule is that any witness who happened upon the scene of an accident can testify that the victim groaned or cried out. There exists a conflict of authority whether spontaneous declarations of a bystander are admissible along with those of a participant in the event.

Declarations against interest form a significant exception to the hearsay rule. Such declarations are admissible as a form of proof, if, at the time uttered, they were against the pecuniary or proprietary interest of the since-deceased or otherwise-unavailable declarant and if the latter had competent knowledge of the matter on which he spoke and had no real motive to misrepresent the facts.[7]

---

[5] "Res gestae" means "things done."

[6] *State v. Messamore,* 639 P.2d 413 (Hawaii App. 1982).

[7] The opposite of a declaration against interest is known in law as a *self-declaration.* The latter type is normally inadmissible when offered in favor of the party by whom made; otherwise, he would be in a position to create evidence for himself in advance of the trial. Although the foregoing is the rule in general, there are instances in which self-serving declarations are admissible—as, for example, where they are part of the res gestae.

By statute in some jurisdictions and under the common law in others, it is possible to put in evidence a writing or report, prepared in the regular course of business, without having to call as witnesses all those who had a hand in the writing. Thus, Rule 4518(a) of the New York Civil Practice Law and Rules[8] reads as follows:

> Any writing or record, whether in the form of an entry in a book or otherwise, made as a memorandum or record of any fact, transaction, occurrence or event, shall be admissible in evidence in proof of that act, transaction, occurrence or event, if the judge finds that it was made in the regular course of any business and that it was the transaction, occurrence or event, or within reasonable time thereafter. All other circumstances of the making of the memorandum or record, including lack of personal knowledge by the maker, may be proved to affect its weight, but they shall not affect its admissibility. The term business includes a business, profession, occupation and calling of every kind.

26-7 Admissions and Confessions Where one of the litigating parties has in the past made some acknowledgment of fact or taken some action which contradicts the position he is adopting at the trial, proof will be received against him about such prior inconsistent behavior, and the proof is said to be in the form of an *admission.* The testimony relating to past conduct is introduced as probative evidence on points in dispute and in an effort to discredit the party involved by showing that he has "changed his stripes." Technically, admissions represent still another exception to the hearsay rule. As is the case with other forms of evidence, the weight to be given a particular admission is for the jury's determination.

26-8 Best-Evidence Rule If it becomes necessary in the course of a lawsuit for one party to prove the contents of a writing, the document itself is the *best evidence* of its terms, and it must be physically produced (unless extenuating circumstances are present, permitting the introduction of secondary evidence). The best-evidence rule aims at eliminating some errors by minimizing reliance upon copies[9] of an original writing or upon recollections of a witness regarding the contents of the document in dispute.

Although the rule pertains to various kinds of writings which are sought to be used for evidential purposes, there are a few unusual circumstances under which the rule is inapplicable. One such exception involves judicial records and public documents. Because removal of such records would result in inconvenience to the public at large and would create the danger of loss, entries in public documents may be proved by means of certified copies.

Apart from the few situations in which the best-evidence rule does not apply, there are extenuating conditions under which production of the original document, otherwise necessary, will be dispensed with and secondary evidence al-

---

[8] McKinney (1963).

[9] The "original" of the document, and not a "copy," is to be produced. But where "duplicate originals" of a contract have been executed, all are treated as "originals," and any one of them may be produced in court. In such a situation, absence of all the duplicate originals would have to be shown to the judge's satisfaction before secondary evidence (copies or oral testimony) would be proper.

lowed. Adequate excuses for inability to produce the original writing include its loss or destruction, and the fact that it is unavailable because it is outside the jurisdiction of the court or is in the possession of the uncooperative adverse party. If one of the foregoing explanations can be shown to pertain, a foundation exists for the introduction of secondary evidence; that is, (1) a copy of the instrument or (2) oral testimony from a witness who remembers what the writing says.

26-9 Parol-Evidence Rule Almost invariably the parties to a written instrument will have engaged in a number of preliminary talks or negotiations which eventually culminate in the execution of the document itself. The *parol-evidence rule* holds that testimony about such prior conversations is inadmissible to contradict or vary the terms of the writing,[10] which speaks for itself. The rule gives effect to the supposed intention of the parties and protects them from the uncertainties of oral testimony.

The parol-evidence rule has its exceptions. For one it is not unusual for the parties to a written instrument to regard it as only part of their contract. In such circumstances, many decisions have permitted the introduction of parol evidence not to vary the terms of the writing but to show the complete understanding of the parties.[11] Parol evidence will be admitted whenever its purpose is to show that no valid contractual obligation ever existed in the first place, perhaps because of want of consideration, or because of fraud in the inception. Everything depends upon the reason for the offer of parol evidence. If the effort is to modify or alter an admittedly valid contract, the evidence is objectionable. If the purpose is to show that the apparent contractual obligation is not binding because the agreement is invalid in law, the extrinsic evidence should be admitted. By the same token, where the validity of a written document had been attacked, parol evidence *sustaining* the contract will not be excluded.

There are a few miscellaneous circumstances in which oral evidence is admissible to help in the interpretation of an instrument in writing. Thus, parol evidence is proper as a means of identifying the parties to the contract and of showing the capacity in which each acts. If the language of the writing is unclear to the court—perhaps technical terms are employed, or words peculiar to a certain local area—explanation by oral testimony is appropriate. And parol evidence of circumstances attending the execution of the written document may be needful to show the intentions of the parties, where identification of such intentions is important and is not made clear in the writing.

The parol-evidence rule precludes testimony in contradiction of a written instrument whose terms and conditions the litigants have put in issue. The rule affects those who were parties to the writing and their privies. Strangers who had

---

[10] The parol-evidence rule does not preclude testimony as to a collateral, independent oral agreement between the parties, even though same relates to the very subject matter as the written contract and was made at about the same time. Similarly, the rule would not operate to keep out oral evidence providing a new agreement (with its own consideration) which purports to replace the written document in suit or to modify it in some fashion.

[11] Parol evidence cannot supply any of the terms of a contract which the applicable statute of frauds requires to be in writing.

no connection with the instrument in suit are not affected by the parol-evidence rule.

26-10   Opinion Evidence   As a general proposition, witnesses are supposed to confine their testimony to facts and avoid expressing conclusions and opinions. It is nonetheless true that a great deal of legitimate *opinion evidence* finds its way into the average trial.

As far as the *ordinary* (that is, nonexpert) *witness* is concerned, his opinion based upon observations will be admitted when the nature of the subject to which the opinion pertains is such that no more satisfactory form of evidence is obtainable. Not infrequently, a witness will have no way, other than by giving what amounts to a conclusion or opinion, of expressing himself about what he saw or heard. Suppose a point in dispute relates to whether a particular sound represented the report of a gun or the backfiring of a car. The lay witness should be entitled to give his opinion, since he cannot be expected to separate the various indications upon which he reached his conclusion from the very conclusion itself. In other words, it is often impossible for a witness, if he is prevented from stating anything in the line of an opinion, to detail all of what he saw or heard and to do it in a fashion which would enable the jury to reach reasonable conclusions. Accordingly, the ordinary witness may express an opinion on a number of subjects which do not call for particular skill or background. The list includes, among others, such matters as quantity, color, weight, estimated age of a person, and identification of an individual by his voice.

An *expert witness* is one who has, through study or experience, acquired such special skill or knowledge on the subject under consideration that he can assist the jury in understanding and resolving points which the untrained layman is unable to handle without guidance. The real distinction between the expert witness and the nonexpert is that the former gives the results of a process of reasoning which can be mastered only by those of special skills, while the nonexpert testifies about subject matter readily comprehended by the ordinary person and gives the results of a reasoning process familiar to everyday life.

The position of the expert witness varies with the type of case. In some situations he will testify only to *facts,* and in others it is proper for him to state his *conclusions* as well. The following quotation from an old New York case[12] does a good job of drawing the dividing line:

> It may be broadly stated as a general proposition that there are two classes of cases in which expert testimony is admissible. To the one class belong those cases in which the conclusions to be drawn by the jury depend upon the existence of facts which are not common knowledge and which are peculiarly within the knowledge of men whose experience or study enables them to speak with authority upon the subject. If, in such cases, the jury with all the facts before them can form a conclusion thereon, it is their sole province to do so. In the other class we find those cases in which the conclusions to be drawn from the facts stated, as well as knowledge of the facts themselves, depend upon

---

[12] *Dougherty v. Miliken,* 162 N.Y. 527, 533–534 (1900).

professional or scientific knowledge or skill not within the range of ordinary training or intelligence. In such cases not only the facts, but the conclusions to which they lead, may be testified to by qualified experts. The distinction between these two kinds of testimony is apparent. In the one instance the facts are to be stated by the experts and the conclusion is to be drawn by the jury; in the other, the expert states the facts and gives his conclusion in the form of an opinion which may be accepted or rejected by the jury. . . .

If the knowledge of the experts consists in descriptive facts which can be intelligently communicated to others not familiar with the subject, the case belongs to the first class. If the subject is one as to which expert skill or knowledge can be communicated to others not versed in the particularly science or art only in the form of reasons, arguments or opinions, then it belongs to the second class.

When a party offers a witness as an expert on some matter in issue, the trial judge must pass upon the necessity for expert testimony[13] and must also decide whether the prospective witness has the special skill or experience to qualify as an expert. Unless the other side concedes the point, the offering party must proceed to establish the standing of his would-be expert, and opposing counsel is entitled to cross-examine about the claimed qualifications.

The opinion of an expert may be founded upon facts derived from his own knowledge or observation—which facts he states to the jury before giving his conclusions[14]—or the opinion may be based upon hypothetical questions embracing a set of assumed facts some of which may have come out in testimony produced earlier in the trial. Often the expert will be asked to explain the reasons for his opinion.

Every expert opinion is supposed to rest upon fact. If a hypothetical question is involved and the facts are of an assumed nature, the weight of the opinion will depend upon the jury's finding whether the assumed facts were ultimately proved. But an attorney may not simply concoct a hypothetical situation to fit the needs of his client's case. Ordinarily, all the facts set forth in the hypothetical question must be supported by at least some evidence produced at the trial prior to the asking of the hypothetical question.

If, on the other hand, the conclusions are based upon the expert's personal knowledge of the facts, the average jury is inclined to have somewhat more confidence in the opinion expressed, though the trustworthiness of the witness is yet open to consideration. The adverse party, of course, is very likely to try to undermine the jury's confidence in "expert number one" through the introduction of contradictory testimony by other experts.[15]

26-11   Depositions   A *deposition* is sworn testimony taken in writing, pursuant to notice, upon oral or written interrogatories, and with opportunity afforded for cross-examination. Sometimes depositions are taken for the purpose of "dis-

---

[13] While the admissibility of expert testimony is for the court to determine, its weight is for the jury.

[14] When the expert has personal knowledge of the facts which form the foundation for his opinion, the court may allow him to state the conclusion without first detailing the underlying facts. However, such facts may then be elicited either on direct examination or on cross-examination of the witness.

[15] The next chapter is devoted specifically to the engineer as an expert witness.

covery" of information material to the issues in the case, and sometimes for the purpose of making certain that a person's testimony will be available at the trial. In this connection there are circumstances in which a prospective witness will be, for good reasons, such as physical disability, prevented from appearing in court. Arrangements may then be made to have his or her testimony recorded under oath on the outside. Examination will be conducted by the respective attorneys, as though the witness were giving evidence in a courtroom. This recorded testimony is termed a *deposition,* and it will be available for reading in open court so that the judge and jury may hear all the questions and answers. Of course, such a reading may not prove as impressive as would the same testimony given in person to the court, but it is certainly better than losing altogether the benefit of what the witness has to say.

Statutory provisions vary with the jurisdiction, and the applicable laws would have to be consulted to determine in what situations a deposition is appropriate, who may take it,[16] and what formalities are to be observed.[17]

Under the typical deposition procedure, the attorneys may object to certain questions as they are put to the witness. There being no judge present at that stage to rule upon the admissibility of the disputed evidence, unresolved issues of this nature will remain to be settled by the judge when, at subsequent time, the deposition is read in court. The official stenographer will have recorded in the deposition transcript and, at appropriate points, the various objections and the reasons presented in support of same so that the court can make rulings as the reading proceeds. Thus, the witness whose deposition is being taken will answer questions for transcript purposes despite opposing counsel's objection, but the judge, if he eventually sustains such objection, will have the answer stricken from the records of the case.

QUESTIONS

1 What is the purpose of evidence?
2 Explain direct evidence.

[16] In California, for example, the deposition of a witness or party may be handled by a judge or clerk of any court, a justice of the peace, a notary public, or a commissioner of the superior court. See California Code of Civil Procedure, §179 (1982).

[17] Thus, New York Civil Practice Law and Rules, §3113(b) (McKinney 1977) deals with "objections," among other considerations:

"The officer before whom the deposition is to be taken shall put the witness on oath and shall personally, or by someone acting under his direction, record the testimony. The testimony shall be recorded by stenographic or other means, subject to such rules as may be adopted by the appellate division in the department where the action is pending. All objections made at the time of the examination to the qualifications of the officer taking the deposition or the person recording it, or to the manner of taking it, or to the testimony presented, or to the conduct of any person, and any other objection to the proceedings, shall be noted by the officer upon the deposition and the deposition shall proceed subject to the right of a person to apply for a protective order. The deposition shall be taken continuously and without unreasonable adjournment, unless the court otherwise orders or the witness and parties present otherwise agree. In lieu of participating in an oral examination, any party served with notice of taking a deposition may transmit written questions to the officer, who shall propound them to the witness and record the answers."

 3 What is circumstantial evidence? When may it be necessary? What are the weaknesses and dangers connected therewith?
 4 Explain (*a*) how evidence may be relevant, (*b*) how it may be irrelevant.
 5 What is the meaning of "weight" of evidence?
 6 What qualities of a witness should be examined when judging his testimony?
 7 What is "opinion" evidence? Under what circumstances is it admissible?
 8 Who determines the weight of evidence?
 9 What is meant by the "best" evidence? What part does the best-evidence rule play in a trial?
10 What is "secondary" evidence? Illustrate.
11 Compare written with oral (parol) evidence as respects the value of each and the role each fills.
12 What is meant by "burden of proof"? Which party assumes it?
13 What is "hearsay" evidence? When is it admissible?
14 What is meant by "competency" of evidence?
15 What is meant by "material" evidence?
16 Define "testimonial" evidence.
17 What is "real" evidence? Illustrate.
18 What is "documentary" evidence? Illustrate.
19 Differentiate between the functions of a judge and a jury.
20 What is the judge's "charge" to the jury?
21 What is meant by "judicial notice"? Illustrate.
22 Who determines the "credibility" of a witness?
23 What is the significance of "conclusive" presumption? How does it operate?
24 What is "rebuttal"? Illustrate.
25 Who passes upon the competency of a prospective witness?
26 What are "privileged communications"? Can the privileged party be forced to divulge such communications in court? May he do so voluntarily?
27 What factors might make a prospective witness incompetent?
28 List several exceptions to the hearsay rule.
29 Illustrate the "res gestae" doctrine.
30 On what basis are "dying declarations" admitted?
31 Explain "declarations against interest," and comment on their admissibility.
32 Illustrate a situation in which a "book entry" would constitute proper evidence.
33 What purpose may admissions and confessions serve in a court case?
34 Indicate the circumstances under which an expert may testify (*a*) as to facts, (*b*) as to conclusions.
35 A witness said: "Mrs. Jones told me that her husband had willed all of his property to her." Do you think that such a statement is admissible as evidence in a suit by other relatives who claimed a share in the estate, since no will made by Mr. Jones was found?
36 What is a deposition? How may one be used in the courtroom?
37 Explain how objections made by the lawyers during the recording of a deposition are handled (*a*) at the time the deposition is made, (*b*) when it is read in the courtroom.

CHAPTER 27

THE ENGINEER AS AN EXPERT WITNESS

27-1 Introduction Engineers are frequently called upon for court appearances in the role of *expert witness,* a term referring to a person whose educational background, training, and professional experience indicate superior knowledge about a particular field of endeavor and serve as the foundation for presumably meaningful conclusions and opinions. Experts are in demand in many types of lawsuits, and engineers who testify a number of times may well find that their reputations and career potential may depend to some extent on how well they conduct themselves on the stand.

Engineers may also be asked for testimony and statements of opinion in (1) arbitration proceedings, (2) commission hearings, (3) legislative committee hearings, (4) conferences pertaining to contract disputes, (5) hearings to establish valuations of property and rates for services, (6) hearings before zoning boards, (7) public or private meetings to consider matters relating to building codes, and (8) many other proceedings which are not held in courts but which are nevertheless of great importance.

Expert witnesses are usually brought in by one or another of the contesting parties, although the sentiment has been expressed that it would be preferable to have nonpartisan experts called by the court itself, either on its own initiative or that of a party. The underlying philosophy seems to be that divorcing expert testimony from the source of compensation therefor would come as close as possible to ensuring the impartiality of the witnesses involved.

While it is generally thought that courts have the inherent power to select experts to give evidence, several jurisdictions have enacted statutes expressly authorizing trial judges to appoint expert witnesses, to fix their remuneration (subject to any statutory maximum), and to treat such as costs of the lawsuit.[1]

27-2 The Function of an Engineer in a Lawsuit Apart from the possibility of being himself a party, an engineer may have one or more of the following functions to perform in connection with a lawsuit:

1 He may be called upon as an expert witness to explain some technical matter or to give his opinion regarding a particular point at issue. His role can be thought of as aiding the court and jury in understanding intricate evidence so that they can make a proper evaluation of the case.

2 He may assist the attorney in the preparation of the cross-examination of witnesses for the opposition, and may also assist during the actual cross-examination itself. He may do this by passing along the attorney's suggestions regarding the questions to be asked and weak points to be attacked in the statements of the opposing witnesses.

3 He may serve as a general assistant to the lawyers for his side in the development of their case and not appear at all in the courtroom. In highly technical cases, in which it is important for the attorney to gain some measure of competence in the technical field itself, this educational and advisory role may be important.

When filling one of the first two capacities, the engineer should be in constant attendance at the proceedings, because what has been said by others may have a significant impact upon what he tries to bring to the attention of the court in his own testimony or indirectly through advice and suggestions to his attorneys.

27-3 Knowledge of Subject Above all things, one should not attempt to act as expert witness regarding a matter about which he is not well informed. If he does so, the opposing lawyers and their engineering advisers will be likely to detect his shortcomings, and the result may be the discrediting of his testimony.

A witness on the stand is fair game for opposing counsel and is in no position to try any bluffing. Erroneous statements open the floodgate. Hence a witness should avoid making broad assertions unless he is sure of his ground. When a witness is asked for a fact rather than an opinion and he cannot answer with assurance, it is best to say, "I do not know."[2]

---

[1] An illustrative provision of the nature is California Evidence Code §730 (West 1966). Section 733, also part of the picture, states that: "Nothing contained in this article shall be deemed or construed to prevent any party to any action from producing other expert evidence on the same fact or matter mentioned in Section 703; but, where other expert witnesses are called by a party to the action, their fees shall be paid by the party calling them and only ordinary witness fees shall be taxed as costs in the action."

[2] There have been cases wherein the supposed "expert" witness actually weakened his side by ill-advised answers on the stand. In one instance, a witness had testified to certain facts. The opposing

It is also obvious that a witness should check his facts before attempting to base any conclusions thereon. For example, an engineer testified that certain equipment and expense would be necessary if a dam were to be built in the river adjoining a particular industrial plant. He offered this opinion without having seen the final plans for the proposed construction. Later on, the plans were presented in court, and they proved to be much different from what he had assumed was the case. His testimony was made to appear ridiculous in the light of the actual situation.

Thorough examination of the situation and the circumstances to be involved in a proposed suit are usually vital before an engineer agrees to serve as an expert witness. Such preliminary investigations may be helpful to the owner and his lawyer as well as the engineer. An example of this is where an owner of a large multistory building planned to bring suit for damages against the contractor because large vertical cracks had occurred in the concrete foundation walls. After making careful measurements, the engineer found that there were no differential settlements of adjacent parts of the foundation, showing that the soil had not yielded unevenly. However, he was convinced that the cracking of the long foundation walls had been caused by the lack of vertical contraction joints which would have been essential in order to allow for horizontal shrinkage and thermal contraction of the concrete. This lack was the responsibility of the designer. The cracks were simply nature's way of showing what should have been provided. Therefore, the engineer, because of his careful investigation, had enabled the owner to avoid the expense of suing the wrong party.

27-4   Commencing a Suit   The attorney for a would-be plaintiff in an action may want the help of an engineer in preparing a formal complaint outlining allegations which plaintiff will be prepared to prove. Preparation of this important document entails a thorough study of the entire situation and of the evidence ostensibly available. Even at this early juncture, the expert advice of engineers and other technicians can help the attorney in deciding what prospective evidence is, and what is not of value in establishing relevant and complicated engineering points.

27-5   Necessity for Honest Advice   When an engineer is assisting an attorney in the study or preparation of a case, she should be absolutely honest and straightforward in giving opinions and advice. The engineer and the attorney have diverse areas of knowledge and must rely upon each other to a considerable extent. Even though the engineer's opinion on matters in her own field of activity may not coincide with what the lawyer wishes to hear, the engineer should nevertheless

---

lawyer then asked him about related things which were really matters handled entirely by another man. In trying to answer without prior study of the points in question, the witness made a statement that he subsequently had to admit was in error. This episode gave the cross-examiner an opportunity to criticize the witness severely and thus to cast doubt upon his credibility.

express her ideas as completely and as convincingly as she can. If any serious flaws in the case can be found, it is better to detect them in advance (even if it means dropping the suit) than to have them brought out by the opposition at the trial.

By way of illustration consider the following. A man built a cinder-block house about ¾ mile from a stone quarry. By the time two years had elapsed, the front porch and one corner of the house had settled and cracked badly. The interior plastered partitions had settled near the center of the structure, and they also were badly cracked. The owner engaged a lawyer to sue the quarry operators for damages, claiming that the cracks were caused by the blasting of rock. The attorney called in an engineer to assist in some research. The engineer made a careful study of the situation and learned that the owner had built the house himself, without engineering or architectural aid and without the assistance of a contractor. The porch and corner that settled had been founded upon newly filled-in soil, the subsequent compaction of which had evidently caused their subsidence. The floor in the center of the house was supported upon a central wooden beam which had shrunk at least ¼ inch in drying out, thus causing the partitions to sag at the center, whereas they were supported upon unyielding masonry walls at the outer edges. When the engineer advised the owner and the attorney of his findings, it was decided to forego the suit because the situation appeared hopeless. The engineer's advice had saved the owner considerable money which would otherwise have been spent in fruitless litigation.

In another instance, the construction of a power dam was, without doubt, going to cause the flooding of a portion of a valuable industrial-plant site. Since direct negotiations between the parties had failed to effect a settlement, the owners of the plant site brought a "friendly" suit against the power company in order to have the court determine the payment which the power company should make to the owners of the property to compensate them for the "damages." The attorneys for the power company called in an engineering firm to design protective works for the site, intending to offer the plant owners the estimated cost of these works as a settlement. The engineers insisted that, although complete protection would cost more than a less effective construction job, they would not present to the court a plan which they could not defend wholeheartedly, because, if their plan could be proved inadequate by the opposition, the power company might be in a very unfortunate position and, as a result, might have to pay even more than the cost of really protective facilities. The lawyers for the power company finally approved this idea. Subsequently, the hearings revealed the wisdom of this course, because the opposing attorneys would have been able to prove that anything less than the measures the engineer had suggested to the power company would constitute inadequate protection for the plant owner. A settlement substantially equal to the estimated cost of the proposed works was ordered by the court.

27-6 Preparation for Taking the Stand It is essential that the engineer who is to act as witness should first become thoroughly familiar with the overall case and particularly with his special part in it. This normally involves close advance cooperation with the attorneys for his side. There should be a definite understanding about what the expert will be asked, what information he is in a position to give, and what his knowledge or experience limitations may be. The expectant witness should also prepare himself as much as he can for cross-examination by imagining temporarily that he is on the opposing side and that he is trying to counteract the very testimony which he plans to give on direct examination. This sort of exercise will help him get ready to meet anticipated attack from the opponents.

27-7   Qualification of Witness as an Expert   When an engineer-witness takes the stand, he (like any other witness) will give his name and address and will be sworn in. Then he will be asked for information about his background and qualifications.[3] The usual questions relate to his education, practical experience, writings, accomplishments in his profession, and technical organizations with which he is connected. It is helpful if the witness can state that he is licensed as a professional engineer (or architect) in one or more states. The public recognizes that the licensed engineer has had to pass suitable examinations to prove his ability in the art and practice of his particular branch of work and that he is bound by codes of ethics to conduct himself in accordance with proper moral and professional principles.

Here are some of the questions asked by an attorney during the "qualification" of an engineer as an expert in a suit involving waterworks.

What is your occupation?
What college did you attend?
What degree or degrees have you received?
Are you a licensed professional engineer?
In what states do you hold a license?
To what professional societies do you belong?
Where and by whom are you employed?
What is the title of your position?
What is the general nature of your duties in that position?
Will you state briefly an outline of your professional experience?
What are some of your other activities in your profession?
Have you written any books on engineering subjects?
If so, what are they?
Have you written any articles for engineering periodicals?
If so, to what subjects do they apply in general?
Have you ever worked upon the construction of waterworks?
Has that work been in the office or in the field?
Please explain in some detail the character of this work and your responsibilities in connection with it.

The apparent qualifications of the expert will have considerable bearing on the

---

[3] Any question as to whether a given witness truly qualifies as an expert is for the court to decide as a preliminary to his substantive testimony.

weight his opinion carries with the trier of the facts. The opposing attorney may challenge the standing of an expert witness on the basis (1) that he has not had adequate training in the particular field concerned, (2) that he is too young to have had satisfactory experience, (3) that he is too old (perhaps retired) and out of touch with the most recent developments in the particular field, (4) that his training is too broad on the one hand or too specialized on the other, or (5) that he is too closely connected with the case to have an unbiased opinion. If at all feasible, the expert chosen to testify should be a recognized authority in his field.

The importance of the relative qualifications of expert witnesses may be demonstrated by considering the jury's problem in judging the weight of opposing testimony given as "opinion" evidence in connection with the failure of a bridge during its erection. The attorney for the plaintiff asked his witness, "In your opinion what was the cause of this failure?" The witness replied, "I believe that the timber piles under the erection bent were not properly braced in a lateral direction. This allowed the erection bent to 'jackknife,' which means that it buckled sidewise at the top of the piles." The second expert witness, called by the opposition, stated in reply to similar questioning that he believed the temporary steel erection frame or post on top of the piles failed, thus causing the structure to fall, and that the piles *were* braced properly. The jury was thus faced with conflicting expert opinions. Since it was impossible, after the structure had fallen, to obtain absolute, factual proof of the cause of the collapse, the jury's finding would depend in large part upon its estimate of the qualifications and credibility of the respective experts.

27-8   Direct Examination   When an expert witness first takes the stand, the attorney for the side which brought about his presence will conduct the so-called *direct examination.* He will ask questions designed to bring out certain points he believes will be relevant and helpful in establishing his case. The substance of what the witness can be expected to say on direct is generally no secret to his attorney. The principal function of the witness will be either (1) to bring out technical information or opinions in support of his side or (2) to present involved facts or expert opinion, the tendency of which is to refute statements made by witnesses for the opposition.

When the expert has not actually observed the facts on which his attorney would have him base an opinion, the lawyer may have to get his conclusions by asking one or more hypothetical questions which will enable the witness, through his answers, to make the desired statements.

A witness should be very careful to make certain that he understands the question before he tries to answer it. He may ask to have the question repeated if he is not sure of what it was. To illustrate: a pumping station was to be installed by the power company for the benefit of the plant owner, and it was to be operated by the latter. The witness had been told in advance that his attorney would ask him if this pumping equipment would be *inexpensive* to operate. Actually the question as asked was: "Will this pumping equipment be *expensive* to operate?" The expert, being a bit inattentive, replied: "Yes, we designed it that way." To the embarrassment of the witness, the lawyer said, "I believe that you misunderstood my question." Then he repeated it, and allowed the witness to correct himself.

To the extent possible, the expert should give his testimony in generally understood language, rather than in highly technical terms. He should aim at making himself clearly understood by the judge and jury and, accordingly, should try to keep his statements (and any illustrations which he uses) within the range of common experience.

There may be considerable difference in the best way for a witness to present engineering matters to a jury or court. As juries are usually composed of nonprofessionals, all information should be given in a way which they can be reasonably expected to understand. If the case is presented before a judge, he will have excellent knowledge of the legal matters but may not grasp the technicalities of engineering as readily; hence the need for simplification is just as great. If the case is being heard by an arbitration board composed of experienced engineers, the witness can be free to present his testimony in technical terms.

Here is one illustration of how a witness presented a technical matter to a lay jury. The case involved some serious diagonal cracking of masonary walls in a large building. The opposition had claimed that the cracking was produced by a wedging action of part of the structure settling with respect to adjacent portions. The witness, having heard this testimony, prepared to counteract it. To do so, he made a wedge of wood, then he placed two bricks side by side on a table in front of the jury. Next, he inserted the wedge vertically between the bricks, causing them to separate horizontally. However, he showed that a vertical force had to be applied to the wedge to cause it to move, that the wedge had to move downward appreciably in order to separate the bricks, and that the wedge had to be in contact with the sides of the bricks because it could not exert a lateral pressure through an air space. He pointed out that no such downward force existed in the structure, that there had been no permanent downward movement of one part of the structure with the respect to the next, and that there was an air space at each crack. The jury could see that the claimed wedge action could not have produced the existing cracking of the walls.

After cross-examination of the expert has intervened, his attorney may find it advantageous to recall him to the stand to explain or contradict points brought out by the opponent. This type of return appearance is called *redirect*.

27-9   Cross-Examination   After completion of direct testimony by a witness, the opposing attorney will conduct his *cross-examination*. Naturally, this attorney's objective is to break down or nullify the effect of the direct testimony given by the witness. The attorney may try in various ways to discredit the "expert" and will likely ask questions intended to confuse him, to get him to contradict his previous statements, and to find any discernible weakness in the position he took on direct examination. Questions regarding the amount of the fee paid to the witness are allowable and should be answered candidly and without embarrassment.

This undergoing of cross-examination can be an ordeal, since the witness is very much alone on the stand.[4] If the cross-examiner can prove, through the interrogated witness or through other witnesses, that even a small part of the direct testimony given by the witness was inaccurate, this circumstance is likely to shake the jury's confidence in such a witness.

The expert should guard against being drawn into an argument with the opposing attorney. He should also endeavor to keep from making irrelevant statements that may becloud the issue or serve as openings for the opposing lawyer to challenge the witness effectively.

The following episode illustrates the role of cross-examination in discrediting a witness. A financial expert had been asked on direct to testify regarding the interest rate that should be used in estimating the capitalized value of certain annual expenses that a manufacturer could expect to incur in the future. He stated that a yield of 5 percent was the maximum obtainable from really prudent investments. Under cross-examination the witness admitted that he was trustee of certain funds and that these monies were currently invested so as to earn interest and dividends ranging from 7 to 8 percent a year. The witness-trustee was thereupon asked if he considered such monies prudently invested, and he naturally had to admit that such was the case. Thus, his previous statement that 5 percent was a maximum return on a "prudent" investment was made to appear foolish.

Absolute honesty under cross-examination is essential even though it may occasion temporary embarrassment. A witness, testifying in a suit for damages, was asked by the opposing attorney if he had seen a certain 24-inch pipeline to which he had supposedly been close. In an attempt to trip the witness, the attorney inquired, "Didn't you see that big pipeline lying there on the ground?" The witness replied, "No, I was examining the topography and water levels to the south of that area." This admission may appear to show a lack of keen observation on the part of the witness, but, if he had made a blind guess at the actual condition and had answered in the affirmative (which is really lying under oath), his honesty and reliability would obviously have been open to question when independent evidence later established that the pipeline was buried underground and could not have been seen.

It is customary for the engineers assisting the opposition to pass notes to their attorneys as the trial proceeds, suggesting troublesome questions that should be asked of a particular expert on cross-examination (or suggesting that one of them should be called to combat some aspect of the other side's evidence). Before the cross-examination of any given expert is finished, the opposing lawyers will be satisfied that they have impaired the effect of his direct testimony or that they cannot succeed in such an endeavor.

If a witness is asked questions which he cannot at once answer completely, it is sometimes possible for him to ask for the privilege of consulting his attorney, other engineers, or available data, in order to make sure that he will be answering correctly. One may also ask for a reading of certain previous testimony in order to

---

[4] Apart, of course, from such help as his attorney can afford through objections to improper questions.

refresh his memory. Such devices are especially useful when complicated figures, or statements made earlier in the trial, are involved.

27-10   Objections   While a witness is under examination by the attorney for his side, opposing counsel will be paying close attention so that he may raise objections to questions he finds improper or to answers which he wishes to have stricken from the record. Under the rules of evidence there are a number of grounds upon which to base possible objections. Should the judge overrule the objection, of course, the question concerned will stand as originally propounded (or, perhaps, as rephrased), and the witness will be directed to make answer.

In a typical case a witness was asked his opinion of the construction work on a certain dam. The opposing lawyer objected, asserting that the witness was not experienced in the design of dams and knew nothing about the one in question. A series of further questions by the first attorney sufficiently established the qualification of the witness respecting the subject under discussion, and the objection was not sustained.

Having a number of objections raised during his appearance may prove very disconcerting to the witness who had planned to make certain statements in reply to anticipated questions. The problem for his attorney becomes one of rephrasing those questions to which valid objections are interposed, so as to bring out the desired information in such form as will meet with the court's approval.

27-11   Conduct in Court   When in the courtroom, a witness should conduct herself with dignity and confidence. It is clearly desirable for an expert witness to give the impression that she knows her business, that she is scrupulously honest, and that she is confident as to the accuracy of her testimony. She should be thoroughly competent and able to defend her opinions. She should be absolutely fair and be able to convince the court of her fairness. Moreover, she should not attempt to be an expert witness for a plaintiff (or defendant, as the case may be) whose general position—at least on the matter about which expert testimony is desired—she cannot wholeheartedly endorse and support.

A witness on the stand ought to be courteous, self-controlled, and well-mannered at all times. Although he may be under nervous tension, he should not reveal this by his actions. And he must not allow himself to become angry or intimidated. If he is sure of his ground, he need not be afraid of opposing counsel because he will realize that he probably knows more about the subject than does the lawyer.

In court a witness should be dressed neatly and in good taste. During questioning, a witness should look directly at his interrogator and occasionally at the judge or jury. He should give short, clear answers. If the opposition lawyer tries to overpower him with rapid-fire questions, the witness may ask for the repetition of each question and may answer each one individually. If the cross-examiner becomes abusive, the witness's attorney may appeal to the court. The witness should not allow opposing counsel to wrest control of the situation from him or demoralize him to the extent that he will make some untrue, inconsistent, or self-contradictory statement.

A prospective expert witness should not be reticent about appearing in court and expressing himself before a group of spectators. He should be mentally prepared for his task and able to think clearly and speak effectively under difficult circumstances.

A witness ought not to blindly accept unproven facts as true or indicate agreement with counsel's statements unless the expert believes them to be correct. He should be on guard against leading questions.[5] When he cannot answer a question with a direct "yes" or "no," he should state that he cannot make an unqualified answer, but would like to answer with an explanation. If the witness is not positive regarding his answer, he can say, "To the best of my knowledge. . . ."

A witness should not try to make his answer cover so many fine points that it will be confusing to the judge and jury. He should try to emphasize the important features, and, if necessary, arrange with his attorney to have the latter ask additional questions which will bring out any further details desired. Generally speaking, it is dangerous for a witness to volunteer more information than the questions require. He may thereby unwittingly give the opposition something to attack or to use as a "red herring" to divert the attention of the judge and jury from the really pertinent information.

The witness should not be unduly partisan when he is on the stand. It is human nature for the expert to experience some feeling in favor of the side for which he is appearing, but it is incumbent upon him to maintain a sense of balance. He should not say anything which is incorrect or misleading. If he inadvertently does so, a correction at the earliest opportunity is indicated. False evidence given with knowledge of its falsity amounts to perjury.

Before and during the actual court proceedings the expert witness will do well to remember that the lawyers on his side are trying, partly through his testimony, to bring out certain points relating to the issues in order to tip the scales in favor of the client they represent. The opposing lawyers, of course, will be endeavoring either to preclude the recording of any damaging expert testimony in the first place or in some fashion to attack its accuracy or value once it has "come in."

27-12   Presentation of Information   When giving evidence in a trial, a witness should remember that the court stenographer is taking down what he says and that the written record must be depended upon for future reference and as the foundation for a possible appeal to a higher tribunal. Therefore, he should be sure to speak slowly and distinctly so that the record will be intelligible. To illustrate a related point, assume that a witness is making reference to a map which shows in red the location of a certain pipeline. The map has previously been placed in evidence and bears a certain exhibit number. Suppose that the witness points to a location on the map and says, "The pipeline is here." Those present in the court-

---

[5] *Leading questions* are those which are so worded as to suggest their own answers. For example: "When you inspected the site, did you notice the fine, water-bearing sand running into the excavation?" This form of query virtually puts words in the mouth of the witness. The question would properly be put in this fashion: "When you inspected the site, what did you see?" Leading questions may be acceptable in direct examination regarding preliminary matters and in cross-examination, but not at other times.

room will understand his testimony, but the record would mean nothing.[6] Actually, the witness should have said something like this: "The pipeline is shown on Exhibit K by the red line extending eastward from the water tank to the northwest corner of the building designated on this map as No. 3."

Furthermore, an expert witness should bear in mind that hearings on engineering matters will doubtless involve technical terminology with which the court stenographer is not familiar. It is the courteous thing for the witness to spell out such words as he thinks necessary, thus making certain they are recorded correctly. When an engineer is trying to explain or illustrate some involved matter to a judge or jury, he should remember that their background and experience are probably much different from his own. What is a commonplace thing to him may be utterly unknown to them. His terminology and examples should be chosen with this general situation in mind.

In one instance, a lawyer asked an expert witness to explain what the latter meant when he said that a certain piece of cast iron "probably failed because of fatigue." The witness replied: "Let us assume that we have a 2 by 10 plank about 18 feet long lying flat on two blocks 15 feet apart. A man who weighs 150 pounds steps on the center of the plank. It sags 3 or 4 inches but does not fail. Now, if the man steps on and off the plank time and time again, it may be that eventually the plank seems to get tired out and finally breaks under the weight of the man whom it supported safely so many times before. In a way, that is what I mean by "fatigue." " The example selected was sufficiently "down to earth" that the average layman could readily see the point.

The use of maps, charts, models, and similar materials is often helpful in illustrating what a witness has to say. However, such maps and the like should be simple, attractively prepared, and readily understood. They should be large enough to be visible at a distance or should be in such form that they can be passed around to all persons entitled to see them. A witness will not make a favorable impression on the court if he tries to illustrate points by reference to visual aids which are not easily seen and understood by the judge, jury, and other interested parties.

The engineer must be sure that all materials which he intends to use in court are correct to the most minute detail and are consistent one with another. In one instance of litigation resulting from a contract, an engineer (expert witness) discovered that there was a reference in the specifications to something that had been removed from the the drawings had been put in the record as exhibits. During a recess, the witness told his attorney about the discrepancy. The latter took the initiative and immediately brought the matter to the court's attention. The opposing attorneys attempted to show that the inconsistency between specifications and drawings was evidence of poor preparation of the contract documents. However, they accomplished little or nothing compared to what they might have achieved if the engineer's lawyer had not voluntarily exposed the situation.

If a witness can use some statements made by the opposition's witnesses to

---

[6] Presumably, the lawyers will move to clarify the record, but the witness can help the situation by intelligent testimony in the first place.

prove his point, this is helpful in that the other side cannot very well attack the validity of their own source of information. The foregoing possible opportunity reveals one of the reasons why a witness may benefit if he attends all the trial sessions preceding the giving of his own testimony. In one case, witness *A* had stated that the building of a proposed dam would cause sufficient flooding of part of the plaintiff's land to destroy its use (and, therefore, sale value) as a site for a future industrial plant or for other structures. He claimed that this flooding would cause a decline of at least $50,000 in the value of the property in question. Witness *B,* appearing for the defendant, showed the court certain records which revealed that the land in question had been flooded many times in the past when severe freshets occurred in the adjacent river. Witness *B* pointed out that, if flooding caused by the proposed dam would injure the sale value of the land, the similar and frequent flooding of this land by the river would cause similar loss of value of the property. Furthermore, since (without the dam) flooding of this land would undoubtedly recur many times in the future by reason of the river alone, the construction of the proposed dam could not be regarded as likely to have an appreciable detrimental effect on the value of the land. Thus, witness *B* was able, in a way, to use the statement by *A* to show that the asserted loss was not a fact.

In many cases it is necessary to present to the court the ultimate results of detailed calculations or even to explain the calculations themselves.[7] It is best to have these sets of figures complete and in proper form that they can be placed in evidence as exhibits. Upon being offered for the record, prospective exhibits have to be submitted for the perusal of the judge and of opposing counsel who will doubtless show them to his engineering experts or other technical advisers on hand.

27-13   Use of Reference Material and of Prior Decisions   Generally speaking, learned treatises or extracts therefrom are inadmissible (under the rule precluding hearsay evidence) when offered to demonstrate the truth of what is said therein. But scientific books and similar reference materials nonetheless play an important part in many courtroom scenes. Thus, such items may be used effectively on cross-examination for the purpose of discrediting or contradicting an opposition witness. Assume, for instance, that such a person is asked if he is familiar with a book on structural design by an X. Y. Smith. With an affirmative answer in hand, the cross-examiner may then inquire whether Smith is thought to be a reputable and capable engineer. Upon receiving a second affirmative response, the lawyer might point out that on page 210 of this published work Smith makes a statement which flatly contradicts the position which the witness has taken on the point in question during his direct examination. The maneuver, of course, is designed to raise serious doubts regarding the validity and accuracy of the witness's testimony.[8]

---

[7] One should be very careful to avoid making hasty mental or written computations while on the stand. Under pressure, a witness may make errors that will later be pointed out by the opposition.

[8] However, during redirect examination, the attorney at whose behest the witness appeared might ask the latter if he is familiar with a book by Q. Z. Jones, who is also admittedly a well-known author in the field of structural design. Thereafter, the attorney may be positioned to bring out the fact that Jones, in his treatise, makes a statement substantiating that given by the witness and opposing that of author Smith.

It is obvious that a witness must be on guard when writings are likely to be used against him. He should be prepared to state (1) whether a particular author is regarded as being a genuine authority on the subject at issue, (2) whether the written material (to which his attention is directed on the stand) is now out of date or really inapplicable to the case, and (3) whether other authors disagree with the one being quoted—with the result that the latter,s statements are perhaps of questionable value.

27-14   Fees   Assuming that the expert witness is testifying at the request of a litigating party and on a scientific or technical matter requiring special learning, skill, or knowledge—and quite likely some research or other extensive preparation—there will probably be an understanding in advance that he is to be paid for his services[9] reasonable remuneration which amounts to something beyond the ordinary witness fees.[10] It is customary for the expert to base his total charges primarily upon the amount of time involved in preparation for his court appearance as well as at the trial itself. The rate normally approximates charges for comparable consulting work. The relative importance of the expert's prospective role in the court proceedings will undoubtedly influence the precise rate. If he figures to be the key person in resolving the principal question at issue, and particularly if his professional reputation is on the line, the measure of his compensation may well reflect this.

QUESTIONS

1 Define the meaning of the term "expert witness."
2 In what kinds of proceedings may an engineer have an important part as an expert witness?
3 Describe in what capacities an engineer may aid in court proceedings.
4 Why may "expert" opinion be admissible at a trial when opinion evidence in general is not? Explain.
5 Explain the difference in an engineer's situation (a) when acting as a witness for one of the litigants and (b) when rendering service strictly as an adviser to one of the attorneys.
6 Illustrate the procedure which may be followed in court in qualifying a witness as an expert.
7 What facts may greatly affect the weight of an expert witness's testimony in court?
8 Explain the difference between direct examination of an expert witness and cross-examination of him.
9 Illustrate the raising of objections by an attorney during the taking of testimony when an expert witness is on the stand.

---

[9] Payment under the usual agreement is made regardless of the outcome of the case. Occasionally an expert witness will have been asked to testify on a "contingent fee basis," meaning that he would be remunerated only in the event the party who called him prevails; this arrangement would obviously give the expert a pecuniary interest in the lawsuit, a situation hardly calculated to enhance an image of impartiality. Incidentally, an engineer (as any other expert witness) should be prepared, if circumstances warrant, to state in open court the understanding concerning his fee.

[10] To be distinguished is the situation in which an individual, though his expertise be great, is asked merely to testify to facts which have come within his observation; such an individual may be subpoenaed and paid the statutory witness fee and travel allowance.

10 How may visual aids be used in connection with the presentation of testimony by an expert witness?

11 Explain when and how references may be used by an expert witness in connection with presentation of his testimony.

12 Can an engineer be *compelled* to give testimony in court? How?

13 If an engineer is asked to serve as an expert witness, should he investigate the case before accepting? Why?

14 What may happen if a witness makes a statement on the stand and then, under cross-examination, has to admit that he erred?

15 How can the court decide upon the relative value of opposing testimony presented by various expert witnesses?

16 Why might an engineer who is to help the attorney for plaintiff in a lawsuit be asked to assist in the preparation of the complaint?

17 If an engineer sees that a case in which he is asked to help an attorney is weak or has weak points in it, what should he do about it?

18 Why is it customary for a lawyer to go over in advance the general character of the questions which he will ask an expert witness on direct examination?

19 In preparation for direct examination, is it proper for the prospective witness himself to suggest questions which his attorney might well ask him? Explain.

20 Can an expert witness generally be recalled to the stand after he has undergone both direct and cross-examination? Why?

21 What is the purpose of making a stenographic copy of all proceedings during a hearing or court session?

22 If an expert witness is not sure that he understands what a question asked him means, what should be do?

23 Can an expert witness, who is on the stand, ask for an opportunity to refresh his memory before answering a specific question?

24 When may an expert witness make use of books to prove technical points he made in his testimony?

25 May an expert witness properly be questioned about his fee?

26 Can book data, used to substantiate statements made by an expert witness, be placed in evidence as exhibits?

27 If a witness refers to computations and figures when giving his testimony, will such data be made a part of the record?

28 On the stand, an expert witness was asked: "Exactly what was the speed of the truck when it crashed through the railing of the bridge?" The witness answered: "I don't know." (*a*) Does that answer discredit the other testimony of the witness? (*b*) Should the witness try to make a specific answer? (*c*) Would the situation be different if the question were: "In your opinion, what was the approximate speed of the truck when it crashed through the railing of the bridge?"

29 Lawyer *A* came to engineer *B* to ask the latter to serve as expert witness. As the result of a discussion of the matter, *B* found that he could not agree with much of *A*'s position and claims. Should *B* accept the job nevertheless? Why?

CHAPTER 28

ARBITRATION OF DISPUTES

28-1 Function of Arbitration The engineer not infrequently is empowered to settle disputes between the contractor and the owner. This authority usually pertains to misunderstandings about the contractor's performance arising from the quality of the materials used or the interpretation of the plans and specifications. Questions of breach of contract lie byond the expertise of the engineer.

Disputes between the contractor and the owner that cannot be settled satisfactorily by the engineer should be resolved, if possible, without resort to lawsuits. To accomplish this it is advisable to establish in advance some agreements as to an arbitration procedure that is to be followed if occasion arises. The stipulation for and description of this procedure are properly a part of the contract documents.

28-2 Definition of Arbitration *Arbitration* is the submission of dispute to a disinterested person or persons for the final decision. Arbitration has been described as:[1]

A substitution, by consent of parties, of another tribunal for tribunal provided by ordinary process of law, and its object is final disposition, in speedy, inexpensive, expeditious and perhaps less formal manner, of controversial differences between parties.

Arbitration can attain its goal of providing final, speedy and inexpensive settlement of disputes only if judicial interference with process is minimized; it is, after all, meant to be a substitute for and not springboard for litigation.

Notice that arbitration is an extrajudicial determination of a controversy. Its

---

[1] *Barcon Associates, Inc. v. Tri-County Asphalt Corp.*, 430 A.2d 214, 86 N.J. 179 (1981).

purpose is to avert court action by either party. Arbitration is a summary, businesslike, and relatively inexpensive method of settling disputes. The decision, of course, may involve many thousands of dollars, and the outcome of the proceedings is thus of great importance to the parties. Arbitration is generally favored by the courts, which will only overturn an award on carefully circumscribed grounds.

28-3   Distinctions   Miscellaneous arrangements which leave to third parties the making of decisions which may be necessary regarding prices, values, quantities, qualities, and amounts are not really submissions to arbitration, and such third persons are not properly called "arbitrators." *Appraisement* is often confused with arbitration. Arbitration presupposes a controversy and is designed to resolve such, whereas appraisement is a stipulated procedure to prevent future disputes and to set a value, a price, or the like.[2] Arbitration is a quasi-judicial procedure usually involving a hearing, whereas appraisement generally implies decisions by a third-party expert upon the basis of his own knowledge and investigation without necessity of formal hearings.

A *referee* differs from an arbitrator in that a reference is appointed by a court as its officer and acts under its direction. An arbitrator is selected by the disputants. If the parties are not willing or able to appoint an arbitrator when time comes to do so, the court will make the appointment, but such an appointment is based upon prior agreement of the parties to arbitrate.

28-4   The Demand or Submission   Most matters are brought to arbitration as a result of the parties having agreed in advance by inserting an arbitration clause in their contract. Such an agreement is commonly enforced under federal or state laws.

Where there is an agreement, the moving party may file a *demand* with the administrating agency or otherwise initiate the proceeding in accordance with the specific terms of the contract. In other cases, it will be necessary for the parties to execute a *submission* to arbitrate after the dispute arises. The agreement to submit should cover everything that is necessary to give the arbitrators the power to make a binding determination.

It is quite customary in agreements to arbitrate as contained in construction contracts, or in connection with existing contract disputes, to refer to the *Construction Industry Arbitration Rules* of the American Arbitration Association (AAA).

28-5   Resort to Arbitration   Arbitration may be resorted to even while work under the contract is progressing. Utilization of this medium avoids both the costs of litigation and the delays that frequently are necessary because of crowded court

---

[2] As it was put in *Sanitary Farm Dairies v. Gammel,* 195 F.2d 106, 113 (8th Cir. 1952): "In general, where parties to a contract, before a dispute and in order to avoid one, provide for a method of ascertaining the value of something related to their dealings, the provision is one for an appraisement and not for an arbitration." See also *Bailey v. Tempore,* 75 Ill.2d 539 (1979).

dockets. Largely because it can proceed with relative speed, the arbitration procedure is especially beneficial if the dispute is such that it has caused or may cause performance under the contract to be suspended pending a settlement. In a situation of that sort, the arbitrator should be authorized to direct that work under the contract is to continue during the arbitration process if such an order seems to be in the best interest of the project.

It is quite possible that arbitration will be needed to settle disagreements that materialize after performance of some portion of the work but prior to final winding up of accounts. It may also happen that disputes arising during constructing are held to be resolved subsequently. Thus, a contractor was making a large excavation strictly accordingly to the plans. One side of the cut caved in and caused serious settlement of some adjoining construction. The engineer and the contractor agreed that the latter should take immediate remedial steps to save the structure, keeping account of all expenses involved. They agreed that the question of responsibility for the failure and thus the question of which side must assume the cost was to be settled by arbitration later on.

Even though a contract does not require arbitration the parties can nevertheless agree somewhere along the line to settle questions in this manner. In other words, they can submit the questions to arbitration even though such action is not mentioned in the contract.

The arbitrator customarily resolves disputes involving such matters as (1) what is or is not included in the contract, (2) the proper interpretation of drawings and specifications, (3) the extent of monetary adjustment for mistakes made and poor workmanship performed by the contractor, (4) claims for liquidated damages, (5) determination of the computed quantities to which the bid unit prices are to be applied, (6) claims for delays, and (7) extras and allowances.

The American Arbitration Association maintains a variety of arbitration rules for the conduct of such proceedings. The *Construction Industry Arbitration Rules* are uniquely applicable to disputes of an engineering nature. The association does not itself act as an arbitrator, nor does it give legal advice. However, for a specified fee it will handle the entire administration of the case: (1) preparing a list of persons considered to be competent to act as arbitrators in the type of situation involved; (2) furnishing rules for the guidance of the arbitrators; (3) appointing an arbitrator or arbitrators, if this is desired; and (4) assisting in various administrative capacities. If the parties elect to conduct the arbitration under these rules, both sides are bound thereby throughout the proceedings.

The association suggests that, if a construction contract is to require arbitration under AAA rules, the following clause should be inserted in such contract:

Standard Arbitration Clause:

Any controversy or claim arising out of or relating to this contract, or breach thereof, shall be settled by arbitration in accordance with the Construction Industry Arbitration Rules of the American Arbitration Association, and judgment ipon the award rendered by the Arbitrator(s) may be entered in any Court having jurisdiction thereof.

Efforts have been made to educate people to the advantage of arbitration so that disputants will resort to this device in lieu of bringing suit. As a result, the federal government and most states have passed legislation designed to facilitate enforcement of arbitration clauses in contracts and to provide judicial support for the arbitration process. By way of illustration, consider the United States Arbitration Act of 1925, as amended.[3] Key provisions of the act were characterized by the Supreme Court[4] in this fashion:

> Section 2 provides that a written provision for arbitration "in any maritime transaction or a contract evidencing a transaction involving commerce. . .shall be valid, irrevocable, and enforceable, save upon such grounds as exists at law or in equity for the revocation of any contract." Section 3 requires a federal court in which suit has been brought "upon any issue referable to arbitration under an agreement in writing for such arbitration" to stay the court action pending arbitration once it is satisfied that the issue is arbitrable under the agreement. Section 4 provides a federal remedy for a party "aggrieved by the alleged failure, neglect, or refusal of another to arbitrate under a written agreement for arbitration," and directs the federal court to order arbitration once it is satisfied that an agreement for arbitration has been made and has not been honored.

Among the other important provisions of the said act is Section 9 which says, in part:

> If the parties in their agreement have agreed that a judgment of the court shall be entered upon the award made pursuant to the arbitration, and shall specify the court, then at any time within one year after the award is made any party to the arbitration may apply to the court so specified for an order confirming the award, and thereupon the court must grant such an order unless the award is vacated, modified, or corrected as prescribed in sections 10 and 11 of this title. If no court is specified in the agreement of the parties, then such application may be made to the United States court in and for the district within which such award was made.

Section 10 permits the federal court to vacate an award for any of a variety of reasons, one such being a finding of "evident partiality or corruption in the arbitrators." In a similar vein, Section 11 lists instances in which an award may properly be modified or corrected, as for example "where there was an evident material miscalculation of figures."

The United States Arbitration Act represents an exercise of the authority constitutionally delegated to Congress to regulate interstate commerce and maritime transactions. In enacting such legislation Congress was providing for those engaged in interstate transactions an expeditious extrajudicial process for settling disputes—a process which contracting parties can avoid entirely by merely refraining from including arbitration provisions in their agreements. A representa-

---

[3] 9 U.S.C. §§1–14 (1970). "This legislation overruled longstanding judicial precedents which had refused to enforce agreements to submit justiciable controversies to arbitration on the grounds that they were contrary to public policy." *National R.R. Passenger Corp. v. Missouri Pacific R.R. Co.,* 501 F.2d 423, 426 (8th Cir. 1974).

[4] *Prima Paint Corp. v. Flood and Conklin,* 388 U.S. 395, 400 (1967).

tive decision interpreting the act, and a given arbitration provision, is *Metro Industrial Painting Corp. v. Terminal Construction Co.*[5] The court held that the particular disputes which had arisen between the parties were appropriate for arbitration under a contract clause which required submission to arbitration of "any question with respect to performance, non-performance, default, compliance or non-compliance, whether on behalf of the Contractor or Subcontractor." The opinion concluded in these words:

> The grievance asserted by petitioners is that respondents failed to meet time schedules, insisted upon performance of duties not required of petitioners under the contract and failed to supply materials for petitioners. In view of the federal policy to construe liberally arbitration clauses, to find that they cover disputes reasonably contemplated by this language, and to resolve doubts in favor of arbitration. . ., it is clear that these disputes can reasonably be said to fall within the category of compliance or non-compliance.

28-6   Disputes Subject to Arbitration   A great many matters, or types of disputes (not involving the commission of a crime),[6] are regarded as proper for settlement by arbitration. Issues submitted can be both factual and legal. Of course, illegal transactions cannot be the subject of arbitration because such transactions are not enforceable.

28-7   Scope of Arbitrator's Authority   A question of the authority of the arbitrators can be decided by those individuals themselves or by a court. An effort will be made to give effect to the intent of the parties. A liberal interpretation of the extent of the arbitrators' powers is usually made by the courts, but any matter not included in the submission cannot properly be settled by the arbitrators.

28-8   Reference to Courts   If the contracting parties have agreed that certain disputes, should they arise, will be submitted to arbitration, the United States Arbitration Act and other modern arbitration statutes provide for enforcement of the agreement to arbitrate. For example, the New York Arbitration Law provides "A party aggrieved by the failure of another to arbitrate may apply for an order compelling arbitration."[7] This is a typical statute requiring courts to enforce an obligation to arbitrate and thereby making the arbitration clause a meaningful part of the contract.

Many of the most widely used construction industry forms refer to these laws. For example, Article 7.10,1, of the AIA (*American Institute of Architects*) Form A-201, entitled *"General Conditions of the Contract for Construction,"* provides that "this agreement so to arbitrate shall be specifically enforceable under the prevailing arbitration law. The award rendered by the arbitrators shall be final, and judgment may be entered upon it in accordance with applicable law in any court having jurisdiction thereof."

---

[5] 287 F.2d 382 (2d Cir. 1961). *cert. denied* 368 U.S. 817. See also *American Home, Assurance Co. v. Vecco Concrete Construction Co.,* 629 F.2d 961 (4th Cir. 1980).

[6] Disputes of a criminal nature are of concern to the general public and are not subject to arbitration.

[7] Art. 75, New York Civil Practice Law and Rules, §7503(a) (McKinney 1980).

Modern arbitration laws give the courts power to enforce arbitration agreements and to exercise supervision of the process. For example, if the agreement fails to designate a method for selecting the arbitrator, either party has the right to apply to a court for a designation. In this connection, the *Construction Industry Arbitration Rules* of the American Arbitration Association, referred to in many of the form contracts used in the construction industry, provide a self-contained arbitration procedure so that no need would normally occur for applying to the courts. This reduces the cost and delay for the parties involved.

If there is not a bona fide agreement to arbitrate, it is possible under modern arbitration legislation to apply to the courts to stay an arbitration.[8]

28-9   Role and Qualifications of an Arbitrator   An *arbitrator* may be defined as a private judge chosen by the parties and endowed by them with power to decide a matter in dispute.

Any disinterested person may act as an arbitrator unless he lacks qualifications prescribed in the agreement (or in some applicable statute). He must be impartial and nonpartisan, and he must not be financially interested in the dispute. He should also be knowledgeable with respect to the general subject matter involved. Engineers, architects, and contractors are, by the nature of their work, in a good position to arbitrate disputes arising from construction contracts, other than those with which they have some connection.

28-10   Selection of an Arbitrator   Ordinarily, when arbitration is provided for, the contract will specify the procedure for selecting the arbitrator or arbitrators.

Perhaps the most common arrangement is to refer the matter to the American Arbitration Association so that administration can proceed under the *Construction Industry Arbitration Rules.* A list of experts in the industry will then be sent to the parties for their consideration so that arbitrators agreeable to both sides can be appointed. The AAA then handles the details of the arbitration process.

A less desirable procedure is for the owner to select one arbitrator, the contractor to choose another, and the two arbitrators to select a third. This system is cumbersome and leads to an occasional impasse in selection of the third arbitrator.

Where three arbitrators are deemed desirable, it is preferable that each arbitrator be entirely impartial. By mutually selecting the arbitrators from a list prepared by an outside agency, the parties can obtain such a panel. Where three neutral arbitrators hear a case, one of them is ordinarily designated as the chairman and may rule on minor procedural matters. Otherwise, a majority decision is necessary unless the arbitration agreement requires unanimity.

28-11   Procedure in Arbitration   If the arbitration agreement stipulates certain rules which are to govern the process, such rules must be followed.

Otherwise, the arbitrators have considerable latitude in their procedure. For

---

[8] See, for example, §7503(b), New York Civil Practice Law and Rules (McKinney 1980).

instance, they need not follow technical rules of evidence, but proceedings must nevertheless be conducted with honesty and fairness. The arbitrators should hold a hearing within a reasonable time after their appointment and can utilize statutory subpoena power to compel the attendance of witnesses.

Both parties have the right to be notified of any pending hearing and to present evidence. In the usual situation, an award made without a full and proper hearing[9] may be invalid unless the losing party waived his right to be heard or refused to attend.

The subject matter and scope of the dispute to be resolved should be made entirely clear to the arbitrators. Their function is to settle the question at issue completely, but their decision should not go to the extent of covering matters beyond the range of the submission.

When a hearing is held before the arbitrators, it is customary for representatives of both contesting parties to present evidence somewhat as though it were a court proceeding. In major cases the lawyers on either side will conduct the presentation of the evidence. Engineers normally act as expert witnesses in the course of the arbitration, and they may help in the preparation of facts and figures to substantiate whatever point of view they represent in the argument. Direct examination and cross-examination of witnesses may take place. The arbitrators will play the role of judges in determining whether proper evidence is or is not relevant and admissible. There may be no fixed procedure for the presentation of evidence and arguments by the opposing parties. Normally, the entire affair is relatively informal. Each side should conduct itself in a dignified manner and each will try to present its case in as favorable a light as possible. The arbitrators are supposed to be given all available data. They may ask for additional material from either or both parties if they think the information offered is inadequate to enable them to determine the matter. When a decision is rendered, it will be final, unless fraud or some other element is present which would support a contention that the award should be vacated or modified.

28-12   The Award   *An award* is the decision or judgment made by the arbitrators. Unless abandoned by both parties, it supersedes any claims which were submitted for arbitration.

An award will usually stand up in a court of law if it represents the honest judgment of the arbitrators. The latter must have acted fairly, reasonably, and within any time limit on their powers. The award must cover every point included in the submission and should fix responsibility for the costs of the arbitration unless the handling of such costs has been the subject of an arrangement between the respective parties.

An award should be complete and definite, thus leaving no room for future doubt or confusion on the points involved. Even though an award may seem excessive, the courts are generally reluctant to invalidate it absent clear indication

---

[9] There may be some cases in which the arbitrators are appointed as experts and specifically authorized to decide the dispute upon the basis of their own knowledge, dispensing with testimony of outsiders.

that the arbitrators exceeded their authority or were guilty of some impropriety. A valid award, made pursuant to a prior agreement to submit to arbitration, can be enforced through the judicial process.

The arbitrators are to reach their decision on the basis of their judgment as to what constitutes a proper and fair conclusion, considering all the data available to them. They need not explain to the parties exactly how or why they arrived at a given determination, though they may wish to do so voluntarily at the mutual request of the parties.

28-13 Allocation of Costs of Arbitration Arbitrators upon occasion serve without compensation, or they may be paid a reasonable sum. Sometimes the amount of compensation is stated in the agreement to arbitrate. *The Construction Industry Arbitration Rules* authorize the AAA to fix an appropriate daily rate and to outline other arrangments for submission to the parties at the outset. The rate of compensation should be fair and adequate in the light of the service rendered. If there are several arbitrators, they should generally be compensated equally.

Unless the contract clarifies the matter, a question may arise about who pays the cost of arbitration proceedings. If the owner is obligated by the contract to pay the arbitration bills, the contractor could cause him considerable expense by insisting upon submitting some petty issue at every opportunity. If the expense of arbitration, according to the terms of the contract, is to be divided equally between the owner and the contractor, this arrangement may help to prevent needless resort to arbitration by either party. It is often desirable to empower the arbitrators, as a part of their decision, to allocate costs; they can then in appropriate cases "penalize" a party who has been unreasonable in bringing to arbitration a minor issue which should have been settled amicably without the intervention of outsiders.

QUESTIONS

1 What is arbitration? What is its purpose?
2 Is an agreement to arbitrate necessarily a part of a contract? If such an agreement is part of a contract, can this bind the parties to submit a future dispute to arbitration before the difference arises? Why might this be desirable?
3 Can an agreement to arbitrate be made after a contract is in operation? How?
4 Does an agreement to arbitrate preclude subsequent resort to court proceedings?
5 What is the difference between (a) arbitration and appraisement? (b) an arbitrator and a referee?
6 What is a "submission," and what does it constitute?
7 Can an agreement to arbitrate be revoked or waived? Can these things be done if the agreement is a part of the signed contract?
8 Who may decide whether a dispute is to be heard by one or by more than one arbitrator?
9 What are the advantages (a) of having one arbitrator? (b) of having two? (c) of having three?

10 How are arbitrators generally chosen?

11 Can arbitration be resorted to only after completion of a contract? Why?

12 What is the American Arbitration Association, and what are its purposes and functions?

13 What is an "award"? How is an award arrived at?

14 Must the reasons for the magnitude and nature of the award be publicized? Why?

15 Will an arbitration award be likely to have any effect in subsequent litigation? Why?

16 From whom do arbitrators get their authority to make an award?

17 Does the submission indicate intended compliance with the future award?

18 What information should an arbitration agreement contain?

19 Is it customary to name the membership of a board of arbitration before a contract is started? Why?

20 Does arbitration have to proceed immediately after a dispute arises? Why?

21 If the arbitrators make a mistake, what can be done about it?

22 Should the contract or agreement to arbitrate specify the allocation of costs for arbitration?

23 A dispute arose between Black, the owner, and Brown, the contractor, over the interpretation of a clause in the specifications. Is this a suitable subject for arbitration? Why?

24 A contract contained a clause requiring arbitration of disputes. During progress of the work, a serious fire occurred. The owner and contractor could not agree upon the size of, and responsibility for, the resultant damages; therefore the owner brought suit against the contractor. Is this proper procedure? Will the court be likely to hear the case?

25 Jones, the contractor, claimed $10,000 compensation from Smith, the owner, because of extra work done under the contract. There was no arbitration clause in the contract. Jones demanded that the dispute be submitted to arbitration. Smith refused. Can Jones force Smith to arbitrate? Why?

26 According to the terms of a certain contract, one member of a board of arbitration is to be selected by the owner, one by the contractor, and one by these two appointees. Is this better than having one man as the sole abribrator? Explain.?

27 What assistance can be rendered by the American Arbitration Association? Will it, as an organization, arbitrate disputes?

GLOSSARY

Technical legal terms and Latin phrases are not abundantly employed in the text. Of those which do appear, many are defined at the point first used and, in order to reduce the length of this glossary, most such definitions are not repeated here.

a fortiori   Literally, "with stronger reason"; thus, given fact *A*, related and more probable fact *B* must likewise exist.

aliquot   A fractional part of something else, supposed to divide the latter without remainder.

alter ego   Another or second self; broadly, a confidant.

assignee   One to whom an assignor transfers property or rights under an assignment.

beneficiary   One who is the recipient of advantage, profit, or other benefits.

breach   Violation of a right, a duty, or of law, either by an act of commission or by nonfulfillment of an obligation. Thus, breach of contract is the unexcused failure to satisfy one's contractual undertaking.

caveat emptor   Let the buyer beware. In other words, the buyer should take pains to discover for himself any obvious defects in an article he is about to purchase.

chattel   An article of personal or movable property.

de facto   In fact.

de jure   Of right; lawful. "De facto" and "de jure" are contrasting terms. The former is used to refer to an action or state of affairs which is essentially without legal justification but which must nonetheless be accepted for all practical purposes.

de novo   Anew. Generally, to retry a case from the beginning.

defeasance clause A condition whose fulfillment defeats the operation of an instrument and renders it null and void.

delectus personae Personal preference, or choice of the person; thus, the right of partners to pass upon prospective members of the firm.

delegata potestas non potest delegari A delegate cannot (further) delegate. Without authorization from his principal, an agent cannot delegate his nonministerial functions and duties to another.

delict A violation of duty, public or private; a wrong.

devisee One to whom realty is given by will.

donee The recipient of a gift.

emancipation A term used primarily with reference to the "setting at liberty" of minors. When the parents, expressly or by implication, surrender the right to the custody and earnings of their child and renounce their parental duties, such child becomes his own master and is thus *emancipated.*

estoppel One who has by his conduct induced another to act in a particular manner is precluded (*estoppedth*) from thereafter adopting an inconsistent position and thus causing injury to such other person. The bar is raised when the party against whom it operates has brought about such conditions as make it inequitable for him to claim a right to which he would otherwise be entitled.

expressio unius est exclusio alterius The express enumeration (as in a list) of certain things (or persons or places) is, by implication, the exclusion of another or others not mentioned.

fee simple The largest estate known to the law; an absolute estate in which the owner is entitled to the entire property with unconditional power to dispose of same during his lifetime and which passes to his heirs if the owner dies intestate.

filum aquae The middle line or thread of a stream, which divides it into two parts and often serves as the boundary between the riparian owners on each side.

fraud Intentionally deceitful practice aimed at depriving another person of his rights or doing him injury in some respect.

grantee One to whom real property is transferred by grant.

gravamen The material part, or gist, of a complaint or charge.

in extremis In the last (that is, the ultimate) illness.

infant A person under the age of legal majority.

infra Below, beneath; often used in this fashion: "page 6, *infra.*"

injunction A writ issued by a court of equity ordering a person to refrain from a given course of action.

in loco parentis Acting in the place of a parent.

in pari delicto In equal fault; equally to blame.

in personam See *in rem.*

in rem Actions *in personam* and those *in rem* are distinguishable on the ground that the former are directed against specific persons and seek personal judgments, while in rem proceedings seek to determine rights in particular property as against all the world. Similarly, rights in personam are those primarily available against certain persons, while rights in rem are good against the world at large.

inter alia Among other things.

intestacy   The circumstance in which a person dies without leaving a valid will to indicate his wishes about the disposal of his property.

jurisprudence   Broadly speaking, the science and philosophy of law and of legal relations.

laches   Unreasonable delay in doing what in law should have been done, or neglect in seeking to enforce a right at a proper time, which delay or neglect has operated to the disadvantage of another party.

lessee   The person to whom a lease is made.

lex non scripta   Unwritten or common law.

lien   A charge (encumbrance) imposed on specific property as security for a debt or other obligation.

liquidated damages   A fixed sum agreed upon, at the outset of a contractual arrangement, as the proper compensation to be paid to the injured party in the event of breach.

malfeasance   An act which the party has no right to do and ought not to do; an illegal act.

mandamus   A court writ issued to compel the performance of a public or official duty owed by the person to whom the order is directed.

ministerial   Pertaining to an action done under the authority of a superior and with relatively little reliance upon the judgment or discretion of the doer.

misfeasance   The performance in a wrongful and injurious manner of an act which might have been done in lawful fashion. Thus, *misfeasance* is the doing of a lawful act in an unlawful manner, as distinguished from *malfeasance,* which is the performance of an act positively unlawful.

modus operandi   Manner of operating.

mores   Usages; folkways; customs.

mortgagee   The person to whom property is mortgaged.

nonfeasance   Failure to do something which ought to have been done.

nullity   Invalidity; nothingness.

parol   Verbal.

per se   By itself.

post hoc   After the fact.

prima facie   On first appearance; at first view.

privity   Successive (or mutual) relationship to the same property rights; a connection between parties.

promisee   The recipient of a promise.

pro rata   Proportionately.

quantum meruit   Literally, "as much as he deserved."

quid pro quo   One thing for another; something for something. The term is used in connection with *contract consideration.*

quo warranto   Literally, "by what authority?"; the writ by which it is sought to recover an office or franchise from the individual or entity in possession thereof. The writ requires the defendant to appear and show by what authority he maintains possession.

ratification   Confirmation; broadly, agreement by *A* (in the face of his right to repudiate) to adopt an act performed for him by *B,* or to affirm an act which *A* himself had performed earlier.

reformation   The correction of defects so as to bring an instrument in line with the actual intent of the parties, which intent had been, through inadvertence or otherwise, improperly expressed in the agreement as originally drawn.

rescission   Annulment; cancellation.

res gestae   Things done; every circumstance and statement associated with and incident to a given transaction or occurrence.

seisin   Possession with intent to claim a freehold interest.

situs   Location; position.

sovereign immunity   A doctrine holding that governments are generally not responsible in damages for their tortious actions.

statute of limitations   A statute limiting the time period in which a lawsuit may be brought.

subrogation   Substituting one person for another in regard to rights, claims, etc.

supra   Above in position; upon. "Page 6, supra" refers the reader to an earlier page in the same publication.

tenement   Broadly, everything of a permanent nature (principally realty) which may be held. Another use of the word is, of course, to refer to a dwelling place.

testator   One who has left a will to control disposition of his property upon his death.

tort   Wrongful behavior (aside from breach of contract) for which a civil action will lie; the unprivileged commission (or omission) of an act whereby another person incurs loss or injury; such breach of duty as results in damage to plaintiff.

tort-feasor   One who commits a tort.

trespass   In its ordinary sense, trespass refers to an unlawful act committed with actual or implied violence and resulting in injury to or physical interference with the person or property of someone else. In a more limited sense, trespass denotes an unauthorized entry upon the real property of another, frequently entailing some measure of damage to such property.

ultra vires   In the law of corporations, an act beyond the scope of corporate powers.

vendee   One to whom a sale (usually of real property) is made; a purchaser.

waiver   Voluntary abandonment, surrender, or relinquishment of a known right or privilege.

INDEX